PRINCIPLES OF
BEHAVIORAL NEUROLOGY

CONTEMPORARY NEUROLOGY SERIES AVAILABLE:

Fred Plum, M.D., *Editor-in-Chief*
J. Richard Baringer, M.D., *Associate Editor*
Sid Gilman, M.D., *Associate Editor*

PRINCIPLES OF
BEHAVIORAL NEUROLOGY

M-MARSEL MESULAM, M.D.

Director, Division of Neuroscience and Behavioral Neurology
Beth Israel Hospital
Professor of Neurology
Harvard Medical School
Boston, Massachusetts

F. A. DAVIS COMPANY • PHILADELPHIA

Printed in the United States of America

Last digit indicates print number:

10 9 8 7 6

NOTE: As new scientific information becomes available through basic and clinical research, recommended treatments and drug therapies undergo changes. The author(s) and publisher have done everything possible to make this book accurate, up-to-date, and in accord with accepted standards at the time of publication. However, the reader is advised always to check product information (package inserts) for changes and new information regarding dose and contraindications before administering any drug. Caution is especially urged when using new or infrequently ordered drugs.

Library of Congress Cataloging in Publication Data

Main entry under title:

Principles of behavioral neurology.

(Contemporary neurology series; 26)
Includes bibliographies and index.
1. Mental illness—Etiology—Addresses, essays, lectures.   2. Neuropsychology—Addresses, essays, lectures.   3. Brain—Diseases—Complications and sequelae —Addresses, essays, lectures.   I. Mesulam, M-Marsel. II. Series.   [DNLM: 1. Mental Disorders.   2. Nervous System Diseases.   W1 C0769N v.26/WL100 P9568]
RC454.4.P73 1985   616.89'071   84-28650
ISBN 0-8036-6151-7

In memory of Norman Geschwind, 1926–1984

PREFACE

One factor that has contributed prominently to the rapid growth and evolution of contemporary behavioral neurology is the distinctly multidisciplinary approach to patient care and research. The phenomenal progress in the basic neurosciences, for example, has provided a source of new direction and enthusiasm for the clinician. In turn, clinical observations are once again beginning to guide basic experimentation, as they had done so fruitfully during the earlier days of both disciplines. Many of the recent developments in the areas of attention, emotion, memory, and dementia owe their impetus to this fertile interaction.

Since the single most fundamental aspect of behavioral neurology is the systematic assessment of mental state, the neuropsychologist has played one of the most important roles in the development of the entire field. The relationship between psychiatry and neurology is also central to behavioral neurology. Although this interaction is not new, it is undergoing a welcome revival. For example, temporolimbic epilepsy is now regularly included in the differential diagnosis of atypical psychiatric problems; some cases of shyness and childhood depression have been linked to a developmental right-hemisphere dysfunction; and it is now generally accepted that substantial alterations of mood, personality, and comportment may arise as the sole manifestation of focal injury not only in the frontal lobes but also in various parts of the right hemisphere. These are only some of the developments that have fueled the intensity of the interchange between psychiatry and neurology.

This is not to say that there is an inexorable movement toward reclassifying all psychiatric diseases within the neurologic nosology. In fact, for the vast majority of patients who seek outpatient psychiatric help, a neurologic approach is no more useful (or desirable) than a chemical analysis of the ink would be for deciphering the meaning of a message. However, new discoveries on the cerebral organization of emotion and personality are prompting the inclusion of neurologic causes into the differential diagnosis of many conditions that have traditionally been attributed to idiopathic psychiatric disorders.

Attitudes that once considered the organic approach as the insensitive sledgehammer of psychiatry and the area of behavior as the soft underbelly of neurology are rapidly changing. Behavioral neurology is now firmly established as a *bona fide* specialty that includes, among other fields of interest, the borderland area between neurology and psychiatry. It also provides an outlook that focuses on the behavioral consequences of almost all neurologic diseases and of many medical conditions that influence brain physiology. The clinical mandate of behavioral neurology is vast and its future is bright. The growth of this field promises to be of major benefit to a large group of patients, who can hope to receive a new understanding for conditions that have not been of central interest to the mainstream of modern medical, neurologic, and psychiatric practices. These developments are also likely to have a considerable impact on the scientific investigation of brain-behavior relationships.

This book aims to provide a background for some of the major areas in behavioral neurology. Chapter 1 contains a survey of anatomic and physiologic principles that guide the interpretation of brain-behavior interactions. Chapter 2 gives an overview of the mental state assessment and its correlations with cerebral damage. This chapter also places a special emphasis on the examination of elderly and demented patients. Chapters 3 through 7 contain in-depth analyses of major behavioral areas: attention, memory, language, affect, and complex perceptual processing. Chapter 8 is somewhat unique in dealing with a single disease process. This appeared justified in view of the remarkable spectrum of psychiatric and endocrinologic conditions that are seen in conjunction with temporolimbic epilepsy. Chapters 9 through 11 survey recent advances in the behavioral application of evoked potentials, regional metabolic scanning, and quantitative computerized tomography.

Each chapter in this book contains a broad range of information, which can provide not only an introductory background for clinicians who are new to the field but also an update for the expert. However, this book will not accomplish its full purpose unless it also proves useful to the basic neuroscientist who is interested in bridging the gap between the experimental laboratory and the human brain.

During the preparation of this book, it has been a pleasure to work with Fred Plum, M.D., Editor-in-Chief of the Contemporary Neurology Series, Sylvia K. Fields, Ed.D., Senior Medical Editor at F. A. Davis, Ann Huehnergarth, Production Editor, and the other F. A. Davis staff members who have directed the production process. I also want to thank my secretary, Leah Christie, who diligently participated in the preparation and editing of almost every chapter in this book.

M-Marsel Mesulam, M.D.

CONTRIBUTORS

D. FRANK BENSON, M.D.
Department of Neurology
UCLA School of Medicine
710 Westwood Plaza
Los Angeles, CA 90024

HOWARD W. BLUME, M.D.
Department of Neurosurgery
Beth Israel Hospital
330 Brookline Avenue
Boston, MA 02215

ANTONIO R. DAMASIO, M.D., Ph.D.
Department of Neurology
University Hospital
Iowa City, IA 52242

NORMAN GESCHWIND, M.D.
Neurology Department
Beth Israel Hospital
330 Brookline Avenue
Boston, MA 02215

RUBEN C. GUR, Ph.D.
Department of Neurology
Graduate Hospital
19th and Lombard Streets
Philadelphia, PA 19146

ROBERT T. KNIGHT, M.D.
Martinez V.A. Hospital
150 Muir Road
Martinez, CA 94553

M-MARSEL MESULAM, M.D.
Neurology Department
Beth Israel Hospital
330 Brookline Avenue
Boston, MA 02215

MARGARET A. NAESER, Ph.D.
Boston V.A. Hospital
150 South Huntington Avenue
Boston, MA 02130

ELLIOTT D. ROSS, M.D.
Department of Neurology
University of Texas Health Science Center
5325 Harry Hines Boulevard
Dallas, TX 75235

DONALD L. SCHOMER, M.D.
Neurology Department
Beth Israel Hospital
330 Brookline Avenue
Boston, MA 02215

JEAN-LOUIS SIGNORET, M.D.
Neurologie et Neuropsychologie
Hopital de la Salpetriere
47 Blvd. de l'Hopital
75013 Paris
France

PAUL A. SPIERS, M.A.
Behavioral Neurology Unit
Beth Israel Hospital
330 Brookline Avenue
Boston, MA 02215

SANDRA WEINTRAUB, Ph.D.
Behavioral Neurology Unit
Beth Israel Hospital
330 Brookline Avenue
Boston, MA 02215

CONTENTS

Chapter 1

PATTERNS IN BEHAVIORAL NEUROANATOMY: ASSOCIATION AREAS, THE LIMBIC SYSTEM, AND HEMISPHERIC SPECIALIZATION*

M-Marsel Mesulam, M.D.

The neurons that make up the human brain display marked regional variations in architecture, connectivity, and transmitter neurochemistry. In this chapter, the relationship of these regional variations to complex behavior is reviewed from the vantage point of behavioral neurology. Little attempt will be made to be inclusive or balanced, since such an approach would commit the reader to many volumes of intricate detail. Instead, the intent is to provide a reorganization of basic information in a manner that highlights patterns rather than specifics.

A great deal is already known about the cerebral correlates of human behavior and a great deal of new information is being discovered with the help of novel technologies, some of which are reviewed in Chapters 9 to 11. However, there are also many aspects of cerebral connectivity and neurochemistry that cannot yet be investigated directly in the human brain. Experimental evidence gathered from studying subhuman primates has been very useful for bridging these gaps. Structural homologies among different species are never simple, always incomplete, and sometimes misleading. Furthermore, many human behaviors, especially those of greatest interest to the behavioral neurologist, have hardly any analogue in other animals. Nonetheless, this chapter draws heavily upon relevant evidence in nonhuman primates when similar information is not yet available in the human. Since the emphasis is on patterns rather than on specific detail and since patterns are likely to remain relatively stable across closely related species, it is unlikely that this reliance on animal data will lead far astray.

*The preparation of this chapter was supported in part by the Javits Neuroscience Investigator Award, the McKnight Foundation Director's Award, and by NINCDS grants NS09211 and NS20285. I am grateful to Leah Christie and Rick Plourde for expert secretarial and photographic assistance.

CEREBRAL CORTEX

Many frameworks have been proposed for subdividing the vast cortical surface of the human brain on the basis of regional variations in the architectonic arrangement of neurons and myelin. These schemes can be divided into two major groups: those that focus on the unique properties of individual regions and those that focus on patterns formed by regions sharing common characteristics. Proponents of the first school have constructed a wide variety of cortical maps, ranging in complexity from the map of Exner,[73] which boasted hundreds of sharply delineated subdivisions, to the more modest and also much more widely accepted ones of Brodmann[33] and of von Economo.[70] The second school is more difficult to identify, since few of its proponents have produced systematic surveys of the entire brain. Members of this second school include theoreticians of brain function such as Broca,[32] Abbie,[1] Filimonoff,[75] Yakovlev,[305] and Sanides.[246] It is the thinking of this second school which guides much of the discussion in this section on cortex. This approach helps to show that the entire cortical mantle can be subdivided into only five subtypes, which display a gradual increase in structural complexity and differentiation. This purely structural typology is then shown to have far-reaching implications for understanding the distribution of neural connectivity and behavioral specialization throughout the brain.

The Corticoid Areas

Although it is not customary to designate the basal forebrain as cortex, it is important to realize that some components of this region are situated directly on the ventral and medial surface of the hemispheric vesicle. By definition, these components could be considered part of the cortical mantle. The three basal forebrain structures that can be included in this group are the septal region, the substantia innominata, and parts of the amygdaloid complex (Fig. 1). These structures contain the simplest and most undifferentiated type of cortex in the entire forebrain. The organization of the constituent neurons is so rudimentary that no consistent lamination can be discerned (Fig. 2A). Even the orientation of dendrites appears random and haphazard. Therefore, these three components of the basal forebrain could be designated as having a *corticoid*, or cortex-like, structure. For the purposes of this chapter, we will consider the entire amygdaloid complex as part of the corticoid structures even though a substantial part of the amygdala has a subcortical nuclear appearance. In fact, all of the corticoid areas have architectonic features that are part cortical and part nuclear.

The Allocortical Structures

The next more complex type of cortex carries the designation of *allocortex*. This type of cortex contains one or two bands of neurons arranged into moderately well-differentiated layers (Fig. 2B). There are two allocortical formations in the brain: the hippocampal formation, which also carries the designation of archicortex; and the piriform or primary olfactory cortex, which is also known as paleocortex.

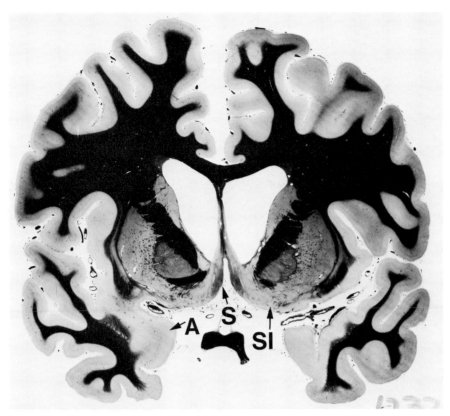

FIGURE 1. Coronal section of human brain stained for myelin. The basal forebrain contains the three corticoid components: the septal area (S), the substantia innominata (SI), and the amygdaloid complex (A). Since they are on the surface of the brain, these three formations are part of the cortical mantle. (Magnification ×1.6.)

Paralimbic Areas (Mesocortex)

The next level of structural complexity is provided by the *paralimbic* regions of the brain, which have also been designated as *mesocortex*. These areas are intercalated between allocortex and isocortex so as to provide a gradual transition from one to the other (Fig. 3). The parts of paralimbic areas that border the allocortex are also known as *periallocortical* or *juxta-allocortical*, whereas the parts closest to the isocortex can be designated as *pro-* or *peri-isocortical*. The demarcations among these regions are never sharp and always include zones of gradual transition.

Allocortical cell layers often extend into the periallocortical component of paralimbic areas (Fig. 3). Several gradual changes in the direction of increased complexity and differentiation occur from the allocortical pole toward the isocortical side of paralimbic regions. These include:

1. Progressively greater accumulation of small granular neurons first in layer IV and then in layer II
2. Sublamination and columnarization of layer III
3. Differentiation of layer V from layer VI and of layer VI from the underlying white matter
4. An increase in intracortical myelin, especially along the outer layer of Baillarger (layer IV)

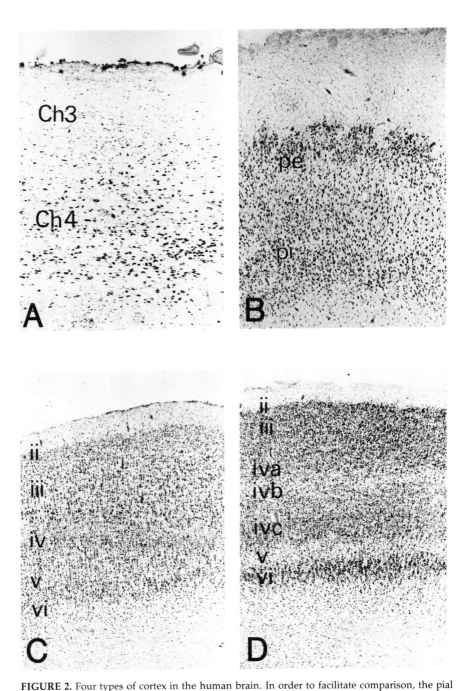

FIGURE 2. Four types of cortex in the human brain. In order to facilitate comparison, the pial surface is toward the top in all four photomicrographs. (*A*) An example of the corticoid type. This is a photomicrograph of the substantia innominata showing the nucleus basalis of Meynert (Ch4) and the more superficial horizontal limb nucleus of the diagonal band (Ch3). The lamination is incomplete, and there is no uniformity in the orientation of neurons. (*B*) An example of allocortex from the subicular portion of the hippocampal formation. There are two distinct layers, an external pyramidal (pe) and an internal pyramidal (pi). Dendrites within each layer have a relatively uniform orientation. (*C*) An example of homotypical isocortex from prefrontal heteromodal regions. There are six distinct layers including differentiated granularity in layer ii and layer iv. (*D*) An example of idiotypic cortex from the striate visual area. There are at least seven layers, many strongly granular. From corticoid to idiotypic cortex there is a gradual increase of cell density and laminar differentiation. (Magnification ×10.)

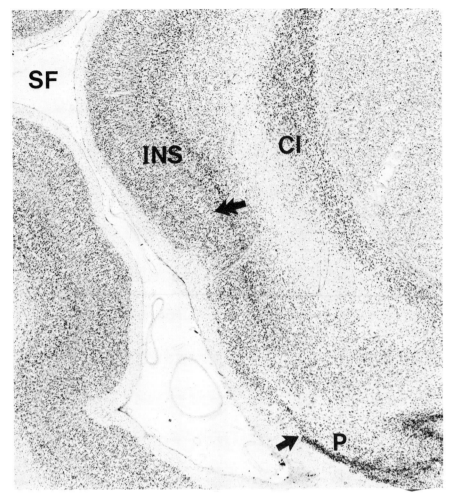

FIGURE 3. Insula of the rhesus monkey. The *single arrowhead* points to the direct continuity between piriform cortex (P) and the insular paralimbic cortex. Adjacent to piriform allocortex, the insula has two to three agranular layers. This is the peri-allocortical sector of the insula. More dorsally, a granular layer IV begins to appear *(double arrowhead)*. There is further differentiation of insular cortex in the dorsal direction toward parietal isocortex. Abbreviations: Cl— claustrum, INS—insula, P—piriform cortex, SF—sylvian fissure. (Magnification ×30.)

In general, the emergence of relatively well-differentiated granular cell layers in layers IV and II, the sublamination of layer III, and the differentiation of layer V from layer VI mark the end of the paralimbic zone and the onset of six-layered homotypical isocortex.

There are five major paralimbic formations in the primate brain: (1) the caudal orbitofrontal cortex; (2) the insula; (3) the temporal pole; (4) the parahippocampal gyrus (which includes the entorhinal, prorhinal, perirhinal, presubicular, and parasubicular areas); and (5) the cingulate complex (which includes the retrosplenial, cingulate, and parolfactory areas). These five paralimbic regions form an uninterrupted girdle surrounding the medial and basal aspects of the cerebral hemispheres (Fig. 4).

The paralimbic belt can be divided into two major groups: the olfactocentric and the hippocampocentric. The olfactory piriform cortex provides the allocortical nidus for the orbitofrontal, insular, and temporo-

PATTERNS IN
BEHAVIORAL
NEUROANATOMY:
ASSOCIATION
AREAS, THE LIMBIC
SYSTEM, AND
HEMISPHERIC
SPECIALIZATION

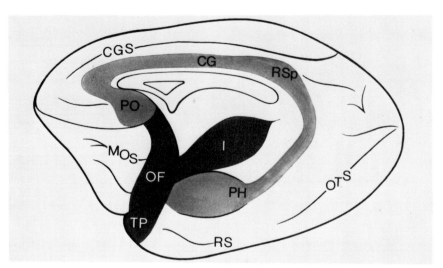

FIGURE 4. The girdle of olfactocentric (black) and hippocampocentric (gray) paralimbic areas in the brain of the rhesus monkey. Abbreviations: CG—cingulate cortex, CGS—cingulate sulcus; I—insula; MOS—medial orbitofrontal sulcus; OF—caudal orbitofrontal cortex; OTS—occipitotemporal sulcus; PH—parahippocampal region; PO—parolfactory area; RS—rhinal sulcus; RSp—retrosplenial area; TP—temporopolar cortex.

polar paralimbic areas. The hippocampus and its induseal rudiment, on the other hand, provide the allocortical nidus for the cingulate and parahippocampal components of the paralimbic brain.[182]

Homotypical Association Isocortex

By far the greatest extent of the cortical surface in the human brain consists of six-layered *homotypical isocortex,* also known as association cortex (see Fig. 2C). Association isocortex can be subdivided into two major types: modality-specific *(unimodal)* isocortex and high-order *(heteromodal)* isocortex.[14,189]

Unimodal association isocortex is defined by three essential characteristics:

1. The constituent neurons are almost exclusively responsive to stimulation in only a single sensory modality
2. The predominant cortical inputs are provided by the primary sensory cortex or by other unimodal regions in that same modality
3. Damage yields modality-specific deficits confined to tasks guided by cues in that modality

The unimodal areas for the three major sensory modalities have been determined experimentally in the brain of the macaque monkey. Such experiments have shown that the superior temporal gyrus is the unimodal auditory association area, that the superior parietal lobule provides the somatosensory unimodal association area, and that the peristriate, midtemporal, and inferotemporal regions provide the unimodal association regions in the visual modality.[14,179,189]

In contrast to unimodal isocortex, the heteromodal component of isocortex is identified by the following characteristics:

1. Even within relatively small areas of cortex, neuronal responses are not confined to any single sensory modality
2. Damage to this type of cortex leads to behavioral deficits that are not modality specific
3. The cortical inputs originate from unimodal areas in more than one modality and/or from other heteromodal areas

Some neurons in heteromodal association areas respond to stimulation in more than one modality, indicating the presence of direct multimodal convergence.[20,28] More commonly, however, there is an admixture of neurons with different preferred modalities.[118,121] Defined in this fashion, heteromodal cortex includes the types of regions that have been designated as high-order association cortex, polymodal cortex, multimodal cortex, polysensory areas, and supramodal cortex.[189]

There are essentially two and perhaps three major heteromodal fields in the brain of the monkey. One is in the prefrontal region, including the anterior orbitofrontal surface and the dorsolateral frontal convexity.[46] The second heteromodal field includes the inferior parietal lobule and extends into the banks of the superior temporal sulcus.[36,118,189,254,255] There may be a third heteromodal region in the posterior part of the ventral temporal lobe.[63,253]

There are some relatively subtle architectonic differences between unimodal and heteromodal areas. In general, the unimodal areas have a more differentiated organization, especially with respect to sublamination in layers III and V, columnarization in layer III, and more extensive granularization in layer IV and especially in layer II. On these architectonic grounds, it would appear that heteromodal cortex is closer in structure to paralimbic cortex and that it provides an intercalated step between paralimbic and unimodal areas.

Idiotypic Cortex

The koniocortex of primary sensory areas and the macropyramidal cortex of the primary motor region constitute unique and highly specialized structures which can be designated as having an *idiotypic* architecture. There are two divergent opinions about these areas. One is to consider them the most basic and elementary components of cortex; the other is to see these areas as the most advanced and highly differentiated components of the cortical mantle.[246] The latter point of view will dominate this discussion. Indeed, the gradual trend toward granularization and increased lamination does appear to reach its climax in koniocortical regions such as the striate area (see Fig. 2D).

The locations of these idiotypic regions are well known. The primary visual area covers the occipital pole and the banks of the calcarine fissure, the primary auditory cortex covers Heschl's gyrus on the supratemporal plane, the primary somatosensory cortex covers the postcentral gyrus, and the primary motor area is located in the precentral gyrus.

A GENERAL SCHEMA FOR PATTERNS OF BEHAVIORAL SPECIALIZATION AND NEURAL CONNECTIVITY IN CORTEX

The preceding discussion shows that the entire hemispheric surface can be subdivided into five essential types of cortex that range from the simplest to the most differentiated. The relative position in this structural hierarchy also determines the topographic relationships among the individual cortical types. For example, paralimbic cortex is always flanked by allocortex at one extreme and by isocortex at the other; unimodal association cortex is always intercalated between primary sensory areas and heteromodal cortex. These sets of relationships lend themselves to the schematic representation of the cortical surface shown in Figure 5. Although this schema arises primarily from architectonic and topo-

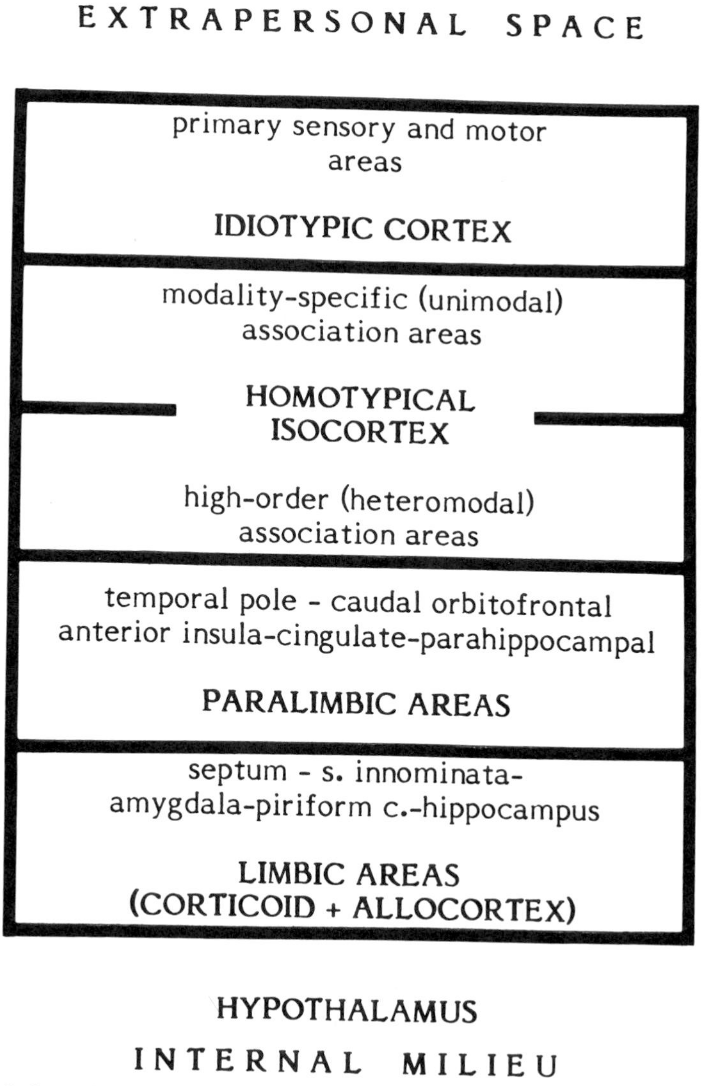

EXTRAPERSONAL SPACE

primary sensory and motor
areas

IDIOTYPIC CORTEX

modality-specific (unimodal)
association areas

**HOMOTYPICAL
ISOCORTEX**

high-order (heteromodal)
association areas

temporal pole - caudal orbitofrontal
anterior insula-cingulate-parahippocampal

PARALIMBIC AREAS

septum - s. innominata-
amygdala-piriform c.-hippocampus

**LIMBIC AREAS
(CORTICOID + ALLOCORTEX)**

HYPOTHALAMUS

INTERNAL MILIEU

FIGURE 5. Cortical zones of the human brain.

graphic considerations, it leads to general principles about the organization of connectivity patterns and the distribution of behavioral specializations.

The corticoid and allocortical areas at one pole of the schema in Figure 5 are also collectively designated as "limbic" structures. Of all cortical areas, these have the closest association with the hypothalamus. Through neural and also hormonal mechanisms, the hypothalamus is in a position to coordinate electrolyte balance, blood glucose levels, basal temperature, metabolic rate, autonomic tone, sexual phase, circadian oscillations, and even immunoregulation.[25,34,199] The hypothalamus is essentially the head ganglion of the internal milieu and also a major generator of drives and instincts that promote the survival of the self and of the species.[273] In keeping with these functions of the hypothalamus, areas in the limbic zone of cortex assume an important role in the regulation of four types of behaviors: memory and learning, the modulation of drive, the affective coloring of experience, and the higher control of hormonal balance and autonomic tone. These specializations of the limbic structures are very much related to the maintenance of the internal milieu (homeostasis) as well as to the associated operations necessary for the preservation of the self and the species.

At the other pole of the schema in Figure 5 lie the most highly specialized primary sensory and motor areas. These are the parts of cortex that are most closely related to the extrapersonal space: sensory input from the environment has its first cortical relay in the primary koniocortical areas while the motor cortex coordinates actions that lead to the manipulation of the extrapersonal world. The adjacent zone of unimodal association cortex provides the neuronal machinery for the subsequent processing of sensory input. Unimodal cortex is therefore very much directed toward the extrapersonal world and closely associated with primary sensory areas.

The two remaining zones of heteromodal and paralimbic formations provide neural bridges that mediate between the needs of the internal milieu and the realities of the extrapersonal world. These zones enable two major types of neural transformation: (1) the further associative elaboration of sensory processing and (2) the integration of this information with drive, affect, and other aspects of mental content. In contrast to the idiotypic and unimodal belts, which are characterized by relatively more "dedicated" and homogeneous neural mechanisms confined to single modalities of information transfer, the intermediate belts of heteromodal and paralimbic cortex represent more "generalized" mechanisms with heterogeneous input-output relationships. This is one reason why no uniform behavioral specialization can be assigned to components of the paralimbic and heteromodal zones.

The schema in Figure 5 is also relevant to patterns of corticocortical interconnections. Components of each zone have *vertical* connections with components of other zones and also *horizontal* connections among each other. Experimental evidence, gathered mostly from the brain of the monkey, shows that the most intense vertical connectivity of an individual cortical area is with zones that are immediately adjacent to it in the schema of Figure 5. Although this schema is most relevant to intracortical connections, it also addresses the interactions between cortex and hypothalamus.

For example, recent anatomic evidence has definitively shown that all types of cortical areas, including association isocortex, receive direct hypothalamic projections.[140,183] However, for the great majority of cortical regions, this hypothalamic input is quite minor. The only exception is provided by the limbic structures. Thus, the septal nuclei, the basal nucleus of the substantia innominata, the amygdaloid complex, the piriform cortex, and the hippocampus are set apart by the presence of substantial hypothalamic connections.[6,181,248,304] This distribution suggests that the hypothalamic influence on cortex is likely to be felt most intensely within the limbic zone. The second major source of vertical connections for limbic structures originates in the paralimbic zone. For example, the amygdala receives what is probably its most extensive vertical cortical input from the insula;[203] the hippocampus from the entorhinal sector of the parahippocampal region;[286,287] and the piriform cortex as well as the nucleus basalis from the olfactocentric paralimbic areas.[180,181] Thus, the two most extensive vertical connections of limbic areas are with the hypothalamus and with paralimbic areas. This is entirely consistent with the position of the limbic zone within the schema of Figure 5.

A similar analysis can be extended to the cortical connections of other zones in Figure 5. Paralimbic areas, for example, have the most extensive vertical connections with limbic and heteromodal areas. This has been demonstrated experimentally in the case of the anterior insula, the parahippocampal formation, and the cingulate complex.[180,182,216,286,287] Relevant information is not yet available for the other components of the paralimbic belt.

With respect to the heteromodal areas, the experimental literature shows that their most extensive vertical connectivity is with components of the paralimbic zone, on the one hand, and with those of the unimodal zone, on the other.[14,129,189,217] The presence of extensive connections between paralimbic areas and heteromodal association cortex has been shown only recently with the help of modern pathway tracing techniques. The demonstration of such connections is of obvious importance, since it provides an anatomic substrate for interactions between motivational factors and extensively preprocessed sensory information.

The other two zones in Figure 5 also display selective patterns of connectivity. Thus, the major vertical connections of unimodal areas occur with primary areas on the one hand and with heteromodal areas on the other, whereas primary areas derive their major cortical connections from unimodal areas.[66,129,217]

In most instances the position of a cortical area in the schema of Figure 5 will only help to identify its *predominant* cortical connectivity patterns. For example, although the amygdala is also known to receive direct input from association isocortex,[3,283] this does not appear to be as substantial as the connections of this limbic structure with the hypothalamus and with paralimbic regions. In some cases, however, there are more absolute distinctions. For example, the primary areas in the more advanced primates do not seem to receive any limbic or paralimbic cortical input. This may ensure that the initial processing of sensory information is done objectively and without being influenced by drive and impulse. Perhaps this is why emotional state does not alter the shape of an object or the pitch of a sound. The adaptive value of this arrangement is evident. In other mammalian species such as the rat, however, this separation may be less complete, since direct connectivity between primary and paralimbic areas have been reported.[289]

Many cortical areas also have horizontal cortical connections with other components of the same zone. These are extremely well developed within the limbic, paralimbic, and heteromodal zones.[14,181,182,189,240] For example, of all the cortical neurons that directly projected to a subsector of prefrontal heteromodal cortex, 26 percent were located within unimodal areas, 13 percent in paralimbic regions, and *61* percent in heteromodal regions.[14] Furthermore, the insula as well as cingulate cortex have interconnections with virtually each of the other paralimbic regions of the brain.[182,216] Although the horizontal connections of unimodal and primary areas may also be substantial in quantity, they have a more restricted distribution. For example, while unimodal regions may receive extensive input from other associaton areas in the same modality, there is essentially no interconnectivity among areas belonging to different modalities. In a similar vein, except for the intimate interconnections between the primary somatosensory and motor areas, there are no neural projections among primary areas belonging to separate modalities. It appears, therefore, that there is a premium on channel width within the limbic, paralimbic, and heteromodal zones, whereas the emphasis is on fidelity within the unimodal and primary zones.

BEHAVIORAL PATTERNS IN HUMAN CORTEX

Although some of the relevant evidence needs to be extrapolated from experimental animals, the entire cortical surface of the human brain can also be subdivided into the five zones shown in Figure 5. In this section, we will see how these zones are distributed in reference to a standard architectonic map of the human brain such as that of Brodmann and how the behavioral specializations of these zones follow the general principles outlined earlier.

Primary Areas (the Idiotypic Zone)

Because of their unique architectonic features, the primary motor and sensory areas are easily located. The primary sensory areas are koniocortical (hypergranular) in architecture, and they constitute the first cortical relay for inputs from the modality-specific thalamic relay nuclei.

Primary Visual Cortex

In the visual modality, the primary cortex (V1) covers the occipital pole and the banks of the calcarine fissure (area 17 in Figure 6). This region is also known as striate cortex, because of the conspicuous myelinated stripe of Gennari (or of Vicq d'Azyr), which is easily detected in layer IV even by naked-eye inspection of unstained specimens. Retinal input is relayed to striate cortex through the lateral geniculate nucleus. The importance of visual input to striate cortex is shown by the fact that fully 70 percent of all its neural input originates from the lateral geniculate nucleus.[66] The entire visual field is faithfully mapped onto striate cortex. The striate area in each hemisphere receives input from the contralateral visual field, and dorsal parts of striate cortex contain a representation of the lower visual field, whereas the ventral parts represent the upper visual field. The central (macular) part of the visual field is mapped onto the most caudal part of striate cortex, whereas the more peripheral parts of the visual field are represented more anteriorly along the calcarine fissure. Except for the representation of the vertical meridian, other parts of striate cortex in the monkey do not have callosal interhemispheric connections.[137] This arrangement may further ensure the specificity of the spatial map.

The physiologic exploration of neurons in primary visual cortex has led to one of the most exciting chapters in neuroscience.[117] It is now well known that these neurons form columnar modules and that they have an exquisite organization of connectivity that leads to the orderly processing of information about the location, distance, size, shape, movement, and color of objects.

In the rhesus monkey, extensive (but almost certainly incomplete) bilateral removals of striate cortex lead to a loss of fine discrimination of stationary objects. However, visuospatial orientation and the ability to reach toward moving peripheral targets remain relatively intact.[59] In humans, partial destruction of geniculostriate structures leads to characteristic visual field deficits within which conscious form perception of still objects is lost. On the other hand, flashing or moving objects can sometimes be detected in the blind field, even though the detection thresholds for such stimuli are about 50 times higher than those in the good field.[15] Some ability for reaching toward visual stimuli in the blind field may also be retained, even though the patient may deny awareness of the stimulus.[222,297]

Primary Auditory Cortex

The primary auditory koniocortex (areas 41 and 42), also known as A1, is located on Heschl's gyrus on the posterior aspect of the supratemporal plane (Fig. 6). This area receives inputs from the medial geniculate body, which is the thalamic relay nucleus in the auditory modality. There is a tonotopic organization in A1 so that the low frequencies are represented more rostrally than are the higher frequencies.[177] Single-unit recordings in monkeys also show that A1 units are sensitive to the location of sound sources.[22] For example, some A1 units are much more active when the animal is required to identify the location of a sound source than when the task is merely the detection of the sound. Neurons in the A1 region of each hemisphere are more likely to have their peak firing rate in response to sounds originating in the contralateral extrapersonal space. However, the spatial map in A1 is nowhere as specific as the map in V1. In monkeys, bilateral ablations of A1, which extend into adjacent asso-

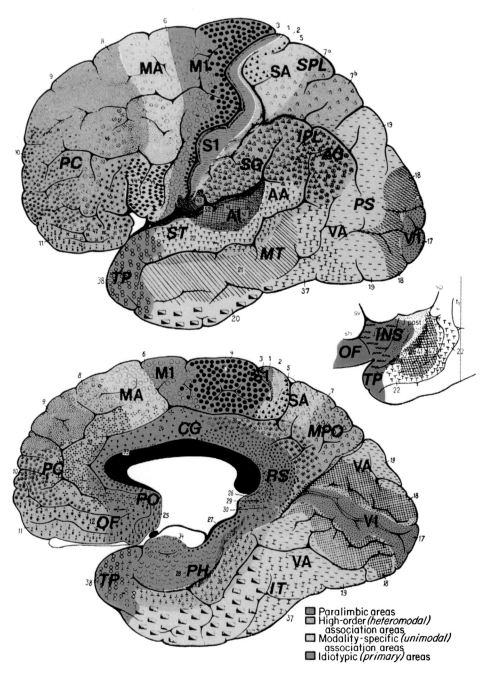

FIGURE 6. Distribution of functional zones in relationship to Brodmann's map of the human brain. The boundaries are not intended to be precise. Much of this information is based on experimental evidence obtained from laboratory animals and needs to be confirmed in the human brain. Abbreviations: AA—auditory association cortex; AG—angular gyrus; A1—primary auditory cortex; CG—cingulate cortex; INS—insula; IPL—inferior parietal lobule; IT—inferior temporal gyrus; MA—motor association cortex; MPO—medial parieto-occipital area; MT—middle temporal gyrus; M1—primary motor area; OF—orbitofrontal region; PC—prefrontal cortex; PH—parahippocampal region; PO—parolfactory area; PS—peristriate cortex; RS—retrosplenial area; SA—somatosensory association cortex; SG—supramarginal gyrus; SPL—superior parietal lobule; ST—superior temporal gyrus; S1—primary somatosensory area; TP—temporopolar cortex; VA—visual association cortex; V1—primary visual cortex.

ciation cortex, lead to deficits in a task that requires the animal to walk toward the source of a sound.[105]

The anatomic organization of the primary auditory cortex is quite different from that of primary visual and somatosensory cortices. First, through multisynaptic pathways that have extensive decussations in the brainstem, the A1 of each hemisphere has access to information from both ears even though the influence of the contralateral ear appears to be stronger. Therefore, unilateral A1 lesions do not lead to contralateral deafness. In fact, such lesions would probably remain undetectable without the assistance of auditory evoked potentials or dichotic listening tasks. In the latter test, the simultaneous delivery of stimuli to both ears yields excessive suppression of the input into the ear contralateral to the A1 lesion (see Chapter 2). Second, in contrast to the primary visual and somatosensory thalamic relay nuclei (which have no substantial connections with the pertinent association areas), the medial geniculate body has major projections not only to A1 but also to the unimodal auditory association areas in the adjacent superior temporal gyrus.[186] Therefore, complete cortical deafness is not likely to arise unless there is bilateral damage both to A1 and to the adjacent auditory association areas.

Primary Somatosensory Cortex

The postcentral gyrus contains the primary somatosensory cortex, S1 (Fig. 6). This area is the major recipient of projections from the ventroposterior thalamic nucleus which, in turn, is the thalamic relay for the ascending somatosensory pathways. Some investigators have convincingly argued that only area 3b should be included in S1, since areas 1 and especially 2 have characteristics more consistent with those of unimodal somatosensory association cortex.[135]

The contralateral half of the body surface is somatotopically mapped onto the S1 region of each hemisphere. The mouth and face areas are represented most ventrally; the hand, arm, trunk, and thigh more dorsally; and the leg and foot medially.[220,237] With the exception of the hand and foot representations, most parts of S1 have callosal connections with homologous parts of S1 in the opposite hemisphere.[137]

In the monkey brain, single-unit recordings show that area 3a is activated by muscle spindle afferents, that areas 3b and 1 are most easily activated by cutaneous input, and that area 2 has a preferential response to joint receptors. Each of these three subdivisions may contain an individual body map.[135] It has also been shown in the monkey that area 2 lesions impair size and curvature discrimination, that area 1 lesions impair texture discrimination, and that both types of deficits arise from area 3b lesions.[42,230] Neurons in S1 are especially involved in *active* tactile exploration. Thus, in one study, 13 of 76 S1 neurons fired only when the monkey's finger was moving but not when it was at rest, even though the same receptive field was being stimulated.[54]

In humans, S1 damage is usually associated with a selective impairment of the so-called cortical sensations, such as two-point discrimination, touch localization, position sense, and stereognosis; whereas touch, pain, and temperature detection may remain relatively preserved.[50,235]

Primary Motor Cortex

The primary motor cortex (M1) is located in the precentral gyrus (Fig. 6) and contains a body representation closely paralleling that of S1.[220] The M1 region can be identified on the basis of three characteristics:

1. The cytoarchitecture stands out by the presence of large pyramids that reach their greatest size in area 4 (Betz cells)
2. The threshold for eliciting movement upon electrical stimulation is lower in M1 than in any other cortical area
3. The M1 region contains the greatest density of neurons giving rise to corticospinal and corticobulbar fibers

The literature contains considerable controversy with respect to the distinction between M1 and "premotor" cortex. Some have argued that only area 4 (the Betz cell zone) deserves to be included in M1. However, there are convincing arguments showing that M1 should include not only area 4 but also the posterior half of area 6, a view that is followed in this chapter.[292,302] Experiments in monkeys show that the M1 output is especially important in controlling the early-recruited portion of the motoneuron pool which is involved in precise fine movements.[72] In keeping with these physiologic characteristics, extensive M1 removals in monkeys lead to relatively subtle deficits, mostly confined to individual finger movements, whereas other movements and especially posture and gait remain relatively intact.[151]

The clinical consequences of M1 lesions in man are poorly understood. Some authorities argue that such lesions may only impair fractionated distal limb movements while leaving tone and proximal muscles intact. Others argue that M1 lesions lead to increased tendon reflexes and to widespread paralysis of entire limbs.[155]

Some forms of apraxia result from an inability to convert verbal command into skilled movements that require fractionated control of distal limb musculature (see Chapters 2 and 5). Such apraxias may result from lesions that interrupt the multisynaptic pathways leading from language areas to M1. The execution of commands aimed at the axial musculature may remain preserved in these patients, since such movements can be coordinated by other parts of the brain.[86] As in S1, the hand and foot representations of M1 have no callosal connectivity.[137] Since this arrangement is likely to promote hemispheric independence of hand control, it may provide at least one essential anatomic substrate for the development of handedness.

Secondary Sensory-Motor Areas

In addition to the primary areas (V1, A1, S1, M1), it is also customary to designate secondary sensory and motor areas. The V2 area is located in peristriate cortex; A2 is just medial to A1; S2 is located in the inner aspect of the parietal operculum and adjacent to the dorsal insula; and M2 (supplementary motor area) is found on the medial surface of the hemisphere, just anterior to M1. These regions have some properties in common with the relevant primary areas; however, they have a more isocortical architecture and a less well-defined topographic representational map. These features suggest that the secondary areas may occupy an intermediate position between primary cortex and unimodal association areas. The physiology of these secondary areas and their role in behavior are not fully understood. Some of the relevant information will be discussed in the sections on unimodal association cortex.

Modality-Specific (Unimodal) Association Areas

The second stage in the cortical analysis of sensory information takes place within modality-specific (unimodal) association areas. The unimodal areas act as obligatory relays for the intercortical transfer of sensory information from primary areas toward other parts of cortex. Consequently, there are two major classes of behavioral deficits that reflect a physiologic disruption of unimodal cortex function. First, lesions directly within a unimodal area (or one that selectively disrupts its input from the relevant primary sensory area) give rise to complex perceptual deficits in that modality, while elementary sensation remains intact. Second, lesions that interfere with specific output channels from unimodal areas deprive selected heteromodal, paralimbic, or limbic regions of information in that sensory modality. This second type of lesion leads to modality-specific *disconnection syndromes*, which are of central importance to behavioral neurology.

Visual Unimodal Association Areas

In the primate brain, the unimodal visual association areas are more extensive than association areas in any other modality. Extrapolation from experiments in monkeys suggests that the peristriate belt (areas 18 and 19), as well as most of the middle and inferior temporal gyri (areas 20, 21, and 37), are occupied by unimodal visual association isocortex (Fig. 6). Experimental observations in monkeys show that a tightly organized multisynaptic cascade of neural connections links V1 to peristriate cortex and then peristriate cortex to successively more anterior parts of the temporal visual association zones.[193] Peristriate cortex depends on V1 for its visual input, and the temporal visual areas depend on peristriate cortex for theirs. This neural arrangement is consistent with physiologic observations showing that progressively more abstract features of visual information become extracted along a gradient that leads from V1 toward the anterior end of the visual unimodal area in the temporal lobe.

Neurons in the peristriate area have firing properties similar to those of neurons in V1. However, they also show regional specialization for analyzing more complex aspects of visual information such as motion, form, binocular disparity, and color.[282] The peristriate belt and adjacent visual unimodal areas have recently been divided into at least five individual regions (V2, V3, V4, VP, and MT), each emphasizing an apparently different aspect of visual information processing. Many of these "areas" have a topographic representation of the visual field although the mapping is not as specific as in V1.

In monkeys, selective (but probably not complete) peristriate ablations lead to deficits of spatial orientation, depth perception, distance judgment, stereoacuity, and hue discrimination thresholds, while leaving stationary object discrimination relatively intact.[51,59] The peristriate component of visual association cortex has four major cortical output pathways: to the contralateral peristriate belt, to the caudal part of the frontal eyefields (area 8), to posterior parietal heteromodal fields, and to the temporal visual association areas. It has been suggested that the outflow from the dorsal part of peristriate cortex (directed mostly to parietal heteromodal fields) is predominantly involved in visuospatial processing, whereas the ventral outflow (directed to temporal visual unimodal areas) plays a major role in pattern identification and visual-limbic interactions.[195]

The second major component of visual association cortex is located within the middle and inferior temporal areas. This region receives its major visual input from peristriate cortex and has four major cortical outputs: to the contralateral temporal visual association areas, to the prefrontal heteromodal region, to the posterior heteromodal fields (especially to the banks of the superior temporal sulcus), and to the paralimbic and limbic areas of the medial temporal lobe. With respect to this last projection, it is interesting to note that the limbic and paralimbic outputs originate mostly from the more medial and anterior (e.g., synaptically most downstream) parts of the temporal visual association cortex.[109,180,181,287]

In the monkey, neurons in these temporal visual association areas respond almost exclusively to visual stimulation and not to stimulation in other sensory modalities.[63] As in striate and peristriate cortex, these neurons are also sensitive to contrast, wavelength, size, shape, orientation, and movement. However, in contrast to peristriate neurons, these neurons have much larger receptive fields that often include the fovea and that are commonly bilateral. These temporal visual neurons also seem to have very idiosyncratic trigger properties in response to specific objects, including faces.[36,96,223] Thus, one process that may take place in this part of the brain is the extraction of complex features from visual information, so that neurons become responsive to individual patterns rather than to isolated stimulus features such as color and location.

It has been suggested that the multiplicity of peristriate visual areas may represent a division of labor for analyzing different aspects of stimulus features and that convergent inputs from these areas to the temporal visual association areas could provide a mechanism for integrating these different features into the perception of an individual object.[62] Thus, during the act of perceiving an object, a unique constellation of peristriate inputs—representing a specific set of the object's visual attributes such as shape, color, and texture—could converge upon single neurons of the temporal visual association areas.[193] Assemblies of these neurons could then provide a central visual representation for that object. The invariant properties of objects could be extracted through a subsequent process of signal averaging. For example, since the same object can be seen in different portions of the visual field and under different lighting conditions, such variations in brightness and location could be averaged out upon multiple exposures, leaving a template that emphasizes the invariant features of the object. Through repeated procedures of convergence and signal averaging, different levels of visual abstraction could be achieved.

The hypothetical sequence of events outlined above is most relevant to object *discrimination* tasks. However, the task of object *recognition* or *identification* also requires that the visual representation interact with other components of mental content, including the individual's past experience. These aspects of visual information processing depend on the second major role of the temporal visual association areas—namely, to relay extensively processed visual information toward heteromodal, paralimbic, and limbic zones of the brain.

As in the case of peristriate cortex, the physiologic function of the temporal visual association areas can be disrupted through two separate mechanisms. Direct damage to this part of the brain or to its input from peristriate cortex may impair the formation of visual templates. On the other hand, the interruption of one or more output pathways may lead to modality-specific disconnections that impair recognition and associative elaboration of visual information, while leaving object discrimination abilities intact. For example, in the monkey, the major behavioral impact of lesions in the anterior parts of the temporal visual association areas seems to occur through the interruption of outflow to limbic-paralimbic areas. Such lesions lead to severe deficits of visual recognition memory, with pattern discrimination abilities remaining intact. On the other hand, extending the ablation to the more posterior aspects of the temporal visual area impairs pattern discrimination abilities as well.[114,193,245]

In the human, electrical stimulation of peristriate cortex often leads to the experience of color, light, shadow, outline, and movement.[220] In volunteer subjects undergoing positron emission tomographic scanning, exposure to white light increased the metabolic activity only of V1, whereas the presentation of a more complex visual scene led to additional metabolic activation of peristriate cortex.[226] This observation is consistent with the notion that peristriate cortex analyzes the more complex aspects of visual information.

Damage to peristrate or temporal visual association areas and to their connections in the human brain would have a number of consequences:

1. Impairment of specialized visual processing and template formation
2. The loss of previously formed templates
3. Interruptions in the relay of visual input to parietal and frontal heteromodal areas, leading to visual-somatosensory, visual-auditory, visual-motor, and visual-verbal disconnections
4. Interruptions in the flow of visual input into limbic and paralimbic regions of the brain

Patients have been reported in whom lesions involving peristriate areas have led to remarkably specific disturbances of either color vision (*achromatopsia*) or movement perception, without essentially disturbing other visual functions.[53,309] Such cases show that as in the monkey, the human peristriate cortex also contains subregions specialized for the perception of color and movement.

Dorsally situated parieto-occipital lesions interfere with peristriate outflow pathways directed toward parietal heteromodal zones where visual-somatosensory interactions are likely to occur. When bilateral, such dorsal lesions may lead to severe visuospatial deficits as part of *Balint's syndrome* (see Chapter 7). The visual reaching and ocular scanning disturbances seen in Balint's syndrome may reflect the interruption of peristriate projections not only to posterior parietal cortex but also to the frontal eye fields. More ventrally situated temporo-occipital lesions may damage the temporal visual association areas, their input from peristriate cortex, or their outflow toward language areas and limbic structures. The clinical manifestations of such lesions include *alexia, visual anomia, pattern discrimination difficulties, modality-specific amnesias,* and other *visual agnosias* (see Chapter 7). Such ventral lesions are also likely to be associated with achromatopsia, since the peristriate region specialized for color perception appears to be ventrally situated (see Chapter 7).

The condition of *object agnosia*, in which the patient can neither name nor recognize an object by sight, is the most severe of these ventral syndromes and involves widespread visual-verbal and visual-limbic disconnections, a destruction of previously formed templates, and probably also the disruption of new visual template formation. The verbal and limbic disconnections are likely to account for the *associative* aspects of object agnosia, whereas the impairment of template formation would introduce an additional *apperceptive* component. *Prosopagnosia*, which might be considered an incomplete form of object agnosia, is based on a similar set of mechanisms, except that it does not include a prominent visual-verbal disconnection. In *visual anomia*, the patient recognizes the object (knows its meaning) but cannot name it until he is given nonvisual cues. The patient with *pure alexia*, on the other hand, has a deficit of visual recognition confined to words and letters, whereas visual naming of objects remains intact (see Chapters 5 and 7). Pure alexia (word blindness) is essentially a visual agnosia for *verbal* material. Visual anomia and pure alexia result from visual-verbal disconnections that arise when the language areas of the left hemisphere are partially deprived of their visual input from unimodal areas. In still other patients, lesions to these temporo-occipital parts of the brain can lead to modality-specific *visual memory disturbances* (see Chapter 4). In such patients, some of whom may have no anomia or agnosia, the mechanism of visual-limbic disconnection predominates. By analogy to the connectivity pattern in the monkey brain, lesions in the anterior temporal visual areas would be more likely to lead to isolated visual-limbic disconnections, whereas more posterior lesions—especially those that also include the adjacent peristriate regions—are likely to result in additional visual-verbal disconnections.

In object agnosia, prosopagnosia, and modality-specific visual amnesia, the cerebral damage is often bilateral. The ventral occipitotemporal areas are particularly susceptible to such bilateral involvement, since they are supplied by the posterior cerebral arteries that arise from a common basilar trunk. Pure alexia, on the other hand, can result from a single left-sided lesion that prevents both the ipsilateral and the contralateral visual association input from reaching the language areas in the left temporoparietal region (see Chapter 7).

While the disconnections that cut off visual information from language areas of the left hemisphere have attracted a great deal of attention, there is much less information about the consequences of lesions that interfere with the visual input to equivalent heteromodal association areas of the right hemisphere. It is conceivable that such lesions may lead to nonverbal modality-specific cognitive disturbances. For example, we saw a highly articulate right-handed man in his 50s who suffered an acute infarction involving much of the right temporal lobe (Fig. 7A). Following the acute period, the only residual elementary disturbance was confined to an incomplete left upper quadrantanopia. However, he bitterly complained of complex visual disturbances. He realized that he could no longer focus awareness on the relevant aspects of the visual environment. For example, while fishing, he could not concentrate on the bobber and would find his attention attracted equally strongly by a bubble or a dead leaf on the stream. His work required him to visit factories and to visually inspect the fire extinguishers, exits, and building materials in order to estimate the risk factors related to insurance premiums. After the cerebral injury, although he had no difficulty identifying individual components, he was no longer able to rapidly integrate visual information with his past experience in order to reach an expert judgment. This did not indicate an overall decrement in mental capacity, since written documents of considerable complexity posed no more difficulty than before the stroke, especially if they were read aloud to him. It appears that this man suffered a permanent impairment in his ability to distribute attention within the visual space and in the ability to integrate visual input with other components of mental activity.

Damage to the human brain almost never respects cytoarchitectonic boundaries. For example, lesions in the temporal lobes commonly extend to the adjacent V1, to the more medially situated limbic and paralimbic areas, or to the more dorsal language areas. Furthermore, nearby fiber bundles such as the inferior longitudinal fasciculus, geniculocalcarine pathways, and the splenium of the corpus callosum may also be involved. This explains why patients with the clinical manifestations of visual association cortex lesions may also have additional features of aphasia, generalized amnesia (secondary to limbic-paralimbic involvement), and visual field deficits (secondary to involvement of V1 or geniculostriate pathways).

Although the clinical deficits that arise from lesions of unimodal visual association areas are decidedly complex and heterogeneous, they all share a modality specificity for tasks that depend on the processing of visual information.

Auditory Unimodal Association Areas

Although the auditory and somatosensory association areas should theoretically lend themselves to an analysis similar to that of visual associ-

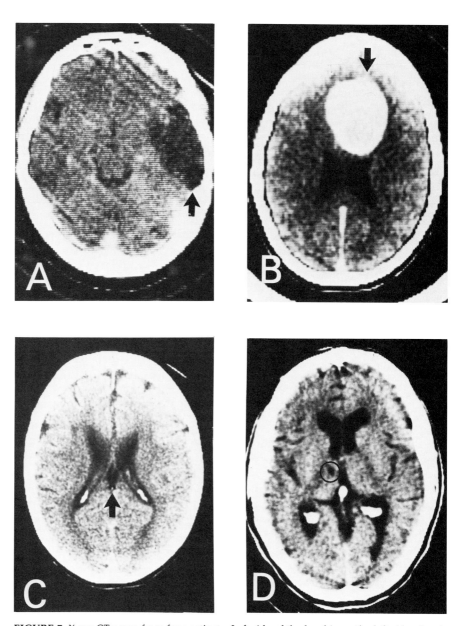

FIGURE 7. X-ray CT scans from four patients. Left side of the head is on the left side of each scan. (*A*) This 54-year-old right-handed man had a right temporal stroke (*arrow*). The lesion included the superior, middle, and inferior temporal cortex. Salient behavioral deficits included impairments in visual processing and in paralinguistic skills. (*B*) For three years before the discovery of this prefrontal meningioma (*arrow*), this 50-year-old ambidextrous woman carried the diagnosis of an atypical depression with hysterical features. There were essentially no elementary neurologic findings, and most cognitive faculties (e.g., memory and language) appeared preserved. She displayed inappropriate and grandiose behavior and many errors of commission in the go–no-go task. Her behavior and her go–no-go performance both improved markedly after removal of the tumor. (*C*) This 83-year-old right-handed woman had a relatively isolated and progressive amnestic state. The CT scan showed a lipoma of the posterior corpus callosum (*arrow*). (*D*) This 55-year-old right-handed man suffered a lacunar stroke to the left thalamus. The lesion appears to have involved both the anterior nuclear group and the medial part of the medialis dorsalis nucleus of the thalamus (*encircled area*). The patient developed an amnestic condition of moderate severity (mostly for retrieval of verbal material) in conjunction with this lesion.

ation areas, there is much less basic and clinical information on these modalities. Therefore, the relevant discussion of auditory and somatosensory association areas is much briefer than that of visual association areas.

The superior temporal gyrus (area 22) contains the auditory unimodal association area (see Fig. 6). Experiments in the monkey brain show that the flow of auditory information follows a hierarchical cascade that shares many characteristics with the flow of visual information. Projections from A1 are directed to the posterior parts of the superior temporal gyrus. This information is then relayed in multiple steps to progressively more anterior parts of the gyrus. It is the most downstream part of auditory association cortex that gives rise to most of the outputs directed toward paralimbic and limbic structures of the temporal lobe. The auditory association area also projects to the prefrontal and temporoparietal heteromodal fields. In monkeys, lesions that destroy auditory unimodal association areas lead to impairments in the *retention* and *discrimination* of auditory frequency and sequences.[52,123] These deficits are likely to reflect auditory-limbic disconnections and disturbances of auditory template formation, respectively.

In humans, electrical stimulation of the temporal lobe results in auditory experience only when the supratemporal plane and superior temporal gyrus are being stimulated.[220] In conscious subjects, auditory stimulation increases the metabolic activation of the superior temporal gyrus.[171,236] By analogy to the connectivity pattern in the monkey brain, it is reasonable to assume that the heteromodal language areas (e.g., parts of Wernicke's area and the inferior parietal lobule) are dependent on the unimodal auditory association cortex for their auditory input. Thus, bilateral lesions in auditory association areas or a strategically situated unilateral left-sided lesion that also interrupts the transcallosal input from the contralateral auditory association area may lead to a complete auditory-verbal disconnection. This is thought to be the mechanism of *pure word deafness,* which is one of the most dramatic phenomena in behavioral neurology (see Chapter 5). Patients with this condition react quite appropriately to environmental sounds (hence, they are not deaf), and they have perfect understanding of written language (hence, they are not aphasic), but they cannot understand or repeat spoken language. Depending on the site of the lesion, there may be variations in the relative contribution of apperceptive and associative components in pure word deafness. The apperceptive component is likely to dominate if A1 and auditory association areas are heavily involved, whereas the associative component is likely to dominate if the damage is predominantly directed to the relevant output channels. Pure word deafness is essentially an auditory agnosia for *verbal* material. In that sense, it should be considered an auditory analogue of pure alexia without agraphia.

The auditory association cortex in the caudal third of the superior temporal gyrus is generally included within the region commonly designated as Wernicke's area (see Fig. 1 in Chapter 5). This is the only modality-specific constituent of Wernicke's area and reflects the fundamental importance of auditory input in the acquisition of language (see Chapter 5). Damage to auditory association cortex can therefore also contribute to the emergence of *Wernicke's aphasia* (perhaps mostly to its word-deaf component). Wernicke's area almost certainly includes additional heteromodal association areas as well, since the language deficit in Wernicke's aphasia is not confined to any single sensory modality.

Lesions in the auditory association cortex of the right hemisphere may lead to *nonverbal auditory agnosias*. Such patients may have difficulty identifying environmental sounds, familiar melodies, and variations in timbre.[172,173,288] As in the case of visual association areas, therefore, lesions to auditory association areas lead to complex but modality-specific deficits.

Somatosensory Unimodal Association Areas

In the monkey, the superior parietal lobule (area 5) is the major recipient of S1 projections (mostly from area 2) and is generally designated as the unimodal somatosensory association area. The sensory responses of approximately 98 percent of the neurons in area 5 are exclusively confined to the somatosensory modality. However, the firing patterns show an intramodal convergence among deep and superficial sensations, thus indicating a higher order of neuronal processing than in S1.[202] There is essentially no information on the behavioral consequences of lesions confined to area 5 in the monkey.

There is a major difficulty in establishing architectonic homologies for this part of the human brain. In the rhesus monkey, essentially the entire superior parietal lobule has been designated as area 5 by Brodmann. In the human, on the other hand, Brodmann's area 5 is confined to the anterior fifth of the superior parietal lobule (see Fig. 6). Brodmann designated the rest of the human superior parietal lobule as area 7. However, in the monkey, area 7 was used by Brodmann to designate the cortex of the *inferior* parietal lobule, an area that contains heteromodal association cortex. It is not clear whether more emphasis should be given to the topographic homologies or to the cytoarchitectonic parallels implied by the Brodmann nomenclature. For the purposes of this chapter, an intermediate approach is taken, and area 5 and the anterior part of area 7 in the superior parietal lobule of the human brain are assumed to constitute a unimodal somatosensory association area (see Fig. 6). The more caudal aspects of area 7, together with the inferior parietal lobule, are considered as components of heteromodal association cortex. (See Chapter 3 for further discussion on the homologies of posterior parietal cortex.)

The behavioral consequences of damage confined to the human somatosensory association areas are poorly understood. It is conceivable that this region may be essential for the finer aspects of touch localization, for active manual exploration, for the synthesis of the body schema, and for the formation of complex somatosensory memories. Furthermore, this region almost certainly provides the somatosensory output destined for posterior parietal heteromodal areas where extensively processed visual and somesthetic information are likely to converge. This convergence may be essential for visuospatial orientation, as well as for the ability to align the body axis with other objects during the process of dressing, sitting in a chair, or getting into a bed. Interfering with the output of this region to parietal heteromodal areas may therefore contribute to the pathogenesis of *Balint's syndrome* and other aspects of *spatial disorientation*.[138] It is conceivable that some lesions could selectively cut off heteromodal language areas from somatosensory association input. Such lesions should theoretically give rise to "agraphesthesia without astereognosis," which could be considered the somatosensory analogue of pure word deafness and pure alexia. Somatosensory input is relayed to the limbic-paralimbic regions of the brain via the insular cortex. In the monkey, an interruption of this relay leads to modality-specific somatosensory learning deficits.[206]

Motor Association Areas

It is possible to consider the cortex anterior to M1 as a motor analogue of the other modality-specific association areas (see Fig. 6). This region is unimodal with respect to output rather than with respect to input. The motor association area has the following characteristics:

1. In contrast to the idiotypic macropyramidal architecture of M1, it has a more isocortical (granular) appearance
2. It provides the major cortical source of association inputs into M1
3. Its lesions yield complex and sometimes category-specific deficits of motor output without primary weakness

In the monkey brain, it is possible to include the rostral part of area 6 and the M2 region (the supplementary motor area) within the group of motor association areas. While the anterior parts of area 8 (frontal eye fields) have characteristics more consistent with a heteromodal association area, the posterior part of this region could also be included within the group of motor association areas, especially in relation to eye movements.

The rostral part of area 6 receives input from several unimodal and heteromodal areas of the brain so that it has access to complex information in all major sensory modalities.[129,217,233,294] This region provides one of the most substantial cortical inputs into M1. This area also contributes a substantial number of descending corticospinal and corticobulbar fibers but the density is less than in M1.[275] Microstimulation of this region leads to movement but the threshold for this is higher than in M1.[294] Although these motor association (or premotor) neurons respond to sensory stimuli, the responses vary according to the movement that will follow. For example, visually responsive neurons in this area show one response to a visual cue that triggers a movement and a different response when the same cue requires the animal to withhold movement.[301] In monkeys that were taught to elevate a lever to obtain reward, a surface-negative slow potential appears over M1 and also over the motor association cortex about 1 second prior to the movement and gradually increases until about 100 milliseconds before the movement.[102] It is conceivable that this slow potential reflects the neural mechanisms that subserve motor intention and initiation.

Monkeys with lesions that include components of motor association cortex develop category-specific and complex deficits of movement in the absence of primary weakness. For example, animals with lesions in area 6 may no longer be able to change the nature of a motor act in response to differential sensory cues.[101] The visual guidance of motor behaviors is also impaired after such lesions.[197] Furthermore, unilateral lesions in the region of the frontal eye fields impair the motor scanning and exploration of the contralateral hemispace (see Chapter 3).

In the human brain, it seems reasonable to include anterior area 6, the supplementary motor region (M2), posterior area 8, and area 44 (a component of Broca's region) within the category of motor association cortex (see Fig. 6). As discussed in Chapter 3, area 8 in humans as well as in monkeys may modulate the motor aspects of exploration and scanning in the process of directed attention. In the left hemisphere, damage that includes area 44 is often associated with *Broca's aphasia* (see Chapters 5 and 11). Although this aphasia has many complex linguistic features, one of its most salient manifestations is an impairment in the generation of speech-related motor programs. This is a category-specific motor impairment, since some of these patients have no difficulty coordinating the same muscle groups in the act of singing or whistling. The clinical effects of small lesions in this region may eventually be resolved to a state of nonaphasic dysarthria and right-sided clumsiness.[196]

Anatomic and physiologic considerations show that M2 is much more like premotor cortex than it is like M1.[294] Despite the voluminous literature, it is still not clear whether there are any long-term motor consequences of supplementary motor area damage other than an impairment of rapid alternating movements.[156] Roland and associates[238] have made the interesting suggestion that the supplementary motor area may provide a covert plan for intended movements. They showed that actual finger movements yielded local bloodflow increases in this part of the brain as well as in the appropriate M1 region. However, when they asked the subject to think about performing the same movement without actually executing it, only the supplementary motor area still showed activation. This region may therefore be important for motor planning. Others have suggested that this area modulates the initiation of motor responses and the ability to sustain motor output. For example, the paucity of speech output in *transcortical motor aphasia* may result from a disconnection between the supplementary motor region and Broca's area.[80]

The concept of a motor association area is complex. The unifying feature of the component areas is that they modulate the sensory guidance, as well as the initiation, inhibition, planning, and perhaps also learning of complex movements.[31] Furthermore, damage to these areas results in category-specific disturbances of movement without generalized weakness, clumsiness, or dystonia.

High-Order (Heteromodal) Association Areas

Within primary and unimodal regions, the analysis of sensory experience remains confined to single modalities. The opportunity for inter-relating the attributes of real events belonging to separate modalities occurs in the subsequent stage of sensory processing within heteromodal regions of the brain. There is a real hierarchy in the processing of sensory information (see Fig. 5). For example, the heteromodal areas in the monkey brain almost never receive their sensory information directly from primary areas. Instead, unimodal association cortex acts as an obligate intermediate relay.[129,189,217]

Within heteromodal areas, the modality specificity of information is lost in favor of intermodal associations. Even the distinction between what is sensory and what is motor is no longer present. For example, many cells in heteromodal areas increase firing not only in response to sensory input but also in phase with motor output.[146,149] It could be argued that during the process of gaining awareness of the world, sense organs are not passive portals for sensory input. Instead, they could be considered to act as tentacles or feelers for the active scanning and updating of a dynamically shifting inner representational map.[68,179] At this level, the building blocks of awareness are as much motor as they are sensory phenomena, and this is clearly reflected in the physiology of heteromodal areas.

At least two essential transformations are likely to occur in heteromodal areas. First, these areas provide a neural template for intermodal associations necessary for many cognitive processes, especially language (see Chapter 5). Second, they provide the initial interaction between extensively processed sensory information and limbic-paralimbic input.[189] Thus, another distinction that is lost in heteromodal areas is that between limbic and nonlimbic. This may initially come as a surprise, since there is a tendency to think of heteromodal cortex as a high-order association area devoted to intellectual processes and hence impervious to limbic impulses. However, the evidence obtained in the macaque brain unequivocally shows that heteromodal areas receive substantial paralimbic input. This anatomic arrangement explains how mood and drive can influence the manner in which the self and the world are experienced and also how thought and experience eventually influence mood.

Damage to heteromodal areas yields a set of complex deficits with cognitive as well as affective components. In lesions of the temporoparietal heteromodal fields, it is the disruption of transmodal integration that gains prominence. On the other hand, the additional affective and motivational components become more conspicuous in lesions of the frontal heteromodal fields.

Temporoparietal Heteromodal Association Areas

In the monkey, the inferior parietal lobule and the banks of the superior temporal sulcus receive inputs from a great variety of unimodal and heteromodal areas as well as from paralimbic regions.[189,216,254] The posterior part of the inferior parietal lobule (area PG or the caudal part of area 7 in the monkey) is somewhat unique in having almost exclusively heteromodal and paralimbic inputs.[189] This region of the monkey brain can thus be considered a supramodal part of cortex that acts as an association area for other high-order association regions of the brain.

Single-unit recordings in the temporoparietal heteromodal areas of the monkey show that neurons responsive to a certain modality are intermingled with those that respond to other modalities, and that some neurons actually respond to multiple modalities.[20,63,118] Many neurons in the inferior parietal lobule have sensory as well as motor response contingencies.[119,202] Furthermore, neurons in the more caudal aspects of the inferior parietal lobule of the monkey are also responsive to the motivational relevance of objects. For example, these neurons may increase firing in response to a behaviorally relevant object, such as food when the monkey is hungry or liquid when thirsty. This increment in discharge declines as soon as the motivational state is altered by the delivery of reward.[166]

Lesions in the temporoparietal component of heteromodal cortex in the monkey brain yield three types of deficits: those that are perceptual-motor; those that depend on transmodal integration; and those that are related to the distribution of attention within the extrapersonal space. For example, after bilateral ablations in these areas, monkeys cannot resolve a complex visual-motor maze, even though they have no difficulties in visual pattern perception or muscle coordination.[225] Similar lesions also lead to impairments in the monkey's ability to determine spatial relationships among objects in the extrapersonal space.[279] Although visual and auditory perception remains intact, damage to these heteromodal fields impairs the ability to perform visuo-auditory compound discriminations.[224] When unilateral, damage to this part of the brain leads to contralateral neglect, a syndrome that may reflect a disruption in the perceptual and motivational representation of the extrapersonal space (see Chapter 3).

The human brain almost certainly contains an analogous temporoparietal heteromodal field (see Fig. 6). This is likely to include areas 39 (the angular gyrus) and 40 (the supramarginal gyrus) of the inferior parietal lobule, the banks of the superior temporal sulcus, and probably also the caudal aspects of area 7 in the superior parietal lobule. One very important constituent of this region is Wernicke's area, which includes the posterior parts of auditory association cortex, heteromodal cortex along the banks of the superior temporal sulcus, and perhaps also some of the adjacent inferior parietal lobule. The transmodal associations that are necessary for processing written and spoken language are thought to be mediated largely through Wernicke's area in the left cerebral hemisphere.[86] Damage to this area gives rise to *Wernicke's aphasia* (see Chapters 5 and 11). The comprehension deficit that is the hallmark of Wernicke's aphasia encompasses all modalities of language input and highlights the heteromodal nature of the underlying brain region. This contrasts with pure alexia and pure word deafness, in which the comprehension deficits reflect the interruption of modality-specific input into Wernicke's area (see Chapters 5 and 7).

When damage spares Wernicke's area but involves the adjacent heteromodal fields of the left hemisphere, especially those in the inferior parietal lobule, then complex combinations of *anomia, alexia, construction deficits, acalculia, dysgraphia, finger identification disturbances*, and *left-right naming difficulties* may be noted, sometimes in the absence of generalized language comprehension deficits. This group of impairments is collectively referred to as the *angular gyrus syndrome*. When only the last four findings appear without the others, the term *Gerstmann syndrome* is used to identify the clinical picture. Both syndromes indicate a breakdown in complex transmodal associations. In contrast to pure alexia, which is a modality-specific disconnection syndrome, the alexia that arises in conjunction with inferior parietal lobule damage is referred to as a "central" alexia, reflecting a breakdown in the relevant transmodal associations (see Chapter 5).

Unilateral lesions in the analogous regions of the right hemisphere lead to characteristic deficits including *dressing difficulties, constructional difficulties, multimodal neglect* of the left hemispace, and *global confusional states* (see Chapter 3). Lesions that predominantly include the dorsal part of these heteromodal fields, especially when bilateral, give rise to complex visuomotor and visuospatial disturbances (e.g., *Balint's syndrome*) and to *global spatial disorientation.*[138] Some of these deficits reflect a breakdown of visual-somatosensory interactions that are likely to take place in these regions. The parietal heteromodal areas may also contain a multimodal representational schema of the extrapersonal world. The disruption of this schema may account for some manifestations of unilateral neglect (see Chapter 3).

In addition to these cognitive and integrative deficits that are distinctly heteromodal with respect to sensory channels, there are also affective components that emerge as a consequence of lesions in this region of the brain. For example, neglect behavior also includes an aspect of motivational indifference directed toward the contralateral hemispace (see Chapter 3). Denial of hemiplegia *(anosognosia)*, which can be seen as part of neglect syndromes, may be another manifestation of this affective component. Patients who develop Wernicke's aphasia may show severe mood alterations ranging from anger to paranoia and indifference (see Chapter 5). There are also patients with lesions in the posterior temporoparietal regions of the right hemisphere whose clinical presentation is dominated by psychiatric and affective disturbances.[98] These clinical features might reflect the disruption of complex sensory-limbic interactions that are likely to occur in this part of the brain.

In the brain of the monkey, anatomic and physiologic observations suggest the presence of an additional heteromodal field in the ventral parts of the temporal lobe.[63,253] It is conceivable that an analogous area may also exist in the human brain (see Fig. 6). Perhaps the *agitated confusional states* that emerge in conjunction with lesions in this general area reflect, at least in part, the involvement of this heteromodal association area and of the adjacent paralimbic regions (see Chapter 3).

Prefrontal Heteromodal Association Areas

The prefrontal region contains the second large heteromodal field in the primate brain. In the monkey, this region receives neural inputs from all unimodal areas and from all the other heteromodal regions. The exact pattern of these inputs and the extent of intermodal overlap show considerable regional variations.[14,129,217] There are also substantial paralimbic inputs from cingulate, caudal orbitofrontal, and insular regions.[14,182,216] Single-unit recordings show a heterogeneity of sensory responses within any given region of the prefrontal heteromodal fields. Some neurons have a single preferred modality but are intermixed with those that respond to another modality; other neurons are truly multimodal and respond to stimulation in more than one sensory modality.[20,28,121]

In prefrontal cortex (as in the temporoparietal heteromodal region) the responses of neurons are even more "abstract" and less dependent on specific sensory dimensions than are those of neurons in unimodal association areas. Many of the visually responsive neurons in this part of the brain have no specificity for color, size, orientation, or movement.[149,191] On the other hand, these neurons are very sensitive to the behavioral significance of sensory information. For example, a neuron that responds briskly to a stimulus when it is associated with reward, may drastically alter its response when the same stimulus becomes associated with an aversive or neutral outcome.[145,274] These response characteristics imply that the prefrontal heteromodal fields may be in a position to associate motivational cues with complex objects and events.

There is also an integration of sensory input with motor output in prefrontal heteromodal cortex. Single-unit recordings suggest that these neurons may participate in the motivational aspects of movement initiation and in the modulation of response inhibition in tasks that require a delay of motor responses.[81,145,146,212] These effects could be mediated, at least in part, through the connections that exist between prefrontal heteromodal cortex and motor association areas.[9]

Ablations in the prefrontal heteromodal fields of monkeys lead to complex deficits that are not confined to any single modality. Depending on their exact location such lesions impair tasks that are dependent on delayed response alternation and also those that require the inhibition of nonrewarded responses in a go–no-go paradigm.[89,123] As mentioned earlier, the posterior part of the frontal eye fields can be included in the group of motor association areas, whereas the anterior parts are components of heteromodal cortex. Bilateral damage of the entire frontal eye fields, including their heteromodal portions, impairs compound multimodal stimulus discrimination and also the ability to vary a conditional motor response according to the nature of a sensory cue.[89,224,266,285] Unilateral lesions to this area result in multimodal neglect for the contralateral extrapersonal space (see Chapter 3). The type of behavioral disturbance that emerges in conjunction with damage to the prefrontal heteromodal fields of the monkey and the firing contingencies of the constituent neurons suggest that this region is a site for complex interactions among multimodal sensory input, motor output, and internal state.

The prefrontal granular cortex in the human brain (areas 45, 46, 47, 8, 9, 10, 11, and 12) almost certainly contains an analogous heteromodal region (see Fig. 6). For example, studies based on regional metabolic activation show that neural responses in these regions are not confined to any single modality of stimulation.[176,236] As in the monkey brain, unilateral lesions in the caudal part of prefrontal cortex lead to *contralateral neglect*, especially if the lesion is in the right hemisphere (see Chapter 3).

The identification of the other behavioral specializations of the human prefrontal cortex remains one of the most enigmatic areas in behavioral neurology. On the one hand, this part of the brain has been associated with the highest integrative faculties of the mind and has been considered the most important factor in the evolutionary success of the human race.[2,30,90] On the other hand, numerous reports have been published in which the authors fail to find any substantial cognitive deficits following considerable bilateral prefrontal lesions.[104,154] At least one explanation for this apparent discrepancy is that the behaviors that are most radically altered by prefrontal lesions are also those that are most difficult to examine in the office or at the bedside.

For example, clinical experience has repeatedly shown that the most dramatic consequences of prefrontal injury are seen in the more complex aspects of comportment and personality (Fig. 7B). Depending on the exact location of the lesion and perhaps also on premorbid personality factors, such patients may show a bewildering variety of behavioral disturbances. Some become puerile, slovenly, inappropriately jocular, grandiose, and irritable; some become apathetic and show a profound slowing of thought processes (abulia); others show an erosion of foresight, judgment, and insight; some others show a dulling of curiosity, vitality, and feeling; still others show an impairment in the abilities for reasoning and abstraction, jump to premature conclusions, and become excessively stimulus-bound. The planning and sequencing of complex behaviors, the ability to pay attention to several components at once, the capacity for grasping the gist of a complex situation, the resistance to distraction and interference, the inhibition of inappropriate response tendencies, and the ability to sustain behavior output for relatively prolonged periods may each become markedly disrupted. In contrast, motor dexterity, perceptual abilities, memory, language, and most other cognitive faculties remain intact. In the office, even the most assiduous testing may reveal surprisingly little. Even when deficits are noted, examiners are often left with the impression that patients could have done much better if only they had tried harder. However, the workplace and home provide the setting for a more rigorous assessment, and patients will usually report that they can no longer function properly because they cannot "concentrate" or because they cannot follow complex instructions (see Chapter 2).

In a most remarkable account, Penfield recorded the effects of a right-sided prefrontal lobectomy, which he was called upon to perform on his own sister.[219] At the end of the acute postoperative period, Penfield notes that his sister's judgment, insight, social graces, and cognitive abilities were quite preserved. However, when he visited her home as a dinner guest, he noticed a diminished capacity for the planned preparation and administration of the meal and a slowing of thinking. While such subtle changes are rather characteristic of unilateral prefrontal lesions, bilateral involvement leads to the more dramatic disturbances of motivation, insight, judgment, and comportment described earlier.

One deficit that is fairly common after prefrontal injury—and one that is relatively easy to demonstrate by standard tests—is a disturbance of certain attentional functions (see Chapters 2 and 3). Patients with this deficit are particularly sensitive to interference and they have great difficulty in inhibiting inappropriate response tendencies (Fig. 7B). This is particularly striking in the go–no-go paradigm, in which such patients emit a substantial number of inappropriate responses (i.e., commission errors) to cues that require them to withhold responding (see Chapter 2).

Except for the go–no-go deficits and the emergence of contralateral neglect, it has been difficult to find animal models for the other aspects of the frontal lobe syndrome. It appears that many of the behaviors that are closely dependent on the integrity of this part of the human brain are either absent or unobservable in the monkey. The one common denominator, however, is that the resultant deficits are not confined to any sensory modality and that many represent a dissolution of purpose, foresight, inhibition, and motivation.

Paralimbic (Mesocortical) Areas and Comments on the Breakdown of the Stimulus-Response Bond

The paralimbic belt encircles the basal and medial aspects of the hemispheres (see Fig. 6). In the human brain, the olfactocentric paralimbic formations include the temporal pole (area 38), the insula (area J), and caudal orbitofrontal cortex (caudal part of areas 11 and 12). The hippocampocentric paralimbic formations include the parahippocampal regions (areas 34, 28, 27, and 35), the retrosplenial area (areas 26, 29, and 30), the cingulate gyrus (area 23, 24, 31, and 33), and the pre- and subcallosal (parolfactory) regions (areas 32 and 25). Architectonically, these areas provide gradual transitions from piriform or hippocampal allocortex to the adjacent isocortical association areas.

The paralimbic areas receive their major cortical sensory information from heteromodal areas. They also have inputs from the synaptically most "downstream" parts of unimodal areas. It is therefore reasonable to assume that paralimbic areas have access to an extensively abstracted overview of the extrapersonal world. This information is then integrated with the limbic inputs that also reach the paralimbic areas. Even to a greater degree than in heteromodal areas, it would appear that paralimbic neurons are in a position to emphasize the behavioral relevance of a stimulus (e.g., edibility, attractiveness) much more extensively than its physical aspects (e.g., size, color, orientation).

This is not to say that heteromodal and paralimbic zones are unique in their responsiveness to the behavioral relevance of stimuli. In fact, there are neurons even in primary sensory cortex and also in unimodal regions that are sensitive to these parameters.[82,116] However, it appears that the relative impact of the task-related parameters gradually increases from primary to unimodal, heteromodal, and paralimbic regions. For example, looking at vertical rather than horizontal bars is likely to have a far greater impact on striate than on visually responsive orbitofrontal neurons. On the other hand, if the behavioral relevance of the vertical bars were to change from session to session, it would be predicted that these changes would influence orbitofrontal neurons more than they would those of striate cortex.

In simpler animal species, there may be little distinction between task and stimulus parameters. The absence of this distinction would seem to be an essential feature of instinctual behavior. For example, a turkey hen with a newly hatched brood will treat every moving object within the nest as an enemy unless it utters the specific peep of its chicks. If a hen is experimentally made deaf it will invariably kill all its own progeny soon after hatching. Furthermore, male sticklebacks (which are red on the ventral side) will attack any key stimulus with a "red below" pattern as if it were a rival.[164] These behaviors indicate instances in which the physical dimensions of a specific stimulus set in motion preprogrammed behavior sequences that are relatively impervious to modification by additional situational cues. These types of behaviors are based on a more rigid stimulus-response bond than is generally found in the behavioral repertoire of primates.

In more advanced animal species, the heteromodal and paralimbic areas that are interposed between the primary sensory-motor regions and limbic mechanisms introduce a synaptic buffer between external reality and internal urges. It is in these parts of the brain that the same stimulus can elicit very different responses, depending on the context. This offers considerable behavioral flexibility in a way that transforms the rigid stimulus-response bond of lower species into the more adaptive set-goal relationship of higher species. With respect to output, these cortical areas may also allow the organism to mobilize not predetermined motor sequences but a plan of action and intention that could lead to one of many suitable motor sequences needed to reach the desirable goal. In a programming board, the paralimbic and heteromodal parts of the brain might be considered analogous to *and gates* and *or gates* interposed between input and output channels. These parts of the brain allow the organism to alter, trigger, or inhibit behavioral goals in keeping with inner needs, the present circumstances of the extrapersonal world, the potential consequences of the contemplated act, and the past experiences of the individual with similar situations.

Although heteromodal and paralimbic areas are both involved in this type of neural interaction, the associative elaboration of perceptual and cognitive processes appears to receive more emphasis in heteromodal areas, whereas the motivational and affective processes may predominate in the paralimbic belt. In keeping with this general outline, the more specific behavioral specializations of the paralimbic belt are generally associated with three realms of function: (1) memory and learning, (2) the channeling of drive and affect, and (3) the higher control of autonomic tone. These areas will be reviewed in this section.

Memory and the Paralimbic Belt

The more severe disorders of memory are usually associated with damage to the hippocampus, amygdala, and subcortical limbic structures (see Chapter 4). However, paralimbic lesions also give rise to memory deficits, perhaps by interrupting the interactions between association isocortex and limbic regions.

In monkeys, just about every major component of the paralimbic belt has been implicated in memory and learning. For example, lesions that involve the paralimbic parts of orbitofrontal cortex and the adjacent parolfactory gyrus give rise to considerable deficits in visual recognition tasks.[194] Furthermore, insular damage yields modality-specific learning deficits in tasks of tactile discrimination.[206] This is quite consistent with the anatomic connectivity patterns, which show that the insula is the major relay of somatosensory information into the limbic system.[180] Anatomic experiments show that the hippocampus receives virtually all of its cortical sensory information through obligatory relays in the paralimbic areas of the parahippocampal gyrus.[284] This may explain why parahippocampal ablations yield memory deficits that are comparable to those of hippocampal lesions.[169] Bilateral damage to the cingulate gyrus has also been shown to yield deficits in tasks that require retention of spatial information.[205,228] In rabbits, physiologic recordings suggest that cingulate neurons may play an important role in learning to avoid aversive stimuli.[83] No similar information is available in primates.

In humans, lesions that damage paralimbic structures often also involve adjacent limbic structures so that the contribution of the paralimbic lesion to the resultant amnesia is difficult to determine. For example, orbitofrontal lesions (often seen in conjunction with anterior communicating artery aneurysms) have been associated with severe amnesias, but limbic areas of the basal forebrain are usually also damaged in these patients (see Chapter 4). Relatively mild and transient memory impairment has been reported after involvement of the anterior cingulate region.[299] A more severe amnestic syndrome has been described in conjunction with tumors or resection of the posterior corpus callosum (Fig. 7C). It is probable that these deficits reflect bilateral injury to the adjacent posterior cingulate and retrosplenial areas. Severe memory disorders are also seen after medial temporal damage that includes the parahippocampal regions. However, the hippocampus is almost always also damaged, so that the role of the paralimbic lesion is difficult to determine.[281]

Drive, Affect, and Paralimbic Areas

Paralimbic areas also play a major role in the affective coloring of experience and in the channeling of drive toward the appropriate extrapersonal targets. In monkeys, lesions that involve paralimbic parts of orbitofrontal cortex dramatically alter the emotional response to novel or threatening stimuli.[40] These animals show enhanced aversive responses to relatively novel objects but a lowered aggressive reaction to humans and to a model snake.[40] It is interesting to note that under similar experimental situations amygdalectomy reduces both aversive and aggressive behavior, showing that orbitofrontal cortex is not simply acting as a passive relay for conveying sensory information into the amygdala. Bilateral ablations in the anterior cingulate region are associated with an enhancement of the startle response. These animals are described as more resistant to handling.[262]

Conspecific affiliative behaviors are also closely associated with the paralimbic belt. In social animals such as monkeys, dominance hierarchies and social bonds rely heavily on specific aggressive and submissive displays, grooming behavior, and vocalizations. Success depends on directing the proper behavior to the proper individual in the proper setting. Conspecific affective and affiliative interactions are severely disrupted after lesions in the paralimbic regions. For example, orbitofrontal and temporopolar damage in monkeys decreases the effectiveness of aggressive encounters and results in a reduction of positive affiliative behaviors, eventually leading to social isolation.[79,143,229]

It is highly probable that the paralimbic areas in the human brain are also involved in channeling drive toward the appropriate object and in imparting affective coloring to thought and perception. The available information, although limited, is consistent with these expectations. For example, electrical stimulation of the anterior cingulate, insula, and parahippocampal regions elicits a number of experiential phenomena such as mood alterations, dreamlike states, feelings of familiarity, and memory flashbacks.[13,88,300] It should be noted, however, that such experiential phenomena are even more closely associated with stimulation of limbic areas, especially the amygdala.

Several additional lines of observation in neurologic patients provide indirect evidence for the participation of the cingulate area in the channeling of affect and drive. For example, the cingulate region may play an essential role in determining the spatial distribution of motivation in the process of directed attention (see Chapter 3). Bilateral anterior cingulate lesions can lead to severe apathy and personality changes.[7] Furthermore, selective and persistent impairments in the emotional intonation of speech have been attributed to lesions in the cingulate area.[132] It is also interesting to note that bilateral cingulotomies are sometimes effective in relieving otherwise intractable depressions and obsessive compulsive syndromes.[12]

Paralimbic areas may also participate in determining the hedonic aspects of sensory experience. Biemond[27] reported a patient in whom a lesion involving the S2 region and the adjacent insula was associated with a dramatic loss of pain perception. The loss of pain sensation in this patient is commonly attributed to the involvement of S2. However, the extension of the lesion into the insula would also have produced a somesthetic-limbic disconnection, since the insula is the major relay of somatosensory inflow into limbic structures. Since pain is an emotion as well as a sensation, this disconnection may explain, at least in part, the loss of pain sensation in Biemond's patient. In this context, it is interesting to note that the insula and other paralimbic areas have higher levels of endogenous opiate receptors than adjacent cortical areas.[161]

Higher Autonomic Control and Paralimbic Areas

Emotional states are associated with specific patterns of autonomic response. It is therefore not surprising that paralimbic areas should also participate in regulating autonomic tone. In monkeys as well as in humans, electrical stimulation in the anterior insula, anterior cingulate gyrus, caudal orbitofrontal cortex and in the temporal pole yields marked and consistent autonomic responses.[45,111,133,134,221,227,259,270,290a] Insular stimulation is more likely to elicit gastrointestinal responses, whereas stimulation in the other areas yields cardiovascular and respiratory changes. Some of these effects can be quite dramatic and include inhibition of gastric peristalsis, respiratory arrest, and blood pressure changes of as much as 100 mm Hg. Even multifocal cardiac necrosis can be obtained when monkeys with no intrinsic cardiovascular disease receive electrical stimulation to orbitofrontal cortex.[100]

It is well known that the major pacemakers and final common pathways for autonomic regulation are located in the hypothalamus, the reticular formation, the dorsal motor nucleus of the vagus, and the solitary tract nucleus. Nonetheless, the paralimbic regions, together with other limbic structures may provide a neural mechanism for the higher control of autonomic activity according to the prevailing mental state. Interactions between autonomic responses and mental states exist in many realms of behavior. For example, cognitive tasks as well as emotional states are associated with specific patterns of autonomic activity. These patterns reflect the difficulty of the task, the type of ongoing information processing, the nature of the emotion, and even the magnitude of its impact upon the individual.[10,74,136,152,187] In medical practice, the interactions between mental stress and autonomic activity are well known. For example, stress can increase blood pressure, promote the formation of ulcers, lead to abnormalities of esophageal motility, and even induce potentially lethal cardiac arrhythmias in the absence of cardiovascular predisposing factors.[48,165] A disturbance in the relationship between mental events and autonomic activity is also thought to be a crucial factor in the pathogenesis of psychosomatic disease.[5] It is conceivable that the integration of mental state with patterns of autonomic activation in these normal and abnormal conditions is strongly influenced by the paralimbic areas of the brain. These parts of the brain may therefore provide a potential anatomic substrate for psychosomatic disease, for "essential" hypertension, and perhaps even for some types of heart disease. MacLean[167] has reached similar conclusions and has proposed the term "visceral brain" as a generic term for a group of regions that include the paralimbic areas.

The pathways through which paralimbic areas influence autonomic responses are almost certainly multisynaptic. The initial relay probably consists of paralimbic projections to the amygdala. Amygdalofugal pathways to the hypothalamus and to the brainstem could then have direct access to autonomic regulation. In the rodent, direct connections exist from the insula to the solitary tract nucleus.[247,258] Such pathways have not been reported in the primate.

Gustation and Olfaction in Paralimbic Areas

In contrast to the auditory, visual, and somatosensory modalities, olfactory and probably also gustatory information is processed predominantly in paralimbic and limbic zones of the brain. The primary olfactory cortex is itself an allocortical limbic structure, intimately related to many paralimbic areas of the brain. The primary gustatory cortex is situated in the frontal operculum, immediately adjacent to the anterior insula.[246] In monkeys, cortical responses to olfactory and gustatory stimulation are most readily obtained in the orbitofrontal and insular components of the paralimbic zone.[271,272,274] Damage to temporopolar, anterior insular, or caudal orbitofrontal cortex in monkeys impairs olfactory and gustatory discrimination behaviors.[11,271,272] In humans, anterior temporal lobectomy may lead to impairments of olfactory memory and olfactory discrimination.[71,231] Very little is known about the effects of cortical damage upon gustatory sensation in humans. The close anatomic relationship between paralimbic regions of the brain and the olfactory-gustatory modalities explains why olfactory and gustatory hallucinations are so common in temporolimbic epilepsy.

It is interesting that olfactory and gustatory sensations are so closely associated with the limbic-paralimbic zones, whereas the other sensory modalities are processed within idiotypic and isocortical areas. Perhaps this is because olfactory and gustatory modalities are essentially chemical senses that might have evolved from mechanisms closely related to the monitoring of the internal milieu.

The Limbic Structures (Corticoid Areas and Allocortex)

The limbic zone provides the state for the most intimate interaction between the products of cortical information processing and hypothalamic impulses. The components of the limbic zone include the septal area, the substantia innominata, the amygdala, the piriform cortex, and the hippocampal formation. These are the only cortical structures that receive major hypothalamic inputs.

The behavioral specializations of these limbic structures are very similar to those of paralimbic regions. However, perhaps because they are synaptically closer to the hypothalamus, lesions in the limbic structures result in much more profound alterations in the entire fabric of memory, drive, affect, autonomic tone, endocrine control, and even immunoregulation.

Septal Nuclei and the Substantia Innominata

This corticoid region of the brain contains a large number of intricately interdigitated cell groups. Four of these cell groups (the medial septal nucleus, the vertical and horizontal limb nuclei of Broca's diagonal band, and the nucleus basalis of Meynert) contain cholinergic cell bodies that provide the major cholinergic innervation for the entire cortical surface. The cholinergic cell bodies within the medial septal nucleus (also designated as Ch1) and those within the vertical limb nucleus of the diagonal band (Ch2) provide the major cholinergic input of the hippocampus, those of the horizontal limb nucleus (Ch3) provide the major cholinergic input into olfactory structures, and the cholinergic neurons of the nucleus basalis (Ch4) give rise to the major cholinergic innervation for all the other cortical zones and for the amygdala.[183] These cholinergic nuclei receive their most substantial neural input from the hypothalamus and from components of the limbic and paralimbic zones.[181] There is indirect evidence that the projections of the Ch4 complex to paralimbic areas may be more intense than are its projections to isocortex.[188] This pattern of connectivity is consistent with the schema shown in Figure 5.

In many animal species, including humans, the integrity of central cholinergic pathways appears necessary for intact memory function.[16]

Since the basal forebrain is the source for most of the telencephalic cholinergic innervation, its role in memory function has attracted a great deal of interest. In rats, septal lesions as well as lesions in the nucleus basalis interfere with passive avoidance learning tasks.[77,97] In humans, the amnestic state that may emerge in patients with anterior communicating artery aneurysms and with septal tumors may, at least in part, reflect the involvement of these cholinergic nuclei (see Chapter 4). In *Alzheimer's dementia*, in which memory loss is probably the single most salient aspect of the clinical picture, there is also a profound loss of Ch4 neurons as well as of cortical cholinergic innervation.[298] However, since many other neural systems are also affected, the extent of the relationship between the Ch4 cell loss and the amnesia in Alzheimer's disease has not been determined with any degree of specificity.

These components of the basal forebrain have also been implicated in other aspects of drive and motivation. For example, septal lesions in rats result in pathologically exaggerated emotional reactivity to novel or threatening stimuli, hyperdipsia, transient hyperphagia, and also alterations in taste preferences.[97] In the monkey, single-unit recordings show that neurons of the nucleus basalis (Ch4) are especially sensitive to motivational variables and to the delivery of reward. These neurons alter their activity when the animal detects an edible object, especially if it is hungry and if the object happens to be a favorite food item.[57,239] This pattern of activity indicates that these neurons may establish complex associations between extrapersonal objects and their motivational value.[181] In this context, it is interesting to note the presence of isolated reports describing spike activity in the septal area of patients with schizophrenia. Furthermore, electrical stimulation of this area apparently leads to pleasurable sensations.[103] Despite a great deal of interest, however, the role of this area in the normal and pathologic aspects of human mental life remains largely unexplored.

The Piriform Cortex

Piriform cortex is the primary cortical relay for incoming olfactory information. In contrast to the other sensory modalities, olfactory information does not have to be relayed through the thalamus in order to reach cortex. In addition to olfactory input, and in keeping with its position in the limbic zone, the piriform cortex also has well-developed hypothalamic, paralimbic, and limbic connections.[6,180,182] The importance of olfactory sensation in sexual, territorial, and feeding behaviors suggests that this region may participate in the modulation of functions that are closely associated with the other specializations of the limbic zone.

Cetacea (e.g., whales and dolphins), which are entirely anosmic, have a well-developed piriform cortex,[124] raising the possibility that this region might have nonolfactory functions even in other mammals. In the cat, for example, piriform cortex lesions induce hypersexuality, whereas stimulation of this region results in a restriction of the trigeminal sensory fields that elicit attack behavior upon hypothalamic stimulation.[29,94] These observations suggest that piriform cortex may have an important role in regulating the direction of drive within the extrapersonal space and that this may be partially independent of its olfactory functions. Thus, piriform cortex may have a dual role both as a sensory area and also as a limbic structure.

The Amygdala

The amygdala has extensive reciprocal connections with the hypothalamus and with other components of the limbic zone, such as the hippocampus and the substantia innominata.[181,189,209,240] It also receives sensory information from unimodal, from heteromodal, and especially from paralimbic areas.[182,283]

The amygdala plays a pivotal role in the channeling of drive and affect. This function was highlighted by Downer's[67] experiments in monkeys. In these experiments, the forebrain commissures including the optic chiasm were sectioned and the amygdala on one side was ablated. As a consequence of these surgical manipulations, one eye but not the other could convey visual information to the intact amygdala. Monkeys usually react to the presence of human onlookers with a characteristic aggressive display. Following surgery, the animal reacted to the sight of onlookers quite placidly when only the eye ipsilateral to the amygdalectomy was uncovered. However, when the onlookers were viewed through the eye ipsilateral to the intact amygdala, the customary aggressive response was triggered. It appears, then, that the amygdala is an essential component for the association of the appropriate emotional response with extrapersonal objects. This is also shown in monkeys with the Klüver-Bucy syndrome, a syndrome that is now thought to arise when the amygdala is deprived of its cortical input.[86,113] These animals show three salient behavioral changes: (1) They indiscriminately initiate sexual activity without regard to the appropriateness of the object; (2) They no longer show the customary aggressive-aversive reaction to their human keepers; and (3) They seem to have lost the ability for visually distinguishing edible from inedible objects so that they keep mouthing all kinds of objects, discarding inedible ones only after buccal inspection. The one common denominator for all of these behaviors is a breakdown in the channeling of drive to the appropriate visual target in the extrapersonal space. It is not the drive that is altered but its association with the proper object. For example, despite excessive mouthing, monkeys with the Klüver-Bucy syndrome are neither hyperphagic nor obese.

Additional evidence shows that the amygdala closely participates in a wide range of behaviors related to drive and affect. For example, radiotelemetered activity from the amygdala of freely moving monkeys engaged in social interactions showed a high rate of responding during sexual and aggressive encounters.[144] In keeping with this observation, amygdalectomy in monkeys leads to dramatic alterations in conspecific aggressive behaviors and in reversals of the individual's position within dominance hierarchies.[241] Amygdalectomy also interferes with the affective vocalizations of monkeys.[131] These observations in monkeys suggest that the amygdala, together with components of the paralimbic belt, may be essential for the interpretation of affective gestures and vocalizations and also for the ability to emit them in the proper context. In the human, the more isocortical parts of the brain, especially in the right hemisphere, also assume a major role in this area of behavior (see Chapter 6).

In the rodent, the amygdala may also participate in the experience of fear. For example, LeDoux and co-workers[158] showed that a neutral auditory tone increased bloodflow mainly in auditory structures. However, when the same tone was conditioned to a fear response, it then elicited bloodflow increases also in the amygdala and hypothalamus. It appears that the human amygdala is also associated with the experience of emotions. For example, during depth recordings in epileptic patients, discharges confined to the amygdala were associated with rage attacks, the experience of strong emotions, and feelings of familiarity.[88,300] In fact, among all the temporal lobe structures, the amygdala discharges were most frequently associated with such affective and experiential phenomena.[88]

As we have argued elsewhere,[182] a healthy mental life depends on a proper balance among experience, thought, behavior, and affect. The discussion in this section and in the preceding one shows that limbic and paralimbic areas in the brain are likely to play a prominent role in establishing this balance. Under normal conditions, the affective coloring of perception and thought is expected to reflect the individual's history, present internal state, and the characteristics of the ongoing mental experience. In patients who have a disease process within limbic and paralimbic regions, there may be severe disruptions in this relationship. This may lead to an unpredictable and incongruous affective coloring of mental activity in a way that may distort the entire fabric of experience.[178] This disruption may account for the remarkable spectrum of acute and chronic psychiatric syndromes ranging from *panic attacks* to *dissociative states, depression,* and *schizophreniform conditions* that have been described in conjunction with temporolimbic epilepsy (see Chapter 8).

The extent to which the amygdala participates in memory and learning is somewhat controversial. In humans, bilateral amygdalectomy does not appear to impair memory.[207,261] On the other hand, it appears that the amnesia resulting from hippocampal damage in humans and monkeys becomes much more severe if there is additional involvement of the amygdala (see Chapter 4).[193] Perhaps this role of the amygdala is related to the well-established observation that stimuli associated with strong emotions are better memorized. It is conceivable that the amygdala plays a major role in establishing an affective link in memorization.

The amygdala also participates in the regulation of autonomic, endocrine, and even immunologic processes. For example, stimulation of the amygdala results in widespread changes of autonomic tone.[8] In some mammals, direct connections have been demonstrated from the amygdala to the dorsal motor nucleus of the vagus and to the solitary tract nucleus.[250] It is not known if analogous pathways exist in the primate brain. The amygdala may have access to autonomic control also through its projections to the hypothalamus via the stria terminalis. As in the case of paralimbic areas, the amygdala may be involved in integrating mental states with autonomic response patterns. The direct participation of the amygdala and of the other limbic-paralimbic structures in autonomic regulation explains the frequent occurrence of autonomic discharges and visceral sensations (e.g., epigastric pain) in patients with temporolimbic epilepsy (see Chapter 8).

Lesions and electrical stimulation of the hippocampus and especially of the amygdala also influence the hypothalamic regulation of endocrine balance.[311] This may be one reason why an increasing number of hormonal problems, ranging from menstrual irregularity to polycystic ovary, infertility, and impotence, are being reported in patients with temporolimbic epilepsy (see Chapter 8). The amygdala as well as other limbic structures such as the hippocampus, the septum, and the diagonal band nuclei also have estrogen and testosterone concentrating neurons.[269] Consequently, in addition to participating in endocrine control, these limbic structures may also be substantially influenced by the hormonal milieu. This offers one plausible mechanism for the sensitivity of mood to hormones.

There is an intriguing possibility that limbic structures may also participate in the immune response. In rats, lesions of the amygdala as well as of the hippocampus and hypothalamus lead to marked alterations in lymphoid cell number and in lymphocyte activation.[34] This may provide at least one anatomic substrate for the putative relationship between mental stress and autoimmune disturbances.

The Hippocampal Formation

The hippocampus is almost entirely dependent on paralimbic areas for its cortical sensory input.[284] Other major connections of the hippocampal formation include those with the hypothalamus and with other components of the limbic system, especially the amygdala and the septal area.[183,240] As discussed earlier, the hippocampus has been implicated in experiential phenomena, endocrine control, and immunoregulation. However, its principal behavioral affiliations are generally considered to be in the realms of memory and learning.

In monkeys, selective hippocampal lesions yield only modest deficits in the retrieval phase of learning.[169] The memory disturbance becomes severe only when the amygdala is also damaged.[193] Cases with isolated hippocampal damage are quite rare in humans. However, combined hippocampal and parahippocampal infarcts frequently lead to severe *amnestic states* even when the amygdala is intact (see Chapter 4). Furthermore, a review of the voluminous literature on this subject shows that damage to the hippocampus is the single most consistent factor in patients who develop an amnestic state in conjunction with medial temporal lesions.

It is unlikely that the amnesia in patients with limbic and paralimbic lesions reflects the actual destruction of memory banks. Instead, it is far more likely that limbic-paralimbic structures, and especially the hippocampus, are necessary for *imprinting* widely distributed neural templates into storage and also for *rekindling* them during retrieval. This may ensure that mental contents with motivational relevance are more likely to be memorized and remembered. According to this mechanism, medial temporal lesions would disrupt the processes of storage and retrieval while leaving the memory banks essentially intact. This is consistent with the phenomenology of amnesia (see Chapter 4).

BASAL GANGLIA: STRIATUM AND PALLIDUM

The importance of the basal ganglia to motor function and to extrapyramidal pathways needs no introduction. Marsden[170] has suggested that "the basal ganglia are responsible for the automatic execution of learned motor plans." However, the basal ganglia have also been associated with complex nonmotor behaviors that are of particular interest to the behavioral neurologist.

The Striatum

The striatum can be divided into four components: the caudate, the putamen, the olfactory tubercle, and the nucleus accumbens. Each striatal component receives cortical input but none projects back to cortex. The caudate and putamen, which are also collectively designated as the *dorsal striatum*, receive cortical input predominantly from association cortex and primary idiotypic areas. The dopaminergic input to these striatal components originates in the pars compacta of the substantia nigra. On the other hand, the cortical input to the olfactory tubercle and nucleus accumbens originates in limbic and paralimbic parts of the brain.[95,210,211] The nucleus accumbens, for example, receives convergent input from the amygdala and the hippocampus. On the basis of this connectivity pattern, the nucleus accumbens and the olfactory tubercle could be designated as the *limbic striatum*.[107] Their dopaminergic innervation originates in the ventral tegmental area of Tsai, which is just medial to the substantia nigra. Dopamine turnover is higher in the limbic striatum than in the neostriatum.[291] Furthermore, the density of cholinergic cell bodies is also higher in the limbic parts of the striatal complex.[184]

All parts of cortex project to the striatum. These corticostriatal projections obey a complex topographic arrangement. One feature of this arrangement is that the input from each cortical area forms multiple patches of axonal terminals within the striatum. Yeterian and Van Hoesen[306] made the interesting suggestion that terminal patches from separate cortical areas are more likely to show partial overlap if the relevant cortical areas are interconnected. This implies that there may be some replication of cortico-cortical interaction patterns within the striatum. Although it is possible that this arrangement largely subserves the control of complex motor output, it is also possible that it reflects the role of the striatum in nonmotor behavioral specializations that are usually attributed to the cortical surface.

With respect to the dorsal striatum, it appears that the caudate may have a lesser role than the putamen in motor control. For example, Künzle[150] has shown that motor cortex projects to the putamen but not the caudate. In the monkey, some caudate neurons seem to respond uniquely to the sight of food but not during the motor act of bar pressing, which leads to the reward.[213] The head of the caudate receives most of its input from dorsolateral prefrontal cortex. It is therefore interesting to note that lesions in the head of the caudate yield deficits that are essentially identical to those that emerge upon ablating prefrontal cortex.[122] This raises the possibility that each striatal region, and especially those in the caudate, may have behavioral specializations that are similar to those of the cortical area(s) from which it receives its major cortical input. It remains to be seen whether any of these specializations is organized within the striatum in a fashion that is largely independent of motor planning.

Lesions in the head of the caudate and in the putamen have been associated with *aphasia* and also with *unilateral neglect* (see Chapters 3 and 5). However, in almost all such cases the adjacent white matter is also involved so that it is not possible to determine whether these deficits reflect damage to the striatum or the adjacent fibers which interconnect cortical areas related to language and attention. We have seen one ambidextrous patient with multi-infarct dementia who also had a substantial infarction in the head of the left caudate. This patient had no motor deficit, aphasia, or amnesia. His mental state deficits were characterized by a severe lack of judgment, insight, and planning. While it is impossible to determine what aspect of his mental state abnormality could be attributed to the striatal lesion, this patient suggests that damage to the head of the caudate may give rise to mental changes that are also seen in conjunction with prefrontal cortex lesions.

Alterations in mental state occur in several basal ganglia diseases. For example, an increasing body of evidence is now indicating the presence of nonmotor cognitive deficits in *Parkinson's disease.* This occurs even in those patients who do not show the cortical plaques and tangles that are characteristic of Alzheimer's disease.[115] It appears, therefore, that there may be a dementia intrinsic to Parkinson's disease. Furthermore, in *Huntington's disease,* in which there are major disturbances of mental state, the striatum is heavily damaged although cortical areas do not harbor any consistent pathology. It is conceivable that the behavioral deficits in both of these conditions reflect striatal involvement. The responsible mechanisms would be a loss of striatal neurons in Huntington's disease and defective striatal dopaminergic innervation in Parkinson's disease. Perhaps the motor deficits in these two conditions are attributable mostly to the putaminal involvement, whereas the behavioral deficits may reflect the involvement of the caudate and limbic striatum.

In rodents, the dorsal striatum (putamen and caudate) appear to have behavioral specializations different from those of the limbic striatum (accumbens and olfactory tubercle). For example, disrupting the dopaminergic pathways from the ventral tegmental area to the limbic striatum of the rat results in locomotor hyperactivity and deficits in passive avoidance learning; whereas interrupting the nigrostriatal pathway to the dorsal striatum yields hypoactivity, bradykinesia, and alimentary disturbances.[160] In the cat, nucleus accumbens stimulation reduces the receptive field for the hypothalamic biting reflex.[91] This role of the nucleus accumbens in the channeling of drive and affect is consistent with its limbic connectivity. There is essentially no specific information on the behavioral specialization of the human limbic striatum even though it is almost certainly involved in Parkinson's disease and perhaps also in Huntington's disease.[17,126] Although the nucleus accumbens has been implicated in the pathogenesis of schizophrenia, the evidence remains circumstantial at best.[268]

The Globus Pallidus

The globus pallidus receives the striatal outflow and projects to thalamic nuclei, the habenula, and other extrapyramidal structures. The globus pallidus of the primate has four easily identified components: (1) the outer (lateral) segment, (2) the inner (medial) segment, (3) the ventral (subcommissural) pallidum, and (4) the pars reticulata of the substantia nigra.

There is essentially no disagreement about the crucial role of the globus pallidus in motor control. In humans, lesions of the globus pallidus are frequently associated with severe extrapyramidal disturbances. In the monkey, neurons of the globus pallidus change firing in close association to push-pull movements.[57] However, the relationship of the globus pallidus to movement may be quite complex and appears to involve substantial sensory-motor integration. For example, local cooling in the area of the globus pallidus in monkeys yields a severe and reversible breakdown of a learned flexion-extension movement, but only when the animal is blindfolded. In the presence of visual input, the deficit is no longer observed.[112]

The medial zone of the inner pallidal segment and also the ventral pallidum have close associations with limbic structures. For example, in contrast to the more lateral and dorsal parts of the pallidum, which receive their striatal input from the caudate and putamen, the ventral pallidum receives its major striatal projections from the nucleus accumbens.[107] Furthermore, a substantial number of ventral pallidal neurons respond to amygdaloid stimulation, and this response is probably mediated through the nucleus accumbens.[307]

In the monkey, the core of the internal pallidal segment projects to the motor thalamus. However, a medial crescent of this pallidal segment projects predominantly to the lateral habenula, which is a structure generally considered to be closely related to limbic regions of the brain.[218] In keeping with this anatomic pattern, some types of pallidal lesions interfere with behaviors generally associated with limbic mechanisms. For example, MacLean[168] showed that damage to the medial globus pallidus of monkeys severely disrupts species-specific sexual display patterns. Thus, the ventral pallidum and the medial portion of the inner pallidal segment could be considered to have preferential limbic affiliations. In humans, carbon monoxide intoxications and infarctions in the territory of the anterior choroidal artery commonly involve these parts of the pallidal complex. However, the behavioral deficits that can be attributed to such lesions have not yet been identified.

The pars reticulata of the substantia nigra is essentially a caudal extension of the globus pallidus. There is evidence suggesting that this portion of the pallidal complex may participate in the programming of saccadic eye movements in response to actual or remembered targets.[110]

THE THALAMUS

The thalamus relays subcortical inputs to cortical areas. Almost all thalamic nuclei have well-developed reciprocal connections with cortex. The one exception is the reticular nucleus, which receives subcortical and cortical input but does not project back to cortex.[127] There is very little interconnectivity among individual thalamic nuclei, so that there is very little interaction at the thalamic level among the different types of information being relayed to cortex. The one exception is provided by the reticular and intralaminar nuclei, which have extensive connections with other thalamic nuclei. Each thalamic nucleus, especially those that are not sensory relay nuclei, has projections to multiple cortical areas. Furthermore, each cortical area is interconnected with many thalamic nuclei. However, most thalamic nuclei have a preferred cortical region as their major projection territory. The bewildering array of thalamic nuclei can be subdivided into several functional groups on the basis of their preferred cortical and subcortical projections (Fig. 8).

It is important to realize that there are substantial lacunes in the understanding of thalamocortical connectivity patterns in the primate

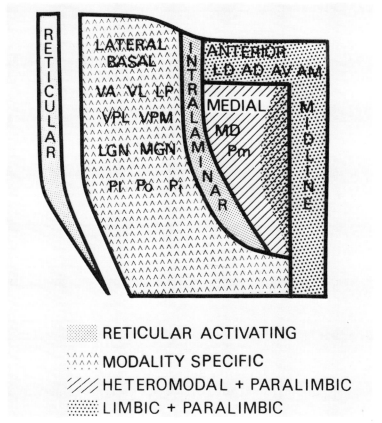

FIGURE 8. A schematic diagram of the four major groups of thalamic nuclei. Abbreviations: AD—anterior dorsal; AM—anterior medial; AV—anterior ventral; LD—laterodorsal; LGN—lateral geniculate; LP—lateroposterior; MD—medialis dorsalis; MGN—medial geniculate; Pi—inferior pulvinar; Pl—lateral pulvinar; Pm—medial pulvinar; Po—oral pulvinar; VA—ventral anterior; VL—ventral lateral; VPL—ventroposterior lateral; VPM—ventroposterior medial.

brain. Furthermore, nuclear boundaries are rarely sharp and the nomenclature is subject to variation. The following account is confined to the more widely accepted nuclear elements in the thalamus. The nomenclature will follow that of the Carpenter and Sutin[43] textbook and will be supplemented, whenever necessary, by the nomenclature of Olszewski.[215]

Primary Nuclei of the Thalamus (Nuclei of Idiotypic Cortex)

The primary relay nuclei are the easiest to identify. The caudal part of the ventroposterior lateral nucleus (VPL_c) and the principal division of the ventroposterior medial nucleus (VPM) receive fibers from the medial lemniscus and quintothalamic tract and constitute the somatosensory relay nuclei of the thalamus. The lateral geniculate nucleus (LGN), which receives the optic tract, and the anterior two thirds of the medial geniculate nucleus (MGN), which receive the brachium of the inferior colliculus, are the primary relay nuclei for the visual and auditory modalities, respectively. These sensory relay nuclei provide the major thalamic inputs of the idiotypic cortical areas S1, V1, and A1. Damage to the VPL_c or to the LGN gives rise to hemihypesthesia and hemianopia, respectively. In contrast to S1 lesions, in which pain sensation is preserved (probably because the VPL_c projection to S2 is spared), VPL_c lesions generally lead to a loss of pain sensation as well. Since inputs from both ears reach the MGN in each hemisphere, unilateral damage to this thalamic nucleus does not lead to contralateral ear deafness. In fact, unilateral MGN lesions may be extremely difficult to detect clinically. The major thalamic input into primary motor cortex (M1) comes from the caudal ventrolateral nucleus (VL_c) and the oral ventroposterior lateral nucleus (VPL_o).[130,249] The behavioral effects of lesions in the thalamic nuclei of motor cortex are poorly understood.

Thalamic Nuclei of Modality-Specific (Unimodal) Association Cortex

In the rhesus monkey, the major thalamic projections to the somatosensory association cortex of the superior parietal lobule (area 5) come from the lateroposterior nucleus (LP) and perhaps also from the oral subdivision of the pulvinar nucleus (P_o).[130] In the visual modality, the nuclei that provide the major projection to visual unimodal association areas include the inferior (P_i) and lateral (P_l) subdivisions of the pulvinar nucleus.[19,39,280] In the auditory modality, the unimodal association region receives its major thalamic input from the MGN and probably also from a ventral rim of the medial pulvinar.[39,186] Thus, the MGN is the source of thalamic projections not only to A1 but also to the auditory association cortex. The motor association cortex receives its major thalamic input from the oral ventrolateral nucleus (VL_o) and from parts of the ventral anterior nucleus (VA).[249]

In the monkey, lesions of the inferior pulvinar disrupt visual discrimination behavior.[44] Except for the implications based on their patterns of cortical connectivity, there is essentially no information about the behavioral specialization of the other modality-specific association nuclei.

Nuclei of the Thalamus That Are Not Modality-Specific: Nuclei of Heteromodal, Paralimbic, and Limbic Cortex

The lateral part of the medial dorsal nucleus (MD) is the major thalamic nucleus for the prefrontal heteromodal fields, whereas the medial pulvinar nucleus (P_m) and parts of the adjacent lateral posterior nucleus (LP) are the major nuclei for the heteromodal fields in the inferior parietal lobule and within the banks of the superior temporal sulcus.[189,276]

The close interactions between cortical heteromodal and paralimbic zones (see Fig. 5) are also reflected in the arrangement of thalamic connectivity patterns. Thus, the MD and P_m nuclei, which are the major nuclei for heteromodal cortical areas, also have extensive paralimbic connections. For example, the medial part of MD (including the magnocellular [MD_{mc}] component) appears to be the major thalamic nucleus for the orbitofrontal paralimbic region while the P_m has reciprocal projections with all components of the paralimbic belt and is probably the major nucleus for the temporopolar paralimbic area.[92,93,204] The MD and P_m also have direct limbic connections. Thus, the medial and magnocellular parts of MD have connections with the amygdala, piriform cortex, and the septal region.[99,106,208] Furthermore, the P_m has also been shown to have reciprocal projections with the amygdaloid complex.[128] Thus, in contrast to the more laterally and basally placed modality-specific nuclei, the MD and the P_m have extensively developed heteromodal, paralimbic, and limbic interconnections.

Another group of dorsally and medially placed nuclei are collectively known as the nuclei of the "anterior tubercle." These nuclei include the anterior thalamic nucleus (including its dorsal [AD], ventral [AV], and medial [AM] components) and the laterodorsal nucleus (LD). They provide the major thalamic connections for the posterior cingulate cortex, for the retrosplenial area, and for some of the parahippocampal paralimbic areas.[163,290] The anterior thalamic nucleus receives the mammillothalamic tract and is therefore an important component of the Papez circuit (see Chapter 4).

A number of nuclei situated close to the thalamic midline are collectively known as "midline nuclei." These include the paratenial, paraventricular, subfascicular, central, and reuniens nuclei. These nuclei have extensive projections with paralimbic areas (e.g., temporal pole and anterior cingulate gyrus) and also with the hippocampal formation.[92,290,303]

The effects of lesions in these nuclei are consistent with their patterns of cortical connectivity (see Fig. 7D). For example, in the rhesus monkey, bilateral MD lesions reproduce deficits in spatial delayed alternation similar to those associated with prefrontal ablations.[120] On the other hand, more medial MD lesions that also involve adjacent midline nuclei lead to visual object recognition deficits similar to those obtained after medial temporal ablations.[4] In humans, Spiegel and associates[263] reported that bilateral MD thalamotomies reduce schizophrenic agitation and intractable anxiety. In other patients, even unilateral lesions in the

PATTERNS IN
BEHAVIORAL
NEUROANATOMY:
ASSOCIATION
AREAS, THE LIMBIC
SYSTEM, AND
HEMISPHERIC
SPECIALIZATION

45

more medial parts of MD and in the anterior tubercle nuclei have been associated with severe *amnestic conditions* (see Fig. 7D).[190,265] In *Wernicke's encephalopathy* involvement of the MD is thought to play a major role in the genesis of the amnestic state (see Chapter 4).

The P_m nucleus is also often involved in Wernicke's encephalopathy and may contribute to the overall behavioral changes. Lesions of the right pulvinar nucleus, including its medial component, have been described in conjunction with contralateral neglect for the left extrapersonal space.[41] Electrical stimulation of the left medial pulvinar has been reported to induce transient anomia.[214] These behavioral relations are consistent with the connections of the P_m with the parietotemporal parts of heteromodal cortex and with limbic-paralimbic structures.

While it is easy to understand the influence of the sensory relay nuclei on cortical function, it is much more difficult to surmise the role of the other thalamic nuclei. It is conceivable that cortico-thalamo-cortical loops in heteromodal, paralimbic, and limbic areas are important for imprinting associative links and for reactivating them under specific behavioral conditions.[193]

The Reticular Nuclei of the Thalamus

The reticular nucleus of the thalamus as well as the intralaminar nuclei (e.g., the limitans, paracentralis, centralis lateralis, centromedian, and parafascicularis) have strong associations with the ascending reticular activating pathways (see Chapter 3). Although the centromedian nucleus is usually included within the intralaminar group, it stands out because of its strong connections with the striatum. In contrast to other thalamic nuclei, which have somewhat more restricted projection zones, the intralaminar nuclei are also known as "diffuse projection nuclei."[125] The putative role of these thalamic nuclei in the control of arousal and attention is discussed in Chapter 3.

THE CONCEPT OF A LIMBIC SYSTEM

In the section on cortex, the allocortical and corticoid regions of the brain were collectively designated as limbic structures (see Fig. 5). The literature also contains frequent reference to a limbic "system." In a now lost manuscript written in the 5th century BC, Empedocles is quoted as having stated that "the nature of God is like a circle of which the center is everywhere and the circumference nowhere." This description is just as applicable to the concept of the limbic system, which has ebbed and welled to fit the preference of individual authors. Although it is impossible to provide a universally acceptable definition of the limbic system, the surveys of cortex, basal ganglia, and thalamus in the preceding sections of this chapter lead to certain criteria for delineating its widest boundaries. The following components can each legitimately be included within the limbic system; together, they define what is probably its most extensive boundaries:

1. The hypothalamus
2. The limbic (allocortical and corticoid) components of cortex

3. The paralimbic cortical belt
4. The limbic striatum (olfactory tubercle and the nucleus accumbens), the limbic pallidum, the ventral tegmental area of Tsai, and the habenula
5. The limbic and paralimbic thalamic nuclei—medial part of MD, P_m, AD, AV, AM, LD, and the midline nuclei

One justification for combining these five groups of structures into a unified system is their intricate anatomic interconnections. Another is their common behavioral affiliations. For example, each component of the limbic system is more concerned with the homeostasis of the internal milieu, with memorization, and with the channeling of drive and affect than with sensory-motor processing. In fact, while lesions in each component of this system may yield major behavioral disturbances, elementary sensory and motor functions usually remain relatively intact. One of the most consistent relations in behavioral neurology is that severe and multimodal amnestic states always imply an involvement of the limbic system. Thus, while it may seem that a great many disease conditions can each result in an amnestic state, they do so only insofar as they damage one or more components of the limbic system (see Chapter 4).

The cortical components of the limbic system also have several common immunologic, physiologic, and neurochemical properties. For example, the herpes simplex virus has a special affinity for these regions and frequently causes severe behavioral disturbances and amnesias, while other cognitive, sensory, and motor functions usually remain intact. This suggests that limbic neurons may share a common membrane antigen that is being recognized by the herpes simplex virus. The limbic and paralimbic parts of the cortex are also particularly susceptible to the development of independent seizure foci and to the phenomenon of kindling (see Chapter 8). Furthermore, these cortical areas are characterized by their high density of cholinergic innervation and opiate receptors.[161,188] These neurochemical features may provide one mechanism for the involvement of the limbic system in memory and in the perception of pain and pleasure. It is also possible that the preferential concentration of cholinergic innervation contributes to the susceptibility of these areas to seizures and kindling.

CHANNEL FUNCTIONS AND STATE FUNCTIONS

Many neural interactions among different parts of the brain occur along fairly discrete pathways. For example, there is a point-to-point connectivity between the lateral geniculate nucleus and striate cortex so that small groups of geniculate neurons are dedicated to small subsectors of striate cortex. Cortico-cortical connections have comparable specificity. In addition to the selective patterns of connectivity that follow the schema in Figure 5, the pathways from one cortical region to another display further areal, columnar, and laminar specificity. In many of these pathways, the area of cortex giving rise to a projection is comparable in size to the area that is the major target of its projections.

Most of the interactions among cortex, basal ganglia, and thalamus display this relatively discrete organization. Damage to these pathways or to their cells of origin is likely to yield specific deficits in a class of behaviors that can be designated as *channel-dependent* functions. For example, patients with damage to the left primary visual cortex have a right homonymous hemianopia but no alexia, since visual information from the right visual association cortex can be transferred across the splenium of the corpus callosum to the contralateral visual association areas and then to the language area of the left hemisphere. If such a patient has additional damage to the splenium, the language areas are selectively deprived of all visual input so that alexia (word blindness) emerges in the absence of other signs of aphasia (see Chapters 5 and 7). Therefore, the syndrome of pure alexia indicates a breakdown of discrete neural channels. Many aspects of memory, language, and complex perceptual-motor tasks belong to this category of channel-dependent functions.

In addition to its more specific connectivity, each cortical area also has more diffusely organized connections. These pathways are arranged in such a way that a relatively small and homogeneous area in the brain projects widely to the neocortex or thalamus. At least seven such pathways can be identified:

1. In contrast to most other thalamic nuclei, the intralaminar group of nuclei projects diffusely to widespread cortical areas.[125]
2. Cholinergic neurons of the septal area and in the substantia innominata (Ch1–Ch4) project to the entire cortical surface with an overlapping topographic arrangement.[183]
3. Neurons in the lateral and medial hypothalamus project to widespread areas of cortex.[183]
4. Serotonergic neurons in the brainstem raphe nuclei project to the entire cortical surface.[200]
5. Cholinergic neurons in the pontomesencephalic reticular formation project to the entire thalamus and, to a lesser extent, to the cortical surface.[185]
6. Noradrenergic neurons in the nucleus locus coeruleus project to the entire cortical surface.[198]
7. Dopaminergic neurons in the substantia nigra and in the ventral tegmental area of Tsai innervate the entire striatum as well as many limbic, paralimbic, and perhaps heteromodal cortical areas.[35]

Each of these pathways is organized such that a relatively small group of neurons is in a position to cause rapid modulations in the information processing *state* of the entire cortex and thalamus. These modulations could influence the efficiency of all neural operations and could markedly bias the affective coloring of experience and the energy content of behavior. Certain aspects of mood, motivation, memory, arousal, and vigilance would seem to reflect such widespread shifts in the mode of neural activity and can therefore be designated as state-dependent functions. In fact, arousal and vigilance are very much influenced by the more diffuse thalamocortical and reticulothalamic pathways (see Chapter 3); some aspects of memory require the integrity of ascending cholinergic pathways (see Chapter 4); and ascending monoaminergic pathways have been implicated in the regulation of mood and motivation and even in setting the signal-to-noise ratio for neural transmission.[38,200]

These diffuse pathways could influence the efficiency of all other channel-dependent functions without necessarily altering the content of what is being transmitted along any individual channel. Each aspect of complex behavior is likely to represent an inextricable interaction between channel-dependent and state-dependent functions. For example, the distribution of attention within the extrapersonal space is a channel-dependent function. However, its efficiency is determined by the state of arousal and vigilance (see Chapter 3).

The distinction between state-dependent and channel-dependent functions, while speculative with respect to mechanism, is of heuristic importance in assessing mental function and in considering strategies for treatment. For example, the state-dependent aspects of arousal, mood, and motivation influence performance in all other channel functions. Therefore, deficits in specific cognitive tasks cannot be attributed to a breakdown in the relevant channel until the examiner has ascertained that the underlying state-dependent functions are relatively intact (see Chapter 2). This can be translated into the practical caveat that a memory disorder may have no localizing significance if the patient is also inattentive. With respect to treatment, a disturbance such as pure alexia, which reflects a breakdown in specific channel functions, is unlikely to respond to pharmacologic agents, since these could not be expected to convey the missing visual information to language areas. On the other hand, since many of the diffuse ascending pathways use a single transmitter and since their predominant function appears to be the modulation of information processing rather than the introduction of new information, it is conceivable that deficits in state-dependent functions could respond to systemic pharmacotherapy. This provides the rationale for treating mood disorders with monoamine agonists and memory disorders with cholinomimetics.

HEMISPHERIC DOMINANCE

One of the most fundamental aspects in the anatomic organization of function is the marked hemispheric differences in behavioral specialization. Whereas primary sensory-motor functions are equally distributed between the two sides of the brain, the control of more complex functions is markedly asymmetrical in organization. Although such laterality effects have also been reported in other animal species,[58] they are subtle, at best, when compared with the asymmetries of behavioral specialization between the two sides of the human brain. Hemispheric asymmetry is probably the most fundamental biologic hallmark of human cerebral evolution. Almost every chapter in this book discusses the mechanisms and manifestations of hemispheric dominance patterns. In this section, an introductory overview is presented so that the contents of the other chapters can be placed in perspective.

Language was the first area of behavior for which hemispheric dominance was demonstrated. As discussed in Chapter 5, the left hemisphere is both necessary and sufficient for linguistic functions in the overwhelming majority of individuals. Thus, countless clinical cases show that damage to the left hemisphere leads to aphasia, whereas equivalent damage to the right side of the brain very rarely does so. Experiments on patients whose two hemispheres have been surgically isolated (split-brain), have repeatedly shown that the left hemisphere contains all the machinery for linguistic function, whereas the right-hemisphere capacity in this area of behavior is rudimentary at best. Furthermore, a large number of experiments in neurologically intact volunteers has shown a left-hemisphere advantage for all language-mediated tasks.[264] In addition to language, clinical observations suggest that the left hemisphere is also specialized for the manipulation of number facts in the process of calculation (see Chapter 5).

Another behavioral asymmetry, which is so obvious in most individuals that it often escapes notice, is the remarkable clumsiness of the left hand in comparison to the right in just about every task requiring fine motor control, from screwing in a lightbulb to throwing a ball. In fact, the words "dexterity" (from Latin, *dexter*, right) and "adroitness" (from French, *droit*, right) are synonyms for skillfulness. These superior skills of the right hand indicate that the left hemisphere is also specialized for fine motor control.

As discussed in Chapter 5, there is considerable insight into the anatomic foundation of language dominance. For example, the supratemporal plane, a region generally included in Wernicke's area, is larger on the left side of the brain in the majority of right-handed individuals.[87] This anatomic asymmetry is noticeable even during the embryonic stages of development.[47] The anatomic basis for the motor aspects of handedness remains to be discovered.

In contrast to the dominant functions of the left hemisphere, which have been appreciated for more than a hundred years, the specializations of the right hemisphere did not gain widespread acceptance until much later. In fact, one can still find reference to the right side of the brain as the "minor" hemisphere. Although there is still considerable debate on detail, it is now generally accepted that the right hemisphere in most dextrals is specialized for at least four major areas of behavior: (1) complex and nonlinguistic perceptual tasks, including face identification; (2) the spatial distribution of attention; (3) emotional behavior; and (4) paralinguistic aspects of communication.

There are two different ways of looking at right-hemisphere specializations. According to one hypothesis, it is conceivable that at some time in the distant phylogenetic past, all complex functions were equally distributed between the two hemispheres of the hominoid brain. The subsequent emergence of linguistic behavior and its concentration within the left hemisphere could gradually have created a dilution effect on the former functions of this hemisphere. Consequently, if the right hemisphere now seems to be specialized in several nonlinguistic tasks, this is merely because the left hemisphere has lost some of its former capacity in those same areas of behavior. Since it must have retained some residues of its former functions, this hypothesis leads to the prediction that lesions of the left hemisphere would also disrupt all nonlinguistic behaviors, and to a greater extent than the disruption of language caused by equivalent right-hemisphere damage. According to an alternate hypothesis, it is conceivable that right-hemisphere specializations may represent truly emergent functions that have evolved simultaneously with the linguistic skills of the left hemisphere. In that case, the influence of left-hemisphere damage on the behavioral specializations of the right hemisphere should be no more than the influence of right-hemisphere lesions on linguistic tasks. The type of evidence required for choosing between these two alternative hypotheses is rarely available, especially since it is quite difficult to find tasks that are entirely free of covert linguistic mediation. However, it may be useful to keep both hypotheses in mind when interpreting the relevant literature on the behavioral specializations of the right hemisphere.

Right-Hemisphere Specialization for Complex Nonlinguistic Perceptual Skills and Face Identification

There is a voluminous and continuously expanding literature showing that patients with right-hemisphere lesions, especially in the posterior aspects of the hemisphere, have a much greater impairment in complex visuospatial tasks than do those with equivalent lesions in the left hemisphere.[60,264] For example, patients with posterior right-hemisphere lesions show marked deficits in a task that requires a visual judgment of line orientation, whereas patients with equivalent left-sided lesions perform at the same level as age-matched controls.[24] Furthermore, visually guided stylus maze performance, the Block Design Test of the Wechsler Adult Intelligence Scale, the tactile judgment of rod orientation, the ability to identify objects presented from an unusual visual perspective, and the ability to detect variations in a familiar melody are also more impaired in patients with right-hemisphere lesions, compared with those with left-hemisphere lesions.[61,139,175,257,293] Although the visuospatial deficits have received more emphasis in the literature, these reports indicate that the right hemispheric specialization for complex perceptual tasks also extends into the auditory and somesthetic modalities. Even memory processes show hemispheric asymmetry along these lines (see Chapter 4). For example, patients with left temporal lobectomies are more impaired in learning verbal material, whereas patients with right temporal lobectomies are more impaired in the memorization of nonlinguistic complex perceptual material.[192]

If right-hemisphere injury leads to a greater impairment of nonlinguistic complex perceptual tasks, then it would be expected that this side of the intact brain is normally more active in the execution of such tasks. Confirmatory evidence comes from several lines of observation. For example, a patient who was preoperatively able to copy a cube almost equally well with either hand showed a much greater postoperative decrement with the right hand after undergoing commissurotomy.[85] Assuming that no other part of the brain was damaged during surgery, the effect

PATTERNS IN
BEHAVIORAL
NEUROANATOMY:
ASSOCIATION
AREAS, THE LIMBIC
SYSTEM, AND
HEMISPHERIC
SPECIALIZATION

51

of the commissurotomy was to isolate the control of each hand to the contralateral side of the brain. The postsurgical superiority of the left hand therefore indicates that the right hemisphere is more specialized in this task. The better presurgical performance of the right hand may have reflected the transcallosal influence that it received from the right hemisphere.

The superiority of the right hemisphere in tasks that require complex perceptual processing has also been demonstrated in neurologically intact subjects. In dichotic listening experiments, for example, there is a consistent right-ear (and therefore left-hemisphere) advantage for words and numbers but a left-ear advantage for pitch and melody identification.[141,260] In tachistoscopic experiments, it has been repeatedly shown that there is a left visual field (and therefore right-hemisphere) superiority for depth perception, spatial localization, and the identification of complex geometric shapes.[141,264,278] Furthermore, the left hand is more accurate in judging the orientation of a rod by palpation.[23] With the availability of methods that allow the mapping of cerebral metabolism in normal subjects, it has been possible to show directly that there is more left-hemisphere activation during a verbal task but more right-hemisphere activation during a nonverbal tonal memory task (see Chapter 10).[171]

The identification of faces is a most complex perceptual task that is also of great biologic importance (see Chapter 7). It has firmly been established that there is a left visual field (and therefore right-hemisphere) superiority in the identification of faces.[142,147,234] Some have argued that this is just another manifestation of the right-hemisphere superiority in complex perceptual transformations, whereas others have argued that this represents an independent area of specialization.[159] As shown in Chapter 7, it is also clear that the left hemisphere has a major part in the process of facial identification, since severe deficits of facial recognition (prosopagnosia) are seen only after bilateral lesions.

A few additional points are worth making about this area of specialization in order to address some of the apparent controversies in the literature. First, it appears that the right-hemisphere superiority in complex perception is maximized if the task is made especially difficult and also if some memorization is required for successful performance.[278] Second, it is necessary to consider the contribution of past experience, individual talent, and also peculiarities in the mode of information processing. For example, while naive listeners show a right-hemisphere superiority in the recognition of melodies, musically experienced individuals—as well as those who use a more analytic approach in organizing the relevant information—may show a left-hemisphere specialization in similar tasks.[26,171] There may also be substantial sex differences in the hemispheric specialization for these functions. For example, the left visual field superiority for complex perceptual tasks is generally of lower magnitude in females.[64,174] Perhaps this is a reflection of a recently reported observation that the splenium of the corpus callosum may be larger in the female brain.[153] It appears, therefore, that the most consistent hemispheric differences in complex perceptual tasks are likely to be obtained in right-handed male subjects who are performing unfamiliar and difficult perceptual tasks that also require some memorization.

Right-Hemisphere Specialization for the Spatial Distribution of Attention

Several lines of evidence that are analyzed in greater detail in Chapter 3 lead to the conclusion that the right hemisphere is also specialized for determining the distribution of attention within the extrapersonal space. According to a model suggested in Chapter 3, the right hemisphere contains the neural machinery for attending to both sides of the extrapersonal space, whereas the left hemisphere contains the machinery for attending to only the contralateral right hemispace. This leads to the emergence of marked contralateral neglect after right-hemisphere injury but not after equivalent left-hemisphere injury. Additional details and implications of this right-hemisphere specialization are discussed in Chapter 3.

Right Hemisphere and Emotion

The inconsistencies in the terminology and obvious difficulties in the experimental manipulation of the relevant behaviors have erected substantial obstacles to the neurologic investigation of human emotions. However, recent observations suggest that there may be marked hemispheric differences in the organization of emotions. First, it has been suggested that each hemisphere introduces a different affective perspective to experience and behavior. Second, it also seems that the right hemisphere may be more important than the left hemisphere for the experience as well as for the expression of emotions (see Chapter 6).

Clinical observations suggest that there may be consistent hemispheric differences in the emotional response to experience. For example, the lack of concern and even the inappropriate jocularity in response to hemiplegia is a striking feature of some right-hemisphere infarcts, although it is almost never seen in patients with left-hemisphere lesions.[84] In keeping with this clinical observation, there is preliminary evidence showing that the right hemisphere is more closely associated with the experience and expression of negative (dysphoric) emotions, whereas the left hemisphere seems to impart a more positive (euphoric) perspective to experience.[65,232,243,251] In the intact brain, the modulation of emotional perspective could reflect a complex balance between these two tendencies. Thus, a right-hemisphere lesion could diminish the negative component of affective valence and lead to inappropriate jocularity. It also seems that these hemispheric differences in emotional attitude originate quite early in life. For example, there is greater EEG activation over the left hemisphere when 10-month-old infants are shown happy faces but not when they are shown faces bearing other types of expression.[55]

There are at least two major difficulties associated with the conclusion that each hemisphere imparts a perspective of different polarity to experience. First, the *expression* of jocularity in left hemiplegics may have very little relationship to their state of *feeling*, since there may be major discrepancies between emotional feeling and expression in patients with brain damage (see Chapter 6). Despite their apparent jocularity, some left hemiplegics may, in fact, be severely depressed. Second, the assumption is usually made that the despondency in response to right hemiplegia is qualitatively equivalent to the jocularity that emerges after left hemiplegia, except that it is of the opposite valence. However, one could also argue that the despondency of the right hemiplegic is an entirely appropriate reaction that need not depend on the disruption of a putative inter-hemisphere affective balance. On the other hand, the expression of jocularity by the patient with left hemiplegia is always inappropriate. This suggests that proper emotional perspective (or its expression) is more likely to be disrupted after right-hemisphere lesions.

In keeping with this implication, there are several lines of investigation that attribute a dominant role in all aspects of emotional expression and experience to the right hemisphere. For example, the ability to express affect through emotional prosody in speech as well as by means of facial expression and gesture is more impaired after right-hemisphere lesions than after damage to the left side of the brain (see Chapter 6). Similarly, a patient's ability to identify the nature of the emotion expressed through prosody and facial expression is also more severely impaired after injury to the right hemisphere.[21,148,277]

As shown in Chapter 6, deficits in the *encoding* (display) of emotional expression are associated with lesions in the more anterior frontal parts of the right hemisphere, whereas more temporoparietal lesions tend to disrupt the *decoding* (understanding) of emotional expressions. Thus, the neural substrate for the encoding and decoding of affect shows a plan of organization that parallels the organization of language in the left hemisphere. In view of what was said in prior sections of this chapter about the relationship between emotion and the limbic system, it may seem inconsistent that difficulties in affective expression should be so closely associated with *nonlimbic* parts of the right hemisphere. This apparent dilemma is resolved in Chapter 6, where it is pointed out that these patients do not necessarily have an impairment in their abilities to *feel* emotion. Furthermore, these encoding and decoding deficits are confined to the expression of relatively subtle affective modulations, whereas outbursts associated with more intense feelings are generally spared (see Chapter 6).

In keeping with these observations on patients with brain lesions, there is also evidence that the right hemisphere of neurologically intact individuals is better equipped for encoding and decoding emotional expressions. Thus, emotional expressions are more accentuated on the left side of the face, which is predominantly controlled by the right hemisphere.[201,244] Furthermore, there is a left visual field advantage in the identification of emotional expressions.[108] This body of evidence leads to a potential paradox. For example, in face-to-face encounters, the more expressive left half of the face is likely to fall within the right visual field of the observer. This would appear to create the setting for an intrinsic inefficiency in the communication of affective states, since the right visual field is analyzed primarily by the side of the brain (left) that has the less-developed abilities for decoding emotional expressions. However, it is also possible that this arrangement could reflect the survival value in making one's emotions less than perfectly obvious.

In addition to the expression of emotions, there is also indirect and preliminary evidence showing that there may be right-hemisphere specialization for the *experience* of emotions. For example, experiments that used the electrodermal response as an index of emotional experience found that patients with right-hemisphere lesions were markedly impaired in their responses to emotional stimuli, whereas those with left-hemisphere injury were not.[310] Furthermore, in a sample of right-handed individuals, emotionally loaded questions elicited a larger number of left-ward eye movements than did neutral questions—a finding that some would interpret as an indication of greater right-hemispheric activation in response to the induced emotional state.[252] In a different and rather remarkable set of experiments, neurologically intact subjects were asked to stimulate themselves into sexual climax while their EEGs were being monitored. The results show that the EEG amplitude during orgasm was greater over the right hemisphere.[49] These results have often been cited as supporting the specialized role of the right hemisphere in the experience of emotions.

In further support of a putative right-hemisphere specialization for emotional behavior, there are indications that affective disease may be more closely associated with right-hemisphere dysfunction. For example, it has been suggested that in temporolimbic epilepsy, left-sided foci are more likely to be associated with ideational disorders, whereas right-sided foci are more likely to be associated with affective disturbances.[18] In individuals with unilateral brain injury, affective disorders are also more common in conjunction with right-sided brain damage.[162] Even in patients with otherwise typical manic-depressive disturbances, the EEG power spectra tend to show greater disturbances over the right hemisphere.[78] In fact, when such patients were given a dichotic listening task, their performance was found similar to the performance of individuals with right temporal lobectomy.[308] Furthermore, neurologic signs indicative of right-hemisphere dysfunction may be seen in depressed children and may disappear when the depression is treated.[37,295]

Even the relationship between affective state and the motor system shows a more intimate linkage in the right hemisphere. For example, the left side of the body appears more responsive to hypnotic suggestion.[76,242] Furthermore, unilateral hysterical paralysis is more frequently encountered in the left side of the body, even in left-handed people.[267] In view of the close interrelation between emotion and autonomic tone, it is interesting to note that there may also be hemispheric asymmetry and perhaps a right-sided dominance in autonomic regulation.[239a,310] However, many of the relevant details remain to be elucidated.

Although the neurologic organization of emotions is of fundamental relevance to behavioral neurology, it is important to emphasize that there are formidable methodologic problems associated with this area of observation. Many of the conclusions listed in this section are preliminary and in need of substantial consolidation. If confirmed, the existence of right-hemispheric specialization for emotion could imply that there is asymmetry directly at the level of the limbic system. Alternatively, it is also conceivable that the asymmetry stems from the nonlimbic parts of the right hemisphere and that these regions have a stronger influence on the pertinent components of the limbic system.

Paralinguistic Aspects of Communication and the Right Hemisphere

The production of phonemes and also word choice, syntax, grammar, and semantics constitute the formal, linguistic aspects of speech. These are the major aspects of communication that are impaired after left-hemisphere injury. However, there is much more to communication, and even to speech, than these formal linguistic features. For want of a better terminology, such additional components can be designated as *paralinguistic* aspects of communication. The encoding and decoding of emotional expression through variations of prosody and facial expression are some of the paralinguistic aspects of communication. As shown in Chapter 6, these behaviors are closely associated with right-hemisphere function. However, there are additional paralinguistic functions that are independent of emotion and that also seem to show a relative right-hemisphere specialization.

For example, in addition to imparting emotional tone, speech prosody is also used to denote emphasis and attitude (see Chapter 6). These aspects of prosody also appear to be under preferential right-hemisphere control.[256,296] An analogous relationship could conceivably also occur in written communication. Although patients with right-hemisphere dysfunction usually have no impairment in spelling or writing grammatically, they could fail to impart the appropriate tone and organization to what they write. Paragraphs might not be organized effectively, and the tone of the prose might not suit the subject matter or the intended audience. For example, it has been shown that lesions in the right hemisphere interfere with the organization of paragraphs.[56]

Another aspect of paralinguistic communication is the comprehension of situational context. Patients with right-hemisphere injury are more severely impaired than are those with left-hemisphere injury in the inference of affective as well as neutral contexts through paralinguistic cues.[21] Furthermore, neurologically intact right-handed individuals show a left-ear advantage in the ability to infer context during the dichotic presentation of sentences that have been altered to render the verbal material indecipherable while keeping the melodic structure intact.[69]

The modulation of verbal output according to contextual demands, the choice of appropriate address (*tu* versus *vous*), and the choice of proper pitch (using a higher frequency when speaking to a child) each constitute additional paralinguistic functions about which there is very little neurologic information. The patient whose CT scan is shown in Figure 7A, for example, was known to be a shy and considerate man before suffering a right temporal infarct. Following the stroke, the patient became uncharacteristically brazen and abrasive. From his hospital room he would keep calling the physician's office and use forms of address and a conversational style that reflected inappropriate familiarity. The same patient also talked excessively, and during a conversation would not take the cue to "yield the floor" to the examiner. Kolb and Taylor[148] have also reported an increased verbal output after right temporal lobectomy. These observations raise the possibility that these additional aspects of paralinguistic communication may also yield to neurologic investigations and that they may closely depend on right-hemisphere mechanisms.[295,296]

It appears, therefore, that the right hemisphere may be specialized not only for the emotional aspects of communication but also for many additional paralinguistic functions. It is conceivable that the "inappropriate" behavior of patients with right-hemisphere injury reflects, at least in part, their marked difficulty in inferring context and in modulating their paralinguistic behaviors to fit this context.

LOCALIZATION OF BEHAVIOR

That the brain is the organ of the mind is no longer a subject for dispute, unless there is a fundamental misunderstanding in the definition of terms. One unresolved question that is likely to remain of central importance to behavioral neurology, however, is the way in which mental function is mapped onto cerebral substance. Divergent opinions about the details of this mapping have led to several theories of localization.

One theory, commonly associated with the equipotentiality principle of Lashley,[157] tends to minimize local variations and favors the view that complex behavior is uniformly distributed throughout the brain. The evidence reviewed in this chapter is quite clearly inconsistent with this position, unless one cares to discuss areas of behavior that are impossible to define neurologically. Thus, writing and facial recognition are most certainly not organized according to the equipotentiality principle, whereas character or intelligence could be.

Another extreme position is based on the assumption that each complex function is coordinated by a specific and dedicated center in the brain.[73] This position is also inconsistent with much of the evidence in this chapter that shows the interdependence among cerebral areas in all phases of information processing. This interdependence suggests that complex behaviors are subserved not by isolated centers but by sets of distributed regions that form integrated cerebral networks based on neural interconnections.[179] This network approach, which has its roots in the writings of many neurologists including Hughlings Jackson and Bastian, offers an intermediate position between the two extremes of the holistic and centrist approaches.

The network approach that was initially proposed to describe the mechanism of directed attention is schematically shown in Figure 9.

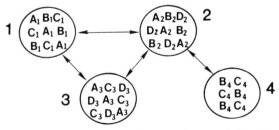

FIGURE 9. A schematic diagram of the network approach to cerebral localization.

Thus, a complex multimodal function A (e.g., spatial distribution of attention) is considered in terms of three component processes: A_1 (representation of the extrapersonal space), A_2 (distribution of motivation), and A_3 (motor schema for scanning and exploration). These behaviors are modulated by sites 1 (parietal heteromodal cortex), 2 (cingulate cortex), and 3 (prefrontal cortex), respectively. The component sites are interconnected and constitute an integrated network subserving function A. Each one of these three sites has additional behavioral specializations (B, C, D) and is therefore also part of intersecting but distinct networks. For example, sites 1, 2, and 4 provide an intersecting network for function B.

The network approach suggests several general principles for the cerebral localization of complex functions:

1. Complex functions are represented within distributed but interconnected sites which collectively constitute an integrated network for that function
2. Each individual cerebral area contains the neural substrate for several sets of behaviors and may therefore belong to several partially overlapping networks
3. Lesions confined to a single region are likely to result in multiple deficits
4. Different aspects of the same complex function may be impaired as a consequence of damage to one of several cortical areas or to their interconnections.

In some complex functions such as directed attention and memory, the behavioral deficits that emerge from damage to separate components of the pertinent network differ from each other only in subtle ways (see Chapters 3 and 4). In other complex functions such as language, lesions to each component of the underlying network give rise to markedly different deficits (see Chapters 5 and 6).

SUMMARY

In this chapter, the entire cortical surface of the brain is divided into five zones that form an orderly succession of architectonic structure, from the simplest to the most differentiated. Highly organized cortical interconnections enable the sequential transfer of information across these five zones. The innermost zone, the one with the strongest hypothalamic connectivity, contains the corticoid and allocortical areas of the brain and is designated as "limbic." This zone is most closely associated with homeostatis, drive reduction, affect, and memory—basic neural mechanisms that are essential for self-preservation and species propagation. The two outermost cortical zones contain the idiotypic and unimodal areas and have the most immediate contact with the extrapersonal world: they process incoming information and coordinate the motor sequences for the manipulation of the environment. The two intermediate zones, containing the heteromodal and paralimbic areas, occupy most of the cortical surface and provide a matrix for the further refinement of incoming information and also for its association with limbic input.

The organization of the striatum, globus pallidus, and thalamus lend themselves to analogous subdivisions. The limbic and paralimbic zones of cortex, together with the hypothalamus and the limbic components of the basal ganglia and thalamus, constitute the limbic system, which is of crucial importance in all aspects of memory, learning, and emotions.

Some pathways originating from discrete groups of neurons are widely distributed throughout the brain. These pathways are associated with state-dependent functions that influence all other behaviors. In contrast, most of the interactions among cortex, basal ganglia, and thalamus are organized in the form of more discrete pathways. These discrete pathways subserve the channel functions of the brain. Such channel functions are localized at the level of integrated cerebral networks.

Superimposed on these patterns of regional brain-behavior relations, there is also a marked hemispheric asymmetry in the functional specialization of the human brain. Whereas the left hemisphere is dominant for language and fine motor skills, the right hemisphere is more specialized for nonverbal perceptual-motor skills, for determining the spatial distribution of attention, for emotional behavior, and for the paralinguistic aspects of interpersonal communications. The anatomic foundations for these hemispheric specializations, especially for those of the right hemisphere, are largely unknown and are likely to offer fertile directions for future research.

REFERENCES

1. ABBIE, AA: *Cortical lamination in a polyprotodont marsupial, Perameles nasuta.* J Comp Neurol 76:509–535, 1942.
2. ACKERLY, S: *Instinctive, emotional and mental changes following prefrontal lobe extirpation.* Am J Psychiatry 92:717–729, 1935.
3. AGGLETON, JP, BURTON, MJ, AND PASSINGHAM, RE: *Cortical and subcortical afferents to the amygdala of the rhesus monkey (Macaca mulatta).* Brain Res 190:347–368, 1980.
4. AGGLETON, JP, AND MISHKIN, M: *Visual recognition impairment following medial thalamic lesions in monkeys.* Neuropsychol 21:189–197, 1983.
5. ALEXANDER, F: *Psychosomatic Medicine: Its Principles and Applications.* WW Norton, New York, 1950.
6. AMARAL, DG, VEAZEY, RB, AND COWAN, WM: *Some observations on hypothalamo-amygdaloid connections in the monkey.* Brain Res 252:13–27, 1982.
7. AMYES, EW, AND NIELSEN, JM: *Clinicopathologic study of vascular lesions of the anterior cingulate region.* Bull Los Angeles Neurol Soc 20:112–130, 1955.
8. ANAND, BK, AND DUA, S: *Circulating and respiratory changes induced by electrical stimulation of limbic system (visceral brain).* J Neurophysiol 19:393–400, 1956.
9. ARIKUNI, T, SAKAI, M, HAMADA, I, AND KUBOTA, K: *Topographical projections from the prefrontal cortex to the post-arcuate area in the rhesus monkey, studied by retrograde axonal transport of horseradish peroxidase.* Neurosci Lett 19:155–160, 1980.
10. AX, AF: *The physiological differentiation between fear and anger in humans.* Psychosom Med 15:433–442, 1953.
11. BAGSHAW, MH, AND PRIBRAM, KH: *Cortical organization in gustation (Macaca mulatta).* J Neurophysiol 16:499–508, 1953.
12. BALLENTINE, HT, JR, LEVEY, BA, DAGI, TF, AND DIRIUNAS, IB: *Cingulotomy for psychiatric illness: Report of 13 years experience.* In SWEET, WH, OBRADOR, S, AND MARTIN-RODRIQUES, JG (EDS): *Neurosurgical Treatment in Psychiatry, Pain and Epilepsy.* University Park Press, Baltimore, 1977, pp 333–353.
13. BANCAUD, J, TALAIRACH, J, GEIER, S, BONIS, A, TROTTIER, S, AND MANRIQUE, M: *Manifestation comportementales induites par la stimulation éléctrique du gyrus cingulaire antérieur chez l'homme.* Rev Neurol (Paris) 132:705–724, 1976.

14. BARBAS, H, AND MESULAM, M-M: *Organization of afferent input to subdivisions of area 8 in the rhesus monkey.* J Comp Neurol 200:407–431, 1981.

15. BARBUR, JL, RUDDOCK, KH, AND WATERFIELD, VA: *Human visual responses in the absence of the geniculo-calcarine projection.* Brain 103:905–928, 1980.

16. BARTUS, RT, DEAN, RL, BEER, B, AND LIPPA, AD: *The cholinergic hypothesis of geriatric memory dysfunction.* Science 217:408–417, 1982.

17. BEAL, MF, BIRD, ED, LANGLAIS, PJ, AND MARTIN, JB: *Somatostatin is increased in the nucleus accumbens in Huntington's disease.* Neurology 34:663–666, 1984.

18. BEAR, O, AND FEDIO, P: *Qualitative analysis of interictal behavior in temporal lobe epilepsy.* Arch Neurol 34:454–467, 1977.

19. BENEVENTO, LA, AND REZAK, M: *The cortical projections of the inferior pulvinar and adjacent lateral pulvinar in the rhesus monkey (Macaca mulatta): An autoradiographic study.* Brain Res 108:1–24, 1976.

20. BENEVENTO, LA, FALLON, J, DAVIS, BJ, AND REZAK, M: *Auditory-visual interaction in single cells in the cortex of the superior temporal sulcus and the orbital frontal cortex of the macaque monkey.* Exper Neurol 57:849–872, 1977.

21. BENOWITZ, LI, BEAR, DM, ROSENTHAL, R, MESULAM, M-M, ZAIDEL, E, AND SPERRY, RW: *Hemispheric specialization in nonverbal communication.* Cortex 19:5–11, 1983.

22. BENSON, DA, HIENZ, RD, AND GOLDSTEIN, MH, JR: *Single-unit activity in the auditory cortex of monkeys actively localizing sound sources: Spatial tuning and behavioral dependency.* Brain Res 219:249–267, 1981.

23. BENTON, AL, VARNEY, NR, AND HAMSHER, K DE S: *Lateral differences in tactile directional perception.* Neuropsychol 16:109–114, 1978.

24. BENTON, AL, VARNEY, NR, AND HAMSHER, K DE S: *Visuospatial judgement.* Arch Neurol 35:364–367, 1978.

25. BENZO, CA: *The hypothalamus and blood glucose regulation.* Life Sci 32:2509–2515, 1983.

26. BEVER, TG, AND CHIARELLO, RJ: *Cerebral dominance in musicians and nonmusicians.* Science 185:537–539, 1974.

27. BIEMOND, A: *The conduction of pain above the level of the thalamus opticus.* Arch Neurol Psychiatr 75:231–244, 1956.

28. BIGNALL, KE, AND IMBERT, M: *Polysensory and cortico-cortical projections to frontal lobe of squirrel and rhesus monkeys.* Electroenceph Clin Neurophysiol 26:206–215, 1969.

29. BLOCK, CH, SIEGEL, A, AND EDINGER, H: *Effects of amygdaloid stimulation upon trigeminal sensory fields of the lip that are established during hypothamically-elicited quiet biting attack in the cat.* Brain Res 197:39–55, 1980.

30. BRICKNER, RM: *An interpretation of frontal lobe function based upon the study of a case of partial bilateral frontal lobectomy.* Assoc Res Nerv Ment Dis Proc 13:259–351, 1934.

31. BRINKMAN, C, AND PORTER, R: *Supplementary motor area and premotor area of monkey cerebral cortex: Functional organization and activities of single neurons during performance of a learned movement.* In DESMEDT, JE (ED): *Motor Control Mechanisms in Health and Disease.* Raven Press, New York, 1983.

32. BROCA, P: *Anatomie comparée des circonvolutions cérébrales: Le grand lobe limbique dans la série des mammifères.* Rev Anthropol 1:384–498, 1878.

33. BRODMANN, K: *Vergleichende Lokalisationlehre der Grosshirnrinde in ihren Prinzipien dargestellt auf Grund des Zellenbaues.* JA Barth, Leipzig, p 324, 1909.

34. BROOKS, WH, CROSS, RJ, ROSZMAN, TL, AND MARKESBERG, WR: *Neuroimmunomodulation: Neural anatomical basis for impairment and facilitation.* Ann Neurol 12:56–61, 1982.

35. BROWN, RM, CRANE, AM, AND GOLDMAN, PS: *Regional distribution of monoamines in the cerebral cortex and subcortical structures of the rhesus monkey: Concentrations and in vivo synthesis rates.* Brain Res 168:133–150, 1979.

36. BRUCE, C, DESIMONE, R, AND GROSS, CG: *Visual properties of neurons in a polysensory area in superior temporal sulcus of the macaque.* J Neurophysiol 46:369–384, 1981.

37. BRUMBACK, RA, STATON, RD, AND WILSON, H: *Right cerebral hemispheric dysfunction.* Arch Neurol 41:248–249, 1984.

38. BUNNEY, WE, JR, AND GARLAND, BL: *A second generation catecholamine hypothesis.* Pharmacopsychiatry 15:111–115, 1982.

39. BURTON, H, AND JONES, EG: *The posterior thalamic region and its cortical projection in New World monkeys.* J Comp Neurol 168:249–301, 1976.

40. BUTTER, CM, AND SNYDER, DR: *Alterations in aversive and aggressive behaviors following orbital frontal lesions in rhesus monkeys.* Acta Neurobiol Exp 32:525–565, 1972.

41. CAMBIER, J, ELGHOZI, D, AND STRUBE, E: *Lésions du thalamus droit avec syndrome de l'hémisphère mineur. Discussion du concept de négligence thalamique.* Rev Neurol 136:105–116, 1980.

42. CARLSON, M: *Characteristics of sensory deficits following lesions of Brodmann's areas 1 and 2 in the postcentral gyrus of Macaca mulatta.* Brain Res 204:424–430, 1981.

43. CARPENTER, MB, AND SUTIN, J: *Human Neuroanatomy.* Williams & Wilkins, Baltimore, 1983.

44. CHALUPA, LM, COYLE, RS, AND LINDSLEY, DB: *Effect of pulvinar lesions on visual pattern discrimination in monkeys.* J Neurophysiol 39:354–369, 1976.

45. CHAPMAN, WP, LIVINGSTON, KE, AND POPPEN, JL: *Effect upon blood pressure of electrical stimulation of tips of temporal lobes in man.* J Neurophysiol 13:65–71, 1950.

46. CHAVIS, DA, AND PANDYA, DN: *Further observations on cortico-frontal connections in the rhesus monkey.* Brain Res 117:369–386, 1976.

47. CHI, JG, DOOLING, EC, AND GILLES, FH: *Gyral development of the human brain.* Ann Neurol 1:86–93, 1977.

48. CLAUSE, RE, AND LUSTMAN, PJ: *Psychiatric illness and contraction abnormalities of the esophagus.* N Engl J Med 309:1337–1342, 1983.

49. COHEN, HD, ROSEN, RC, AND GOLDSTEIN, L: *Electroencephalographic laterality changes during human sexual orgasm.* Arch Sex Behav 5:189–199, 1976.

50. CORKIN, S, MILNER, B, AND RASMUSSEN, T: *Somatosensory thresholds.* Arch Neurol 23:41–58, 1970.

51. COWEY, A: *Sensory and non-sensory visual disorders in man and monkey.* Phil Trans R Soc Lond B 298:3–13, 1983.

52. COWEY, A, AND DEWSON, JH: *Effects of unilateral ablation of superior temporal cortex on auditory sequence discrimination in Macaca mulatta.* Neuropsychology 10:279–289, 1972.

53. DAMASIO, AR, AND DAMASIO, H: *Localization of lesions in achromatopsia and prosopagnosia.* In KERTESZ, A (ED): *Localization in Neuropsychology.* Academic Press, New York, 1983, pp 417–428.

54. DARIAN-SMITH, I, SUGITANI, M, HEYWOOD, J, KORITA, K, AND GOODWIN, A: *Touching textured surfaces: Cells in somatosensory cortex respond both to finger movement and to surface features.* Science 218:906–909, 1982.

55. DAVIDSON, RJ, AND FOX, NA: *Asymmetrical brain activity discriminates between positive and negative affective stimuli in human infants.* Science 218:1235–1247, 1982.

56. DELIS, DC, WAPNER, W, GARDNER, H, AND MOSES, JA JR: *The contribution of the right hemisphere to the organization of paragraphs.* Cortex 19:43–50, 1983.

57. DELONG, MR: *Activity of pallidal neurons during movement.* J Neurophysiol 34:414–427, 1971.

58. DENENBERG, VH: *Hemispheric laterality in animals and the effects of early experience.* Behav Brain Sci 4:1–49, 1981.

59. DENNY-BROWN, D, AND CHAMBERS, RA: *Physiological aspects of visual perception. I. Functional aspects of visual cortex.* Arch Neurol 33:219–227, 1976.

60. DERENZI, E: *Disorders of Space Exploration and Cognition.* Wiley, Chichester, UK, 1982.

61. DERENZI, E, FAGLIONI, P, AND VILLA, P: *Topographical amnesia.* J Neurol Neurosurg Psychiatr 40:498–505, 1977.

62. DESIMONE, R, FLEMING, J, AND GROSS, C: *Prestriate afferents to inferior temporal cortex: An HRP study.* Brain Res 184:41–55, 1980.

63. DESIMONE, R, AND GROSS, CG: *Visual areas in the temporal cortex of the macaque.* Brain Res 178:363–380, 1979.

64. DIMOND, SJ: *Sex differences in brain organization.* Behav Brain Sci 3:215–263, 1980.

65. DIMOND, SJ, FARRINGTON, L, AND JOHNSON, P: *Differing emotional response from right and left hemisphere.* Nature 261:690–692, 1976.

66. DOTY, RW: *Nongeniculate afferents to striate cortex in macaques.* J Comp Neurol 218:159–173, 1983.

67. DOWNER, DE C JL: *Interhemispheric integration in the visual system.* In MOUNTCASTLE, VB (ED): *Interhemispheric Relations and Cerebral Dominance.* Johns Hopkins Press, Baltimore, 1962, pp 87–100.

68. DROOGLEEVER-FORTUYN, J: *On the neurology of perception.* Clin Neurol Neurosurg 81:97–107, 1979.

69. DWYER, JH III, AND RINN, WE: *The role of the right hemisphere in contextual inference.* Neuropsychology 19:479–482, 1981.

PATTERNS IN
BEHAVIORAL
NEUROANATOMY:
ASSOCIATION
AREAS, THE LIMBIC
SYSTEM, AND
HEMISPHERIC
SPECIALIZATION

61

70. Economo, C von: *The cytoarchitectonics of the human cerebral cortex.* Oxford University Press, London, 1929.

71. Eichenbaum, H, Morton, TH, Potter, H, and Corkin, S: *Selective olfactory deficits in case H.M.* Brain 106:459–472, 1983.

72. Evarts, EV, Fromm, C, Kroller, J, and Jennings, VA: *Motor cortex control of finely graded forces.* J Neurophysiol 49:1199–1215, 1983.

73. Exner, S: *Untersuchungen über Localisation der Functionen in der Grosshirnrinde des Menschen.* W Braumuller, Vienna, 1881.

74. Fenz, WD, and Epstein, S: *Gradients of physiological arousal in parachutists as a function of an approaching jump.* Psychosom Med 29:33–51, 1967.

75. Filimonoff, IN: *A rational subdivision of the cerebral cortex.* Arch Neurol Psychiatr 58:296–311, 1947.

76. Fleminger, JJ, McClure, GM, and Dalton, R: *Lateral response to suggestion in relation to handedness and the side of psychogenic symptoms.* Br J Psychiatry 136:562–566, 1980.

77. Flicker, C, Dean, RL, Watkins, DL, Fisher, SK, and Bartus, RT: *Behavioral and neurochemical effects following neurotoxic lesions of a major cholinergic input to the cerebral cortex in the rat.* Pharmacol Biochem Behav 18:973–981, 1983.

78. Flor-Henry, P: *On certain aspects of the localization of the cerebral systems regulating and determining emotion.* Biol Psychiatr 14:677–698, 1979.

79. Franzen, EA, and Myers, RE: *Neural control of social behavior: Prefrontal and anterior temporal cortex.* Neuropsychologia 11:141–157, 1981.

80. Freedman, M, Alexander, MP, and Naeser, MA: *Anatomical basis of transcortical motor aphasia.* Neurology 34:409–417, 1984.

81. Fuster, J, Bauer, RH, and Jervey, JP: *Cellular discharge in the dorsolateral prefrontal cortex of the monkey in cognitive tasks.* Exper Neurol 77:679–694, 1982.

82. Fuster, JM, and Jervey, JP: *Inferotemporal neurons distinguish and retain behaviorally relevant features of visual stimuli.* Science 212:952–955, 1981.

83. Gabriel, M, Foster, K, Orona, E, Saltwick, SE, and Stanton, M: *Neuronal activity of cingulate cortex, anteroventral thalamus and hippocampal formation in discriminative conditioning: Encoding and extraction of the significance of conditional stimuli.* Prog Psychol Physiol 9:125–231, 1980.

84. Gainotti, G: *The relationships between emotions and cerebral dominance: A review of clinical and experimental evidence.* In Gruzelier, J, and Flor-Henry, P (eds): *Hemisphere Asymmetries of Function in Psychopathology.* Elsevier, North-Holland, 1979, pp 21–34.

85. Gazzaniga, MS, and LeDoux, JE: *The Integrated Mind.* Plenum Press, New York, 1978.

86. Geschwind, N: *Disconnection syndromes in animals and man.* Brain 88:237–294, 1965.

87. Geschwind, N, and Levitsky, W: *Human brain: Left-right asymmetries in temporal speech region.* Science 161:186–187, 1968.

88. Gloor, P, Olivier, A, Quesney, LF, Andermann, F, and Horowitz, S: *The role of the limbic system in experiential phenomena of temporal lobe epilepsy.* Ann Neurol 12:129–144, 1982.

89. Goldman, PS, and Rosvold, HE: *Localization of function within the dorsolateral prefrontal cortex of the rhesus monkey.* Exper Neurol 27:291–304, 1970.

90. Goldstein, K: *The significance of the frontal lobes for mental performances.* J Neurol Psychopathol 17:27–40, 1936.

91. Goldstein, JM, and Siegal, J: *Stimulation of ventral tegmental area and nucleus accumbens reduce receptive fields for hypothalamic biting reflex in cats.* Exper Neurol 72:239–246, 1981.

92. Gower, EC, and Mesulam, M-M: *Some paralimbic connections of the medial pulvinar nucleus in the macaque.* Neurosci Abstr 4:75, 1978.

93. Gower, EC, and Mesulam, M-M: *Thalamic connections of paralimbic cortex in the temporal pole of the macaque.* Neurosci Abstr 5:75, 1979.

94. Green, JD, Clemente, CD, and DeGroot, J: *Rhinencephalic lesions and behavior in cats.* J Comp Neurol 108:505–536, 1957.

95. Groenewegen, HJ, Becker, NEHM, and Lohman, AHM: *Subcortical afferents of the nucleus accumbens septi in the cat, studied with retrograde axonal transport of horseradish peroxidase and bisbenzimide.* Neurosci 5:1903–1916, 1980.

96. Gross, CG, Rocha-Miranda, CE, and Bender, DB: *Visual properties of neurons in inferotemporal cortex of the macaque.* J Neurophysiol 35:96–111, 1972.

97. Grossman, SP: *Behavioral functions of the septum: A re-analysis.* In DeFrance, JF (ed): *The Septal Nuclei.* Plenum Press, New York, 1976, pp 361–422.

98. GUARD, O, DELPY, C, RICHARD, D, AND DUMAS, R: *Une cause mal connue de confusion mentale: le ramollissemant temporal droit.* Rev Med 40:2115–2121, 1979.

99. GUILLERY, RW: *Afferent fibers to the dorso-medial thalamic nucleus in the cat.* J Anat 93:403–419, 1959.

100. HALL, RE, AND CORNISH, K: *Role of the orbital cortex in cardiac dysfunction in unanesthesized rhesus monkey.* Exper Neurol 56:289–297, 1977.

101. HALSBAND, U, AND PASSINGHAM, R: *The role of premotor and parietal cortex in the direction of action.* Brain Res 240:368–372, 1982.

102. HASHIMOTO, S, GEMBA, H, AND SASAKI, K: *Analysis of slow cortical potentials preceding self-paced hand movements in the monkey.* Exper Neurol 65:218–229, 1979.

103. HEATH, RG: *Studies in Schizophrenia.* Harvard University Press, Cambridge, MA, 1959.

104. HEBB, DD: *Man's frontal lobes.* Arch Neurol Psychiatr 54:10–24, 1945.

105. HEFFNER, H, AND MASTERTON, B: *Contribution of auditory cortex to sound localization in the monkey (Macaca mulatta).* J Neurophysiol 38:1340–1358, 1975.

106. HEIMER, L: *The olfactory connections of the diencephalon in the rat.* Brain Behav Evol 6:484–523, 1972.

107. HEIMER, L, AND WILSON, RD: *The subcortical projections of the allocortex: Similarities in the neural associations of the hippocampus, the piriform cortex and the neocortex.* In SANTINI, M (ED): *Golgi Centennial Symposium. Proceedings.* Raven, New York, 1975, pp 177–193.

108. HELLER, W, AND LEVY, J: *Perception and expression of emotion in right-handers and left-handers.* Neuropsychologia 19:263–272, 1981.

109. HERZOG, AG, AND VAN HOESEN, GW: *Temporal neocortical afferent connections to the amygdala in the rhesus monkey.* Brain Res 115:57–69, 1976.

110. HIKOSAKA, O, AND WURTZ, RH: *Visual and oculomotor functions of monkey substantia nigra pars reticulata. III. Memory-contingent visual and saccade responses.* J Neurophysiol 49:1268–1284, 1983.

111. HOFFMAN, BL, AND RASMUSSEN, T: *Stimulation studies of insular cortex of Macaca mulatta.* J Neurophysiol 16:343–351, 1953.

112. HOREL, J, MEYER-LOHMANN, J, AND BROOKS, VB: *Basal ganglia cooling disables learned arm movements of monkeys in the absence of visual guidance.* Science 195:584–586, 1977.

113. HOREL, JA, AND MISANTONE, LJ: *The Klüver-Bucy Syndrome produced by partial isolation of the temporal lobe.* Exper Neurol 42:101–112, 1974.

114. HOREL, JA, AND PYTKO, DE: *Behavioral effect of local cooling in temporal lobe of monkeys.* J Neurophysiol 47:11–22, 1982.

115. HORNYKIEWICZ, O, AND KISH, SJ: *Neurochemical basis of dementia in Parkinson's disease.* Can J Neurol Sci 11:185–190, 1984.

116. HUBEL, DH, HENSON, CO, RUPERT, A, AND GALAMBOS, R: *"Attention" units in the auditory cortex.* Science 129:1279–1280, 1959.

117. HUBEL, DH, AND WIESEL, TN: *Brain mechanisms of vision in the brain.* In *The Brain, A Scientific American Book.* WH Freeman & Co, San Francisco, 1979, pp 84–97.

118. HYVARINEN, J: *Regional distribution of functions in parietal association area 7 of the monkey.* Brain Res 206:287–303, 1981.

119. HYVARINEN, J, AND PORANEN, A: *Function of the parietal associative area 7 as revealed from cellular discharges in alert monkeys.* Brain 97:673–692, 1974.

120. ISSEROFF, A, ROSVOLD, HE, GALKIN, TW, AND GOLDMAN-RAKIC, PS: *Spatial memory impairments following damage to the mediodorsal nucleus of the thalamus in rhesus monkeys.* Brain Res 232:97–113, 1982.

121. ITO, S-I: *Prefrontal unit activity of macaque monkeys during auditory and visual reaction time tasks.* Brain Res 247:39–47, 1982.

122. IVERSEN, SD: *Behavior after neostriatal lesions in animals.* In DIVAC, I, AND OBERG, RGE (EDS): *The Neostriatum.* Pergamon Press, Oxford, 1979.

123. IVERSEN, SD, AND MISHKIN, M: *Comparison of superior temporal and inferior prefrontal lesions on auditory and non-auditory tasks in rhesus monkeys.* Brain Res 55:355–367, 1973.

124. JACOBS, MC, MORGANE, PJ, AND MCFARLAND, WL: *The anatomy of the brain of the Bottlenose Dolphin (Tursiops truncatus). Rhinic lobe (Rhinencephalon) I. The paleocortex.* J Comp Neurol 141:205–272, 1971.

125. JASPER, HH: *Functional properties of the thalamic reticular system.* In *Brain Mechanisms and Consciousness.* Charles C Thomas, Springfield, IL, 1954, pp 374–401.

126. JAVOY-AGID, F, AND AGID, Y: *Is the mesocortical dopaminergic system involved in Parkinson disease?* Neurology 30:1326–1330, 1980.

PATTERNS IN
BEHAVIORAL
NEUROANATOMY:
ASSOCIATION
AREAS, THE LIMBIC
SYSTEM, AND
HEMISPHERIC
SPECIALIZATION

63

127. JONES, EG: *Some aspects of the organization of the thalamic reticular complex.* J Comp Neurol 162:285–308, 1975.

128. JONES, EG, AND BURTON, H: *A projection from medial pulvinar to the amygdala in primates.* Brain Res 104:142–147, 1976.

129. JONES, EG, AND POWELL, TPS: *An anatomical study of converging sensory pathways within the cerebral cortex of the monkey.* Brain 93:793–820, 1970.

130. JONES, EG, WISE, SP, AND COULTER, JD: *Differential thalamic relationships of sensory-motor and parietal cortical fields in monkeys.* J Comp Neurol 183:833–881, 1979.

131. JURGENS, U: *Amygdalar vocalization pathways in the squirrel monkey.* Brain Res 241:189–196, 1982.

132. JURGENS, W, AND VON CRAMON, D: *On the role of the anterior cingulate cortex in phonation: A case report.* Brain Lang 15:234–248, 1982.

133. KAADA, BR: *Cingulate, posterior orbital, anterior insular and temporal pole cortex.* In MAGOUN, HW (ED): *Neurophysiology.* Waverly Press, Baltimore, 1960, pp 1345–1372.

134. KAADA, BR, PRIBRAM, KH, AND EPSTEIN, JA: *Respiratory and vascular responses in monkeys from temporal pole, insula, orbital surface and cingulate gyrus.* J Neurophysiol 12:348–356, 1949.

135. KAAS, JH: *What, if anything, is SI? Organization of first somatosensory area of cortex.* Physiol Rev 63:206–231, 1983.

136. KAHNEMANN, D, TURSKY, B, SHAPIRO, D, AND CRIDER, A: *Pupillary, heart rate and skin resistance changes during a mental task.* J Exper Psychol 79:164–167, 1969.

137. KAROL, EA, AND PANDYA, DN: *The distribution of the corpus callosum in the rhesus monkey.* Brain 94:471–486, 1971.

138. KASE, CS, TRONCOSO, JF, COURT, JE, TAPIA, JF, AND MOHR, JP: *Global spatial disorientation.* J Neurol Sci 34:267–278, 1977.

139. KERTESZ, A: *Right hemisphere lesions in construction apraxia and visuospatial deficit.* In KERTESZ, A (ED): *Localization in Neuropsychology.* Academic Press, New York, 1983, pp 455–470.

140. KIEVIT, J, AND KUYPERS, HGJM: *Basal forebrain and hypothalamic connections to frontal and parietal cortex in the rhesus monkey.* Science 187:660–662, 1975.

141. KIMURA, D: *The asymmetry of the human brain.* Sci Am 228:70–78, 1973.

142. KLEIN, D, MOSCOVITCH, M, AND VIGNA, C: *Attentional mechanisms and perceptual asymmetries in tachistoscopic recognition of words and faces.* Neuropsychologia 14:55–66, 1976.

143. KLING, A, AND STEKLIS, HD: *A neural substrate for affiliative behavior in nonhuman primates.* Brain Behav Evol 13:216–238, 1976.

144. KLING, A, STEKLIS, HD, AND DEUTSCH, S: *Radiotelemetered activity from the amygdala during social interactions in the monkey.* Exper Neurol 66:88–96, 1979.

145. KOJIMA, S: *Prefrontal unit activity in the monkey: Relation to visual stimuli and movements.* Exper Neurol 69:110–123, 1980.

146. KOJIMA, S, MATSUMURA, M, AND KUBOTA, K: *Prefrontal neuron activity during delayed-response performance without imperative GO signals in the monkey.* Exper Neurol 74:396–407, 1981.

147. KOLB, B, MILNER, B, AND TAYLOR, L: *Perception of faces by patients with localized cortical excisions.* Can J Psychol 37:8–18, 1983.

148. KOLB, B, AND TAYLOR, L: *Affective behavior in patients with localized cortical excisions: Role of lesion site and side.* Science 214:89–91, 1981.

149. KUBOTA, K, TONOIKE, M, AND MIKAMI, A: *Neuronal activity in the monkey dorsolateral prefrontal cortex during a discrimination task with delay.* Brain Res 183:29–42, 1980.

150. KÜNZLE, H: *Bilateral projections from precentral motor cortex to the putamen and other parts of the basal ganglia: An autoradiographic study in Macaca fascicularis.* Brain Res 88:195–209, 1975.

151. KUYPERS, HGJM; *Anatomy of descending pathways.* In BROOKS, VB (ED): *Handbook of Physiology–The Nervous System II.* American Physiological Society, Bethesda, MD, 1981, pp 597–666.

152. LACEY, JI: *Somatic response patterning and stress: Some revisions of activation theory.* In APPLEY, MH, AND TRUMBULL, R (EDS): *Psychological Stress.* Appleton-Century-Crofts, New York, 1967, pp 14–37.

153. LACOSTE-UTAMSING, C DE, AND HOLLOWAY, RL: *Sexual dimorphism in the human corpus callosum.* Science 216:1431–1432, 1982.

154. LANDIS, C: *Psychology.* In METTLER, FA (ED): *Selective Partial Ablation of the Frontal Cortex.* Paul B Hoeber, New York, 1949, pp 492–496.

155. Laplane, D, Talairach, J, Meininger, V, Bancaud, J, and Bouchareine, A: *Motor consequences of motor area ablations in man.* J Neurol Sci 31:29–49, 1977.

156. Laplane, D, Talairach, J, Meininger, V, Bancaud, J, and Orgogozo, JM: *Clinical consequences of corticectomies involving the supplementary motor area in man.* J Neurol Sci 34:301–314, 1977.

157. Lashley, KS: *Brain Mechanisms and Intelligence.* University of Chicago Press, 1929.

158. LeDoux, JE, Thompson, ME, Iadecola, C, Tucker, LW, Reis, DJ: *Local cerebral blood flow increases during auditory and emotional processing in the conscious rat.* Science 221:576–578, 1983.

159. Leehey, S, Carey, S, Diamond, R, and Cahn, A: *Upright and inverted faces. The right hemisphere knows the difference.* Cortex 14:411–419, 1978.

160. Le Moal, M, Galey, D, and Cardo, B: *Behavioral effects of local injection of 6-hydroxydopamine in the medial ventral tegmentum in the rat. Possible role of the mesolimbic dopaminergic system.* Brain Res 88:190–194, 1975.

161. Lewis, ME, Mishkin, M, Bragin, E, Brown, RM, Pert, CB, and Pert, A: *Opiate receptor gradients in monkey cerebral cortex: Correspondence with sensory processing hierarchies.* Science 211:1166–1169, 1981.

162. Lishman, WA: *Brain damage in relation to psychiatric disability after head injury.* Br J Psychiatr 114:373–410, 1968.

163. Locke, S, and Kerr, C: *The projection of nucleus lateralis dorsalis of monkey to basomedial temporal cortex.* J Comp Neurol 149:29–42, 1973.

164. Lorenz, K: *Evolution and Modification of Behavior.* University of Chicago Press, Chicago, 1965.

165. Lown, B, Temte, JV, Reich, P, Gaughan, C, Regestein, Q, and Hai, H: *Basis for recurring ventricular fibrillation in the absence of coronary heart disease and its management.* N Engl J Med 294:623–629.

166. Lynch, JC: *The functional organization of posterior parietal association cortex.* Behav Brain Sci 3:485–499, 1980.

167. MacLean, PD: *Psychosomatic disease and the visceral brain: Recent developments bearing on the Papez theory of emotion.* Psychosom Med 11:338–353, 1949.

168. MacLean, PD: *Effects of lesions of globus pallidus on species-typical display behavior of squirrel monkeys.* Brain Res 149:175–196, 1978.

169. Mahut, H, Moss, M, and Zola-Morgan, S: *Retention deficits after combined amygdalo-hippocampal and selective hippocampal resections in the monkey.* Neuropsychologia 19:201–225, 1981.

170. Marsden, CD: *The mysterious motor function of the basal ganglia: the Robert Wartenberg lecture.* Neurology 32:514–539, 1982.

171. Mazziotta, JC, Phelps, ME, Carson, RE, and Kuhl, DE: *Tomographic mapping of human cerebral metabolism: Auditory stimulation.* Neurology 32:921–937, 1982.

172. Mazzucchi, A, Marchini, C, Budai, R, and Parma, M: *A case of receptive amusia with prominent timbre perception defect.* J Neurol Neurosurg Psychiatr 45:644–647, 1982.

173. McFarland, HR, and Fartrin, D: *Amusia due to right temporoparietal infarct.* Arch Neurol 39:725–727, 1982.

174. McGlone, J: *Sex differences in human brain asymmetry: A critical survey.* Behav Brain Sci 3:215–263, 1980.

175. Meerwaldt, JD, and Van Harskamp, F: *Spatial disorientation in right-hemisphere infarction.* J Neurol Neurosurg Psychiatr 45:586–590, 1982.

176. Melamed, E, and Larsen, B: *Cortical activation pattern during saccadic eye movements in humans: localization by focal cerebral blood flow increases.* Ann Neurol 5:79–88, 1979.

177. Merzenich, MM, and Brugge, JF: *Representation of the cochlear partition on the superior temporal plane of the macaque monkey.* Brain Res 50:275–296, 1973.

178. Mesulam, M-M: *Dissociative states with abnormal temporal lobe EEG: Multiple personality and the illusion of possession.* Arch Neurol 38:176–181, 1981.

179. Mesulam, M-M: *A cortical network for directed attention and unilateral neglect.* Ann Neurol 10:309–325, 1981.

180. Mesulam, M-M, and Mufson, EJ: *Insula of the old world monkey. Part III. Efferent cortical output and comments on function.* J Comp Neurol 212:38–52, 1982.

181. Mesulam, M-M, and Mufson, EJ: *Neural inputs into the nucleus basalis of the substantia innominata (Ch4) in the rhesus monkey.* Brain 107:253–274, 1984.

182. Mesulam, M-M, and Mufson, EJ: *The insula of Reil in man and monkey.* In Jones, EG, and Peters, AA (eds): *Cerebral Cortex.* New York, Plenum Press, 1985, in press.

183. MESULAM, M-M, MUFSON, EJ, LEVEY, AI, AND WAINER, BH: *Cholinergic innervation of cortex by the basal forebrain: Cytochemistry and cortical connections of the septal area, diagonal band nuclei, nucleus basalis (substantia innominata) and hypothalamus in the rhesus monkey.* J Comp Neurol 214:170–197, 1983.

184. MESULAM, M-M, MUFSON, EJ, LEVEY, AI, AND WAINER, BH: *Atlas of cholinergic neurons in the forebrain and upper brainstem of the macaque based on monoclonal choline acetyltransferase immunohistochemistry and acetylocholinesterase histochemistry.* Neurosci 12:669–686, 1984.

185. MESULAM, M-M, MUFSON, EJ, WAINER, BH, AND LEVEY, AI: *Central cholinergic pathways in the rat: An overview based on an alternative nomenclature (Ch1-Ch6).* Neurosci 10:1185–1201, 1983.

186. MESULAM, M-M, AND PANDYA, DN: *The projections of the medial geniculate complex within the Sylvian Fissure of the rhesus monkey.* Brain Res 60:315–333, 1973.

187. MESULAM, M-M, AND PERRY, J: *The diagnosis of love sickness: Experimental psychophysiology without the polygraph.* Psychophysiology 9:546–551, 1972.

188. MESULAM, M-M, ROSEN, AD, AND MUFSON, EJ: *Regional variations in cortical cholinergic innervation: Chemoarchitectonics of acetylcholinesterase-containing fibers in the macaque brain.* Brain Res 311:245–258, 1984.

189. MESULAM, M-M, VAN HOESEN, GW, PANDYA, DN, AND GESCHWIND, N: *Limbic and sensory connections of the inferior parietal lobule (Area PG) in the rhesus monkey: A study with a new method for horseradish peroxidase histochemistry.* Brain Res 136:393–414, 1977.

190. MICHEL, D, LAURENT, B, FOYATIER, N, BLANC, A, AND PORTAFAIX, M: *Etude de la mémoire et du langage dans une observation tomodensitométrique d'infarctus thalamique paramedian gauche.* Rev Neurol (Paris) 138:533–550, 1982.

191. MIKAMI, A, ITO, S, AND KUBOTA, K: *Visual response properties of dorsolateral prefrontal neurons during visual fixation task.* J Neurophysiol 47:593–605, 1982.

192. MILNER, B: *Hemispheric specialization: Scope and limits.* In SCHMITT, FO, AND WARDEN, FG (EDS): *The Neuroscience: Third Study Program.* MIT Press, Cambridge, MA, 1974, pp 75–89.

193. MISHKIN, M: *A memory system in the monkey.* Phil Trans R Soc Lond B 298:85–92, 1982.

194. MISHKIN, M, AND BACHEVALIER, J: *Object recognition impaired by ventromedial but not dorsolateral prefrontal cortical lesions in monkeys.* Soc Neurosci Abstr 9:29, 1983.

195. MISHKIN, M, UNGERLEIDER, LG, AND MACKO, KA: *Object vision and spatial vision: Two cortical pathways.* Trends in Neuroscience 6:414–417, 1983.

196. MOHR, JP, PESSIN, MS, FINKELSTEIN, S, FUNKENSTEIN, HH, DUNCAN, GW, AND DAVIS, KR: *Broca aphasia: Pathological and clinical.* Neurology 28:311–324, 1978.

197. MOLL, L, AND KUYPERS, HGJM: *Premotor cortical ablations in monkeys: Contralateral changes in visually guided behavior.* Science 198:317–320, 1977.

198. MOORE, RY, AND BLOOM, FE: *Central catecholamine neuron systems: Anatomy and physiology of the norepinephrine and epinephrine systems.* Ann Rev Neurosci 2:113–168, 1979.

199. MOORE-EDE, MC, CZEISLER, CA, AND RICHARDSON, GS: *Circadian timekeeping in health and disease.* N Engl J Med 309:469–476, 1983.

200. MORRISON, JH, AND MAGISTRETTI, PJ: *Monamines and peptides in cerebral cortex.* Trends in Neuroscience 6:146–151, 1983.

201. MOSCOVITCH, M, AND OLDS, J: *Asymmetries in spontaneous facial expressions and their possible relation to hemispheric specialization.* Neuropsychologia 20:71–81, 1982.

202. MOUNTCASTLE, VB, LYNCH, JC, GEORGOPOULOUS, A, SAKATA, H, AND ACUNA, A: *Posterior parietal association cortex of the monkey: Command functions for operations within extrapersonal space.* J Neurophysiol 38:871–908, 1975.

203. MUFSON, EJ, MESULAM, M-M, AND PANDYA, DN: *Insular interconnections with the amygdala in the rhesus monkey.* Neuroscience 6:1231–1248, 1981.

204. MUFSON, EJ, AND MESULAM, M-M: *Thalamic connections of the insula in the rhesus monkey and comments on the paralimbic connectivity of the medial pulvinar nucleus.* J Comp Neurol 227:109–120, 1984.

205. MURRAY, EA, AND MISHKIN, M: *A further examination of the medial temporal lobe structures involved in recognition memory in the monkey.* Soc Neurosci Abstr 9:27, 1983.

206. MURRAY, EA, NAKAMURA, RK, AND MISHKIN, M: *A possible cortical pathway for somatosensory processing in monkeys.* Soc Neurosci 6:654, 1980.

207. NARABAYASHI, H, NAGAO, T, SAITO, Y, YOSHIDA, M, AND NAGAHATA, M: *Stereotaxic amygdalectomy for behavior disorders.* Arch Neurol 9:1–16, 1963.

208. Nauta, WJH: *Neural associations of the amygdaloid complex in the monkey.* Brain 85:505–520, 1962.

209. Nauta, WJH, and Haymaker, W: *Hypothalamic nuclei and fiber connections.* In Haymaker, W, Anderson, E, and Nauta, WJH (eds): *The Hypothalamus.* Charles C Thomas, 1969, pp 136–210.

210. Newman, R, and Winans, SS: *An experimental study of the ventral striatum of the golden hamster: I. Neuronal connections of the nucleus accumbens.* J Comp Neurol 191:167–192, 1980.

211. Newman, R, and Winans, SS: *An experimental study of the ventral striatum of the golden hamster: II. Neural connections of the olfactory tubercle.* J Comp Neurol 191:193–212, 1980.

212. Niki, H, and Watanabe, M: *Prefrontal and cingulate unit activity during timing behavior in the monkey.* Brain Res 171:213–224, 1979.

213. Nishino, H, Ono, T, Fukuda, M, Sasaki, K, and Muramoto, K-I: *Single-unit activity in monkey caudate nucleus during operant bar pressing feeding behavior.* Neurosci Lett 21:105–110, 1981.

214. Ojemann, GA, Fedio, P, and VanBuren, JM: *Anomia from pulvinar and subcortical parietal stimulation.* Brain 91:1968.

215. Olszewski, J: *The Thalamus of the Macaca Mulatta.* S Karger, Basel, 1952.

216. Pandya, DN, Van Hoesen, GW, and Mesulam, M-M: *Efferent connections of the cingulate gyrus in the rhesus monkey.* Exp Brain Res 42:319–330, 1981.

217. Pandya, DN, and Kuypers, HGJM: *Cortico-cortical connections in the rhesus monkey.* Brain Res 13:13–36, 1969.

218. Parent, A, and Bellefeuille, L de: *Organization of efferent projections from the internal segment of globus pallidus in primate as revealed by fluorescence retrograde labeling method.* Brain Res 245:201–213, 1982.

219. Penfield, W, and Evans, J: *The frontal lobe in man: A clinical study of maximum removals.* Brain 58:115–133, 1935.

220. Penfield, W, and Jasper, H: *Epilepsy and the Functional Anatomy of the Human Brain.* Little, Brown & Co, Boston, 1954.

221. Penfield, W, and Faulk, ME: *The insula. Further observations on its function.* Brain 78:445–470, 1955.

222. Perenin, MT, and Jeannerod, M: *Visual function within the hemianopic field following early cerebral hemidecortication in man–I. Spatial localization.* Neuropsychologia 16:1–13, 1978.

223. Perrett, DI, Rolls, ET, and Caan, W: *Visual neurons responsive to faces in the monkey temporal cortex.* Exp Brain Res 47:329–342, 1982.

224. Petrides, M, and Iversen, SD: *The effect of selective anterior and posterior association cortex lesions in the monkey on performance of a visual-auditory compound discrimination test.* Neuropsychologia 527–537, 1978.

225. Petrides, M, and Iversen, SD: *Restricted posterior parietal lesions in the rhesus monkey and performance on visuospatial tasks.* Brain Res 161:63–77, 1979.

226. Phelps, ME, Kuhl, DE, and Mazziotta, JC: *Metabolic mapping of the brain's response to visual stimulation: studies in humans.* Science 211:1445–1448, 1981.

227. Pool, JL, and Ransohoff, J: *Autonomic effects on stimulating rostral portion of cingulate gyri in man.* J Neurophysiol 12:385–392, 1949.

228. Pribram, KH, Wilson, WA, Jr, and Connors, J: *Effects of lesions of the medial forebrain on alternation behaviors of rhesus monkeys.* Exper Neurol 6:36–47, 1962.

229. Raleigh, MJ, Steklis, HK, Ervin, FR, Kling, AS, and McGuire, MT: *The effects of orbito-frontal lesions on the aggressive behavior of vervet monkeys (Cercopithecus aethiops sabaeus).* Exper Neurol 66:158–168, 1979.

230. Randolph, M, and Semmes, JH: *Behavioral consequences of selective subtotal ablations in the postcentral gyrus of Macaca mulatta.* Brain Res 70:55–70, 1974.

231. Rausch, R, Serafitinides, EA, and Crandall, PH: *Olfactory memory in patients with anterior temporal lobectomy.* Cortex 13:445:452, 1977.

232. Reuter-Lorenz, P, and Davidson, RJ: *Differential contributions of the two cerebral hemispheres to the perception of happy and sad faces.* Neuropsychologia 19:609–613, 1981.

233. Rizzolatti, G, Scandolara, C, Gentilucci, M, and Camarda, R: *Response properties and behavioral modulation of 'mouth' neurons of the postarcuate cortex (area 6) in macaque monkeys.* Brain Res 255:421–424, 1981.

234. Rizzolatti, G, Umilta, C, and Berlucci, G: *Opposite superiorities of the right and left cerebral hemispheres in discriminative reaction time to physiognomical and alphabetical material.* Brain 94:431–442, 1971.

235. ROLAND, PE: *Astereognosis.* Arch Neurol 33:543–558, 1976.

236. ROLAND, PE: *Cortical regulation of selective attention in man. A regional cerebral blood flow study.* J Neurophysiol 48:1059–1078, 1982.

237. ROLAND, PE, AND LARSEN, B: *Focal increase of cerebral blood flow during stereognosis testing in man.* Arch Neurol 33:551–558, 1976.

238. ROLAND, PE, LARSEN, B, LASSEN, NA, AND SKINHOJ, E: *Supplementary motor area and other cortical areas in organization of voluntary movements in man.* J Neurophysiol 43:118, 1980.

239. ROLLS, ET, SANGHERA, MK, AND ROPER-HALL, A: *The latency of activation of neurones in the lateral hypothalamus and substantia innominata during feeding in the monkey.* Brain Res 164:121–135, 1979.

239a. ROSEN, AD, GUR, RC, SUSSMAN, N, GUR, RE, AND HURTIG, H: *Hemispheric asymmetry in the control of heart rate.* Soc Neurosci Abstr 8:917, 1982.

240. ROSENE, DL, AND VAN HOESEN, GW: *Hippocampal efferents reach widespread areas of cerebral cortex and amygdala in the rhesus monkey.* Science 198:315–317, 1977.

241. ROSVOLD, HE, MIRSKY, AF, AND PRIBRAM, K: *Influence of amygdalectomy on social behavior in monkeys.* J Comp Physiol Psychol 47:173–178, 1954.

242. SACKEIM, HA: *Lateral asymmetry in bodily response to hypnotic suggestion.* Biol Psychiatr 17:437–447, 1982.

243. SACKEIM, HA, GREENBERG, MS, WEIMAN, AL, GUR, RC, HUNGERBUHLER, JP, AND GESCHWIND, N: *Hemispheric asymmetry in the expression of positive and negative emotions: Neurological evidence.* Arch Neurol 39:210–218, 1982.

244. SACKEIM, HA, GUR, RC, AND SAUCY, MC: *Emotions are expressed more intensely on the left side of the face.* Science 202:434–436, 1978.

245. SAHGAL, A, AND IVERSEN, SD: *Categorization and retrieval after selective inferotemporal lesions in monkeys.* Brain Res 146:341–350, 1978.

246. SANIDES, F: *Functional architecture of motor and sensory cortices in primates in the light of a new concept of neocortex evolution.* In NOBACK, CR, AND MONTAGNA, W (EDS): *The Primate Brain.* Appleton-Century-Crofts, New York, 1970, pp 137–208.

247. SAPER, CB: *Convergence of autonomic and limbic connections in the insular cortex of the rat.* J Comp Neurol 210:163–173, 1982.

248. SAPER, CB, SWANSON, LW, AND COWAN, WM: *The efferent connections of the anterior hypothalamic area of the rat, cat and monkey.* J Comp Neurol 182:575–600, 1978.

249. SCHELL, GR, AND STRICK, PL: *The origin of thalamic inputs to the arcuate premotor and supplementary motor areas.* J Neurosci 4:539–560, 1984.

250. SCHWABER, JS, KAPP, BS, HIGGINS, GA, AND RAPP, PR: *Amygdaloid and basal forebrain direct connections with the nucleus of the solitary tract and the dorsal motor nucleus.* J Neurosci 2:1424–1438, 1982.

251. SCHWARTZ, GE, AHERN, GL, AND BROWN, S-L: *Lateralized facial muscle response to positive and negative emotional stimuli.* Psychophysiology 16:561–571, 1979.

252. SCHWARTZ, GE, DAVIDSON, RJ, AND MAER, F: *Right hemisphere lateralization for emotion in the human brain: Interactions with cognition.* Science 190:286–288, 1975.

253. SELTZER, B, AND PANDYA, DN: *Some cortical projections to the hippocampal area in the rhesus monkey.* Exper Neurol 50:146–160, 1976.

254. SELTZER, B, AND PANDYA, DN: *Afferent cortical connections and architectonics of the superior temporal sulcus and surrounding cortex in the rhesus monkey.* Brain Res 149:1–24, 1978.

255. SELTZER, B, AND PANDYA, DN: *Converging visual and somatic sensory input to the intraparietal sulcus of the rhesus monkey.* Brain Res 192:339–351, 1980.

256. SHAPIRO, BE, AND DANLY, M: *The role of the right hemisphere in the control of speech prosody in propositional and affective contexts.* Brain Lang, in press.

257. SHAPIRO, BE, GROSSMAN, M, AND GARDNER, H: *Selective musical processing deficits in brain damaged populations.* Neuropsychologia 19:161–169, 1981.

258. SHIPLEY, MT: *Insular cortex projection to the nucleus of the solitary tract and brainstem visceromotor regions in the mouse.* Brain Res Bull 8:139–148, 1982.

259. SHOWERS, MJC, AND LAUER, EW: *Somatovisceral motor patterns in the insula.* J Comp Neurol 117:107–116, 1961.

260. SIDTIS, JJ: *Predicting brain organization from dichotic listening performance: Cortical and subcortical functional asymmetries contribute to perceptual asymmetries.* Brain Lang 17:287–300, 1982.

261. SMALL, IF, HEIMBURGER, RF, SMALL, JG, MILSTEIN, V, AND MOORE, DF: *Follow-up of sterotaxic amygdalotomy for seizure and behavior disorders.* Biol Psychiatr 12:401–411, 1977.

262. SMITH, WK: *The results of ablation of the cingular region of the cerebral cortex.* Fed Proc 3:42–43, 1944.

263. SPIEGEL, EA, WYCIS, HT, FREED, H, AND ORCHINIK, C: *The central mechanism of emotions.* Am J Psychiatr 108:426–432, 1953.

264. SPRINGER, SP, AND DEUTSCH, G: *Left Brain, Right Brain.* WH Freeman & Co, San Francisco, 1981.

265. SQUIRE, LR, AND MOORE, RY: *Dorsal thalamic lesion in a noted case of human memory dysfunction.* Ann Neurol 6:503–506, 1979.

266. STAMM, JS: *Functional dissociation between the inferior and arcuate segments of dorsolateral prefrontal cortex in the monkey.* Neuropsychologia 11:181–190, 1973.

267. STERN, DB: *Handedness and the lateral distribution of conversion reactions.* J Nerv Ment Dis 164:122–128, 1977.

268. STEVENS, JR: *Psychomotor epilepsy and schizophrenia: A common anatomy?* In BRAZIER, MAB (ED): *Epilepsy: Its Phenomena in Man.* Academic, New York and London, 1973.

269. STUMPF, WE: *Anatomical distribution of steroid hormone target neurons and circuitry in the brain.* In MOTTA, M (ED): *The Endocrine Functions of the Brain.* Raven, New York, 1980, pp 43–49.

270. SUGAR, O, CHUSID, JG, AND FRENCH, JD: *A second motor cortex in the monkey (Macaca mulatta).* J Neuropathol Exp Neurol 7:182–189, 1948.

271. TANABE, T, IINO, M, AND TAKAGI, SF: *Discrimination of odors in olfactory bulb, pyriform-amygdaloid areas and orbitofrontal cortex of the monkey.* J Neurophysiol 38:1284–1296, 1975.

272. TANABE, T, YARITA, H, IINO, M, OOSHIMA, Y, AND TAKAGI, SF: *An olfactory projection area in orbitofrontal cortex of the monkey.* J Neurophysiol 38:1269–1283, 1975.

273. THOMPSON, RF: *Foundations of Physiological Psychology.* Harper & Row, New York, 1967.

274. THORPE, SJ, ROLLS, ET, AND MADDISON, S: *The orbitofrontal cortex: Neuronal activity in the behaving monkey.* Exp Brain Res 49:93–115, 1983.

275. TOYOSHIMA, K, AND SAKAI, H: *Exact cortical extent of the origin of the corticospinal tract (CST) and the quantitative contribution to the CST in different cytoarchitectonic areas. A study with horseradish peroxidase in the monkey.* J Hirnforsch 23:257–269, 1982.

276. TROJANOWSKI, JQ, AND JACOBSON, S: *Areal and laminar distribution of some pulvinar cortical efferents in rhesus monkey.* J Comp Neurol 169:371–392, 1976.

277. TUCKER, DM, WATSON, RT, AND HEILMAN, KM: *Discrimination and evocation of affectively intoned speech in patients with right parietal disease.* Neurology 27:947–950, 1977.

278. UMILTA, C, BAGNARA, S, AND SIMION, F: *Laterality effects for simple and complex geometrical figures, and nonsense patterns.* Neuropsychologia 16:43–49, 1978.

279. UNGERLEIDER, LG, AND BRODY, BA: *Extrapersonal spatial orientation: The role of posterior parietal, anterior frontal, and inferotemporal cortex.* Exper Neurol 56:265–280, 1977.

280. UNGERLEIDER, LG, AND PRIBRAM, KH: *Inferotemporal versus combined pulvinar-peristrate lesions in the rhesus monkey: Effects on color, object and pattern discrimination.* Neuropsychologia 15:481–498, 1977.

281. VAN BUREN, JM, AND BORKE, RC: *The mesial temporal substratum of memory.* Brain 95:599–632, 1972.

282. VAN ESSEN, DC, AND MAUNSELL, JHR: *Hierarchical organization and functional streams in the visual cortex.* Trends in Neuroscience 6:370–375, 1983.

283. VAN HOESEN, GW: *The differential distribution, diversity and sprouting of cortical projections to the amygdala in the rhesus monkey.* In BEN-ARI, Y (ED): *The Amygdaloid Complex.* Elsevier, New York, 1981, pp.77–90.

284. VAN HOESEN, GW: *The parahippocampal gyrus.* Trends in Neuroscience 5:345–350, 1982.

285. VAN HOESEN, GW, VOGT, BA, PANDYA, DN, AND MCKENNA, TM: *Compound stimulus differentiation behavior in the rhesus monkey following periarcuate ablations.* Brain Res 186:365–378, 1980.

286. VAN HOESEN, GW, PANDYA, DN, AND BUTTERS, N: *Cortical afferents to the entorhinal cortex of the rhesus monkey.* Science 175:1471–1473, 1972.

287. VAN HOESEN, GW, ROSENE, DL, AND MESULAM, M-M: *Subicular input from temporal cortex in the rhesus monkey.* Science 205:608–610, 1979.

288. VIGNOLO, LA: *Auditory agnosia.* Philos Trans R Soc Lond B 298:49–57, 1982.

289. VOGT, BA, AND MILLER, MW: *Cortical connections between rat cingulate cortex and visual motor and postsubicular cortices.* J Comp Neurol 216:192–210, 1983.

290. Vogt, BA, Rosene, DL, and Pandya, DN: *Thalamic and cortical afferents differentiate anterior from posterior cingulate cortex in the monkey.* Science 204:205–207, 1979.

290a. Wall, PD, and Davis, GD: *Three cerebral cortical systems affecting autonomic function.* J Neurophysiol 14:508–517, 1951.

291. Walsh, FX, Thomas, TJ, Langlais, PJ, and Bird, ED: *Dopamine and homovanillic acid concentrations in striatal and limbic regions of human brain.* Ann Neurol 12:52–55, 1982.

292. Walshe, FMR: *On the "syndrome of the premotor cortex" (Fulton) and the definition of the terms "premotor" and "motor" with a consideration of Jackson's views on the cortical representation of movements.* Brain 58:49–80, 1935.

293. Warrington, EK: *Neuropsychological studies of object recognition.* Philos Trans R Soc Lond B 298:15–33, 1982.

294. Weinrich, M, and Wise, SP: *The premotor cortex of the monkey.* J Neurosci 2:1329–1345, 1982.

295. Weintraub, S, and Mesulam, M-M: *Developmental learning disabilities of the right hemisphere: Emotional, interpersonal and cognitive components.* Arch Neurol 40:463–468, 1983.

296. Weintraub, S, Mesulam, M-M, and Kramer, L: *Disturbances in prosody: A right hemisphere contribution to language.* Arch Neurol 38:742–744, 1981.

297. Weiskrantz, L, Warrington, EK, Sanders, MD, and Marshall, J: *Visual capacity in the hemianopic field following a restricted occipital ablation.* Brain 97:709.-728, 1974.

298. Whitehouse, PJ, Price, DL, Clark, AW, Coyle, JT, and DeLong, MR: *Alzheimer disease: Evidence for selective loss of cholinergic neurons in the nucleus basalis.* Ann Neurol 10:122–126, 1981.

299. Whitty, CWM, and Lewin, W: *A Korsakoff Syndrome in the post-cingulectomy confusional state.* Brain 83:648–653, 1910.

300. Wieser, HG: *Depth recorded limbic seizures and psychopathology.* Neurosci Biobehav Rev 7:427–440, 1983.

301. Wise, SP, Weinrich, M, and Mauritz, K-H: *Motor aspects of cue-related neuronal activity in premotor cortex of the rhesus monkey.* Brain Res 260:301–305, 1983.

302. Woolsey, CN, Settlage, PH, Meyer, DR, Sencer, W, Pinto Hamuy, T, and Travis, AM: *Patterns of localization in precentral and "supplementary" motor areas and their relation to the concept of a premotor area.* Assoc Res Nerv Ment Dis 30:238–264, 1950.

303. Wyss, JM, Swanson, LW, and Cowan, WM: *A study of subcortical afferents to the hippocampal formation in the rat.* Neuroscience 4:463–476, 1979.

304. Wyss, JM, Swanson, LW, and Cowan, WM: *Evidence for an input to the molecular layer and the stratum granulosum of the dentate gyrus from the supramammillary region of the hypothalamus.* Anat Embryol 156:165–176, 1979.

305. Yakovlev, PI: *Pathoarchitectonic studies of cerebral malformations. III. Arrhinencephalies (Holotelencephalies).* J Neuropathol Exp Neurol 18:22–55.

306. Yeterian, EH, and Van Hoesen, GW: *Cortico-striate projections in the rhesus monkey: The organization of certain cortico-caudate connections.* Brain Res 139:43–63, 1978.

307. Yim, CY, and Mogenson, GJ: *Response of ventral pallidal neurons to amygdala stimulation and its modulation by dopamine projections to nucleus accumbens.* J Neurophysiol 50:148–161, 1983.

308. Yozawitz, A, Bruder, G, Sutton, S, Sharpe, L, Gurland, B, Fleiss, J, and Costa, L: *Dichotic perception: Evidence for right hemisphere dysfunction in affective psychosis.* Br J Psychiatr 135:224–237, 1979.

309. Zihl, J, Von Cramon, D, and Mai, N: *Selective disturbance of movement vision after the bilateral brain damage.* Brain 106:313–340, 1983.

310. Zoccolotti, P, Scabini, D, and Violani, C: *Electrodermal responses in patients with unilateral brain damage.* J Clin Neuropsychol 4:143–150, 1982.

311. Zolovick, AJ: *Effects of lesions and electrical stimulation of the amygdala on hypothalamic-hypophyseal regulation.* In Eleftheriou, BE (ed): *The Neurobiology of the Amygdala.* Plenum, New York, 1972, pp 643–683.

Chapter 2

MENTAL STATE ASSESSMENT OF YOUNG AND ELDERLY ADULTS IN BEHAVIORAL NEUROLOGY*

Sandra Weintraub, Ph.D., and M-Marsel Mesulam, M.D.

This chapter provides initial guidelines for structuring the assessment of mental state, for choosing suitable instruments to measure specific behavioral symptoms, and for the appropriate interpretation of the results. General guidelines on interpretation are discussed in this introductory section. More specific comments on individual test instruments follow in subsequent sections.

The assessment of mental state is usually organized according to discrete mental faculties such as memory, language, and visuospatial skills. This provides a convenient checklist for reviewing behavioral systems that are relevant to daily existence. However, it is important to realize that these traditional categories may well reflect abstractions on the part of the clinician rather than realities at the level of brain organization. For example, the examination of brain-damaged patients clearly shows that mental functions that appear unitary from the psychologic point of view (e.g., memory, comprehension, affect) may be far from unitary at the level of neural organization. Following brain damage, some patients could have language comprehension for whole body movements but not for limb movements (see Chapter 5); other patients may learn a new motor skill even though they may have no conscious memory of the experience (see Chapter 4); still other patients may behave as if deeply depressed even though they may feel no sadness (see Chapter 6). In neurologic practice, the traditional psychologic classification of behavior may

*The preparation of this chapter was supported in part by the Javits Neuroscience Investigator Award, by NINCDS grants NS 09211 and NS 20285, and by Alzheimer's Disease Research Center grant 1P50 AG05134-01. We thank Dr. Edith Kaplan for providing a copy of the Stroop Test.

We are grateful to Della Grigsby, Leah Christie, and Rick Plourde for expert secretarial and photographic assistance.

therefore need to be modified in a way that is more consistent with the organization of mental faculties within the brain.

In order to make meaningful brain-behavior correlations, it is also useful to identify the most elementary disturbance that accounts for failure in a complex task. For example, a patient who is unable to do the simplest mental calculation may demonstrate superb numerical skills when allowed to use paper and pencil. This suggests that the initial failure in mental calculations may have been the result of impaired concentration rather than of a genuine acalculia. Alternatively, the patient may be aphasic, and calculation errors may indicate a deficit not at the level of arithmetic reasoning but rather at the level of naming and writing numbers correctly. In still other patients with unilateral spatial neglect, there may be such a profound inability to align numbers properly that the outcome of the calculation may be erroneous even if the individual numerical operations are correct. It is only after all these contributing processes have been eliminated that the examiner can diagnose acalculia, a deficit in the manipulation of numbers and related concepts.

Just as performance of a single task can be impaired through many different mechanisms, several seemingly independent cognitive tasks may be disrupted because of the impairment of a single component process common to all. For example, patients with a diminished attention span or poor motivation may show secondary deficits in several complex cognitive tasks, ranging from mental calculations to constructions and memory (see Chapter 3). It would be an error to conclude that this pattern of performance implies the independent involvement of multiple cognitive faculties and therefore the presence of numerous lesions in the various anatomic regions specialized for each.

In general, each mental task is based on an input channel, an intermediate processing stage, and an output channel. The clinician must be very cautious before reaching conclusions about the intermediate processing stage in patients who have an impairment of the input or output channels. For example, it may be impossible to know if a mute patient has aphasic speech, if an aphasic patient has abnormal thought processes, or if an extremely inattentive patient has retentive memory.

These complexities in the functional organization of behavior must be reconciled with parallel complexities at the level of neural organization. For example, very similar behavioral deficits may arise as a consequence of lesions in entirely different parts of the brain. Thus, left hemispatial neglect has been observed after lesions in the thalamus, in the parietal lobe, and in the frontal cortex (see Chapter 3). One explanation for this is that complex behaviors such as directed attention may be organized at the level of cerebral networks, which contain several distributed but interconnected components. Damage to any of the individual components or even to their interconnections may give rise to similar deficits. In such instances, one must rely on a cluster of "neighborhood signs" in order to reach a more precise anatomic localization. For example, if unilateral neglect is accompanied by hemiplegia but not by hemisensory deficits, then the responsible lesion is more likely to be in the frontal lobe. On the other hand, if neglect is accompanied by left hemianopsia or by loss of two-point discrimination on the left, then it is more likely to be a consequence of a parietal lesion.

An additional point needs to be considered in making use of the existing literature for the purposes of brain-behavior correlations. Consider a hypothetical study in which patients with lesions in the left temporal lobe are compared with those with lesions in the right temporal lobe. Such a study may establish, with great statistical reliability, that patients with left temporal lobe lesions are more likely to develop a material-specific verbal amnesia. There is a tendency to assume that this result implies the presence of a left temporal lobe lesion in all patients with this type of amnesia. This is obviously incorrect. The study has merely shown that in a group of patients, all of whom have proven lesions of the temporal lobe, an amnesia confined to verbal material could be used to predict the laterality of the lesion. This elementary point often escapes attention and leads to misleading conclusions.

The importance of confining inference to the legitimate population cannot be stressed too strongly. For example, much of what we know about the behavioral asymmetries between the two sides of the brain has been learned from patients who develop specific patterns of cognitive deficits as a consequence of single lesions acquired in adulthood (see Chapter 1). Such patterns of mental state deficits have considerable validity for predicting which side of the brain is damaged in adults with acquired unilateral lesions. Individuals with a neurologic disease of very early onset (e.g., epilepsy) may have a very different brain organization. In fact, the Wada test has demonstrated that strict left-hemisphere dominance for speech is much less frequent in right-handed patients with epilepsy who have evidence of early left brain injury than in those with no such evidence.[85] Therefore, the patterns of deficits that are useful in determining the laterality of lesions in adults with stroke or brain tumor may lead to unreliable conclusions about the laterality of a chronic epileptogenic focus. Similarly, evidence derived from neuropsychologic tests is, by itself, of questionable validity for determining the putative anatomic distribution of cerebral dysfunction in schizophrenia, depression, or normal aging.

The behavioral neurologist is concerned with the relationship between brain structure and complex behavior in all realms of experience. However, the actual examination of mental state disproportionately stresses those functions that are easily testable. For example, a great deal of emphasis is placed on drawing, calculating, and language, since these are amenable to convenient assessment. On the other hand, a vast realm of vital human behaviors including curiosity, strength of will, foresight, and many aspects of problem-solving ability remain relatively untapped. Even functions such as judgment, insight, and reasoning can be assessed only in the most rudimentary fashion. Part of the difficulty is that these functions are indeed extremely complex and do not easily lend themselves to laboratory or bedside examination. This probably accounts for much of the current perplexity in understanding the functions of the frontal lobes and of the right hemisphere.

In the course of assessment, the patient may produce some responses that are obviously abnormal from any frame of reference. For example, the performance of an aphasic patient who calls an ashtray a "trash tin" does not require comparison with normative values obtained from a population of neurologically intact individuals. The same holds true for a patient who fails to name only objects placed in his left hand in the absence of any sensory deficits. However, less-dramatic perturbations may be more difficult to interpret. For example, a patient may be able to multiply 2 × 19 but may fail to carry out a process of long division. Another patient may have no paraphasias during spontaneous speech but on the Boston Naming Test[53] may fail the items at the upper levels of difficulty. Quantitative assessment with standardized tests becomes necessary in these instances. If the results show that the patient's performance falls 1 to 2 standard deviations below the norms for age- and education-matched controls, one can hypothesize that an abnormality exists in that function and therefore in the cerebral mechanisms that subserve that function. However, in reaching such a conclusion one makes the assumption that the particular function was normal in the past. Such assumptions are often fraught with uncertainty especially when the deficits are mild or moderate in severity. The importance of past experience becomes of special relevance in the examination of gifted persons who come with complaints of mild but bothersome difficulties in the area of their talents. In such cases, although test scores in all revelant areas may be in the normal range or even better, this level of activity may still represent a significant deterioration compared with past performance. There is no remedy to this dilemma, which is often frustrating to both patient and examiner.

The choice of test intruments is a decision that needs to be made individually for each case. There is no best test for a single mental function or a specific battery of tests that is applicable to all circumstances. If the clinician were to rely on a test battery that would be suitable for all occasions, its administration would be extremely cumbersome and time-consuming. For practical reasons, it is customary to start with a broad survey of cognitive functions, using reliable tests of intermediate difficulty. Then, as deficits are encountered the focus of the examination can be narrowed to explore the specific problem areas in much greater detail.

The approach to testing should be flexible so that it can respond to the level of functioning of the individual patient. This avoids a floor or ceiling effect that would make it impossible to detect change over time. For example, if only the serial threes task is used to assess concentration, then the task will be too easy for some patients and too difficult for others. When the serial threes task is too difficult, the examiner should ask the patient to do serial twos, serial ones, or perhaps even count forward by twos and so on until a task is found that can be done with about 75 percent accuracy. If the serial threes task is too easy, then the examiner can ask the patient to do serial sevens or even to recite the alphabet in reverse order until a similar level of performance is identified. This method of successive approximation for finding a task of moderate difficulty introduces greater flexibility and allows subtle changes over time to be detected.

The need for flexibility in testing is not limited to the level of difficulty. Patients may come to medical attention with puzzling complaints concerning functions for which no tests exist. Consider the patient who comes to the examination with the chief complaint of a loss of the ability for mental visualization. There are no standard batteries to test for mental imagery and the clinician may have to improvise indirect tests of this function. For example, the patient could be asked to learn two word lists—one that contains words of high imagery value (e.g., volcano, chair, pen, candle) and another that contains words of low imagery value (e.g., pride, then, about). Failure to recall the high imagery list but not the other may provide at least some indirect evidence for the patient's capacity for mental visualization. Most complex deficits require improvisation in testing procedures. How else could one differentiate pure word deafness from Wernicke's aphasia or optic aphasia from object agnosia? In still other contexts, a finding such as a left upper quandrantanopia may raise additional questions about the integrity of the right temporal lobe, and the clinician may then have to improvise a method for testing the more complex behavioral specializations of this region. Clearly, the clinician needs to be intimately familiar with the structure of the brain as well as with the organization of behavior in order to plan and interpret the testing of mental state.

From an intellectual point of view, it may be attractive to search for the neuronal circuitry that subserves the memorization of the Rey-Osterrieth figure.[79,91] However, patients do not come to medical attention because they cannot reproduce a complex figure. They seek help because of difficulties, perceived either by themselves or by significant others, in the execution of certain functions of daily living. Therefore, a very important task of the mental state examination is to relate the specific difficulties in testing to the deficits in daily living. This is an area that is frequently neglected. The behavioral neurologist spends a great deal of time examining the patient at the bedside or in the neuropsychologic laboratory but not at home or at work where the activities of daily living are occurring. In some patients, there may be surprisingly little correspondence between the results of the examination and the functional capacity in daily living activities.[114] This lack of correspondence may arise because routine skills performed in a familiar environment are not disrupted as easily as are abstract tasks performed in unfamiliar surroundings. Alternatively, some patients with a disturbance of motivation may perform better on standardized tests than in daily living activities. In this case, the structure and encouragement of the testing milieu may compensate for the lack of spontaneity and foresight that disrupts performance at home. It is hoped that future research will focus on the validity of office tests for predicting performance in daily living activities and on the development of systematic methods for observing and scoring the behavior of the patient at home and at work. Such controlled observations may also offer the opportunity to assess behaviors such as planning, sequencing, and judgment that are not easily demonstrated in the laboratory.

In behavioral neurology, the "history of the present illness" is invariably complex and is in need of considerable reinterpretation. The history given by family members, for example, may differ from that given by the patient, especially if the patient's disease interferes with insight. Furthermore, few patients or family members can be expected to make the distinction between a memory disorder and a word-finding deficit. It is the task of the mental state examination to dissect these complaints by systematic observations that can then be related to brain physiology and therefore to specific diagnostic entities.

In view of the multiple challenges associated with the assessment of mental state and its interpretation, it appears desirable to approach this task with the spirit of an explorer, the cautious flair of a detective, and the training of a neuroscientist. Flexibility is essential, without relying on predetermined test batteries that leave no room for improvisation. Rigid predictions of lesion sites based on single test results should be abandoned in favor of correlations between *patterns* of behavioral deficits and complex cerebral *networks*. Established principles in the psychologic and anatomic organization of behavior should guide these correlations. The ultimate goal is to determine the salient deficits in behavior, their interactions, their impact on daily living activities, their relevance to the structural integrity of the nervous system, and their implications for diagnosis and management. As the subsequent sections of this chapter will show, these goals are most often only partially fulfilled.

ASSESSMENT OF MENTAL FUNCTIONS (TABLE 1)

Wakefulness, Arousal, and the Attentional Matrix (Concentration, Vigilance, Perseverance, Response Inhibition)

The level of wakefulness and arousal can range from deep coma to anxious hyperalertness. This level can be defined operationally by the intensity of stimulation required to elicit a response from the patient. In coma or stupor, for example, intense stimulation, often of a noxious nature, is required to provoke a response. This contrasts with the normal state of wakefulness in which the patient is responsive even to the subtlest of cues. Terms such as "comatose," "stuporous," "drowsy," "alert," and "hyperalert" describe, in ascending order, the levels of wakefulness and arousal. It is important to include level of wakefulness in reporting the mental state examination. The interpretation of other cognitive deficits is substantially limited if the state of arousal is abnormal.

In Chapter 3, attentional functions are divided into two broad categories. One category, generally associated with the function of the ascending reticular pathways and of the frontal lobe, is responsible for maintaining an overall attentional tone or matrix. Words such as "vigilance," "concentration," and "perseverance" are used to describe the positive aspects of this attentional matrix. A disturbance of the attentional matrix is said to result in impersistence, perseveration, distractibility, increased vulnerability to interference, and a peculiar inability to inhibit immediate but inappropriate response tendencies. The testing of these aspects of attention is discussed in this section.

The Continuous Performance Test[94] is useful as a multipurpose test of vigilance despite its low levels of sensitivity and specificity. The task is to attend to a series of visually presented and randomly occurring letters over a 10-minute interval and to respond whenever a target stimulus appears. Rosvold and co-workers[94] reported that normal adults correctly identified close to 90 percent of simple targets (i.e., whenever the letter "A" appears) and 80 percent of complex targets (i.e., the letter "A" only

TABLE 1. Mental State Tests

WAKEFULNESS, AROUSAL, AND THE ATTENTIONAL MATRIX
  A. Tests of Concentration Span and Vigilance
    1. Continuous Performance Test, 76
    2. Digit Span Test, 81
    3. Corsi Block Test, 81
  B. Tests of Perseverance
    1. Serial Recitation Tests, 79
    2. Word List Generation, 79
  C. Tests of Resistance to Interference and Response Inhibition
    1. Trail-Making Test, 79
    2. Stroop Test, 79
    3. Alternating Sequences Test, 80
    4. Go–No-Go Paradigm, 81
MOOD
  A. Self-Administered
    1. Minnesota Multiphasic Personality Inventory, 82
    2. Beck Inventory, 82
  B. Examiner-Administered
    1. Hamilton Depression Rating Scale, 82
MEMORY
  A. Laboratory Tests
    1. Retrograde Memory Tests, 83
    2. Wechsler Memory Scale, 83
    3. Rey Auditory Verbal Learning Test, 84
    4. Rey-Osterrieth Complex Figure Test, 84
    5. Buschke-Fuld Selective Reminding Procedure, 85
    6. Brown-Peterson Technique, 85
    7. Paradigm of Sternberg, 89
  B. Bedside or Laboratory Tests
    1. Three Words–Three Shapes Test, 86
    2. Drilled Word Span Test, 88
LINGUISTIC ASPECTS OF LANGUAGE
  A. Boston Diagnostic Aphasia Examination, 90,92
  B. Western Aphasia Battery, 90
  C. Peabody Picture Vocabulary Test, 91
  D. Token Test, 91
  E. Boston Naming Test, 91
  F. Gates-MacGinitie Reading Tests, 92
CALCULATION ABILITIES
  A. Boston Diagnostic Aphasia Examination–Supplementary Part, 94
  B. Wide Range Achievement Test–Arithmetic Subtest, 94
  C. Wechsler Adult Intelligence Scale–Arithmetic Subtest, 94
COMPLEX PERCEPTUAL TASKS
  A. Hooper Visual Organization Test, 96
  B. Judgment of Line Orientation, 97
  C. Poppelreuter Task, 97
  D. Gollin Incomplete Figures, 97
  E. Seashore Tests of Musical Talent, 97
CONSTRUCTIONAL TASKS
  A. Wechsler Adult Intelligence Scale–Block Design Subtest, 98
  B. Rey-Osterrieth Complex Figure Test, 99
  C. Bedside Tests
    1. Copying a cube, daisy, house, 99
    2. Drawing a clock, 100
SPATIAL DISTRIBUTION OF ATTENTION
    1. Bilateral Simultaneous Stimulation, 101
    2. Verbal and nonverbal stimuli Cancellation Tests (Fig. 10), 101
    3. Blindfolded manual exploration (palpation), 103
PARALINGUISTIC ASPECTS OF COMMUNICATION
  A. Discrimination, repetition and production of affective and nonaffective prosody, 104
  B. Profile of Nonverbal Sensitivity (PONS) Test, 104
REASONING AND ABSTRACTIONS
  A. Wechsler Adult Intelligence Scale–Comprehension Subtest, 105
  B. Gorham Proverbs Test, 105
  C. Wechsler Adult Intelligence Scale–Similarities Subtest, 105

TABLE 1. Mental State Tests (*continued*)

Several tests mentioned in the text are listed above according to the mental function(s) each test emphasizes. Numbers refer to pages in the text where the tests are described. This classification is intended to provide general guidelines. As discussed in the text few, if any, of these tests are specific to single mental functions.

when followed by the letter "X"). Brain-damaged subjects identified an average of 80 percent and 60 percent of simple and complex targets, respectively. Evidence for the validity of this paradigm as a test of attention comes from the finding that performance is impaired in the population of children with hyperactivity,[108] a syndrome clinically defined by the salience of attentional deficits. The stimuli of the Continuous Performance Test can be taperecorded for auditory presentation and can be available for standardized administration even at bedside. When the need arises, it can be modified in several directions. For example, the speed of administration can be varied, and the speed of the patient's response can be monitored as additional variables. Furthermore, it would be possible to quantify the vulnerability to interference by taping the stimuli over a background of conversational or musical sounds. Performance on this test is not influenced by intelligence,[25] thus eliminating a problem that plagues the interpretation of many other mental state tests.

This paradigm may be particularly useful in assessing changes of vigilance in acute confusional states. For example, a 67-year-old man with pulmonary empyema identified only 50 percent of target stimuli on a 5-minute auditory letter detection task. One month following drainage of the empyema and antibiotic treatment, his score exceeded 85 percent. The improvement in his test score paralleled his improvement on other tests of attention and in daily living activities.

The Continuous Performance Test is a relatively nonspecific test of the attentional matrix. A poor test result may indicate an impairment in the basic machinery of vigilance and concentration. However, a patient could also score poorly because of impersistence, distractibility, or an inability to inhibit inappropriate responses to the nontarget stimuli. These additional aspects of attentional behavior can be assessed somewhat more specifically with the help of other tests.

Impersistence, for example, can be assessed with the help of tests of perseverance that require the patient to sustain behavior over a prolonged period of time. The simplest assessment of this function could be done by asking the patient to continue tapping one finger on the table. Tests of serial recitation (e.g., serial sevens[48]) and of word-list generation can also be used to detect impersistence. However, Smith[100] reported a study in which 58 percent of highly educated normal adult subjects made errors on the serial subtraction of seven from one hundred. Therefore, minor errors in this task do not necessarily indicate an attentional impairment. Other serial recitation tasks that can be used include serial threes, serial ones, reciting months of the year in reverse order, and reciting the alphabet in reverse order.

Word-list generation is another measure of the ability to sustain behavior. The task is to generate words that are either related categorically to each other (e.g., animals) or that begin with a specified letter. Spreen and Benton[103] measured the generation of words beginning with the letters F, A, and S each over a 60-second interval. Normal adults with a high-school education can produce an average of 36 words over the 3 minutes.[64,103] A lesser degree of performance may indicate an inability to sustain behavior over time. Naturally, it is first necessary to determine that poor performance is not due to an underlying aphasia. Successful performance in the serial recitation and word-list generation tasks indicates that patients are capable of perseverance and that they can sustain behavior over time. In addition to perseverance, these more complex tasks also measure the overall power of concentration as well as the vulnerability to distraction. Sometimes patients who have difficulty sustaining behavioral output may also show generalized mental slowing and delayed response times. These are important components of the abulic state that is occasionally observed following frontal lesions.

The susceptibility to interference and the inability to inhibit immediate but inappropriate response tendencies are assessed more specifically by tests that call for competing response tendencies. One such test is a modification of the Stroop procedure.[15,20,107] First, a response tendency to read words is established by having the patient read, as quickly as possible, 100 color words (red, blue, green), arranged in random sequence and printed in black and white. Next, a second response tendency to name colors is established by having the patient read 100 colored dots (red, blue, green). In the third, interference condition, the subject is presented with 100 color words, each of which is printed in a color (red, blue, or green) other than the one spelled by the letters (Fig. 1, *top*). The task is to name the color of the letters and to inhibit the equally strong tendency to read the word that they spell. The time to complete each task is recorded. Normative data are available for children and adults.[20]

The Trail-Making Test[88] can also be used to measure response inhibition and vulnerability to interference. On Part A, the patient must draw a line connecting a series of randomly arrayed numbers in numerical sequence (1-2-3, and so on). On Part B, numbers are intermixed with letters and the task is to draw a line connecting numbers and letters in alternating sequence so that the connecting line goes from 1 to A to 2 to B to 3 to C, and so on. Figure 2 *(left)* illustrates Part B of the Trail-Making Test completed by a patient with a left frontal arteriovenous malformation, demonstrating an impairment in response inhibition. Performance of the Trail-Making Test is influenced by intelligence[89] and age.[23] Age-appropriate norms are available for test interpretation.[33]

The Alternating Sequences Test[67] provides another measure of the ability to inhibit inappropriate responses. The examiner produces a short segment of alternating m's and n's in cursive writing or alternating triangles and squares. The patient is asked to copy the segment and then to continue in the same pattern until the end of the page is reached. Subsequent to the copying phase, many patients with attentional impairments show a tendency to repeat one of the letters or shapes rather than alternating them with the other member of the pair (see Chapter 3, Fig. 1). This reveals an inability to inhibit immediate response tendencies and also the emergence of perseveration. The Alternating Sequences Test can also be administered in the motor modality.[66] The examiner asks the patient to make a sequence of three consecutive hand movements. The disruption of the proper sequence and a tendency to repeat the same movement indicate a failure of this test.

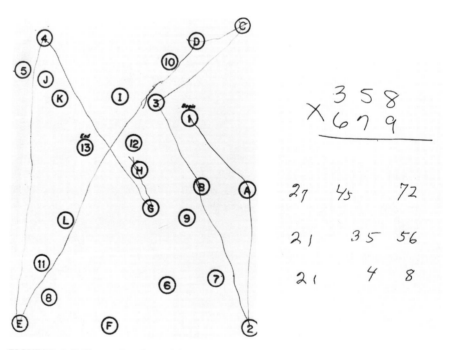

FIGURE 2. *Left*: The results of a modified version of the Trail-Making Test, Part B, from a patient with a left frontal arteriovenous malformation. The task is to alternate between numbers and letters. The patient begins correctly (1-A-2-B-3-C. . . .) but then shows an interference or perseveration effect (C-D-E-4-G-H). *Right*: Acalculia in a 72-year-old patient with a moderate dementia of the Alzheimer type. Number identification and multiplication table values are relatively intact. However, the patient has lost the operations necessary for complex multiplication.

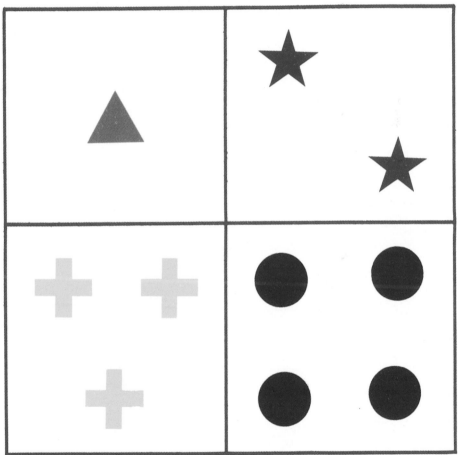

GREEN RED BLUE RED GREEN BLUE RED RED GREEN BLUE

BLUE RED BLUE GREEN RED BLUE GREEN RED BLUE RED

GREEN BLUE GREEN BLUE RED GREEN RED BLUE GREEN RED

BLUE RED BLUE GREEN RED BLUE GREEN RED BLUE RED

FIGURE 1. *Top*, The Stroop interference condition.[15,20,107] The patient must inhibit the tendency to read "Green, red, blue . . . " and instead must say "Red, blue, red . . . ," which corresponds to the colors in which the words are written. (From Broverman,[15] with permission.)

    Bottom, The four test cards of the Wisconsin Card Sort Test.[42] The cards are placed in a horizontal row in front of the patient. See text for explanation. (From Grant and Berg.[42] Copyright © 1980 Wells Printing Company, with permission.)

Response inhibition can also be tested with the "go–no-go" paradigm, which has been used extensively in the clinical setting as well as in the animal research laboratory to assess the effects of brain lesions. A simple test of this function is to ask the patient to place the hand on the table and to raise the index finger in response to a single tap while holding still in response to two taps. The examiner produces the taps by hitting a pencil to the undersurface of the table so as not to give visual cues. Patients with impaired attention, especially those with frontal lobe lesions, cannot inhibit raising the finger in response to the no-go signal (two taps). These inappropriate "errors of commission" are also seen in monkeys with prefrontal cortex lesions (see Chapter 1).

Serial recitation tasks (e.g., serial sevens), the word-list generation test,[103] the Stroop Test,[15,20,107] the Trail-Making Test,[88] and the Alternating Sequences Test[66,67] are all dependent on sustained behavior and are therefore impaired in patients who have a primary deficit in perseverance. The digit-span test,[68,111] on the other hand, places a lesser premium on sustained behavior and can be used as an alternate measure of attention. The digit span is determined by having the patient repeat a string of digits read aloud at the rate of one per second. First, the forward span is assessed by asking the patient to repeat the string of digits in the same order as recited by the examiner. Then backward span is measured by asking the patient to reverse the sequence presented by the examiner. The average number of digits normal adults can repeat in forward sequence is 6 ± 1.[102] This number coincides with Miller's[72] speculation regarding the limitations of immediate information processing capacity. The discrepancy between the backward and forward spans normally does not exceed 2 digits. Digit span is often classified as a test of "short-term memory." However, it should be considered a test of attention span, especially since patients with severe amnestic states have normal digit spans. The digit span is influenced by education and age.[12,51,113] These factors must be considered when interpreting the patient's performance. The digit span is mediated through verbal channels and is therefore not a good test of attention in aphasic patients.

An analogous assessment of nonverbal attention span can be achieved with the Corsi Block Test.[75] Nine wooden blocks are mounted on a board in an irregular array. The examiner points to a gradually increasing number of blocks and the patient is subsequently asked to reproduce the sequence of pointing in the same order or in reverse order. The normal nonverbal attention span assessed in this fashion should be numerically equal to the verbal span assessed by digit repetition.

Although there is a great deal of overlap among the tests described herein, they each emphasize slightly different components of the attentional matrix. For heuristic reasons, these tests can be divided into three groups: (1) those that primarily test concentration span and vigilance—Continuous Performance Test,[94] digit span,[68,111] and the Corsi Block Test;[75] (2) those that primarily test perseverance—serial recitation tests, word-list generation test,[103] finger tapping test; (3) those that primarily test the resistance to interference and response inhibition—Trail-Making Test,[88] the Stroop Test,[15,20,107] the Alternating Sequences Test,[66,67] and the Go–No-Go paradigm. There is some practical value in assessing these various components of attention separately. For example, some patients with frontal lobe damage may have a selective impairment of response inhibition, whereas other patients with a different kind of frontal lesion may have the most difficulty in sustaining a steady output of behavior. On the other hand, metabolic encephalopathies usually affect not only each of these two components but also vigilance and concentration.

Although cumbersome, the assessment of attention is one of the most essential steps in the mental state examination. One reason for this is that attentional functions provide the essential background upon which all the other cognitive faculties operate. Therefore, cognitive impairments in an inattentive patient have less implication for localization than if they were to occur in an attentive and alert individual.

Mood and Motivation

Mood and motivation, like wakefulness and attention, are essential to all other mental faculties (see the discussion on state functions in Chapter 1). The level of motivation can be assessed by both observation and testing. The degree of cooperation, the ability to sustain effort during task performance, and the extent to which encouragement is required to complete a task are all observable aspects of motivation. Motivation and the perseverance aspect of attention overlap greatly so that patients with motivational impairments do poorly on the tests of perseverance listed earlier.

Specific disturbances in motivation are described as apathy and, in the most extreme circumstances, as abulia. The examiner's role in eliciting behavior is nowhere more important than in the examination of the abulic or apathetic patient. The patient may frequently reply "I don't know," offer truncated responses, or take an inordinate amount of time to initiate a response. Unless the examiner allows sufficient time for a response or follows up on these responses with encouragement or even prompting, there may be an underestimation of the patient's cognitive abilities.

The mood or affect of the patient may be determined during the course of the clinical examination with the help of traditional psychiatric methods of observation or with the help of standard questionnaires. Self-administered questionnaires such as the Minnesota Multiphasic Personality Inventory[47] or the Beck Inventory[6] rely on the ability of the patient to read and understand the test items. These demands often exceed the limitations of the brain-damaged patient. Furthermore, these tests are not standardized on specific brain-damaged populations. The Hamilton Depression Rating Scale[46] eliminates the patient as a rater, since it is completed by the examiner during a clinical interview. This scale has been used quite successfully in a population of demented patients.[60] There are also visual analog mood scales that are relatively free of verbal mediation.[32] The patient is asked to indicate a point between the two extremes of a line, one end indicating happiness and the other sadness. Such scales can be quite sensitive to temporal variations of mood.

In the examination of motivation and affect, apathy must be distinguished from depression, and the experience of emotion must be distinguished from its expression (see Chapter 6). In patients with pseudobulbar palsy or certain types of right-hemisphere lesions, for example, the expression of affect may be impaired while its experience may remain intact. Furthermore, some types of frontal lobe lesions or some metabolic encephalopathies may lead to severe apathy (abulia) in the absence of underlying depression.

Most of the attentional tests described in the preceding section are very sensitive to disturbances in mood and motivation. Therefore, when patients perform poorly on such tasks, the clinician must attempt to differentiate the contribution of poor attentional tone from the contribution of disturbances in mood and motivation.

Memory

Memory loss is never complete. Some patients may have a retrograde amnesia (a loss of memory for remote events) as well as an anterograde amnesia (deficit for learning new material), while others may have mostly the anterograde component; some patients may have a specific deficit for learning verbal material but not nonverbal material; some amnesias may be modality-specific, while others may differentially affect individual stages of learning such as registration, storage, or retrieval; finally, even in the most severe amnestic states, perceptual and motor learning may remain relatively intact (see Chapter 4). The examination of memory must therefore be sufficiently broad to detect these patterns without being overwhelmingly lengthy.

Retrograde (remote) memory is tested informally by questioning the patient about significant past public events. When necessary, more quantitative information can be obtained with one of the several extensive batteries that are available. These are based on recognition of photographs of famous personalities,[1] knowledge of public events,[98] and recall of popular television programs.[104] These tests include items that have come to peak public attention at different times during the past several decades, so that it is possible to estimate when they were stored in memory. The retrograde memory loss of Korsakoff's amnesia is characterized by a gradient in which events that occurred closer in time to the onset of the amnesia are recalled least well (Ribot's Law).[1,98] This pattern of retrograde amnesia differs from that of patients with dementias of Huntington's or Alzheimer's disease in whom recall of past events is disturbed for all the decades with a less-definite time gradient.[2,117] Despite the availability of standardized tests of remote memory there are substantial difficulties in their interpretation, since it is impossible to assess the extent to which the patient has acquired the information for which recall is being tested.

The evaluation of new learning (anterograde memory) requires a comparison between verbal and nonverbal learning and also an assessment of the major stages in memorization. Several test instruments are currently being used for this purpose, although none is comprehensive. The clinician needs to decide which instruments to use and how to extract the desired information from the patient's performance.

The Wechsler Memory Scale[112] is the most widely used clinical test of memory. Several shortcomings need to be considered when interpreting the memory quotient (MQ) that can be derived with this instrument.

First, the Wechsler Memory Scale contains not only tests of learning but also tests of orientation and attention. Therefore, the MQ is a mixed score that does not specifically reflect memory functions. Second, only immediate recall is tested, and there is no measure of retention of information over time. This is quite critical, since many amnesic patients can do quite well on immediate recall but have the most difficulty after a delay interval, particularly if they are distracted by some other activity during that time. Third, the verbal and nonverbal tests are combined in the overall MQ so that differences between these two aspects of memory are obscured.

These considerations have prompted several modifications. For example, a 30-minute delayed recall condition can be introduced following the immediate recall in the logical memory (story recall) and visual reproduction (recall of designs) subtests.[3,95] After a 30-minute delay, individuals in the 30–40-year age range retain close to 90 percent of the information that they reproduced during the immediate recall condition for both the verbal and nonverbal subtests.[95] With advancing age, there is a decline of immediate recall that is more marked for nonverbal than for verbal material.[51,119] At the age of 65, the amount of retention from the immediate to the 30-minute delay condition remains at 90 percent for the nonverbal information but drops to 63 percent in the verbal subtest.[119] The amount of information retained shows no further decline between the ages of 65 and 80. However, since immediate recall also declines with age, the absolute amount of information that is being retained is smaller for older individuals, even when they show 90 percent retention after 30 minutes.

For the patient with superior premorbid intellectual achievement, the examiner may wish to administer more challenging tests of memory. Two instruments are commonly used for this purpose. The Rey Auditory Verbal Learning Test[92] provides a complex test of verbal learning, whereas the Rey-Osterrieth Complex Figure Test[79,91] provides a test of approximately equivalent difficulty for assessing nonverbal learning. The Rey Auditory Verbal Learning Test contains two lists of 15 words each. First, the number of test items that can be immediately recalled is assessed over five repeated presentations of list 1. Then list 2 is presented once and immediate recall is tested. During recall of list 2, the patient may report items belonging to list 1. These errors of intrusion are thought to represent a process of proactive interference. Following recall of list 2, the patient is asked once again to recall the items of list 1. The extent to which the new learning of list 2 disturbs the retention of the previously learned list 1 provides a measure of retroactive interference. Normative values are available for young and elderly adults at different occupational levels.[64,92] Young adults learn an average of 14 words on list 1 by the fifth trial; individuals over the age of 70, an average of 10. When recall is not perfect, recognition is tested by having the patient identify the words from list 1 in a short paragraph that contains the test words as well as nontest words. Delayed recall of the lists can also be tested, but there are no normative values available for this procedure.

In the Rey-Osterrieth Complex Figure Test,[79,91] the patient is first asked to copy a complex design containing 36 scorable elements. Adequate copying of the Rey-Osterrieth Figure also indicates that errors during subsequent reproduction from memory are not due to constructional difficulties or to the patient's failure to attend to the stimulus. Normal adults can correctly reproduce an average of 22 elements of the design on immediate recall.[64] There are no normative values for delayed recall but results from studies using a comparable design, the Taylor Figure, suggest that little information is lost by normal individuals over delays as long as an hour.[118]

The Buschke-Fuld selective reminding procedure[17] is recently enjoying a good deal of popularity, especially in drug trials for memory improvement. A 12-word list is presented over a series of 12 trials or until the list is successfully recalled on three consecutive trials. The entire list is presented only on the first trial. Subsequently, only those words not recalled on the immediately preceding trial are presented. Words that are remembered on two consecutive recall trials without "reminding" are presumed to be in long-term storage. Words recalled only after reminding are assumed to be retrieved via short-term retrieval mechanisms. Following the 12 learning trials, standardized lists are available to see if the subject can recognize by multiple choice the test items he fails to recall. This maneuver makes it possible to assess the role of retrieval difficulties. The selective reminding procedure is a convenient research tool even if the theoretical assumptions underlying the measurement of storage and retrieval require a great deal of additional clarification.

Recall in most patients with amnesia is exquisitely sensitive to the effects of distraction. If there is no intervening activity during a delay period, the patient with amnesia may be able to retain information by means of internal rehearsal. Distraction can be introduced informally by engaging the patient in another activity during the delay. The Brown-Peterson technique [16,82] provides systematic distraction during the delay interval. The stimuli to be remembered consist of letters or words presented three at a time and followed by varying delay intervals ranging from 0 to 18 seconds. During the delay intervals, the subject is asked to count backward from a specified number. This procedure has been used extensively to study memory functions in Korsakoff's amnesia and Huntington's disease[18] and to measure treatment effect of cholinergic agents in Alzheimer's disease.[22]

As described in Chapter 4, even the most severe amnesic state may still be compatible with a relatively preserved ability to learn new motor skills. The pursuit rotor test,[21] which can quantitate motor learning, can easily be added to the clinical assessment when the need arises. Furthermore, the standard test batteries put a premium on the auditory and visual modalities, while the somatosensory and especially the olfactory and gustatory modalities are usually neglected. In some patients it may be of special interest to test olfactory and gustatory memory, especially since these two modalities are very closely related to the limbic system (see Chapter 1).

Even in the neuropsychology laboratory, the assessment of memory is a difficult task. The bedside assessment of this function presents even greater challenges. The standard method of asking the patient to remember four words or a short story provides very limited information at best.

Nonverbal memory is frequently not assessed and the role of attention and language are usually not considered. We have developed two simple methods that are amenable to use at the bedside as well as in the neuropsychologic laboratory and that dissect several different components of the learning process.

One of these, the Three Words–Three Shapes Test, is based on three simple designs and three words with low imagery value as test stimuli (Figs. 3 to 5). The patient is first asked to copy the six test stimuli without being told that memory is to be tested. Immediately after copying, the stimuli are removed and the patient is asked to reproduce them from memory. This provides a measure of incidental recall. Patients who have complete incidental recall are then ready to proceed to tests of delayed memory. Patients who show any impairment in the incidental recall stage are allowed a 30-second examination of the six stimuli, this time being told that their memory for these will be tested. Following this second presentation (the first study period), immediate recall is tested. Patients who reproduce correctly five or all six of the items are then ready to enter the delayed retention period. Patients who do not succeed in doing so are allowed repeated study periods either until the criterion of five or six items is reached, or until five study periods have elapsed. Delayed recall is tested after 5 minutes, and again after 15 and 30 minutes. During the delay period, the patient is engaged in other distracting tasks. After the 30-minute delayed recall condition, multiple-

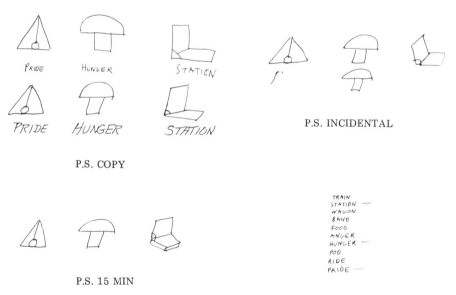

FIGURE 3. The Three Words–Three Shapes Test in a 65-year-old man with a left frontal infarction. The stimuli are first drawn on an 8½-inch by 11-inch sheet of paper by the examiner (*top, upper left*). The patient copies them immediately underneath. In the incidental memory phase (*upper right*), there is a selective difficulty with the words. The patient cannot reproduce the words despite repeated study trials (not shown). Following 15 minutes of delay, he shows adequate retention of the nonverbal stimuli (*lower left*). At that time, despite a failure to reproduce the words, he correctly recognizes them by multiple choice from a list provided by the examiner (*lower right*). This indicates intact nonverbal memory but a retrieval deficit for verbal learning. This patient is likely to do quite well with verbal memory if he is given cues.

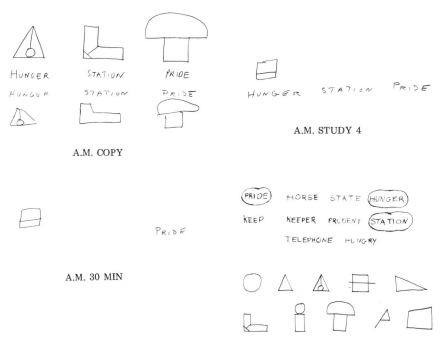

FIGURE 4. The Three Words–Three Shapes Test performance in a 72-year-old woman with dementia. The patient's reproduction of the test stimuli is acceptable (*upper left*). She could reproduce essentially none of the stimuli by incidental recall (not shown). She needed four study periods to reproduce the words by immediate recall (*upper right*), but she could not reproduce the shapes even after the fifth study (not shown). Following a 30-minute delay, two of the words show retention decay (*bottom left*). At that time, when the examiner gave her a list of 10 words and 10 shapes, she correctly recognized all three words but none of the shapes (*bottom right*). This patient shows a dramatic block of incidental memory, sluggish registration as well as temporal decay of retrieval for verbal material and a storage deficit for nonverbal material. Overall, her memory is worse for nonverbal material.

choice recognition of the test items is assessed by presenting the patient with a list of 10 words and 10 shapes containing the test stimuli.

There are several advantages offered by this method of testing. First, incidental memory can be differentiated from volitional (rote) memory. The recall of much of daily experience is based on incidental (contextual) memory. It is conceivable that some patients suffer an impairment only of incidental memory. This could lead to complaints of memory loss even though extensive testing in the neuropsychologic laboratory might reveal no abnormalities, since most standard batteries rely on rote memory. Second, the procedure of drilling by repeated study periods is designed to overcome the learning difficulties based on attentional deficits. The number of study trials needed to reach criterion is a measure of the effectiveness of *registration* and *acquisition*. These two stages of new learning are particularly sensitive to attentional and motivational difficulties. Normal adults under 65 years of age typically learn all six stimuli after one study period. Over the age of 65, two or more study periods may be necessary to reach criterion.

The method of drilling to criterion also allows the standardization (across patients with widely differing abilities for acquisition) of the

MENTAL STATE
ASSESSMENT OF
YOUNG AND
ELDERLY ADULTS IN
BEHAVIORAL
NEUROLOGY

87

F.G. STUDY 1   F.G. STUDY 3

F.G. STUDY 4   F.G. 15 MIN

FIGURE 5. The Three Words–Three Shapes Test in a 71-year-old woman with a presumptive diagnosis of Alzheimer's disease. Her copying of the stimuli was acceptable (not shown). She could reproduce none of the stimuli by incidental recall. Immediate recall kept improving slowly over four study periods (*upper left, upper right,* and *lower left*). However, once registration had taken place, there was relatively little decay of memory traces over time (*lower right*). In this patient, the bottleneck in memory functions appears to be at the stage of *registration.* Memory problems are likely to play a minor role if she is allowed repeated exposure to the relevant material.

amount of information for which retention and recall are being tested. An impairment of recall and recognition following the delays indicates a specific deficit in the *retention* of information over time. If free recall is impaired after the delay but multiple-choice recognition is accurate, then it is possible to conclude that the memory difficulty is predominantly at the stage of *retrieval.* In addition to identifying difficulties at specific phases of the learning process, another major advantage of this test is its ability to identify patients whose learning deficits are specific for verbal or nonverbal items. The Three Words–Three Shapes Test can be modified by increasing the number of items to be learned and also by increasing the delay period.

We have also developed the Drilled Word Span Test, which minimizes the role of attention and which allows the adjustment of the difficulty level for each patient. The patient is presented with a list of unrelated words. The number of items in the list is one less than the patient's forward digit span. The list is read aloud, and the patient is asked to recall the words without intervening delay. The list is read repeatedly until it is recalled correctly on three consecutive trials (criterion). Delayed recall is tested at three points in time: after 60 seconds without distraction, after a second interval of 60 seconds during which the patient is given a serial threes task, and then after 3 minutes filled with distracting activities. Failure to retrieve the full list on either of the 60-second recall trials is followed by repeated drilling of the list to criterion. If free recall is defective at the 3-minute interval, recognition by multiple choice is tested. The number of trials taken to reach criterion is a measure of registration and acquisition. The effect of distraction can be tested by comparing the two 60-second recall trials. Retention over time is measured by comparing the two intervals (60 seconds and 3 minutes) with distraction. Table 2 presents the results of this test obtained from two patients, one with a presumptive diagnosis of Alzheimer's disease and the other suffering an acute confusional state associated with a metabolic encephalopathy. The patterns of performance are distinctly different. A true amnesic disorder is confirmed in the patient with dementia whose recall suffers as a function of distraction and of time. In contrast, it can be seen that retention and recall are normal in the patient with a metabolic encephalopathy despite the difficulty of the initial acquisition due to the attention deficits.

The Three Words–Three Shapes Test and the Drilled Word Span Test are easily used at the bedside. Even though they may take slightly more time than the traditional assessment of memory, the additional information that is obtained can be useful in the differential diagnosis.

Most memory tests require an all or none response: the answer is either incorrect or correct. The paradigm of Sternberg[105] differs from this approach. The subject is asked to learn sets of stimuli containing different numbers of items. The subject is then presented with one probe stimulus at a time and asked to decide whether or not it is a member of the learned set. Sternberg[105] demonstrated that decision time increases as a function of set size so that it takes longer to reach a decision if the test stimulus is a member of a larger set. Since the comparisons are among sets of different size, variables such as mood, motivation, attention, aphasia, and reaction time—factors common to all set sizes—do not influence the results. It has been shown, for example, that the prolongation of response time with increasing set size is more pronounced in elderly subjects.[4] In addition to providing a novel measure of memory process, the objectivity

TABLE 2. The Drilled Word Span Test

	Learning Trials					Delayed Recall			Multiple Choice	
	1	2	3	4	5	60″ND	60″D	3′D	Correct	False Positive
Patient A:*										
Apple	+	+	+	+	+	+	+	+	+	
Shoe	−	−	+	+	+	+	+	+	+	
Horse	−	−	+	+	+	+	+	+	+	(None)
Truck	+	+	+	+	+	+	+	+	+	
Patient B:†										
Apple	+	+	+			+	+	+	+	Cheese
Shoe	+	+	+			+	−	+	+	Apron
Horse	+	+	+			+	+	−	−	
Truck	+	+	+			+	+	−	+	

*Patient A was a 67-year-old man with an acute confusional state secondary to a metabolic encephalopathy. Deficits in attention impeded acquisition, but delayed recall was intact and not affected by distraction once the words had been acquired.
†Patient B was a 56-year-old woman with Alzheimer's disease. Acquisition was rapid and recall unimpaired following a delay of 60 seconds without distraction. However, with distraction, and over time, recall diminished. In addition, the patient failed to recognize one list word and identified two nonlist words on multiple-choice recognition testing.
ND = No distraction; D = Distraction; + = Correct; − = Incorrect.

MENTAL STATE
ASSESSMENT OF
YOUNG AND
ELDERLY ADULTS IN
BEHAVIORAL
NEUROLOGY

of this paradigm, its flexibility and its adaptibility to computerized administration make it particularly attractive.

In order to ascertain that poor performance on any one of the tests described in this section is due to a specific memory disorder, it is essential to eliminate other factors that can contribute to poor performance. Sometimes, patients can be so inattentive or so poorly motivated that memory testing becomes impossible. Some patients have such severe constructional difficulties that nonverbal memory can only be tested by multiple-choice recognition. A similar problem may exist for verbal memory in the case of aphasic patients. If the constructional difficulties or aphasia is sufficiently severe, then even multiple-choice recognition may be affected so that material-specific memory testing may not be possible. In the more severely impaired patient we often find it necessary to test memory by hiding three or four of the patient's personal belongings (e.g., wallet, watch, ring) around the examination room and then testing for recall of the hidden objects and their locations after variable intervals with and without distraction. The ability of the patient to retrieve the objects immediately after they are hidden ensures that he has paid sufficient attention to the task.

Linguistic Aspects of Language

The three major linguistic components of language are word formation, word choice, and grammar. Aphasia refers to a disturbance of these components in spoken or written communication. The clinical findings that characterize the various aphasic syndromes are described in Chapter 5. The identification of these syndromes is based on the evaluation of spontaneous speech, oral language comprehension, repetition, naming, reading, and writing. The Boston Diagnostic Aphasia Examination[40] and the Western Aphasia Battery[57] contain standardized tests for examining each of these functions.

The quantity and grammatical content of speech production as well as the presence of dysarthria and paraphasias are assessed by observing the patient's spontaneous speech production. The importance of these variables to the fluent-nonfluent classification of aphasias is described in detail in Chapter 5. Essentially, nonfluent and dysgrammatical speech is associated with anterior lesions (e.g., Broca's and transcortical motor aphasias), whereas fluent paraphasic speech is associated with more posterior lesions (e.g., Wernicke's, conduction, and transcortical sensory aphasias). The examination of language repetition is also extremely important for classifying the different clinical syndromes of aphasia. In testing repetition, it is important to use sentences such as "If she were here, I could have seen her," since aphasic patients have the most difficulty repeating small grammatical words. The repetition of tongue twisters such as "Irish constabulary" or "hippopotamus" is useful in bringing out dysarthria and palilalia but not for assessing aphasic difficulties. Some patients will fail repetition tasks because the length of the sentence exceeds their word span. This should be attributed to attentional deficits, not to aphasia.

The bedside testing of language comprehension is described in Chapter 5. Comprehension may need to be tested along several individual channels. The most commonly tested channels include auditory speech comprehension of "yes-no" questions, comprehension of written language, comprehension of commands directed to individual limb movements, and comprehension of whole body commands. A different combination of comprehension channels is affected in different types of aphasia. There are standardized tests of comprehension that provide a somewhat more quantitative assessment of this function. The Peabody Picture Vocabulary Test[29] contains items at several levels of difficulty to test comprehension of single words. "Yes-No" questions such as "Does March come before April?" or "Is a coffee pot alive?" can be used to test sentence length comprehension. Patients with Wernicke's and transcortical sensory aphasias do poorly in these tests. The Token Test[27] provides a more difficult test of comprehension. In this test geometric forms differing in shape, size, and color are manipulated by the patient in response to instructions from the examiner (e.g., "Put the small yellow circle behind the large red square"). Several versions of this test including a convenient short form[103] are available. Although this test provides a sensitive assessment of comprehension deficits, its results need to be interpreted cautiously. For example, some patients may fail on the basis of apraxic disturbances. Furthermore, some patients with Broca's aphasia may have a selective comprehension disturbance for complex grammatical structure and may fail this test even though their comprehension in most other language tests may be perfect. The Token Test is particularly useful in identifying the earliest signs of subtle comprehension deficits, especially in patients with the slowly progressive aphasia syndrome.[71]

Naming is impaired in almost all subtypes of aphasia. An initial probe of naming ability can easily be done at the bedside by presenting the patient with common objects and asking the patient to name them as well as their component parts. However, in order to assess change over time, it may be necessary to use a more quantitative measure. The Boston Naming Test[53] is a sensitive standardized instrument consisting of 60 line drawings for words of graded difficulty. Normative data are provided for children and adults at different educational levels. The sensitivity of the test makes it a useful instrument for patients with mild to moderate impairment who might seem unimpaired in a traditional bedside testing of naming functions.

The Boston Naming Test also makes some provisions for differentiating misnaming due to aphasic disturbances from misnaming due to perceptual difficulties. When a patient fails to identify a picture correctly, a "stimulus cue" is presented to aid recognition (e.g., "photographers or surveyors use it" for a picture of a tripod). This cue may be sufficient to allow for the correct word to be retrieved. In the event that the word is not retrieved, even though there is evidence that the object is recognized, a "phonemic cue" is given to provide acoustic information about the word itself (e.g., "tri . . ." for tripod). The patient who fails to name because of misperception may benefit from stimulus cues; the patient who recognizes the object but has a true word retrieval deficit may benefit from phonemic cues. The Boston Naming Test as well as most bedside naming tests almost exclusively rely on the visual modality for input. The importance of other modalities in the assessment of naming is discussed in Chapters 5 and 7.

Reading and writing lend themselves to convenient informal bedside assessment. Quantitative assessment of these functions can be obtained with the Boston Diagnostic Aphasia Examination[40] and the Gates-MacGinitie Reading Tests.[35] The reading tests employ a multiple-choice format, eliminating complex oral and motor responses. It is important to test the reading ability for common nouns separately from that for short grammatical words (see Chapter 5). Patients with Broca's aphasia, for example, may show a selective reading impairment for grammatical words. Special strategies may need to be used in this assessment, since the patient with Broca's aphasia may have very severe impairments of speech output during reading aloud. We usually provide the patient with a list of 10 words such as "house, table, it, alligator, she, then, book, was, coat, them." The examiner reads words at random (some of which are on the list and some of which are not) and the patient is asked to point to the one that was just read. In severely apraxic patients, the examiner may also need to do the pointing and the patient is asked to signal in the affirmative or negative (by nodding or by "yes-no" answers). On such tests, patients with Broca's aphasia do very poorly in matching the heard sound of grammatical words with their written forms. No such difficulty is shown with common nouns.

Writing is disturbed in most of the aphasia syndromes, the errors in writing typically paralleling the errors in spoken language. An isolated disturbance of writing, however, may be observed in patients in acute confusional states. In this instance, it is the mechanical rather than the linguistic aspects of writing that are most often disturbed. Aphasic writing only with the left hand may be observed in right-handed patients with callosal lesions. The motor components of writing can be dissected from the linguistic components by comparing spontaneous writing to writing with a typewriter or with anagram letters.

Praxis

The word "apraxia" has been used to describe a large number of phenomena including dressing difficulties, oculomotor scanning difficulties, and also impairment of skilled movement. The various problems associated with this nomenclature are discussed in Chapter 7. In the strictest sense, the term "apraxia" describes an impairment of skilled movement to command or imitation even though the patient understands what is requested and has no primary motor deficit that interferes with the performance of the movement. This description is consistent with Liepmann's[65] category of ideomotor apraxia. This is the sense in which the term "apraxia" is used in this section.

The examination for apraxia must consider several factors: (1) the body part performing the movement (face, limb, whole body); (2) the nature of the command (verbal versus model for imitation); and (3) the nature of the movement being performed (symbolic or conventional gestures, representation of object usage, nonrepresentational postures). Buccofacial apraxia is common in patients with anterior lesions who have nonfluent or Broca-type aphasias. In response to commands such as "Stick out your tongue," "Cough," or "Show me how you would blow out a match," the patient may demonstrate an inappropriate movement, or simply repeat the command. This latter phenomenon is a dramatic finding in many patients with Broca's aphasia.

In testing limb praxis it is important to emphasize the movements that depend on fine distal actions of the foot and hand. These movements are more vulnerable to apraxia, perhaps because they are more dependent on the integrity of the pyramidal tract. Limb movements can be disturbed bilaterally or, in the case of a right hemiplegia, they can be impaired in the nonhemiplegic left arm. This phenomenon is also known as sympathetic dyspraxia. In assessing limb praxis, symbolic and conventional gesturing can be tested with such commands as "Salute," "Thumbs up," "Wave goodbye," and "Beckon someone to come over to you." The representation of object use can be demonstrated by asking the patient to pretend to use a toothbrush, comb, or scissors. A more subtle form of apraxia is the tendency to substitute a body part for the object rather than positioning the hand in a way that is compatible with holding the imaginary implement. Thus, a patient might rake his fingers through his hair to represent the use of a comb, rub his teeth with his index finger to represent a toothbrush, and use the index and middle fingers to represent the blades of a scissors. This is called the "body-part-as-object" response.[39] When a patient fails to carry out the desired movement, the examiner should perform several comparable movements including the correct one. If the patient can identify the correct movement, then the examiner can conclude that the patient has understood the command. This is a necessary prerequisite for diagnosing an apraxia.

Some patients who fail to emit a symbolic gesture or represent object usage upon verbal command may imitate the relevant movement quite well. These patients demonstrate the presence of a verbal-motor disconnection. On the other hand, some patients are apraxic not only to verbal command but even to imitation. In such cases, one wonders whether there is a direct impairment of "motor engrams."

It has been argued that apraxia is a disorder only for motor output that has a symbolic content. The imitation of nonrepresentational gestures eliminates the symbolic component of movement. The patient is asked to imitate hand and arm postures that have no communicative content, e.g., placing the back of the hand behind the ipsilateral ear or placing the hand vertically to one side of the nose. Presumably, these types of tasks should remain unimpaired in patients with ideomotor apraxia, even if they cannot imitate movements related to symbolic gesture and object use. In our experience, the value of testing nonrepresentational gestures is limited.

In addition to the face and limbs, it is also important to test praxis involving the whole body in response to commands such as "Stand up," "Turn around and sit down," or "Take the stance of a boxer." These commands as well as those directed at eye opening and closure may be perfectly preserved in severely aphasic and otherwise apraxic patients.

The preservation of whole body and axial movements upon command suggests that these may be mediated by different nonpyramidal pathways.

The term "ideational apraxia" has been used by Liepmann[65] to describe an impairment in the performance of a complex activity requiring a sequence of movement with real objects. The act of lighting a cigarette or taking out a pen and beginning to write provide reasonably good tests of this ability. This ability is most often impaired in acute confusional states and in severe dementia. Such impairments probably represent basic attentional disturbances that interfere with the coherence of motor output (see Chapter 3).

Calculations

Calculation abilities can be assessed either at the bedside or with the help of standardized quantitative measures. An assessment of basic calculations (addition, subtraction, multiplication, and division) is provided in the supplementary part of the Boston Diagnostic Aphasia Examination.[40] Normative values for age and educational level are provided for individuals up to 85 years of age. The arithmetic subtest of the Wide Range Achievement Test[52] provides a more extensive examination and includes normative data for ages 5 to 64 years. This may be desirable for use with patients who have a superior premorbid ability in this area. The arithmetic subtest of the Wechsler Adult Intelligence Scale[68,111] contains numerical word problems that are delivered orally with time constraints on responses.

In each of these tests, the patient may fail for a variety of reasons. For example, a patient who is unable to do mental arithmetic as required by the Wechsler Adult Intelligence Scale or by the bedside examiner may be able to perform the same operations when given pencil and paper. This indicates that there is an attentional disturbance rather than a primary involvement of calculation abilities. Patients with right-hemisphere lesions fail calculations because of spatial deficits.[49] For example, in operations that require the use of a pencil and paper, these patients may neglect the numbers on the left side or they may align the numbers incorrectly. Aphasia and alexia may also interfere with calculations.[8] Some patients with these deficits may make excellent use of an electronic calculator.

Even in the absence of these contributing factors, some patients have specific calculation deficits. The ability to manipulate numbers in the process of calculation entails several component processes.[63] First, knowledge of basic table values must be accessible to the patient (i.e., $2 \times 6 = 12$; $9 - 3 = 6$; $5 + 8 = 13$; and so on). Second, the patient must understand the meaning of "multiplication," "subtraction," "addition," "division" and of the symbols that are used for these operations. Third, the sequence of operations necessary for carrying out the relevant computation and such techniques as "borrowing" or "carrying" must be preserved. Patients with acalculia have complex combinations of deficits in all three of these areas (see Fig. 2, *right*). In assessing calculations, it is useful to have the patient perform multiplication (e.g., 3×38), long division (e.g., 7310 ÷ 34), and also a simple word problem. Acalculia is closely associated with lesions of the inferior parietal lobule (especially the angular gyrus) of the left hemisphere (see Chapter 5).

Right/Left Orientation

The symptom of right/left disorientation has attracted a great deal of interest because of its association with Gerstmann's syndrome. As described in Chapter 5, when the cluster of dysgraphia, acalculia, finger identification difficulties, and right/left disorientation occurs in isolation, this is designated as Gerstmann's syndrome and is frequently associated with damage to the left inferior parietal lobule.

There are several standard batteries with normative data for assessing verbally mediated right/left discrimination.[10,40] The complexity of the commands is gradually increased. The patient is first asked to point to his own right and left and then to the examiner's right and left. Then the patient is asked to follow complex commands, such as "Put your left hand on your right eye" or "Point with your left hand to my right ear." Before concluding that the patient has a specific difficulty with this task it is necessary to ascertain that this is not due to an underlying aphasia. Right/left disorientation should be included in the list of deficits only if it occurs either in the absence of aphasia or if its severity is out of proportion to the aphasia.

Even though failures in the tasks described above lead to the designation of right/left disorientation, it is important to realize that this is an unfortunate terminology. What these tests really show is that the patient has a specific naming disorder for right and left. There is no indication from these tests that the patient has actually a problem with egocentric spatial orientation. It is possible to test egocentric spatial orientation with the paradigm of delayed spatial alternation, a task that is well standardized in the experimental literature. Our initial impression is that patients with the right/left naming disturbances of Gerstmann's syndrome do not have right/left spatial disorientation as tested by delayed alternation.

Finger Naming

The naming of fingers has also received prominence because of its association with Gerstmann's syndrome. Essentially, the examiner points to specific fingers on the patient's hand and asks the patient to produce the appropriate names for the fingers. The same procedure can be carried out by having the patient name the examiner's fingers or the fingers on a photograph of a hand. As in the case of right/left orientation, a specific finger-naming disturbance is identified only if it occurs in the absence of aphasia or out of proportion to it.

Complex Perceptual Tasks

Much of the cortical surface is devoted to the transformation of sensory signals into complex percepts. As discussed in Chapter 1, there is a right-

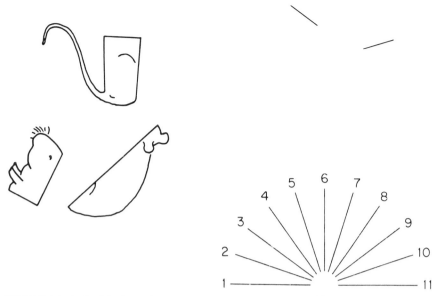

FIGURE 6. *Left*: Card 22 from the Hooper Visual Organization Test.[50] The patient needs to put together the fragmented pieces mentally and to come up with the correct response—a rat or a mouse. A frequent type of incorrect response is given when the patient reaches a premature conclusion by taking into consideration only the upper part and describing the stimulus as a pipe or a watering can. (Copyright © 1983 by Western Psychological Services, 12031 Wilshire Blvd., Los Angeles, CA 90025; reprinted by permission.)

Right: Stimulus card and test array from the Benton Judgment of Line Orientation Test.[10] The patient has to say which of the 11 test lines (*bottom*) match the orientation of the two stimuli on *top*. The correct responses are 3 and 10. (From Benton et al.[10] Copyright © 1983 by Oxford University Press, Inc.; reprinted by permission.)

hemisphere specialization for most nonverbal complex perceptual tasks, including facial recognition. Therefore, many tasks of complex perceptual processing, especially those that are relatively free of verbal mediation, have been used to assess the integrity of the right hemisphere.

The Hooper Visual Organization Test[50] is commonly used for this purpose. It consists of 30 cards containing simple drawings of objects presented in fragmented form (Fig. 6, *left*). The task is for patients to assemble the fragments in their mind's eye and then to name the whole object. Young adults obtain an average score of 25;[50] however, in individuals over the age of 70 a score of 22 may be considered normal.[76] There are several possibilities for errors on the Hooper test. The patient may be unable to name the object simply because of an aphasia; this possibility must be ruled out, since the performance would no longer have any implication for perceptual abilities. A second cause for failure is the tendency to jump to premature conclusions. In other words, the patient may give an answer based on only one of the fragments without taking the time to integrate all of them. This could then lead to poor performance not on a perceptual basis as much as on the basis of faulty reasoning ability. A third reason for failure is that there may be an elementary perceptual deficit in the visual modality so that the basic sensory input may be faulty. This obviously needs to be ruled out. Finally, the performance may be poor because of a high-order deficit in the ability to mentally reorganize complex perceptual input. It is this last type of difficulty that is most closely associated with dysfunction in the right hemisphere. The examiner should analyze the type of errors carefully before concluding that poor performance in this test implies high-order perceptual deficits.

The Judgment of Line Orientation[10] is another task that is sensitive to complex perceptual disturbances. The patient is required to match the orientation of a test line to another line in an array of 11 lines (Fig. 6, *right*). Benton and associates[10] reported that performance was particularly poor in patients with damage to the posterior part of the right hemisphere. This visual skill is preserved until the age of 70; over 70, 11 percent of normal individuals may demonstrate defective performance, defined as a score less than 25 out of 30.[9] This test is relatively free of verbal mediation and can conveniently be used for bedside assessment. A similar test with even greater specificity for right-hemisphere damage can be given in the somatosensory modality by asking the blindfolded patient to adjust two hinged rods in a way that duplicates the angle of the test rods examined by palpation.[70]

Another more traditional test of perceptual processing is the Poppelreuter Task[83] of superimposed line drawings of common objects. The patient is asked to identify each individual object. Using a variation of this test, DeRenzi and Spinnler[26] found that patients with right-hemisphere lesions were impaired on this task compared with those with left-hemisphere lesions. Naturally, the presence of an underlying aphasia that may also interfere with the naming of objects must be ruled out before interpreting the results of this test.

More subtle disturbances of perceptual processing can be detected on tasks in which visual information is incomplete or ambiguous. Warrington[110] noted that patients with right parietal lesions had difficulty identifying the Gollin Incomplete Figures.[38] In this test, the patient is given an incomplete picture of an object. The amount of visual information is gradually increased in subsequent pictures. The performance on this test shows how well the patient can infer the entire picture from partial information. Warrington[110] also demonstrated that patients with right parietal lesions had difficulty identifying objects that had been photographed from an unusual perspective (e.g., a teapot viewed from above; Fig. 7, *left*). Figure-ground and contour discrimination were normal in these patients, suggesting that the observed deficits represent a failure at the level of high-order perceptual classification.

Patients with damage to auditory areas in the right hemisphere may have complex deficits in nonverbal auditory processing. Such lesions may interfere with timbre and pitch discrimination and with memory for tonal patterns.[73] The Seashore Tests of musical talent[97] provide measures of these auditory discrimination functions. The identification of environmental sounds could also be considered as a complex nonverbal auditory task. Tests have been described in which recordings of environmental sounds are played for the patient who must then point to one of several pictures depicting the object that could give rise to such a sound.[101] This technique can easily be adapted for bedside use by employing objects such as keys, combs, crumpling paper, slamming a door, and so forth. The examiner should avoid asking the patient to label the sound and

FIGURE 7. *Left*: A teapot viewed from an unusual perspective. Correct identification probably requires a mental rotation into a more familiar viewpoint. Patients with right-hemisphere dysfunction find this task difficult.

Right: Item 8 from the block design subtest of the Wechsler Adult Intelligence Scale.[68,111] The patient is shown the stimulus card in the upper left. The correct solution on the upper right shows how the nine blocks should be placed together to reproduce the design in the stimulus card. The performance in the bottom left is from a 55-year-old right-handed man with a right temporal lesion and shows the dramatic difficulties including the loss of contour. The performance in the lower right shows the more subtle deficits, mostly in internal detail, of a patient with a left-hemisphere lesion. (Reproduced by permission from the Wechsler Adult Intelligence Scale. Copyright © 1955 by the Psychological Corporation; all rights reserved.)

instead test recognition by multiple-choice selection of the object responsible for the sound, since a naming deficit could otherwise interfere with proper identification.

Complex perceptual processing is a very important area to assess, even though it is commonly neglected at the bedside examination of mental state.

Constructional Tasks

Constructional abilities rely both on complex perceptual and skilled motor functions. They are evaluated by such tasks as drawing, assembling puzzles, and constructing designs from blocks or sticks. These skills are almost automatically associated with the right hemisphere, but the tasks that measure them are so complex that they may also be failed by patients with left-hemisphere damage. Paterson and Zangwill[80] described qualitative differences in the errors made by patients with unilateral hemispheric lesions on drawing tasks. Patients with right-hemisphere lesions were noted to employ a "piecemeal" approach, drawing segments in an isolated and disorganized manner or omitting parts of the model. In contrast, patients with left-hemisphere lesions tended to preserve the configurational features of the model while leaving out details or adding superfluous lines. Thus, performance on tests of constructions must be considered not only in light of accuracy but also with respect to the nature of the errors.

Block design tasks are commonly used to assess constructional abilities. The block design subtest from the Wechsler Adult Intelligence

Scale[68,111] presents complex designs that must be constructed with colored cubes (Fig. 7, *right*). Perceptual and motor factors play a role in this task, which requires that a patient first examine the model and then match his own production to the model. This task has demonstrated interesting differences in the contributions made by the right and left hemispheres. The types of errors made by the right and left hands of commissurotomized patients gives evidence that the right hemisphere contributes to the overall configuration of the design, whereas the left hemisphere contributes primarily to the analysis of internal details.[36] In keeping with this observation, patients with right-hemisphere lesions show an inability to maintain the essential contour or framework of block designs, whereas patients with left-hemisphere damage have difficulty with internal detail (Fig. 7, *right*).[54]

The Block Design Test, in its standard administration, calls for time limits. However, response time is generally slower in brain-damaged patients than in normal individuals, and time limits may thus penalize the patients who can perform the task if allowed to proceed at a slower pace. It would seem preferable to administer this test without time limits if the examiner wishes to assess constructional skills specifically.

Test scores in the block design task may be low after damage to either hemisphere, but the types of errors made after lesions in the right side of the brain are much more dramatic. Similarly, severe deficits in three-dimensional block construction are much more common in patients with right-hemisphere damage.[10] In this task, the patient is given either an exact model or a photograph of what to construct.

The Rey-Osterrieth Complex Figure Test[79,91] is discussed in a previous section as a test of nonverbal memory. However, in the copying condition, this same test can be used to gather information on constructional abilities. Performance on this test is especially vulnerable to right-hemisphere damage. Patients with such lesions show a characteristic piecemeal approach to copying the figure (Fig. 8).

Traditional bedside tests of constructional ability include asking the patient to copy a cube, a daisy, or a house. These may bring out the more

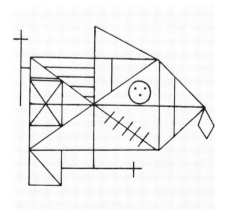
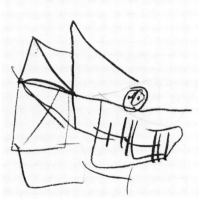

FIGURE 8. *Left*: The Rey-Osterrieth Figure.[79,91] *Right*: Copy of the figure by a 23-year-old man with a lifelong history of learning disability. Slowness and clumsiness of the left limbs during the performance of motor tasks suggested early damage to the right hemisphere of the brain.

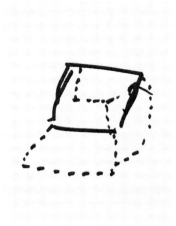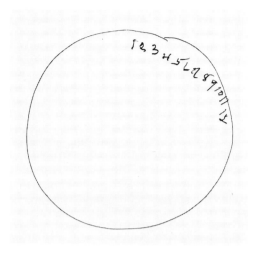

FIGURE 9. *Left*: Drawing of a cube by a 62-year-old, left-handed man with a right frontoparietal infarction. In addition to an element of left neglect, this drawing shows a profound distortion of three dimensionality.

Right: The clock drawing of an 85-year-old, right-handed demented man with no evidence of spatial neglect on cancellation tests or on bilateral simultaneous stimulation. This performance indicates poor planning, not left hemineglect. The difficulty is in the spacing of numbers within the context of the whole circle. If he were given a task that required a counterclockwise sequence, his drawing would have left the right side empty.

severe constructional deficits and are also useful in detecting left-hemispatial neglect (Fig. 9, *left*). In the more subtle cases of constructional disturbances, the patient may do quite well in reproducing two-dimensional figures.

Another traditional bedside test of construction is the drawing of a clock. We have found it very useful for the examiner to provide the patient with a relatively large circle (essentially covering half the surface of an $8\frac{1}{2}$-inch by 11-inch sheet of white paper). The patient is then asked to put in the numbers as if it were the face of a clock. One dramatic type of error is to leave out all the numbers in the left side of the drawing, as seen in conjunction with the syndrome of left-hemispatial neglect. An example of a clock drawn by a patient with severe left-hemispatial neglect is shown in Figure 2 of Chapter 3. However, a clock with nothing on the left side may also reflect poor planning of space rather than hemispatial neglect. An example of this, shown in Figure 9, *right*, is taken from a patient with dementia. The numbers are crowded on the right side not because of left-hemispatial neglect but because the patient fails to plan the appropriate spacing between consecutive numbers. Age and education both influence drawing ability. Goodglass and Kaplan[40] have provided normative values for adults up to the age of 85 on tests of spontaneous drawing and copying of six simple objects, including a clock and a three-dimensional cube.

The various tests of construction do not all assess identical mechanisms. For example, a patient could succeed at block designs but fail on drawing tasks, and vice versa.

Dressing

Impairment of the ability to dress oneself, the so-called dressing apraxia, may be observed in the context of right posterior parietal lesions or bilateral lesions in this region. The localizing aspect of the phenomenon and the objections to designating it as an apraxia are discussed in Chapter 7. If the deficit is exclusively for dressing the left side of the body, then this could be considered as part of the neglect syndrome. In the more severe instances, the deficit encompasses both sides of the body. These patients are unable to align the axis of the clothing to the axis of the body. This type of dressing difficulty reflects complex visuospatial deficits. Other but usually less-severe dressing difficulties may also emerge in patients who have impairments in the ability to sequence complex activities. However, those patients rarely have problems dealing with one piece of clothing at a time, even when the task is made difficult. A simple bedside test is to take away the patient's shirt or blouse and then to return it upside down and with one sleeve inside-out. The examiner can then observe the realignment of the garment with respect to the body axis, the rearrangement of the clothing, and the patient's ability to properly introduce the arms into the garment.

Spatial Distribution of Attention

In an earlier section of this chapter, the testing of the attentional matrix is described. Another very important aspect of attention is its distribution within the extrapersonal space. This function is very sensitive to right-hemisphere damage. Such lesions may give rise to dramatic neglect for the left side of the extrapersonal space. Analogous neglect for the right hemispace is much less common (see Chapter 3).

There are essentially two major types of tests for hemispatial neglect: those based on bilateral simultaneous stimulation and those requiring a systematic scanning of the extrapersonal space. Bilateral simultaneous stimulation in the somatosensory, auditory, and visual modalities is part of the traditional neurologic examination. Extinction during these tasks, especially if it occurs in more than one sensory modality, can be accepted as an indication of sensory neglect. Considerations that are pertinent to the interpretation of bilateral simultaneous stimulation tasks are reviewed in Chapter 3.

Some of the motor manifestations of neglect may emerge in the testing of reading, writing, and constructions. For example, there may be a failure to read the left half of sentences (hemialexia). The patient may leave a very wide margin on the left while writing, may not copy the left side of figures and may leave the left side of a clock empty (see Chapter 3).

We have developed a test of scanning which can be used for evaluating the motor aspects of hemispatial neglect (Mesulam and Weintraub, unpublished). The four test forms consist of random and structured arrays of verbal and nonverbal stimuli, as shown in Figure 10. The

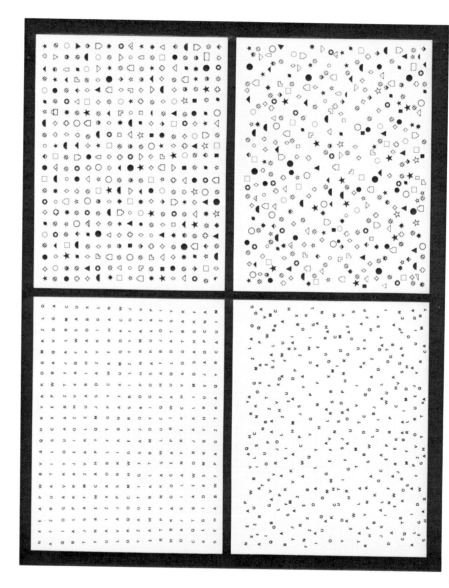

FIGURE 10. The four forms of the Cancellation Test. Each form can be administered in an 8½ × 11 or 11 × 14 format. The letter "A" and the open circle with a line slanted toward the right are the targets. There are fifteen targets in each quadrant of the page. The performance of patients with neglect on this test is shown in Chapter

stimuli are presented on an 8½- by 11-inch or 11- by 14-inch sheet of paper placed directly in front of the patient, who is not allowed to move the test form. Patients are asked to circle all the targets they can find ("A" for the verbal form and the open circle crossed by a single slanted line for the nonverbal array). Each quadrant contains 15 target items. The task is performed with colored pencils, a different color being handed to the patient after the identification of 10 targets or after a specified period of time. The color coding allows the examiner to note the spatial progress of the search over time. Normal adults under the age of 50 can complete each of the four test conditions within 2 minutes without errors. Over the age of 50, the omission of one target from each field could still be considered within normal limits. Over the age of 80, as many as four targets may go undetected in each visual field. Normal adults conduct a systematic search beginning on the left and proceeding to the right in horizontal or vertical rows even in the random arrays.

Data obtained from patients with focal unilateral vascular lesions demonstrate the potential usefulness of these cancellation tasks in detecting neglect behavior. Patients with right-hemisphere lesions begin the search either in the center or on the extreme right side of the page. Those with severe neglect rarely proceed to the left side of the page. Detection of the geometric shapes is less accurate than detection of letters in these patients. More targets are neglected when they are randomly distributed than when they are structured (see Chapter 3). However, the scanning strategy tends to be erratic regardless of the stimulus array. Finally, in patients with subtle or resolving neglect, this test reveals inattention even when other comparable tasks (such as the cancellation of lines) are performed normally. Therefore, this procedure allows a very sensitive assessment of hemispatial neglect in a way that also reveals some of the underlying mechanisms (see Chapter 3).

Patients with left-hemisphere lesions show a distinctly different pattern of performance on this task. Even in patients with a right hemianopia, the right side of the page rarely shows evidence for neglect, unless the examination has been carried out in the acute phase of the stroke. In these patients, detection of geometric shapes tends to be faster than detection of letters. Patients with left-sided lesions employ the systematic scanning strategy characteristic of normal subjects even when the array is random.

Another motor aspect of directed attention is manual exploration. To test this, as described in Chapter 3, the patient can be blindfolded and asked to search by palpation for a small object placed on the table in front of the patient. Alternatively, the patient may be presented with a large board containing objects such as paperclips, keys, and nails taped on the board in random fashion and asked to locate a specific object by palpation. Before using the second paradigm, it is necessary to ascertain that the patient does not have an aphasia or astereognosis. Patients with right-hemisphere lesions and left-hemispatial neglect show defective manual exploration on the left side with either the left hand or the right hand, whereas exploration is much more systematic and rapid with either hand in the right side of space (see Chapter 3).

MENTAL STATE
ASSESSMENT OF
YOUNG AND
ELDERLY ADULTS IN
BEHAVIORAL
NEUROLOGY

103

Paralinguistic Aspects of Communication

As discussed in Chapters 1 and 6, communication entails much more than the formal linguistic aspects of word choice and grammar. These additional components include the encoding and decoding of facial expressions, gestures, and other paralinguistic aspects of communication.

The bedside examination of affective and attitudinal prosody, facial expressions, and gestures is described in detail in Chapter 6. We have also used standardized taperecordings for testing the discrimination, repetition, and evoked production of affective and nonaffective speech prosody.[115,116] In testing the affective component of prosody, the subject is given a simple declarative sentence and asked to utter it with a specified emotional content. Alternatively, the examiner could utter the sentence and ask the patient to identify the underlying emotion. For nonaffective prosody, the patient is first given a basic set of information such as "The man walked to the grocery store, the woman rode to the shoe store." Then, a series of questions are asked, such as "Who walked to the grocery store?" The examiner then determines if the patient has appropriately placed the emphasis (stress) on the word "man" in giving the answer (e.g., "The *man* walked to the grocery store"). It is also possible to test the identification of facial expression with standardized sets of photographs depicting the same person displaying different emotions or attitudes. Normal adults perform exceedingly well on all of these measures, although the spontaneous production of affective prosody requires a certain amount of theatrics, which may be mildly embarrassing to some.

The Profile of Nonverbal Sensitivity (PONS) Test[93] measures the ability to infer interpersonal content (e.g., a scolding versus saying farewell) from prosody, facial expression, and body posture. This test has been validated as a measure of sensitivity to nonverbal communication and has been standardized in a large population of subjects. Poor performance on this test is particularly common to patients with right-hemisphere injury.[7] The original and abbreviated versions of this test are available in film and video format for administration to individuals or groups of patients. One possible disadvantage of this test is its considerable demands for sustained concentration, perceptual abilities, and reading comprehension.

Additional paralinguistic aspects of communication, which are described in Chapters 1 and 6, can be examined informally by observing the interactions of the patient with the examiner and with the staff members. Performance in these paralinguistic behaviors is particularly sensitive to right-hemisphere damage.

Reasoning and Abstractions

The integrity of reasoning and abstraction can be measured by tasks that explore the ability to manipulate old knowledge and to solve novel problems in which analogic reasoning or the extraction of abstract principles is necessary. It is important to assess these skills using both verbal and nonverbal responses. An aphasic patient who fails verbal reasoning tasks may be able to demonstrate sophisticated reasoning if tested by means of nonverbal responses.

Reasoning that draws upon previous experience and knowledge of conventional standards is the focus of the comprehension subtest of the Wechsler Adult Intelligence Scale.[68,111] The patient is asked to reason about dilemmas such as how best to respond to noticing fire and smoke in a movie theater. These tasks do not require novel or creative problem-solving approaches. Instead, they draw upon stored reservoirs of socialized and somewhat automatic thinking. Responses to these questions are often used as evidence for the integrity of judgment. However, it is necessary to point out that patients who give the expected verbal answer in the laboratory may still yell "Fire" when they smell smoke at the theater. It seems that these tests assess the ability for reasoning more than the ability to act with judgment in a real situation.

Proverb interpretation is a standard test of reasoning in the mental state examination. Common proverbs (e.g., "Strike while the iron is hot") do not require abstract thinking, since their meaning is usually highly overlearned. Unfamiliar proverbs (e.g., "The good is the enemy of the best") are preferable, since they cannot be interpreted without resorting to novel ways of thinking. The Gorham Proverbs Test[41] measures both spontaneous and multiple-choice proverb interpretation. The multiple-choice administration is particularly useful for the language-impaired patient if reading comprehension is intact.

The ability to discern similarities is a measure of abstract reasoning. The similarities subtest of the Wechsler Adult Intelligence Scale[68,111] provides a standardized measure of this skill. Naturally, if a patient fails on this test it is important to ensure that this is not due to an underlying aphasia. Elderly patients may tend to interpret the instruction to look for similarity as a request for physical comparison. However, if the instruction is restructured (e.g., "What do a _____ and a _____ have in common?" or "Tell me one thing that is true about a _____ and a _____"), then more abstract responses can be stimulated.

Nonverbal categorical abstraction can be tested with various object- and form-sorting tasks.[37] The task is to sort stimuli that "go together." Sorting of objects may be done according to function, constituent material, and so on. Sorting of forms may use color, shape, or size as the criterion. Once the initial sorting is made, the patient is then asked to regroup the stimuli according to some other criterion. This yields a measure of mental flexibility.

The Visual-Verbal Test[30] is a compact instrument in which abstraction and shifting are tested with cards containing four stimuli. In each card, a group of three stimuli share one attribute and an overlapping group of another three stimuli share another (Fig. 11, *left*). Concept formation and mental flexibility are being evaluated by these methods. Although many aphasic patients fail this test, we have also seen some aphasic individuals who do quite well. The test may therefore provide a probe for investigating the dissociation of thought from language.

The Wisconsin Card Sort Test[42] introduces the elements of hypothesis testing to the basic sorting paradigm. Four stimulus cards are placed horizontally in front of the patient. The cards contain geometric forms

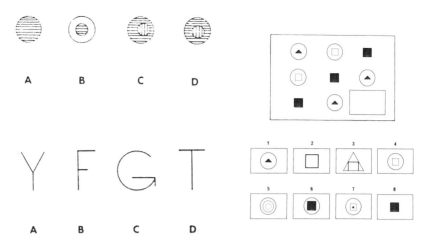

FIGURE 11. *Left*: Cards 13 and 29 from the Visual-Verbal Test. The patient is asked to identify two groups of three stimuli, each group containing a common feature. In Card 13 (top) one group of three is made up of stimuli A, C and D because they each have horizontal lines running through the larger circle. The second group is made up of B, C and D because of the small inner circle. In Card 29 (bottom) one group is made up of stimuli A, B and D because they have no curved lines. The second group is made up of B, C and D because each contains a horizontal line segment. This test assesses mental flexibility. (From Feldman and Drasgow.[30] Copyright © 1959 by Western Psychological Services, 12031 Wilshire Blvd., Los Angeles, CA 90025; reprinted by permission.)

Right: A problem matrix similar to those on the Raven Progressive Matrices. The patient is asked to identify which of the bottom eight stimuli fits the blank region on top. The correct answer is 4. Although the response requires minimal verbal output, it is heavily mediated by verbal reasoning.

that differ from one another in number, shape, and color. The first card contains a red triangle, the second two green stars, the third three yellow crosses, and the fourth four blue circles (see Fig. 1, *bottom*). Sixty-four test cards, each containing from one to four triangles, stars, crosses, or circles in red, green, yellow, or blue are presented one at a time, and the patient is required to sort them in association with one of the four stimulus cards. Each test card has stimuli of only one geometric shape and a single stimulus color. The patient is not apprised of the principle for sorting but instead must deduce the principle from the examiner's response to each placement (i.e., "right" or "wrong").

One possible basis for sorting the cards is by color. In that case, the patient is expected to place cards that contain red stimuli below the first card, green below the second, and so on, regardless of the shape of the stimuli or how many there are on the card. If the sorting principle shifts to shape, then only cards containing triangles are placed below the first card, regardless of the color or number of triangles on each card. Finally, when number is the sorting principle, all single stimuli, regardless of shape and color, are placed below the first card.

After 10 consecutive cards have been correctly sorted, the examiner shifts the principle of sorting (e.g., from color to shape) without forewarning, and the patient must accordingly shift to this new sorting strategy guided only by reinforcement for a correct response. Perseverative errors occur when the patient continues to sort by a principle that is no longer correct. Patients with frontal lobe lesions achieve fewer categories and make more perseverative errors than patients with lesions elsewhere in the brain.[74] Conflicting results have been reported for the relative importance of the left and right frontal lobes in the performance of this task.[28,74] Furthermore, since patients who cannot inhibit interfering response tendencies do quite poorly, performance on the Wisconsin Card Sort Test is also sensitive to the attentional matrix of the patient.

Verbal analogic reasoning is required on the Abstraction subtest of the Shipley Institute of Living Scale.[99] The task is to complete verbal series such as "AB BC CD D_". There are 20 items arranged in order of increasing difficulty. A conceptual quotient is obtained by comparing the abstraction score with the vocabulary score. The drawback of this test is that there are no age-appropriate normative data for middle-aged and elderly adults. However, the combined abstraction and vocabulary scores can yield an estimated IQ for which there are age-appropriate values.[81] The IQ estimated on this basis can be compared with the observed IQ on other measures in order to judge the specificity of analogic reasoning impairments in comparison with more general measures of cognitive functions. In interpreting the results of this test, it is important to ascertain that an underlying aphasia is not principally responsible for observed failures.

The Raven Progressive Matrices[86] is a test of analogic reasoning that has been standardized for adults and that does not necessitate verbal responses. The Colored Progressive Matrices,[87] originally designed for children, is a simpler version that is also appropriate for the brain-damaged patient with moderate functional impairment. The test items consist of visual stimulus arrays with missing parts. The patient is given both appropriate and inappropriate choices for completing the test stimulus. Early items on the test stress perceptual factors, since they require completion of complex designs. Later items employ several designs arranged in a 2 by 2 or 3 by 3 matrix, in which one or more attributes are systematically varied (Fig. 11, *right*). The patient must identify the one design that conforms to the principle of variation. There is a mistaken impression that the Raven Matrices provides a "test of nonverbal reasoning." In fact, this is a "reasoning task based on nonverbal responses." The reasoning itself undoubtedly contains a great deal of verbal mediation. The Raven Progressive Matrices is a useful test of reasoning in aphasic patients who have difficulty with verbal output. The interpretation is reasonably straightforward if performance is adequate despite the aphasia. The score on the test reflects performance not only on the reasoning items but also on items that measure perceptual abilities. Therefore, it is often difficult to interpret the behavioral and structural implications of an abnormal score. It may be useful to score the perceptual and reasoning items separately.

It is very difficult to design tests that specifically assess nonverbal reasoning abilities. Perhaps complex nonverbal perceptual processing tasks, such as the Hooper Test[50] and the identification of objects presented from an unusual angle, also measure aspects of visuospatial reasoning.

Factors that can influence performance on reasoning tasks include attention, language, memory, and visual-perceptual skills. When one of these is impaired, the examiner must decide which of the available tests of reasoning will be least likely to be affected by the deficit. For example, the Wisconsin Card Sort Test[42] and similar instruments may be inappropriate for amnesic patients, since the sorting criteria may be forgotten.

Planning and Sequencing

An essential aspect of mental function is the ability to plan and sequence the individual components of complex goal-directed activities. Patients with frontal lobe damage and those with confusional states may show marked impairment in this area of behavior. Such impairments influence the execution of several daily living activities such as cooking, driving, and dressing. One patient may fail to coordinate the individual steps of a recipe; another may falter in starting the car or in negotiating a set of traffic lights; still another patient may put on items of clothing in an inappropriate sequence. Patients with deficits in these areas may also show a generalized difficulty in following directions and instructions. There are very few formal tests of planning and sequencing abilities. The Alternating Sequences Test (p. 80) and the drawing of a clock (p. 100) assess some aspects of sequencing and planning. Liepmann's[65] ideational apraxia may also be conceptualized as a planning and sequencing deficit in the motor area (see p. 94). The Porteus Maze Test[84] can also be used to assess the patient's ability to plan a strategy for negotiating visual mazes of variable complexity.

Judgment, Insight, Appropriateness

Judgment and insight are very difficult to test objectively. As mentioned above, the common practice of testing judgment by asking patients how they would respond in an ambiguous situation is inadequate. Kohlberg's[61] studies of moral development have illustrated that there are distinctions among knowing what to do, knowing why you are doing it, and actually acting in a real situation. Thus, a patient may be able to describe what should be done but, when faced with the real situation, might act in an entirely different manner that is clearly inappropriate. At present, the most reliable way to assess judgment is by direct observation of behavior or by interviewing family members carefully about the patient's behavior in situations requiring judgment.

Insight is usually judged by the patient's awareness and understanding of the illness. The denial or minimization of disease seen with right-hemisphere lesions is a specific example of poor insight. Socially inappropriate behavior may also reflect impairments of judgment and insight. The patient who strikes up an inappropriately lively conversation with strangers in a waiting room or who strolls into the examiner's office without knocking is providing obvious signs of loss (or disregard) of the rules for socially appropriate conduct. Social inappropriateness as well as poor judgment and insight may be prominent in patients with lesions of the right hemisphere or the frontal lobes.

Family members sometimes report that the patient has lost the ability to grasp the gist of a complex situation or to catch the punch-line of a joke. This probably reflects difficulties in the group of mental faculties that is also essential for insight, judgment, and appropriateness. There are no formal tests for this function. Perhaps some indirect means of assessment could be devised by asking the patient to decipher the humorous message of a cartoon sequence.

TESTING LOBES AND HEMISPHERES

The surface of the brain is divided into the frontal, parietal, occipital, and temporal lobes. As shown in Chapter 1, with the possible exception of the occipital lobe, each lobe of the brain is extremely heterogeneous with respect to functional specialization. Nevertheless, there is a widespread tendency to structure the assessment of mental state according to individual lobes. This approach serves the heuristic purpose of organizing the mental state examination according to anatomic landmarks.

Temporal Lobe

The temporal lobes contain auditory cortex, visual association cortex, and the medially situated limbic and paralimbic structures. The left temporal lobe also contains Wernicke's area. Therefore, the tests of naming and language comprehension described in this chapter and in Chapter 5 can be used to assess the integrity of the left temporal lobe and of its connections. Homologous parts of the right temporal lobe can be tested with the help of the nonverbal complex auditory tasks described earlier (e.g., identification of environmental sounds and discrimination of and memory for tones, timbre, and pitch). The integrity of temporal limbic and paralimbic regions can be tested with the help of verbal memory tests for the left temporal lobe and nonverbal memory tests for the right. Bilateral lesions that directly damage the visual association areas of the temporal lobe or their connectivity with other parts of the brain result in very complex deficits that include visual amnesia, visual anomia, prosopagnosia, and object agnosia. The anatomic correlates of these syndromes are reviewed in Chapters 1, 4, and 7.

For the anatomic reasons listed in Chapter 1, unilateral lesions in auditory association cortex may be extremely difficult to detect clinically unless one uses the Dichotic Listening Test. Two sets of three digits are simultaneously presented, one to each ear. The task is to report the digits heard in each ear. Even normal subjects may fail to report all six digits: those who are right-handed typically do better in reporting digits heard by the right ear (approximately 3 to 10 percent higher scores than for the left ear).[58] This normally higher suppression of left-sided input probably reflects the dominance of the left hemisphere for speech perception. Kimura[59] observed alterations of digit perception as a function of the side of temporal lobectomy. Thus, alterations in the normal pattern of digit perception can be used as a measure of temporal lobe dysfunction. For example, patients with right temporal lobe damage may show an exaggerated form of the normal right-ear advantage, whereas right-ear advantage may disappear altogether or be replaced by left-ear advantage in patients with left temporal lobe damage.

The presence of an upper quadrantanopia is a helpful neighborhood sign indicating the presence of temporal lobe damage.

MENTAL STATE
ASSESSMENT OF
YOUNG AND
ELDERLY ADULTS IN
BEHAVIORAL
NEUROLOGY

109

Occipital Lobe

The entire occipital lobe is occupied by the primary visual area and by visual association cortex. Therefore, the manifestations of damage to this part of the brain are confined to tasks in the visual modality. These are discussed in detail in Chapters 1 and 7. Depending on the selectivity of the lesion, occipital lobe damage is associated with visual field cuts, color perception deficits, difficulties in the perception of movement, alexia, and visuospatial disorientation (see Chapter 7).

Parietal Lobe

The parietal lobe contains the somatosensory areas and the high-order association cortex of the posterior parietal region. Tests sensitive to left parietal lobe damage include naming, reading, writing, calculations, finger identification, and left/right naming. These tests have been described earlier. Equivalent lesions in the right hemisphere give rise to dressing difficulties, constructional difficulties, hemispatial neglect, and neglect of the body surface. The testing for these deficits is also described earlier. Additional manifestations of parietal lobe damage, especially when bilateral, include Balint's syndrome as well as spatial and topographic disorientation (see Chapter 7).

Frontal Lobe

As discussed in Chapter 1, much of frontal lobe function remains elusive. Patients with selective frontal lobe damage usually show two features. First, functions attributed to the other three lobes of the brain are relatively preserved in these patients. Second, there are deficits in judgment, insight, mental flexibility, reasoning, abstraction, planning, sequencing, and in testing of attentional tone—especially those tests that depend on response inhibition and on the ability to sustain behavioral output. Lesions that include the orbital (basal) part of the frontal lobes may also lead to memory disturbances.

Left and Right Hemispheres

The two hemispheres of the human brain have distinctly different behavioral specializations. In most right-handed individuals, tests of linguistic function, ideomotor praxis, left/right identification, finger identification, calculation, and verbal memory are most sensitive to left-hemisphere damage. On the other hand, tests of dressing, constructions, complex nonverbal perceptual skills, spatial neglect, paralinguistic competence, and nonverbal memory are most sensitive to right-hemisphere injury.

Assessment of Handedness

Some individuals perform all skilled manual functions (writing, combing, throwing a ball, using a scissors) with the right hand. Others may write with the right hand but use the left hand for certain skilled movements such as throwing a ball or combing. There is a tendency to classify such individuals as right-handed, since questions about activities other than writing are rarely asked during routine mental state assessments. However, these individuals with mixed handedness may have a considerably different pattern of hemispheric specialization when compared with those who are fully right-handed. Specialized questionnaires are available for computing a quantitative index of handedness when such information becomes desirable.[5,78]

WHICH BATTERY TO USE

The armamentarium of the clinician is full of test batteries. Some are specialized for individual cognitive functions and include the Boston Diagnostic Aphasia Examination,[40] the Boston Naming Test,[53] the Western Aphasia Battery,[57] and the Wechsler Memory Scale.[112] Others are specialized for specific patient populations (e.g., the dementia rating scales). Still others are more general and include the Halstead-Reitan Neuropsychological Tests[45,90] and the Wechsler Adult Intelligence Scale.[68,111] Each of these batteries has the merit of broad-spectrum assessment and standardization. They can be used successfully for population studies or for the assessment of change over time. However, the adoption of such batteries as the only mode of assessment would considerably jeopardize the flexibility of the clinician who may want to alter the testing strategy according to the clinical presentation of the individual patient.

TESTING THE ELDERLY AND THE DIAGNOSIS OF DEMENTIA

Several structural changes occur in the brain as a function of aging. These include a decrease in neuronal density as well as a decrease in the overall weight and volume of the brain.[56] Although there is a general tendency to assume that these reflect degenerative or involutional changes, it is also important to realize that the dendritic branching complexity of neurons increases from the fifth to the seventh decade.[56] Therefore, the behavioral relevance of the structural changes described with age should be interpreted cautiously.

With increasing age, individuals perform less well on tests of reaction time, memory, cognitive flexibility and perhaps also visuospatial skills.[11] Longitudinal studies generally demonstrate that decline occurs at different rates for different cognitive functions, that the decline is generally mild, and that there is a great deal of individual difference.[96] The main message to be derived from aging research is that age-related cognitive changes are relatively minor and do not interfere with the ability to lead an independent existence. When independence is threatened by behavioral changes, a dementing illness should be suspected. Dementia can be defined as a progressive deterioration of previously acquired mental abilities to the point at which the individual is no longer able to perform the daily living activities that are appropriate for his age and background.

The assessment of mental state is probably the single most important phase in the medical diagnosis of dementia in the elderly. The clinician must be able to distinguish changes that occur in the normal course of aging from those due to a disease process. Age-appropriate norms for the elderly have been provided for a number of neuropsychologic tests (see pages 84, 96, and 97). On the Wechsler Adult Intelligence Scale,[68,111] the typical aging pattern consists of relative stability in the verbal IQ and a progressive decline in performance IQ.[14] Since the performance subtests are timed, however, it has been suggested that the observed decline in the performance IQ may be an artifact of slower responses with aging. However, Storandt[106] noted that while untimed scores are higher than timed scores in individuals between 65 and 75, they still remain lower than scores obtained by younger age groups.

Benton and co-workers[9] administered a series of tests to normal subjects ranging in age from 65 to 85 and divided the subjects into 5-year age groups. Performance of each 5-year age group was then compared with the performance of a standard sample of normal adults from 16 to 65 years of age. The percentage of subjects in each 5-year age group that failed in comparison with the standard sample was computed for each individual test. It can be seen from Table 3 that the incidence of decline is different for different cognitive functions. Age-related decline of performance is particularly marked on the Visual Retention Test and the Digit Sequence Test, both of which depend on memory function. In contrast, tests of orientation, simple digit span, and estimation of line orientation remain relatively stable with aging. This study has several important implications for the clinical assessment of dementia in the elderly. For example, if an elderly patient shows impaired performance on several of the tests that remain stable with age (e.g., orientation, digit span), then the likelihood of a dementing disease is enhanced. On the other hand, if a patient performs well even on those tasks that show the most frequent age-related decline, then the likelihood of a widespread dementing process diminishes. Another significant finding was that 33 percent of individuals in the 80- to 85-year age group performed like the younger standard sample in all of the tests. This type of observation,

TABLE 3. Age-Related Changes in Performance on Neuropsychologic Measures: Relative Frequency of Defective Performances

	Test	65–69 yrs (n = 28)	70–74 yrs (n = 62)	75–79 yrs (n = 35)	80–84 yrs (n = 37)
1.	Temporal Orientation	0%	2%	6%	5%
2.	Digit Span	4%	2%	6%	8%
3.	Digit Sequence	18%	10%	14%	22%
4.	Word Association	4%	2%	6%	11%
5.	Logical Memory	4%	3%	3%	11%
6.	Associate Learning	0%	0%	6%	11%
7.	Visual Retention	7%	16%	40%	38%
8.	Facial Recognition	0%	10%	14%	14%
9.	Line Orientation	0%	2%	11%	8%

From Benton et al.[9] Copyright © 1981 Swets and Zeitlinger B. V., Lisse; reprinted by permission.

together with some longitudinal studies,[96] very strongly suggests that a certain subgroup (on the order of 33 percent) of elderly patients may show no or essentially very little cognitive change during much of senescence. This implies that mental decline is not a necessary outcome of age. Therefore, findings that are otherwise compatible with normal aging may, in certain individuals, reflect the beginning of dementia. The differentiation of early dementia from the normal aging pattern poses a formidable challenge to the clinician. (Chapters 10 and 11 describe new developments in neuroimaging, which could assist the clinician in this task.)

Since most dementing illnesses affect relatively widespread areas of the brain, it is customary to use various batteries that offer a composite score of cognitive functioning. Although this practice tends to obscure the more precise correlations between behavior and cerebral physiology, it offers a rapid and reliable assessment of overall functioning. Such batteries are especially useful in patients with more advanced dementia, who may be unable to perform the more specialized tests described earlier. However, such batteries may not be applicable to those with mild and early dementia, for whom more extensive and specialized tests may need to be used to avoid a ceiling effect.

Numerous short mental status tests are available for use with elderly patients. The Mini Mental State[31] provides cutoff scores for normal elderly subjects, patients with dementia and those with psychiatric illness. A positive correlation has been reported between the Blessed Dementia Scale, another short mental state test, and the density of senile plaques in the brains of patients with Alzheimer's disease.[13] This scale now enjoys wide use in clinical and pharmacologic studies of dementia. These brief screening instruments provide a convenient tool for detecting major cognitive alterations. However, they are too insensitive for use with patients who have milder deficits.

One test instrument that holds promise as a useful diagnostic tool is the Mattis Dementia Rating Scale.[69] It contains five subtests (Attention, Initiation and Perseveration, Constructions, Conceptual, and Memory) covering a broad range of mental functions. Normative data have recently been collected for community-based elderly individuals with a mean age of 75.[76,77] The average score obtained by this age group is 137/144 with a standard deviation of 7. Split-half reliability studies show that it is possible to divide this test into two short comparable forms that can be used for test-retest purposes.[34] The Mattis scale does not specifically examine language functions or calculations, but this can be addressed by adding the Boston Naming Test[53] and calculation testing to the evaluation. A profile of cognitive strengths and weaknesses can then be determined by comparing performance on tests of attention, language, memory, visuospatial skills, and reasoning. The Mattis scale scores correlated with measures of cerebral bloodflow in patients with pathologically verified Alzheimer's disease.[19]

Dementia is quite prevalent in old age, and Alzheimer's disease is the single major dementing illness in this age group.[55] Since there are no specific laboratory tests for Alzheimer's disease or for similar dementing illnesses, the clinician is often faced with the task of having to make diagnostic decisions based on the mental state assessment. This is a difficult task, since few symptoms are specific to an individual degenerative dementing illness and since there is a large overlap with the manifestations of other and more easily treatable causes of dementia (such as metabolic encephalopathy, depression, and space-occupying lesions).

Even when the diagnosis of Alzheimer's disease is made on the basis of strict criteria, there is a great degree of clinical variability, especially in the early stages of the disease. In addition to clinical heterogeneity, there is also considerable variability in the distribution of the lesions and in the extent to which different transmitter systems are involved. For example, the distribution of plaques and tangles in Alzheimer's disease may vary from one patient to another.[56] Furthermore, while it seems that most patients with Alzheimer's disease have a major deficit of cholinergic transmission,[24] there are additional but less-universal disturbances of dopaminergic, noradrenergic, serotonergic, and somatostatinergic pathways; the extent of involvement of these additional transmitter pathways varies from patient to patient.[109]

It is conceivable that these clinical and pathologic differences merely reflect chance variations of an underlying unitary process. On the other hand, it is also conceivable that there are many subtypes of Alzheimer's disease and that each subtype reflects a different cluster of biologic alterations. In that case, it may be quite important to identify these subtypes clearly, since they may each respond to different types of treatments.

It is also possible that certain clinical profiles in patients who fit the overall picture of dementia are inconsistent with a diagnosis of Alzheimer's disease or that they have specific implications for prognosis. For example, there are some patients with a very slowly progressive aphasic condition who do not show, at least for many years, other signs of cognitive decline.[71] These patients show a particularly indolent progression of the disease. It may turn out that this condition has a pathophysiology other than that of the traditional clinical forms of Alzheimer's disease. These considerations emphasize the need for a careful mental state assessment in evaluating dementia.

Another very important aspect of the mental state assessment is to differentiate reversible from irreversible causes of dementia. The important alternatives to progressive dementias include metabolic encephalopathy (e.g., confusional state), depression, and focal brain disease (e.g., stroke, tumor). The hallmarks of the confusional state are described in Chapter 3. The presence of a salient attentional disorder in these patients helps the differential diagnosis. However, it is important to realize that some types of intrinsic brain disease, including Alzheimer's disease, can also result in confusional states (see Chapter 3). The contribution of depression to mental state deterioration is difficult to rule out, especially since elderly patients may lack the customary affective and vegetative signs of depression. The presence of language disturbances, quite common in most cases of Alzheimer's disease, is extremely rare in depression. Depressed patients are said to have memory difficulties mostly at the stage of registration, whereas patients with Alzheimer's disease also have an impairment of retention and retrieval. In many instances, however, the distinction between depression and dementia may be so difficult that the clinician may decide to embark on an empirical trial of antidepressant treatment.

Normal pressure hydrocephalus, subdural hematoma, and tumors in the frontal lobe or in the corpus callosum may give rise to the clinical picture of dementia (see Fig. 7B and 7C in Chapter 1). Modern neuroimaging methods can rapidly provide the correct diagnosis in these cases. However, small multiple infarcts that are difficult to demonstrate by available neuroimaging methods may also be associated with the clinical picture of a gradually progressive dementia.[44] Some patients with multi-infarct dementia may even have clinical profiles that are indistinguishable from those of Alzheimer's disease. The Hachinski Ischemic Scale[43] contains 13 factors that are useful in the differentiation of these two conditions: a score above 6 is usually indicative of a multi-infarct dementia. Multifocal EEG slowing and risk factors for cerebrovascular disease also favor the diagnosis. Although multi-infarct dementia is not reversible, it has important implications for the prevention of additional strokes.

It is also important to assess daily living activities and to correlate specific areas of impairment with specific disturbances in mental state. Only a few scales of daily living activities are applicable to the assessment of mildly impaired demented individuals who are living at home. The Instrumental Activities of Daily Living[62] is one such scale, which requires family members to rate routine activities according to the extent to which assistance from others is required for their performance by the patient. Most patients with early symptoms of dementia, especially when the salient problems are in the areas of memory, language, and visuospatial skills, demonstrate impaired performance on formal testing in the office even before significant change is observed in routine home or work activities. However, patients with deficits in functions that are commonly attributed to the frontal lobe may score quite well on cognitive tests even when daily living activities may be severely disrupted.

The patient with superior premorbid intellectual capacity or special talents poses a unique dilemma in the evaluation of early dementia. This is especially true when such an individual comes to medical attention with self-perceived deficits in mental functioning. Many of the standard tests yield a ceiling effect, since they do not match the competence of such individuals. These patients have to be tested with more challenging instruments such as the Graduate Record Examination, the Miller Analogies Test, and the Medical College Aptitude Test. The change of performance over time in these difficult tasks may provide the most important information. A detailed interview with family members or even with colleagues may yield extremely useful information. However, in these individuals even the most extensive evaluation may yield no objective finding.

GUIDELINES FOR THE BEDSIDE EVALUATION OF MENTAL STATUS IN BEHAVIORAL NEUROLOGY

In this section, some guidelines are presented for an examination that can be used during the initial office or bedside assessment of patients. This assessment is expected to lead to a preliminary diagnostic formulation and also to identify, when necessary, areas that should be explored in greater detail by formal neuropsychologic examination. Except in the case of the most difficult patients who are either markedly abulic or severely aphasic, this bedside examination should take from 30 to 45 minutes, once the examiner has acquired some initial experience.

History

The examination of mental state yields the most information when guided by a complete history and physical examination. Since time restrictions would rarely permit all of this information to be collected in a single visit, it is often necessary for the clinician to examine the relevant records before the initial contact with the patient. In many instances, the history is much more reliable if obtained from someone other than the patient; the details of this history can then be compared with those provided by the patient personally. It is important to establish the chief complaint, the mode of onset, and the temporal course of the condition. It is also essential to understand in what way the patient's daily living activities have been impaired as a consequence of the present illness. This historic information will generate certain hypotheses that the clinician will have to test during the examination. The patient's level of education and employment history should also be noted, since they alert the examiner to the level of task difficulty appropriate for the examination.

Initial Observations

Patients' insight and judgment with respect to their condition can be ascertained while obtaining the history of the present illness, by comparing the patients' versions of the history with those of third parties and medical records. Aspects of motivation, cooperation, concentration, the ability to sustain spontaneous behavior, and the presence of aphasia can also be observed during the initial minutes of the interview. If attention or language is grossly impaired, the examination may need to be curtailed. Improvisations in testing to circumvent these deficits, however, can be considered at this time (e.g., testing spatial orientation in the aphasic patient by asking him to point to his location on a map). The patient's display of affect, its range, its appropriateness, and whether it is sustained or prone to abrupt changes should also be noted. The structure of thought processes (circumstantial, tangential, idiosyncratic) is also assessed in this phase.

Initial Probe of Memory Functions

In order to establish a relaxed atmosphere for the examination, testing can begin with informal questions about orientation, current news events, and important recent events in the patient's life. If this information is accurate, a generalized memory deficit is unlikely. The anomic or abulic patient may fail to answer spontaneously questions about current information. In this instance, multiple-choice recognition should be used.

More Detailed Memory Testing

If knowledge of current events or orientation is severely impaired on the basis of the initial probe, the memory task should be simplified. Three of the patient's personal belongings can be hidden around the examining room, making sure that the patient has immediate recall of where they are hidden. Delayed recall is then tested after 5 to 10 minutes. The Drilled Word Span Test (see page 88) is slightly more difficult and can be used when appropriate; but if it is administered, digit span testing should precede it in order to determine the word span. If the memory impairment does not appear to be severe, the Three Words–Three Shapes Test (see page 86) may be administered as the only test of memory. During delay intervals for the memory tests, the time can be used to assess attentional capacity.

Testing the Attentional Matrix

An initial probe of intermediate difficulty is the serial threes test. If serial threes are too difficult, the patient can be asked to recite the months in reverse order. This can be done in about 20 seconds by young and elderly individuals. If this task is too difficult, then the days of the week in reverse order or simply counting backwards from 20 by ones can be tested. If serial threes prove too simple, the patient can be asked to recite serial sevens. If this is in turn too easy, recitation of the alphabet in reverse order can be tested. Young adults can perform this task within 60 seconds with, at most, one error.

Digit span (forward and backward) should also be assessed, if this has not already been done as part of the Drilled Word Span procedure.

Testing of Language Functions

If a patient is able to convey his meaning without circumlocution and to understand the examiner during informal conversation, then it is unlikely that a formal aphasia examination will yield useful information. If speech is obviously aphasic, then the language examination precedes testing of all other cognitive functions and occupies the majority of examination time.

Following the assessment of spontaneous speech output, the language testing continues with naming of common objects (e.g., the eraser and tip of a pencil, the crystal and band of a wristwatch, the receiver and dial of a telephone) and of simple geometric shapes. Repetition can be tested with the use of a complex phrase, such as "No ifs, ands, or buts." Comprehension can be tested by asking questions such as "If the tiger was killed by the lion, which animal is dead?" or "Does March come after April?" A brief writing sample should be obtained by dictation. Reading comprehension can be tested with the help of a newspaper paragraph. A brief screening for apraxia can also be done by asking the patient to demonstrate how to blow out a match and how to use a hammer or comb, first with one hand and then the other. Naming of left and right body parts on the patient's own body and on that of the examiner, as well as finger naming can also be tested at this time. If any one of these stages in the aphasia testing reveals major difficulties, then a more detailed examination can be conducted as outlined in this chapter (see pages 90 to 92) and also in Chapter 5.

Testing of Calculations

Mental arithmetic is first tested with a problem of moderate difficulty (e.g., "Multiply 9 × 13"). If the task is failed, the patient is given paper and pencil. If the task is too simple, a more complex calculation (such as "What is 6 percent of $150?") can be given. It is important to avoid asking multiplication table values (e.g., 9 × 7), since these assess memory for learned values rather than the ability to compute.

Testing the Distribution of Directed Attention Within the Extrapersonal Space

Bilateral simultaneous stimulation in at least two modalities is a test for extinction. The cancellation of random letters (see page 101) tests the motor aspects of directed attention.

Tests of Construction

The patient can be asked to put numbers inside a large circle (5- to 6-inch diameter) as if it were the face of a clock. The examiner draws the circular outline. This procedure facilitates detection of poor spatial planning and unilateral neglect (see page 100).

The patient is then asked to copy a three-dimensional cube. It is important to use the solid cube and not the "transparent" Necker cube, since the latter can be drawn by a trick that is usually learned in elementary school and does not require the understanding of three-dimensional perspective. Design number 6 from the block design subtest of the Wechsler Adult Intelligence Scale[68,111] can be used as an additional probe of constructional skills. If this design and the cube copying are performed well, it is unlikely that there are major visuomotor deficits.

Tests of Perceptual Processing

Items 26 to 30 from the Benton and associates[10] Judgment of Line Orientation Test and cards 6, 9, 10, 16, and 22 from the Hooper Visual Organization Test[50] can be used to assess complex perceptual processing.

Tests of Abstraction and Mental Flexibility

The ability to abstract similarities can be tested with the following items, which range from easy to difficult: "orange and banana," "fish and boat," and "poem and statue." Patients with severe deficits in abstractions may describe differences or insist that there are no similarities between the two items in each pair.

Proverb interpretation can be tested with an unfamiliar proverb, such as "Hunger is the best gravy" or "The good is the enemy of the best."

Abstraction and mental flexibility can be tested using items 6, 10, 13, 14, and 29 from the Visual-Verbal Test.[30] For the patient who is aphasic, items from the Raven Matrices[86,87] are more appropriate.

Tests of Resistance to Intrusion and Interference

The interference condition of the Stroop Test[15,20,107] can be administered to assess the patient's vulnerability to interference. Half the items can be administered at the beginning of testing and half at the end, so that it may be possible to detect whether or not there is a learning effect on task performance.

Motor response inhibition can be tested with a go–no-go task (see page 81). The patient's forearm should be exposed during this task, since even the slightest muscle flicker in the forearm can be counted as an error of commission. Response inhibition can also be tested with written sequences (alternating rectangles and triangles) or motor sequences (successive alternations of hand positions on a table).[66,67]

The Ability to Sustain Behavioral Output

The F-A-S Test[103] is a good test of sustained behavioral output, particularly for patients who are not aphasic.

Formal Neurologic Testing

It is extremely important to complete the assessment of the nervous system by performing a formal neurologic examination of cranial nerves, sensorimotor function, and coordination.

REFERENCES

1. ALBERT, MS, BUTTERS, N, AND LEVIN, J: *Temporal gradients in the retrograde amnesia of patients with alcoholic Korsakoff's disease.* Arch Neurol 36:211, 1979.
2. ALBERT, MS, BUTTERS, N, AND BRANDT, J: *Patterns of remote memory in amnesic and demented patients.* Arch Neurol 38:495, 1981.
3. ALBERT, MS, AND KAPLAN, E: *Organic implications of neuropsychological deficits in the elderly.* In POON, LW, FOZARD, JL, CERMAK, LS, ARENBERG, D, AND THOMPSON, LW (EDS): *New Directions in Memory and Aging.* Lawrence Erlbaum Associates, Hillsdale, NJ, 1980, p 403.
4. ANDERS, TR, FOZARD, JL, AND LILLYQUIST, TD: *Effects of age upon retrieval from short-term memory.* Dev Psychol 6:214, 1972.
5. ANNETT, M: *The binomial distribution of right, mixed and left handed.* Q J Exp Psychol 19:327, 1967.
6. BECK, AT, WARD, CH, MENDELSOHN, M, MOCK, J, AND ERBAUGH, J: *An inventory for measuring depression.* Arch Gen Psychiatry 4:561, 1961.
7. BENOWITZ, LI, BEAR, DM, ROSENTHAL, R, MESULAM, M-M, ZAIDEL, E, AND SPERRY, RW: *Hemispheric specialization in nonverbal communication.* Cortex 19:5, 1983.
8. BENSON, DF, AND DENCKLA, MB: *Verbal paraphasia as a source of calculation disturbance.* Arch Neurol 21:96, 1969.

MENTAL STATE
ASSESSMENT OF
YOUNG AND
ELDERLY ADULTS IN
BEHAVIORAL
NEUROLOGY

119

9. Benton, AL, Eslinger, PJ, and Damasio, R: *Normative observations on neuropsychological test performances in old age.* J Clin Neuropsychol 3:33, 1981.

10. Benton, AL, Hamsher, K de S, Varney, N, and Spreen, O: *Contributions to Neuropsychological Assessment.* Oxford University Press, New York, 1983.

11. Birren, JE, and Schaie, KW (eds): *Handbook of the Psychology of Aging.* Van Nostrand Reinhold, New York, 1977.

12. Black, FW, and Strub, RL: *Digit repetition performance in patients with focal brain damage.* Cortex 14:12, 1978.

13. Blessed, G, Tomlinson, BE, and Roth, M: *The association between quantitative measures of dementia and of senile change in the cerebral grey matter of elderly subjects.* Br J Psychiatry 114:797, 1968.

14. Botwinick, J: *Intellectual abilities.* In Birren, JE, and Schaie, KW (eds): *Handbook of the Psychology of Aging.* Van Nostrand Reinhold, New York, 1977, p 580.

15. Broverman, DM: *Dimensions of cognitive style.* J Pers 28:167, 1960.

16. Brown, J: *Some tests of the decay theory of immediate memory.* Q J Exp Psychol 10:12, 1958.

17. Buschke, H, and Fuld, PA: *Evaluating storage, retention, and retrieval in disordered memory and learning.* Neurology 24:1019, 1974.

18. Butters, N: *The clinical aspects of memory disorders: Contributions from experimental studies of amnesia and dementia.* J Clin Neuropsychol 6:17, 1984.

19. Coblentz, JM, Mattis, S, Zingesser, LH, Kasoff, SS, Wisniewski, HM, and Katzman, R: *Presenile dementia.* Arch Neurol 29:299, 1973.

20. Comalli, PE, jr, Wapner, S, and Werner, H: *Interference effects of Stroop color-word test in childhood, adulthood, and aging.* J Gen Psychol 100:47, 1962.

21. Corkin, S: *Acquisition of motor skill after bilateral medial temporal lobe excision.* Neuropsychologia 6:225, 1968.

22. Corkin, S: *Some relationships between global amnesias and memory impairments in Alzheimer's disease.* In Corkin, S, Davis, KL, Growdon, JH, Usdin, E, and Wurtman, RJ (eds): *Alzheimer's Disease: A Report of Progress in Research.* Raven Press, New York, 1982, p 149.

23. Davies, ADM: *The influence of age on Trail Making Test performance.* J Clin Psychol 24:96, 1968.

24. Davies, P, and Maloney, AJF: *Selective loss of central cholinergic neurons in Alzheimer's disease.* Lancet 2:1403, 1976.

25. DeRenzi, E, and Faglioni, P: *The comparative efficiency of intelligence and vigilance in detecting hemisphere damage.* Cortex 1:410, 1965.

26. DeRenzi, E, and Spinnler, H: *Visual recognition in patients with unilateral cerebral disease.* J Nerv Ment Dis 142:515, 1966.

27. DeRenzi, E, and Vignolo, LA: *The Token Test: A sensitive test to detect disturbances in aphasics.* Brain 85:665, 1962.

28. Drewe, EA: *The effect of type and area of brain lesion on Wisconsin Card Sorting Test performance.* Cortex 10:159, 1974.

29. Dunn, LM: *Expanded Manual for the Peabody Picture Vocabulary Test.* American Guidance Service, Minneapolis, 1970.

30. Feldman, MJ, and Drasgow, J: *The Visual-Verbal Test.* Western Psychological Services, Los Angeles, 1959.

31. Folstein, MF, Folstein, SE, and McHugh, PR: *"Mini-mental state." A practical method for grading the cognitive state of patients for the clinician.* J Psychiatr Res 12:189, 1975.

32. Folstein, MF, and Luria, R: *Reliability, validity, and clinical application of the visual analogue mood scale.* Psychol Med 3:479, 1973.

33. Fromm-Auch, D, and Yeudall, LT: *Normative data for the Halstead-Reitan Neuropsychological Tests.* J Clin Neuropsychol 5:221, 1983.

34. Gardner, R, Oliver-Muñoz, S, Fisher, L, and Empting, L: *Mattis Dementia Rating Scale: Internal reliability study using a diffusely impaired population.* J Clin Neuropsychol 3:271, 1981.

35. Gates, AI, and MacGinitie, WH: *Gates-MacGinitie Reading Tests.* Teachers College Press, Columbia University, New York, 1969.

36. Geschwind, N: *Specializations of the human brain.* Sci Amer 241:180, 1979.

37. Goldstein, KH, and Scheerer, M: *Abstract and concrete behavior: An experimental study with special tests.* Psychol Monogr 53:No 239, 1941.

38. Gollin, ES: *Developmental studies of visual recognition of incomplete objects.* Percept Mot Skills 11:289, 1960.

39. GOODGLASS, H, AND KAPLAN, E: *Disturbance of gesture and pantomime in aphasics.* Brain 86:703, 1963.

40. GOODGLASS, H, AND KAPLAN, E: *The Assessment of Aphasia and Related Disorders,* ed 2. Lea & Febiger, Philadelphia, 1983.

41. GORHAM, DR: *A proverbs test for clinical and experimental use.* Psychol Rep 1:1, 1956.

42. GRANT, DA, AND BERG EA: *The Wisconsin Card Sort Test Random Layout: Directions for administration and scoring.* Wells Printing Co., Inc., Madison, Wisconsin, 1980.

43. HACHINSKI, V: *Cerebral blood flow: Differentiation of Alzheimer's disease from multi-infarct dementia.* In KATZMAN, R, TERRY, RD, AND BICK, KL (EDS): *Alzheimer's Disease: Senile Dementia and Related Disorders.* Raven Press, New York, 1978, p 97.

44. HACHINSKI, V, LASSEN, NA, AND MARSHALL, J: *Multi-infarct dementia.* Lancet 2:207, 1974.

45. HALSTEAD, WC: *Brain and Intelligence.* University of Chicago Press, Chicago, 1947.

46. HAMILTON, M: *A rating scale for depression.* J Neurol Neurosurg Psychiatry 23:56, 1960.

47. HATHAWAY, SR, AND MCKINLEY, JC: *The Minnesota Multiphasic Personality Inventory Manual.* Psychological Corporation, New York, 1951.

48. HAYMAN, M: *Two minute clinical test for measurement of intellectual impairment in psychiatric disorders.* Arch Neurol Psychiatr 47:454, 1942.

49. HÉCAEN, H, ANGELERGUES, R, AND HOULLIER, S: *Les variétés cliniques des acalculies au cours des lésions rétrorolandiques: Approche statistique du problème.* Rev Neurol 105:85, 1961.

50. HOOPER, HE: *The Hooper Visual Organization Test Manual.* Western Psychological Services, Los Angeles, 1958.

51. HULICKA, IM: *Age differences in Wechsler Memory Scale scores.* J Gen Psychol 109:135, 1966.

52. JASTAK, JF, AND JASTAK, S: *The Wide Range Achievement Test.* Western Psychological Services, Palo Alto, California, 1978.

53. KAPLAN, E, GOODGLASS, H, AND WEINTRAUB, S: *The Boston Naming Test.* Lea and Febiger, Philadelphia, 1983.

54. KAPLAN, E, PALMER, EP, WEINSTEIN, C, BAKER, E, AND WEINTRAUB, S: *Block design: A brain-behavior based analysis.* Paper presented at the International Neuropsychological Society, Annual European Meeting, Bergen, Norway, 1981.

55. KATZMAN, R: *The prevalence and malignancy of Alzheimer's disease.* Arch Neurol 33:217, 1976.

56. KEMPER, T: *Neuroanatomical and neuropathological changes in normal aging and dementia.* In ALBERT, ML (ED): *Clinical Neurology of Aging.* Oxford University Press, New York, 1984, p 9.

57. KERTESZ, A: *Aphasia and Associated Disorders.* Grune & Stratton, New York, 1979.

58. KIMURA, D: *Cerebral dominance and the perception of verbal stimuli.* Can J Psychol 15:166, 1961.

59. KIMURA, D: *Some effects of temporal lobe damage on auditory perception.* Can J Psychol 15:156, 1961.

60. KNESEVICH, JW, MARTIN, RL, BERG, L, AND DANZIGER, W: *Preliminary report on affective symptoms in the early stages of senile dementia of the Alzheimer type.* Am J Psychiatry 140:233, 1983.

61. KOHLBERG, L: *Essays on moral development.* Vol 1. *The philosophy of moral development.* Harper & Row, San Francisco, 1981.

62. LAWTON, MP, AND BRODY, EM: *Assessment of older people: Self-maintaining and instrumental activities of daily living.* Gerontologist 9:179, 1969.

63. LEVIN, HS, AND SPIERS, PA: *Acalculia revisited.* In HEILMAN, KM, AND VALENSTEIN, E (EDS): *Clinical Neuropsychology,* ed 2. Oxford University Press, New York, in press.

64. LEZAK, MD: *Neuropsychological Assessment,* ed 2. Oxford University Press, New York, 1983.

65. LIEPMANN, H: *Das Krankheitsbild der Apraxie.* Monatsschr f Psych u Neurol 8:15, 102, 181, 1900.

66. LURIA, AR: *Higher Cortical Functions in Man.* Basic Books, New York, 1966.

67. LURIA, AR: *Human Brain and Psychological Processes.* Harper & Row, New York, 1966.

68. MATARAZZO, JD: *Wechsler's Measurement and Appraisal of Adult Intelligence,* ed 5. Williams & Wilkins, Baltimore, 1972.

69. MATTIS, S: *Mental state examination for organic mental syndromes in the elderly patient.* In BELLAK, L, AND KARASU, TE (EDS): *Geriatric Psychiatry.* Grune & Stratton, New York, 1976, p 77.

70. MEERWALDT, JD: *The Rod Orientation Test in Patients with Right-Hemisphere Infarction.* Roceddavids, Alblasserdam, Netherlands, 1982.

71. MESULAM, M-M: *Slowly progressive aphasia without generalized dementia.* Ann Neurol 11:592, 1982.

72. MILLER, GA: *The magical number seven, plus or minus two: Some limits on our capacity for processing information.* Psychol Rev 63:81, 1956.

73. MILNER, B: *Laterality effects in audition.* In MOUNTCASTLE, VB (ED): *Interhemispheric Relations and Cerebral Dominance.* Johns Hopkins Press, Baltimore, 1962, p 177.

74. MILNER, B: *Effects of different brain lesions on card-sorting.* Arch Neurol 9:90, 1963.

75. MILNER, B: *Interhemispheric differences in the localizaiton of psychological processes in man.* Br Med Bull 27:272, 1971.

76. MONTGOMERY, K, AND COSTA, L: *Neuropsychological test performance of a normal elderly sample.* Paper presented at the International Neuropsychological Society Meeting, Mexico City, Mexico, 1983.

77. MONTGOMERY, K, AND COSTA, L: *Concurrent validity of the Mattis Dementia Rating Scale.* Paper presented at the International Neuropsychological Society Meeting, Lisbon, Portugal, 1983.

78. OLDFIELD, RC: *The assessment and analysis of handedness: the Edinburgh inventory.* Neuropsychologia 9:97, 1971.

79. OSTERRIETH, P: *Le test de copie d'une figure complexe.* Arch Psychol 30:206, 1944.

80. PATERSON, A, AND ZANGWILL, OL: *Disorders of visual space perception associated with lesions of the right cerebral hemisphere.* Brain 67:331, 1944.

81. PAULSON, MJ, AND LIN, TT: *Predicting WAIS IQ from Shipley-Hartford scores.* J Clin Psychol 26:453, 1970.

82. PETERSON, LR, AND PETERSON, MJ: *Short-term retention of individual items.* J Exp Psychol 58:193, 1959.

83. POPPELREUTER, W: *Die psychischen Schädigungen durch Kopfschuss im Kriege 1914/16.* Verlag von Leopold Voss, Leipzig, 1917.

84. PORTEUS, SD: *Porteus Maze Test.* Palo Alto, Pacific Books, 1933.

85. RASMUSSEN, T, AND MILNER, B: *Clinical and surgical studies of the cerebral speech areas in man.* In ZULCH, KJ, CREUTZFELD, O, AND GALLAGHER, GC (EDS): *Cerebral Localization.* Springer Verlag, New York, 1975, p 238.

86. RAVEN, JC: *Guide to the Standard Progressive Matrices.* The Psychological Corporation, New York, 1956.

87. RAVEN, JC: *Guide to using the Coloured Progressive Matrices.* The Psychological Corporation, New York, 1956.

88. REITAN, RM: *Validity of the Trail-Making Test as an indication of organic brain damage.* Percept Mot Skills 8:271, 1958.

89. REITAN, RM: *Correlations between the Trail-Making Test and the Wechsler-Bellevue Scale.* Percept Mot Skills 9:127, 1959.

90. REITAN, RM, AND DAVISON, LA: *Clinical Neuropsychology: Current Status and Applications.* Hemisphere, New York, 1974.

91. REY, A: *L'examen psychologique dan les cas d'encéphalopathie traumatique.* Arch Psychol 28:No 112, 1941.

92. REY, A: *L'examen clinique en psychologie.* Presses Universitaires de France, Paris, 1970.

93. ROSENTHAL, R, HALL, JA, ARCHER, D, DiMATTEO, MR, AND ROGERS, PL: *The PONS Test Manual.* Irvington Publishers, New York, 1979.

94. ROSVOLD, HE, MIRSKY, AF, SARASON, I, BRANSOME, ED, JR, AND BECK, LH: *A continuous performance test of brain damage.* J Clin Consult Psychol 20:343, 1956.

95. RUSSELL, RW: *A multiple scoring method for assessment of complex memory functions.* J Consult Clin Psychol 43:800, 1975.

96. SCHAIE, KW (ED): *Longitudinal Studies of Adult Psychological Development.* The Guilford Press, New York, 1983.

97. SEASHORE, CE, LEWIS, D, AND SAETVEIT, DL: *Seashore Measures of Musical Talents,* rev ed. The Psychological Corporation, New York, 1960.

98. SELTZER, B, AND BENSON, DF: *The temporal pattern of retrograde amnesia in Korsakoff's disease.* Neurology 24:527, 1974.

99. SHIPLEY, WC: *Institute of Living Scale.* Western Psychological Services, Los Angeles, 1946.

100. SMITH, A: *The serial sevens substraction test.* Arch Neurol 17:78, 1967.

101. Spinnler, H, and Vignolo, L: *Impaired recognition of meaningful sounds in aphasia.* Cortex 2:337, 1966.

102. Spitz, HH: *Note on immediate memory for digits: Invariance over the years.* Psychol Bull 78:183, 1972.

103. Spreen, O, and Benton, AL: *Neurosensory Center Comprehensive Examination for Aphasia.* Neuropsychological Laboratory, Department of Psychology, University of Victoria, Victoria, BC, 1969.

104. Squire, LR, Chace, PM, and Slater, PC: *Retrograde amnesia following electroconvulsive therapy.* Nature 260:775, 1976.

105. Sternberg, D: *High-speed scanning in human memory.* Science 153:652, 1966.

106. Storandt, M: *Age ability level and methods of administering and scoring the WAIS.* J Gerontol 32:175, 1977.

107. Stroop, JR: *Studies of interference in serial verbal reactions.* J Exp Psychol 18:643, 1935.

108. Sykes, DH, Douglas, VI, Weiss, G, and Munde, KK: *Attention in hyperactive children and the effect of methylphenidate (Ritalin).* J Ch Psychol Psychiatry 12:129, 1971.

109. Terry, R, and Katzman, R: *Senile dementia of the Alzheimer type: Defining a disease.* In Katzman, R, and Terry, R (eds): *The Neurology of Aging.* FA Davis, Philadelphia, 1983, p 51.

110. Warrington, EK: *Neuropsychological studies of object recognition.* Philos Trans R Soc Lond B 298:15, 1982.

111. Wechsler, D: *Manual for the Wechsler Adult Intelligence Scale.* The Psychological Corporation, New York, 1955.

112. Wechsler, D, and Stone, CP: *Wechsler Memory Scale (Manual).* The Psychological Corporation, New York, 1945.

113. Weinberg, J, Diller, L, Gerstman, L, and Schulman, P: *Digit span in right and left hemiplegics.* J Clin Psych 28:361, 1972.

114. Weintraub, S, Baratz, R, and Mesulam, M-M: *Daily living activities in the assessment of dementia.* In Corkin, S, Davis, KL, Growdon, JH, Usdin, E, and Wurtman, RJ (eds): *Alzheimer's Disease: A Report of Progress in Research.* Raven Press, New York, 1982, p 189.

115. Weintraub, S, and Mesulam, M-M: *Developmental learning disabilities of the right hemisphere: Emotional, interpersonal and cognitive components.* Arch Neurol 40:463, 1983.

116. Weintraub, S, Mesulam, M-M, and Kramer, L: *Disturbances in prosody: A right hemisphere contribution to language.* Arch Neurol 38:742, 1981.

117. Wilson, RS, Kazniak, AW, and Fox, JH: *Remote memory in senile dementia.* Cortex 17:41, 1981.

118. Wood, FB, Ebert, V, and Kinsbourne, M: *The episodic-semantic memory distinction in memory and amnesia: Clinical and experimental observations.* In Cermak, L (ed): *Human Memory and Amnesia.* Lawrence Erlbaum Associates, Hillsdale, NJ, 1982, p 167.

119. York Haaland, K, Linn, RT, Hunt, WC, and Goodwin, JS: *A normative study of Russell's variant of the Wechsler Memory Scale in a healthy elderly population.* J Consult Clin Psychol 51:878, 1983.

MENTAL STATE
ASSESSMENT OF
YOUNG AND
ELDERLY ADULTS IN
BEHAVIORAL
NEUROLOGY

123

Chapter 3

ATTENTION, CONFUSIONAL STATES, AND NEGLECT*

M-Marsel Mesulam, M.D.

If the brain had infinite capacity for information processing, there would be little need for attentional mechanisms. Imagine an ultra wide–angle lens with an infinite depth of field; the entire visual environment could be captured without any need for focusing or moving the camera. Needless to say, such lenses do not exist; one must aim the camera in a certain direction and also choose a limited focal range. The relationship between a photograph and the total visual space is somewhat analogous to the complex relationship between the contents of awareness and the vast quantities of available information.

During wakefulness, the individual is bombarded by a great many sensory signals from the outside and from within. The inexhaustible supply of memory traces and thought sequences generated by the brain itself provide an additional source of potential stimuli. It is fair to assume that the quantity of available information far exceeds the individual's information processing capacity, so that only segments of the stimulus space can be dealt with at any given time. Furthermore, the part of the stimulus space that is most relevant for achieving goals of immediate importance (and these goals could range from food scavenging to resolving deep ethical dilemmas) keeps shifting from moment to moment in a manner that reflects the inner needs of the individual, the dictates of the environment, and the experience gained in the past. There is, therefore, a need for postulating a neural mechanism that regulates the focus of consciousness. The word "attention" is used as a generic term to designate a family of

*This chapter is based on the Third Annual Denny-Brown Lecture, delivered on October 20, 1983, at the 800th meeting of the Boston Society for Neurology and Psychiatry.

The preparation of this chapter was supported in part by NINCDS grants NS-09211, NS-20285, and the Javits Neuroscience Investigator Award. I am grateful to Leah Christie and Rick Plourde for expert secretarial and photographic assistance.

such hypothetical mechanisms that selects the part of the stimulus space that is to capture the center of awareness while holding the other stimuli, which have now become potential sources of distractibility, at bay—at least temporarily.

The work of attention is part of everyday experience. The ability to focus on one out of many simultaneous conversations in a cocktail party setting or the detection of infrequent blips on a radar screen are paradigmatic activities showing attention at work. Orientation, exploration, concentration, and vigilance are positive aspects of attention, whereas distractibility, impersistence, confusion, and neglect reflect attentional deficits. It is not difficult to see why Ferrier,[41] James,[70] and Sherrington,[126] among others, could have singled out attention not only as the climax of mental integration but also as the most important prerequisite for the manifestation of intellectual and reflective powers.

Defined in this fashion, it is quite evident that attention could not possibly represent a unitary phenomenon. Furthermore, in contrast to terms such as aphasia, apraxia, or even amnesia, which have, at least potentially, technical definitions, terms used to define attentional deficits such as distractibility, confusion, and neglect are borrowed from the lay dictionary and are more resistant to strict usage. On these grounds alone one could argue that attention is no more suitable a subject for behavioral neurology than is thrift or guile. Nonetheless, attention and its disorders are inextricably lodged into the working vocabulary of neurology and psychiatry. It is therefore useful to approach this subject as if it represented a unitary faculty, not so much out of scientific rigor as out of clinical convenience.

From the vantage point of psychologic processes, the boundaries of attention intersect those of consciousness, arousal, affect, motivation, memory, and perception. The influence of wakefulness, arousal, affect, and motivation on attention are intuitively obvious. Although some might consider arousal and attention virtually synonymous, this can be debated, since attentional deficits can occur in perfectly awake individuals and since extreme levels of arousal—as in pain or terror—may impair the effectiveness of attention. Attention may also influence perceptual operations. It is common experience among neurologists that sensory examinations are virtually impossible to perform on inattentive subjects. Furthermore, alterations in critical sensory fusion thresholds can arise not only from disturbances along sensory pathways but also from variations in the level of vigilance. Intact attention is also an important requisite for many cognitive processes, especially memory. The relevance of these interactions to the testing of mental state is discussed in Chapter 2.

The study of attention has lent itself to a multitude of dichotomies (e.g., tonic-phasic, diffuse-selective, primary-secondary). In this chapter, the overall process of attention is considered as a composite of two major operations:

1. A *matrix* or *state* function, which regulates the overall information processing capacity, detection efficiency, focusing power, vigilance level, resistance to interference, and signal-to-noise

ratio. This aspect is clearly related to the concept of tonic attention and is generally associated with neural mechanisms in the reticular activating system.

2. A *vector* or *channel* function, which regulates the direction and target of attention in any one of the behaviorally relevant spaces (e.g., extrapersonal, mnemonic, semantic, visceral, and so forth). This aspect of attention is more akin to selective attention and is generally associated with more rostral elements of the neuraxis, especially the neocortex.

This *physiologic* dichotomy is largely blurred at the *behavioral* level, since most attentional behaviors eventually represent an interaction between these two components. Some behaviors, such as the detection of blips on a radar screen, are heavily dependent on the matrix aspect; whereas others, such as the ability to focus on one out of many simultaneous conversations in the environment, appear to place a greater premium on the vector aspect of attention and the related cortical mechanisms.

In the clinical practice of behavioral neurology, there are two relatively common conditions in which attentional deficits emerge as the most salient features of the clinical picture. These are the confusional states and unilateral neglect. In confusional states, the entire matrix of attention is severely impaired. In unilateral neglect, there are dramatic deficits in one vector aspect of attention—namely, the distribution of attention within the extrapersonal space. In this chapter, these two conditions are described from a clinical point of view, and a discussion of relevant topics in the biology of attention follows.

CONFUSIONAL STATES AS DISORDERS OF ATTENTION

Clinical Picture

Attentional disturbances are part of everyday experience. The behavior that follows a sudden awakening from sleep by an unexpected telephone call is a fairly common example. Although the individual may give the external appearance of being awake, it may be difficult to focus on the telephone conversation, to avoid distractibility, or to organize thought and speech coherently. It may even be difficult to recruit all the information necessary for proper orientation in time and space. What should have been simple actions, such as using a pen to write down a message, suddenly requires undue effort and may result in a great deal of fumbling and perhaps even in a few attempts at writing with the wrong end of the pen. Finding the correct word or controlling immediate but inappropriate response tendencies may prove impossible. If such an individual were administered tests that place a premium on attentional functions, the performance could be quite unflattering. Such an individual can be said to be in an acute confusional state. Naturally, no medical intervention becomes necessary, since either voluntary mental effort or a few more hours of sleep will effectively reverse all of these difficulties.

Acute confusional states may also arise as pathologic alterations of mental state, especially in patients with a wide variety of metabolic encephalopathies. In fact, it is fair to say that confusional states constitute the single most common mental state disturbance that most physicians will see. Acute confusional states are also known as delirium, organic psychoses, or acute organic brain syndromes. Since the terms "delirium" and "psychosis" have unwarranted psychiatric connotations and since the term "brain syndrome" is singularly uninformative, it is preferable to use the term "confusional state" for this group of disorders.

A confusional state can be defined simply as a change of mental state in which the *most salient deficits* occur in overall attentional tone. This does not mean that attentional deficits necessarily emerge in isolation. Indeed, patients in acute confusional states commonly have additional cognitive and behavioral disturbances. Some of these are secondary to the attentional difficulties, whereas some others may be independent. It is also important to realize that not all patients with attentional disturbances can automatically be described as being in a confusional state. For example, patients with the amnesic form of Alzheimer's disease commonly also have attentional difficulties. However, they cannot be said to exhibit a confusional state since the salient feature is amnesia rather than inattention. It should also be stressed that although most acute confusional states occur in the context of a metabolic encephalopathy, the two conditions are not synonymous. For example, some metabolic-toxic encephalopathies may lead to paranoid, delusional, hallucinatory states in which the psychotic disturbance—rather than inattention—becomes the most salient feature of the clinical picture. The term "toxic psychosis" may be more appropriate in such instances. Acute confusional states may also result from conditions other than metabolic encephalopathy, such as subdural hematoma or even stroke. The central role of the attentional disorder in confusional states has been emphasized in many clinical descriptions of this condition.[1,16,37,92] In lay usage, confusion often indicates disorientation. Many patients in confusional states are, in fact, also disoriented. However, this is not a necessary feature of the condition, and it is possible to see patients in confusional states who maintain orientation. It is the salience of the attentional deficit rather than the presence of disorientation that is the sine qua non for the diagnosis of a confusional state.

The clinical picture of a patient in an acute confusional state is familiar to most physicians. There are usually no focal neurologic signs of a motor or sensory nature, with the possible exception of a coarse tremor, myoclonus, or asterixis. Attentional deficits arise at several levels of behavior. Vigilance is defective in intensity as well as in selectivity. Attention either wanders aimlessly or is suddenly focused with inappropriate intensity, even if for only a fleeting moment, on an irrelevant stimulus that becomes the source of distractibility. Thought and skilled movement also become vulnerable to interference, impersistence, and perseveration. The patient may volunteer that "concentration" and "thinking straight" require great effort. The stream of thought loses its coherence because of the frequent intrusions by competing thoughts and sensations. Skilled-movement sequences, even those as automatic as dialing the telephone or using eating utensils, lose their coherence and show signs of disintegration, perseveration, and impersistence. Performance in attentional tasks such as the Digit Span Test, the Stroop Interference Test, and the Alternating Sequences Task (see Chapter 2) is impaired. When asked to recite the months of the year in reverse order, the patient may say, "December, November, October, September . . . October, November, December, January," showing the inability to withhold the more customary response tendencies. This clinical description highlights the three cardinal features of confusional states: (1) disturbance of vigilance and heightened distractibility, (2) inability to maintain a coherent stream of thought, and (3) inability to carry out a sequence of goal-directed movements.

Difficulties in additional aspects of mental function are also common in confusional states. Perceptual distortions may lead to illusions and even hallucinations. The patient is often, but not always, disoriented and shows evidence of faulty memory. Mild anomia, dysgraphia, dyscalculia, and constructional deficits are common. Judgment may be faulty, insight appears blunted, and affect is quite labile with a curious tendency for facetious witticism.[16,92] Some of these deficits are probably secondary to attentional difficulties. For example, if the patient is allowed sufficient drilling during the acquisition stage of a learning task, memory improves. Calculations that appear devastated when tested mentally may prove to be quite accurate when the patient is allowed the use of a pencil and paper. Other deficits of mental state, on the other hand (e.g., poor judgment and hallucinations), may be affected independently by the underlying pathogen. It is important to realize, however, that in confusional states these additional deficits are, by definition, of lesser importance than are the attentional difficulties.

Some confusional states are characterized by apathy; others, especially when related to alcohol, barbiturate, or opiate withdrawal, lead to extreme agitation. In their more severe forms, confusional states may lead to stupor and coma. This gives rise to the widely held opinion that confusional states are merely disorders of wakefulness and arousal. However, in the early stages of most confusional states attention is impaired out of proportion to the drowsiness, suggesting that the mechanisms of wakefulness and of attention need not overlap completely. Two other characteristics of confusional states are the rapid fluctuation of mental state that may occur from one hour to the next and the rather typical nocturnal exacerbation. The clinical picture of confusional states would easily lend itself to staging. No such systematic effort has yet been made.

Differential Diagnosis

It is important to differentiate confusional states from four other conditions: Korsakoff's amnesia, Wernicke's aphasia, depression, and dementia. With the possible exception of the most acute phase, patients with Korsakoff's amnesia have an excellent attention span, which sharply contrasts with that of patients in confusional states. Vigilance may be impossible to assess in patients with Wernicke's aphasia, and the speech output could give the impression of an incoherent stream of thought. However, the salience of paraphasias in spontaneous speech and the attentiveness with which the patient performs commands aimed at axial musculature (roll over, sit up, close your eyes) differentiates these patients from those with confusional states. Depressed patients also have attentional disturbances, but the changes in affect dominate the clinical picture. In contrast, the patient in a confusional state may also be transiently sad and tearful, but this is not sustained over time and the affect lacks associative depth. Dementia is merely a descriptive term that refers to a heterogeneous group of patients who have in common a progressive deterioration of cognitive and behavioral faculties. Some patients with dementia have attentional deficits as the most salient deficit of their mental state and can therefore be considered to be in a chronic confusional state. On the other hand, one could also consider acute confusional states as reversible dementias.

Causes and Mechanisms

The causes of confusional states can be divided into six major groups: (1) toxic-metabolic encephalopathies (2) multifocal brain lesions, (3) head trauma, (4) epileptic seizures, (5) space-occupying lesions, and (6) focal brain lesions.

Confusional states are most commonly caused by *toxic-metabolic encephalopathies.* The adequate function of the central nervous system obviously depends on the metabolic integrity of its constituent neurons and glia. Any condition that interferes with the nutritional requirements, acid-base balance, or electrolyte environment of these cells could interfere with nervous function. It is therefore not surprising that metabolic disturbances ranging from renal insufficiency to hepatic failure, anemia, endocrinopathies, hyperglycemia, anoxia, acidosis, alkalosis, and so forth, may each cause an encephalopathy. Withdrawal from alcohol, barbiturates, or opiates, as well as the intake of various psychoactive drugs including analgesics, hypnotics, sedatives, tranquilizers, neuroleptics, antidepressants, and even antihypertensives, can also cause a toxic encephalopathy.

In some of the toxic and metabolic encephalopathies the common denominator appears to be an interference with neurotransmitter action.[38] Toxins and drugs that interfere with cholinergic transmission are particularly apt to produce confusional states. This effect is widely recognized in the case of anticholinergic drugs used for parkinsonism. It is important to realize that a variety of medications including neuroleptics, antidepressants, and antihistamines also have marked anticholinergic activity and that this may be one reason they may cause confusional states. Since many elderly, psychiatric, or depressed patients receive combined treatment with antidepressants and neuroleptics and since antiparkinsonian drugs may be added to the regimen to prevent the extrapyramidal effects of neuroleptics, it is easy to see how such patients may be subjected to considerable anticholinergic effects. In the setting of a surgical service of a general hospital, Tune and associates[143] found that seven of eight patients who developed a postoperative confusional state had serum anticholinesterase activity higher than 1.5 picomolar atropine equivalents, whereas only four of the 17 patients who were not in a confusional state had levels in that range. Interference with central cholinergic pathways may therefore be a major mechanism for the emergence of confusional states. The importance of central cholinergic pathways to vigilance and arousal are reviewed in the section on the biology of the attentional matrix, later in this chapter.

Multifocal brain disease such as those seen in degenerative abiotrophies (e.g., Alzheimer's disease), meningitis, encephalitis, anoxia, vasculitis, disseminated intravascular coagulation,[21] and fat emoblism[35] can result in confusional states—especially in the acute period. These conditions are characterized by a myriad of small tissue lesions spread throughout the brain. A confusional state may be seen in patients with *head trauma*, either as part of a concussion syndrome or even as a fixed and chronic sequel. Patients with *epilepsy* may develop confusional states either postictally or in the course of complex partial seizures.[83] Patients with *space-occupying lesions*, especially subdural hematoma, may come to medical attention in a confusional state. Finally, a number of *focal brain lesions*, usually acquired as part of a cerebrovascular accident, can also result in confusional states. These include unilateral lesions in the parahippocampal-fusiform-lingual gyri on either side of the brain and infarcts of posterior parietal and inferior prefrontal regions in the right hemisphere.[64,88,96]

It is remarkable that so many different processes could lead to a similar clinical picture. In the case of seizures and head trauma, it is likely that the attentional difficulty results from a structural or electrical interference with ascending reticular pathways. A similar mechanism may well account for the confusional states in patients with subdural hematoma, in which supratentorial mass effects may result in brainstem compression.

One could ask why there should be such a predilection for attentional disturbances in patients with metabolic-toxic-multifocal brain disease in contrast, for example, to the relative rarity of salient aphasic and amnesic syndromes in this group of conditions. Perhaps attention has the least "safety factor," so that it emerges as the most salient cognitive difficulty whenever there is widespread disease of the nervous system. In other words, inattention could be the characteristic outcome of "minimal brain dysfunction" as long as this dysfunction is randomly distributed.

It could also be argued that attentional tone is not a unitary faculty, that it cannot be tested independently, and that its presence is merely inferred by inefficient performance in other cognitive tasks. Therefore, brain diseases that randomly interfere with the efficiency of all tasks, ranging from digit span to sequences of skilled movements, without severely disrupting any one of these individually would give the illusion that "attention" is impaired. This is an unlikely possibility, since performance in these additional tasks can be improved in confusional states by manipulating their attentional components, whereas such improvement is not obtainable when these functions are selectively impaired. For example, dyscalculia in a patient in a confusional state is frequently overcome by allowing the use of pencil and paper; however, a similar maneuver has no effect on the dyscalculia of Gerstmann's syndrome.

Although the relationship between confusional states and metabolic-toxic-multifocal conditions implies that attention might be a widely distributed function, the cases of confusional states that follow focal infarctions imply that attentional processes are also subject to anatomic localization even at the level of the neocortical mantle. For example, lesions in the high-order (heteromodal) association cortex of the right frontal or parietal lobes can cause confusional states that may sometimes be indistinguishable from those caused by metabolic-toxic-multifocal brain disease.[96] Even the toxic-metabolic encephalopathies may be affecting the brain quite selectively. For example, alcohol as well as anesthetic agents have their greatest depressant effect on the reticular formation and on high-order association cortex.[67,109] One feature common to high-order association cortex as well as to the reticular formation is that each stands at the end of a polysynaptic chain of information processing. It is conceivable that functions that depend on the most polysynaptic chains of information processing will be most vulnerable to toxic-metabolic encephalopathy and that the psychologic manifestation of this selective involvement may be a disorder of attention. It may initially seem that this explanation cannot apply to the case of multifocal brain disease in which the lesions are randomly distributed. However, on probablistic grounds alone one could argue that the longer polysynaptic chains will contain a greater cumulative effect of randomly distributed lesions. With respect to cortex, both randomly distributed brain disease and toxic-metabolic encephalopathies could therefore have their greatest impact on the physiologic function of high-order association areas. This explanation reconciles the apparent disparity of focal infarcts in high-order association areas causing clinical deficits that are similar to those seen after toxic-metabolic encephalopathies and multifocal brain involvement.

Detection, Vulnerability, and Clinical Course

The behavioral neurologist frequently comes in contact with patients in confusional states. It is therefore useful to consider several additional clinical points. An acute confusional state may sometimes be the presenting clinical sign of an underlying systemic disease. For example, we have seen patients in whom the clinical diagnosis of a confusional state prompted a thorough medical investigation that uncovered occult pulmonary empyema, heart failure, cholecystitis, meningitis, or even spinal epidural abscess. Patients in a confusional state may not give a coherent history and they may show an insensitivity to pain, which could interfere with such clinical signs as rebound tenderness or neck stiffness. When the underlying agent is unclear, an extensive search should be undertaken with appropriate blood tests, radiologic procedures, and lumbar puncture. Otherwise, reversible disease processes can be missed. When a pronounced confusional state is present and a toxic-metabolic encephalopathy is suspected, an electroencephalogram (EEG) may be helpful. If the EEG is free of slowing, a toxic-metabolic cause can virtually be ruled out and another etiology should be considered.

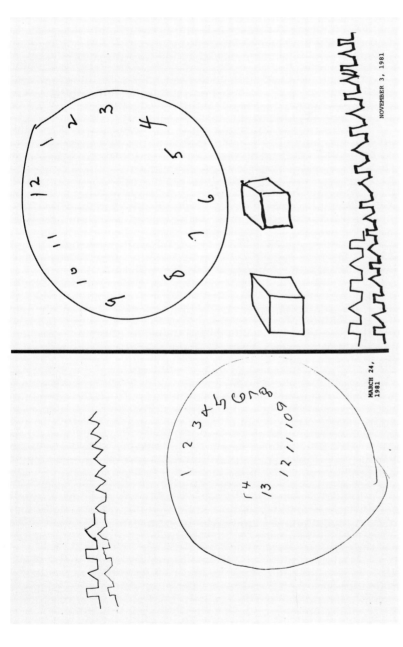

FIGURE 1. An 83-year-old man developed a confusional state in conjunction with congestive heart failure. The confusional state was severe and interfered will all daily living activities. *Left:* Performance on the day of presentation (March 24, 1981) shows some characteristic signs of a confusional state. In drawing a clock he shows poor planning of space and marked perseveration. The tracings above the clock show his performance in the alternating sequences task. The very top tracing is the examiner's model. The patient is asked to copy it and then to continue with the same pattern until he reaches the end of the page. The patient performs reasonably well during the copying phase. However, he subsequently cannot resist the immediate response tendency to repeat the segment just drawn. This results in a series of perseverative triangles instead of the alternating square-triangle pattern. *Right:* The congestive heart failure responded to treatment and disappeared in a matter of a few days. However, the improvement of the confusional state was gradual and took approximately 6 months to reach a plateau. By that time, he was driving and successfully performing customary daily living activities. His performance 7 months later (November 3, 1981) in the alternating sequences and in drawing a clock is much improved. He can even copy a cube reasonably well, whereas the task had been impossible during the acute episode.

133

There are great individual variations in the susceptibility to confusional states.[19] In general, the elderly and those with pre-existing brain disease—especially dementia—seem more vulnerable to developing acute confusional states in response even to mild metabolic stresses.

For patients in toxic-metabolic confusional states, removal of the underlying cause results in improvement of the mental state. In young patients without prior brain disease, this improvement can be rapid and dramatic. In the elderly, however, improvement may not start for days and may last for months (Fig. 1). Furthermore, some of these patients may never regain a fully normal mental state despite the prolonged improvement. It is not clear why recovery can take so long after the presumptive offending agent has been removed or why it is sometimes incomplete. Perhaps these patients suffer a certain degree of irreversible neural damage. It is also conceivable that individuals who show this pattern of extreme vulnerability and indolent recovery may have a certain component of prior brain disease (perhaps degenerative dementia) for which they have partially compensated. This does not, however, alter the basic conclusion that many of the superimposed cognitive deficits of the confusional state are reversible even if the patient may not subsequently regain a completely normal mental state.

BIOLOGY OF THE ATTENTIONAL MATRIX

General Remarks

Approximately a hundred years ago, Ferrier[41] and then Bianchi[7] observed that bilateral frontal lobe lesions in rhesus monkeys severely impaired attentiveness and curiosity. These reports did generate considerable interest in the relationship of cerebral cortex to the process of attention. However, the work of Dempsey and Morison in 1942[24] on the recruiting response and that of Moruzzi and Magoun in 1949[102] on EEG desynchronization rapidly shifted emphasis to the brainstem. Although these workers did not directly conclude that the brainstem was the seat of attention or of consciousness, such an implication subsequently dominated the literature for many years and continues to influence current thinking. For example, the landmark experiments of Broadbent[12] and Hernandez-Peón[61,62] suggested that selective attention was exerted by a peripheral filter that, under the control of the brainstem reticular activating system, inhibited unattended messages at the receptor level. These experiments have since attracted sharp criticism. For example, it has been shown that complex semantic and affective components of a verbal message strongly influence the effectiveness of selective attention in dichotic listening experiments.[98,142] Since peripheral receptors could not possibly identify such complex aspects of the message, the filter clearly needed to be placed much more rostrally in the neuraxis, perhaps at cortical levels. Today, even though vestiges of a predominantly subcortical theory of attention can occasionally still be detected, there is an emerging sense that neocortex, thalamus, and brainstem are inextricably linked in the modulation of attention and that the more complex aspects of this function are executed predominantly by neocortical mechanism.

The Reticular Activating System

Soon after the introduction of EEG recording, several independent lines of investigation established that the spontaneous electrical rhythms recorded by the EEG were very sensitive to alterations in the levels of consciousness and attention.[108] High-voltage slow waves were associated with drowsiness and certain sleep states, whereas desynchronized fast activity occurred during arousal, excitement, attentiveness, and rapid eye movement (REM) sleep. It is now generally accepted that the pacemakers for the EEG rhythms are in the brainstem reticular formation and that their effect upon cortex is largely mediated by the thalamus.[99,102] A great deal of subsequent research in this area has addressed two major questions: (1) What is the relationship between the surface EEG and the activity of subcortical pacemakers? and (2) What is the relationship between these pacemakers and behavioral states, especially attention? Although there has been considerable progress with respect to the first question, many unresolved issues are associated with the second.

Steriade and co-workers[44,136–138] found that neurons in the midbrain reticular core and in the intralaminar thalamic nuclei tend to have higher firing rates during states of EEG desynchronization (waking and REM sleep) than during slow-wave sleep. The firing rate of these neurons decreases just before the appearance of the first EEG spindles during the transition from waking to sleep. An increase in the firing of the same neurons precedes EEG desynchronization and its associated behavioral waking. Increased activity of the midbrain reticular neurons is also correlated with a facilitation of transthalamic sensory transmission and an increased depolarization of cortical output neurons. These changes conceivably could enhance not only the impact of sensory input upon cortex but also the cortical readiness for efferent responses. Thus, the midbrain reticular core is in a position to act as an essential pacemaker for the surface EEG and perhaps also for the associated behavioral states. This notion is very much strengthened by the well-known fact that bilateral midbrain reticular core lesions in man and animals leads to permanent states of stupor and coma.[112] However, the relationship between states of attention and midbrain reticular core activity is far from simple. For example, unit activity in the midbrain reticular core is high not only during wakefulness but also during REM sleep. Some have argued that REM sleep is characterized by intense attentiveness to internal rather than external stimuli. This is obviously a difficult hypothesis to test. A more cautious conclusion is that the activation of the midbrain reticular core is necessary but not sufficient for wakefulness and attention.

The intensity of the attentional tone varies widely during wakefulness. Only a handful of studies have addressed the relationship between midbrain reticular core activity and variations of attentional states during the period of behavioral wakefulness. Goodman,[46] who trained monkeys to press a bar in response to a visual cue, found that the shortest reaction times occurred when the average multi-unit activity in the mesencephalic reticular formation was within a certain range. This range was above the level of activity seen during slow-wave sleep but below the level seen in states of extreme arousal such as startle. However, not all reaction times that occurred while the midbrain activity level was within this range were necessarily short. These experiments suggested that there may be an inverted-U relationship between reticular activity and short reaction times and that reticular formation activity within the optional range is necessary but not sufficient for short reaction times. Since short reaction times may reflect heightened attentiveness, these experiments are relevant to the relationship between the neurons of the midbrain reticular formation and attention during wakefulness. Ray and co-workers[114] recorded the activity of neurons in the pontomesencephalic reticular formation of monkeys engaged in a fixed foreperiod go–no-go task. Some of these neurons showed systematic changes in activity 543 ±246 milliseconds prior to stimulation and returned to control levels after stimulation. The suggestion was made that these neurons could participate in an anticipatory set. Taken together, these studies suggest that the neurons of the brainstem reticular core may influence not only the shift from sleep to wakefulness but also the level of attentional tone even when wakefulness remains constant.

The brainstem reticular formation has many anatomic and neurochemical properties consistent with its involvement in modulating the attentional matrix for the whole forebrain. The brainstem reticular core is not an undifferentiated neuronal swamp; rather, it is composed of a number of subnuclei each with individual architectonic features, although the boundaries among subnuclei are not sharp. Reticular neurons receive collaterals from a large number of ascending and descending pathways in a way that would enable these neurons to integrate a wide spectrum of neural information.[120] Another important aspect of these nuclei is that they provide major ascending cholinergic and monoaminergic pathways to thalamic and neocortical targets.

The pedunculopontine and laterodorsal tegmental nuclei of the upper brainstem contain cholinergic cell bodies that provide the principal cholinergic innervation for thalamic nuclei and also a minor portion of the cholinergic input for the cortical mantle.[94] Acetylcholine could be considered a neuromodulator that makes neurons more responsive to the other inputs they receive.[74] Therefore, these ascending cholinergic pathways that originate in the brainstem reticular core may provide a mechanism for modulating the excitability of widespread regions of the thalamus and cortex in a way that can influence the overall information processing capacity of the nervous system. In fact, many effects of reticular formation stimulation upon cortex and thalamus are mediated, at least in part, by acetylcholine.[86,133,140,141] In animals, anticholinergic agents produce slow-wave synchronization of the EEG without, however, resulting in perceptible alterations of wakefulness.[146] In humans, anticholinergic agents can result in severe confusional states in which the entire attentional matrix is disturbed.[143]

The pontomesencephalic brainstem also provides noradrenergic and serotonergic innervation to cortex and thalamus. The serotonergic pathways originate from the nuclei of the midline raphe. The noradrenergic input originates from the nucleus locus coeruleus. The neurons of this nucleus have been shown to fire slowly during drowsiness and slow-wave sleep, although they fire in rapid bursts during REM.[17] The suggestion has been made that noradrenalin increases the postsynaptic evoked response relative to spontaneous activity, thus enhancing the signal-to-noise ratio in neural transmission.[100] Agents that increase central noradrenergic activity, such as dextroamphetamine and methylphenidate, have been shown to enhance attentiveness in cognitive tasks.[113] In animals, the interruption of the noradrenergic pathways from the nucleus locus coeruleus to neocortex impairs the ability to ignore irrelevant stimuli, thus increasing distractibility.[85] This body of neurophysiologic, anatomic, and pharmacologic evidence shows that the brainstem reticular formation and adjacent nuclei are in a unique position to influence the overall attentional matrix.

Corticoreticular Interactions: The Thalamic Relay

The setting of attentional tone is not under the monopoly of the brainstem reticular core. The thalamus and neocortex have crucial roles in this process. The most coherent framework for analyzing the relationship between cortex and the reticular activating system was proposed by Sokolov.[130] Based on an original observation by Pavlov, Sokolov described an "orienting reflex," which includes an increase in the level of arousal (EEG desynchronization, galvanic skin response, and so forth), an alignment of sensory receptors (turning of head and eyes) toward the source of stimulation, and a facilitation of neural responses to sensory stimuli. Sokolov postulated that this reflex required an interaction between neocortical and reticular components, the relative contribution of each component varying according to the nature of the stimulus. In the case of painful stimuli, for example, arousal can be elicited directly through the stimulation of the reticular core by collaterals of ascending pathways, without much need for cortical participation. On the other hand, the orienting response to a novel stimulus or to a significant word requires initial identification of novelty and significance at neocortical levels. Descending impulses from the cortex are subsequently sent to the reticular formation, which then becomes mobilized to generate the appropriate attentional and arousal tone by influencing thalamic and cortical activity. As repeated exposure to an event decreases its novelty value, an inhibitory signal originates in cortex and dampens the level of reticular core activation, thus leading to habituation. Sokolov's arguments were very much strengthened by a massive body of evidence showing that cortical lesions markedly interfered with the habituation of the orienting response.[82,130] For example, Sharpless and Jasper[125] found that the habituation of evoked responses in the mesencephalic reticular formation of cats was very dependent on corticofugal pathways, especially as the stimulus increased in complexity.

The thalamus acts as the major relay between cortex and the reticular formation. The intralaminar thalamic nuclei receive inputs from the brainstem reticular formation and then relay this information to all parts of neocortex. The reciprocal feedback control of cortex upon these ascending transthalamic pathways is mediated at least in part through the reticular nucleus of the thalamus, which receives cortical inputs and relays them to the other thalamic nuclei.[122,129,156] It should be pointed out that the brainstem core and the nucleus reticularis thalami share the designation "reticular" by accident rather than by design. According to Meyer,[97] the thalamic reticular nucleus had already been designated the "stratum reticulatum" by Arnold in 1838, whereas the brainstem core received the designation "formatio reticularis" by Deiters in 1865 and by Forel in 1877—without any implication of common function.

The reticular nucleus of the thalamus has an inhibitory influence on thalamic nuclei, including the sensory relay nuclei. The frontal lobes, and probably many other areas of cortex, can inhibit thalamocortical transmission by activating the reticular nucleus. Stimulation of the mesencephalic reticular formation, on the other hand, exerts an inhibitory effect on the thalamic reticular nucleus. This effect, which is mediated by acetylcholine, facilitates thalamocortical transmission. These characteristics suggest that the reticular nucleus may act as an attentional valve for regulating ascending thalamocortical transmission according to the integrated influence from the cortex and the reticular core.

It is important to realize that neocortex projects to all other thalamic nuclei as well, so that the reticular nucleus is not the only path through which cortex can modulate the effects of the ascending reticular pathways. The one unique property of the reticular nucleus is that it is the only thalamic nucleus that receives cortical input without sending direct projections back to cortex.[72] The physiologic significance of this anatomic feature remains to be deciphered.

Cortex, Frontal Lobes, and Attention

Although cortical attentional functions are usually attributed to high-order polymodal association areas, even primary sensory cortex participates in this process. For example, Hubel and co-workers[66] found that some neurons in primary auditory cortex reacted to sound only when the animal appeared to pay attention to its source. However, the influence of attentional variables on neural activity is relatively minor in primary areas. For example, in area 3b of the primary somatosensory cortex (S1) in monkeys, Hyvarinen and associates[69] found that the behavioral relevance of a vibratory stimulus influenced responses in only 8 percent of the sampled neurons. In keeping with this principle, the effect of attention on sensory-evoked potentials is least pronounced over the primary sensory areas of cortex (see Chapter 9). Furthermore, there is a gradual decrease of evoked responses during habituation in widespread areas of cortex but not in the primary sensory areas where the evoked responses remain unaltered.[125]

Modality-specific (unimodal) association areas are also known to participate in attentional processes. For example, single-unit recordings from the inferotemporal visual association cortex of monkeys during a delayed matching-to-sample task showed that some neurons had preferential responses to individual colors only in those trials in which color was a behaviorally relevant stimulus and therefore the focus of attention.[42] Furthermore, monkeys with lesions in this same region show learning deficits which appear to be caused by an inability to ignore irrelevant visual information.[131] A similar condition may also occur in the human. In Chapter 1, for example, a patient was described in whom excessive visual distractibility occurred in conjunction with an infarct of the visual association areas of the right temporal lobe (see Chapter 1, Fig. 7). Thus, damage to unimodal association areas may lead to modality-specific attentional deficits.

The more generalized and complex aspects of attention are coordinated at the level of heteromodal (polymodal) association cortex.[90,96] Heteromodal association areas are regions that are not devoted to any single modality and that receive convergent input not only from several sensory association regions but also from limbic-paralimbic areas (see Chapter 1). These heteromodal areas are sensitive to the more abstract features of incoming information and also to its motivational relevance. There are at least three heteromodal association fields in the primate brain: the prefrontal cortex, the posterior parietal cortex, and the ventral temporal lobe.[95,124] It is interesting that acute infarcts, even when unilateral, in any one of these three heteromodal fields (but not elsewhere) give rise to generalized attentional deficits in the form of confusional states.[64,88,96] In the case of the prefrontal and posterior parietal heteromodal fields, the confusional state occurs almost always after the occurrence of right-sided lesions. The implications of this for the hemispheric specialization of attention are reviewed in the section on hemispheric dominance in this chapter. As discussed earlier, heteromodal cortical areas may also be more susceptible to toxic-metabolic encephalopathies and this may explain, at least in part, why a confusional state is a common outcome in such conditions. In addition to this influence on overall attentional tone, heteromodal cortical areas are also involved in modulating directed sensory attention. This is discussed in the following sections.

The frontal lobe seems to be the one cortical region most intimately related to the maintenance of the attentional matrix (see Chapters 1, 2, and 9). In a series of experiments based on determinations of regional cerebral blood flow, subjects were given pairs of auditory, visual, and somatosensory stimuli.[119] The subjects were simultaneously exposed to all three modalities in each trial, but they were asked to ignore two of the channels while making a difficult sensory discrimination in only one of the modalities. Three conclusions emerged. First, the primary and unimodal areas for an individual channel were activated even in trials in which information from that modality was to be ignored. Second, the activation in the modality-specific (primary and unimodal) cortical areas for the channel in which the discrimination was being made was greater than in the other modality-specific areas. Third, a part of the prefrontal heteromodal cortex showed preferential activation during these trimodal attention tasks, regardless of the modality in which the discrimination was being made. These observations show that attentional filtering occurs at a quite advanced stage of information processing, since unimodal areas belonging to the modalities that are not attended continue to be active. This undoubtedly introduces much greater flexibility than if sensory influx were to be barred at a more peripheral level. Furthermore, there appears to be a region in the superomesial prefrontal cortex (probably part of the frontal heteromodal fields) that may be essential for the differential tuning of the attentional matrix across the entire multimodal sensorial space.[119]

Several additional lines of evidence highlight the importance of the frontal lobe to attention. For example, two attention-related components of the evoked potential, the P300 wave and the contingent negative variation, are closely related to frontal lobe mechanisms (see Chapter 9). The contingent negative variation is a surface negative potential that may reflect the intensity of readiness that a task-related warning stimulus elicits. This potential is most readily recorded over the frontal cortex.[10,18,118] Furthermore, the P300 component that is elicited in response to novel stimuli is markedly attenuated in patients with frontal lesions (see Chapter 9). These observations suggest that the frontal lobe may play a major role in the response to complex and novel stimuli (i.e., curiosity).

In addition to many other behavioral alterations, patients with frontal lobe lesions commonly show deficits in virtually all tests that are known to be sensitive to attentional disturbances. These patients may show distractibility, perseveration, impersistence, a heightened susceptibility to intrusion, and an inability to withhold immediate but inappropriate response tendencies. Performance on the Stroop Interference, Go–No Go, and the Trail-Making tests deteriorates, probably on the basis of the attentional components in these tasks (see Chapter 2). The overall span of concentration also declines and the patient may have a diminished digit span. The role of the frontal lobe in digit span was demonstrated by Risberg and associates,[115] who showed that the greatest increase in regional blood flow during a reverse digit span task occurred over the frontal-prefrontal areas. The observations reviewed in this section suggest that the frontal lobes may be particularly important in the regulation of the overall attentional tone.

UNILATERAL NEGLECT AS A DISORDER OF DIRECTED ATTENTION

Clinical Picture

An important vector aspect of attention is the ability to direct the focus of awareness toward behaviorally relevant sensory events in the extrapersonal space. Severe disturbances in this aspect of attention arise in the form of unilateral neglect in certain patients with brain damage. Since this phenomenon is more common and more severe in patients with lesions in the right side of the brain, the following clinical description will focus on the population of patients who develop neglect for the left side. Unilateral neglect probably constitutes one of the most dramatic

FIGURE 2. This figure shows two characteristic manifestations of unilateral neglect. *Left:* A spontaneous writing sample from a 55-year-old right-handed man who developed left unilateral neglect following a right temperoparietal infarct. The margin on the left side of the sheet is uncommonly wide and indicates the presence of neglect. *Right:* This clock was drawn by a 47-year-old right-handed airline pilot who developed left unilateral neglect following a subarachnoid hemorrhage. In addition to a tendency for perseveration, the clock drawing shows marked neglect of the left side.

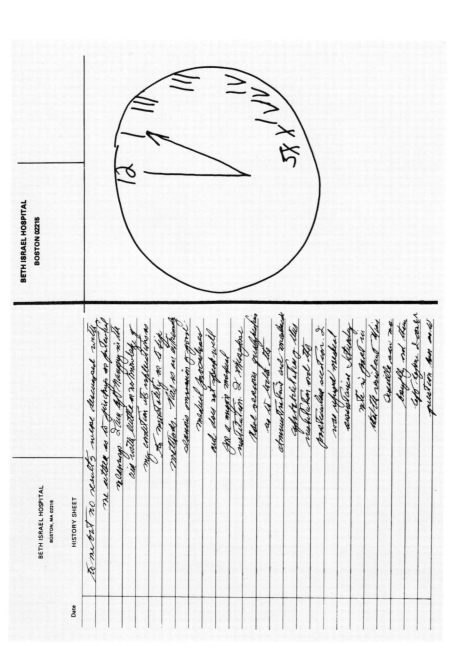

occurrences in clinical neurology. When the neglect is severe, the patient may behave almost as if one half of the universe had abruptly ceased to exist in any meaningful form. One patient may shave, groom, and dress only the right side of the body; another may fail to eat food placed on the left side of the tray; another may omit to read the left half of each sentence; still another may fail to copy detail on the left side of a drawing and may show a curious tendency to leave an uncommonly wide margin on the left side of the paper when asked to write (Fig. 2). When the examiner approaches the bed from the neglected left side and starts a conversation, the patient may briskly orient to the right side and initiate a fruitless but determined search for the source of the voice. In other patients, the unilateral neglect is much more subtle and might not be detected by observation of spontaneous behavior alone. In these cases, it may be necessary to use specialized bedside maneuvers such as bilateral simultaneous stimulation or the random letter cancellation task described in Chapter 2.

Since many patients with unilateral neglect may also have hemianopia, hemihypesthesia, or hemiparesis, the question is often raised whether the neglect merely reflects a combination of elementary sensory motor deficits. This possibility is easily dismissed on the basis of several arguments. First, it has been shown that the extent of sensory loss does not correlate with the severity of neglect.[15,43,55,153] In fact, some patients with intact visual fields may still show neglect, whereas patients with hemianopia do not necessarily neglect the blind field, as long as they can freely move their head and eyes (Fig. 3). Hemiparesis is not an essential factor either, since patients show neglect for the left side even when they are asked to use the intact right limb during tasks based on manual exploration. Since unilateral neglect may occur in the absence of any perceptible limitation of eye movements, gaze paresis is not a necessary component either. It can be concluded, then, that unilateral neglect is not a deficit of seeing, hearing, feeling, or moving but one of looking, listening, touching, and searching. It appears that this phenomenon represents a fundamental disturbance in a vector aspect of attention—namely, in the spatial distribution of directed sensory attention. Unilateral neglect is a composite phenomenon that can be analyzed from the vantage point of complex perceptual, motor, and motivational aspects.

Perceptual Aspects of Unilateral Neglect

In patients with unilateral neglect, sensory events within the left extrapersonal hemispace appear to lose their impact on awareness, especially when competing events are taking place in the other half of the extrapersonal world. The most obvious demonstration of this occurs when patients who respond perfectly well to unilateral stimulation from either the left or the right side consistently ignore the stimulation on the left under conditions of bilateral simultaneous stimulation. Especially when it occurs in more than one modality, the phenomenon of sensory extinction suggests that the left hemispace is neglected, even though the brain receives the relevant elementary sensory information.[147]

FIGURE 3. This figure shows the performance of two different patients on the random letter cancellation task. An 8 × 10 sheet of paper containing 15 As in each quadrant is placed directly in front of the patient, who is then asked to check or encircle all the As without moving the sheet of paper. *Top:* A 59-year-old right-handed woman suffered a left-sided stroke that left her with a dense right homonymous hemianopia. Despite the blind right hemifield, she does not miss any targets on the right. *Bottom:* A right-handed woman in her 70s had an infarct in the right frontal region. She developed a hemiparesis. Visual field testing did not reveal any hemianopia. However, she has marked neglect for targets on the left. These two patients demonstrate that there is no obligatory relationship between hemianopia and unilateral neglect.

Extinction can be subtle or severe, and it can be tested in a great many different ways by varying the input or output parameters. The severity of neglect can lend itself to quantification by measuring the critical difference between the size, loudness, duration, and so on, of the two stimuli which is necessary for overcoming the extinction. In some patients, the tendency for extinction may be so powerful that the mere presence of ambient visual input on the right may elicit left neglect. In such cases, even unilateral stimuli on the left may appear to be ignored, raising the possibility of a hemianopia rather than hemineglect. It may be useful to test these patients in a darkened room where brief flashes of light can be presented as stimuli. Since this eliminates ambient stimulation, it may allow the examiner to make a more definitive distinction between neglect and blindness.

In some patients with unilateral neglect, it also seems as if sensory events in the non-neglected right hemispace have an excessive impact on awareness. Some patients may keep orienting toward the right as if they were abnormally distracted by sensory events within that hemispace. Posner and co-workers[111] have shown that patients with left unilateral neglect have excessive difficulty disengaging attention from targets in the right hemispace when the task requires a leftward shift in covert orientation. There is much less difficulty with such covert leftward shifts when disengagement from a stimulus on the right side of the target is not required. It is therefore conceivable that one mechanism that underlies extinction and perhaps other aspects of neglect is the abnormal magnetism exerted by the right-sided stimulus.

It is important to realize that extinction is not necessarily always a manifestation of a hemispatial attentional deficit. For example, patients with damage to the corpus callosum have left-ear extinction for words during dichotic listening tasks.[132] In some cases, somatosensory extinction may reflect a subtle sensory loss on the side that is being extinguished.[36] Therefore, when extinction in a single modality is the only relevant finding, one needs to be very cautious before concluding that this is a manifestation of inattention or neglect.

The sensory components of unilateral neglect may well indicate that there has been a shift in the intracerebral representation of the extrapersonal space.[8,90] Under normal circumstances, it is reasonable to assume that each segment of the extrapersonal space has potentially equal representation in the brain and that there are synaptic mechanisms that flexibly alter the valence of those segments that harbor relevant events. For example, the representation for the space behind the body is relatively muted in the course of most daily activities. However, while driving a car this part of the extrapersonal space acquires as much behavioral relevance as any other segment of the environment. It is conceivable that under such circumstances the inner representation is altered in a manner that devotes more synaptic activity to the space behind the body, requiring constant updating through the rear-view mirror. Such task-related shifts in the inner representation are of obvious behavioral value. In patients with left unilateral neglect, it appears that the inner representation becomes permanently skewed toward the right hemispace without the possibility of flexible leftward shifts. For example, when patients with

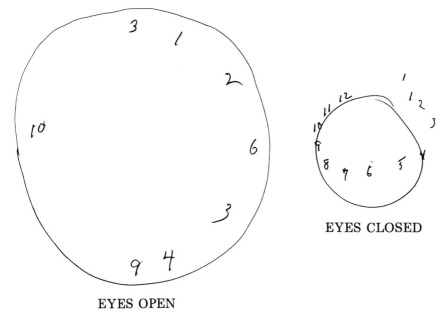

EYES CLOSED

EYES OPEN

FIGURE 4. A 60-year-old right-handed man had a right frontoparietal stroke that left him with left unilateral neglect and severe aphasia. The clock on the left was drawn with the eyes open and shows neglect of the left. The clock on the right was drawn with the eyes closed and shows a marked reduction of the neglect.

left neglect are asked to close their eyes and point toward the body midline, they usually point right of midline.[51] This may well illustrate the shrinkage of the left hemispace in the context of the representation for the entire extrapersonal space. It is interesting that aphasic patients with comparable lesions in the left hemisphere do not show this bias in pointing, suggesting that this behavior may be specific to the neglect syndrome and not a generalized outcome of all comparable brain lesions.[51]

The formation of an internal representation for the extrapersonal world probably requires the complex integration of input from all sensory modalities. It is therefore conceivable that some manifestations of neglect reflect a disturbance at the level of this multimodal integration itself. For example, we have seen several patients whose neglect in drawing a clock improved when the task was performed with the eyes closed, under proprioceptive guidance alone (Fig. 4).

Motor Aspects of Unilateral Neglect

The effective distribution of attention also requires a strategy for orienting, scanning, and searching within the extrapersonal space. These motor aspects of attention are also deficient in patients with unilateral neglect. There is a reluctance to scan and explore the left hemispace even in the absence of obvious gaze paresis. This scanning bias is probably responsible, at least in part, for the left hemialexia seen in these patients. The contribution of scanning difficulties to neglect can be demonstrated with the aid of a letter cancellation task. When the letters are randomly oriented, the individual must impose a systematic scanning strategy for

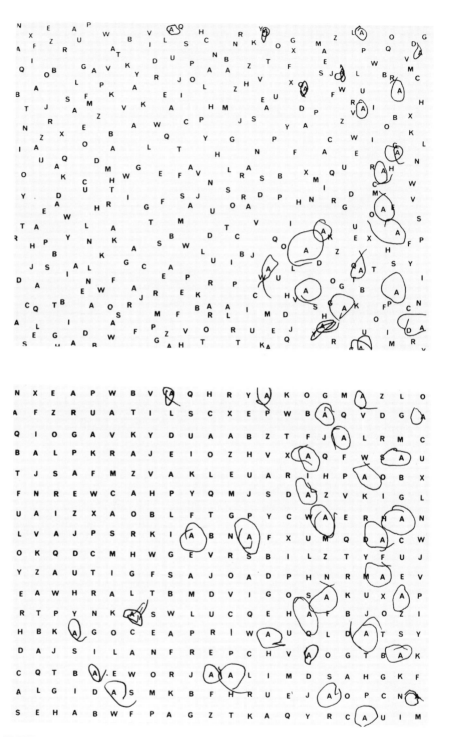

FIGURE 5. This shows the random letter cancellation task performance of a 40-year-old man who sustained a severe head injury that damaged widespread frontoparieto-occipital regions of the right hemisphere. The accident occurred 7 years prior to this testing. *Top:* Performance in the random letter cancellation task shows severe left neglect. *Bottom:* When the same number of letters are arranged in an orderly fashion, performance improves.

effective performance. Most normal individuals start at the upper left corner of the random array and systematically scan in horizontal rows or vertical columns without spatial bias. In this same task, patients with left hemineglect miss most of the targets on the left side of the page (Fig. 5). However, if the letters are arranged into rows and columns, performance frequently improves and even some targets left of the midline begin to be detected accurately (Fig. 5). This improvement occurs probably because the ordering of the letters places a lesser premium on systematic scanning, thus indirectly showing that scanning deficits contribute to neglect behavior.

Patients with left unilateral neglect also show difficulties with the manual exploration of the environment. If such patients are blindfolded and asked to search manually for small objects on a table placed in front, performance is intact on the right side of the table not only with the right hand but also with the left hand in cases where there is no left hemiparesis.[90] However, manual search becomes haphazard and ineffective on the left side of the table even when the intact right hand is being used. These observations suggest that there are complex, high-order deficits for exploratory behavior within the left hemispace regardless of the limb that is being used. This also implies that at certain stages of neural integration, the motor programs are organized not according to the muscle groups that are being activated but according to the hemispace within which the movement is to be discharged.[2,57,90]

The left and right somatocentric hemispaces are separated by a sagittal plane, which passes through the midline of the body. In the visual modality in which the axis of gaze need not be aligned with the body axis, there is often an interaction between neglect for hemifield and neglect for hemispace. For example, when the eyes look to the right, the part of the right hemispace that falls within the left visual field is also neglected. This effect can be demonstrated clinically. For example, when the random letter cancellation task is placed in front of the patient, the eyes usually start by fixating straight ahead or even slightly left of midline, and the targets right of this initial fixation are detected accurately as the eyes move rightward. However, the bias in the internal representation of the test sheet, the fact that more and more of the test sheet becomes neglected as the eyes move rightward, and the reluctance to scan leftward collectively result in a failure to return fixation all the way back to the starting point. The second fixation point is therefore slightly rightward of the first. This rightward drift of the starting point gives rise to the characteristic diagonal line of demarcation between the area of neglect and that of accurate detection (Fig. 6, top). This interaction between the sensory and motor aspects of this syndrome provides another explanation for the failure to eliminate neglect just by placing the test sheet on the right side of the patient. Although this may normalize performance at the beginning, the diagonal effect becomes gradually established, since parts of the test sheet enter the left hemifield as scanning proceeds rightward. Asking the patient to look to the left is also of limited benefit, since a similar diagonal will soon be established as the leftward excursion of the eye becomes gradually more limited.

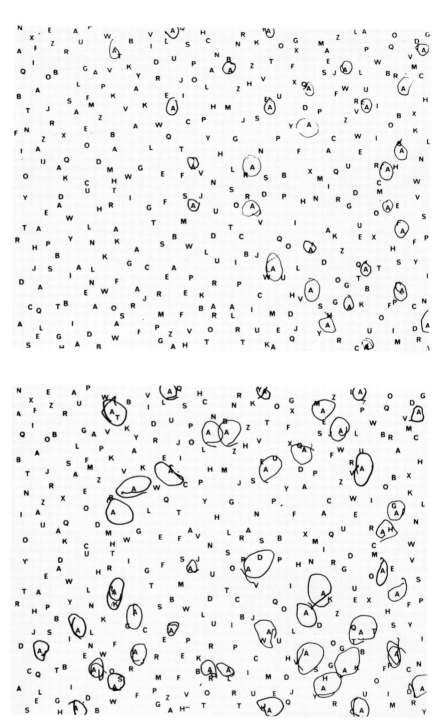

FIGURE 6. Same patient as in Figure 5, but during a different testing session, approximately a year later. *Top:* The performance in the random letter cancellation task shows the diagonal effect. *Bottom:* Following the performance shown on the top, the patient was promised 1 penny for each correct detection. The bottom figure shows how this incentive diminishes the neglect. Performance a few minutes later was back to baseline, so that practice effects are unlikely to account for this difference.

Motivational Aspects of Unilateral Neglect

It is intuitively evident that a major role of any attentional mechanism is to focus on behaviorally relevant events. In the normal individual, each segment of the extrapersonal space is potentially capable of harboring events of motivational importance. Patients with unilateral neglect, on the other hand, behave not only as if nothing were actually happening in the left hemispace but also as if nothing of any importance could be expected to occur there. This motivational aspect of neglect can be demonstrated indirectly with the letter cancellation task. For example, a patient with severe left neglect in the letter cancellation task showed marked improvement in detecting targets on the left when he was provided a reward of one penny for each accurate detection (Fig. 6, *bottom*). Although there are several alternative explanations for this improvement, it is also possible to attribute it to a reward-induced enhancement in the behavioral relevance of the left hemispace. This observation implies that part of the neglect behavior could arise from a motivational devaluation of sensory events on the left.

Causes and Course of Unilateral Neglect and Relationship to Other Components of the Right-Hemisphere Syndrome

Unilateral neglect arises as a consequence of focal intraparenchymal lesions of the brain or as an ictal manifestation in patients with seizures.[52] The location of these lesions and their relevance to the biology of attention are discussed in the following section on the anatomy of unilateral neglect. Metabolic-toxic encephalopathy, subdural hematoma, head injury, multifocal brain disease, which are the major causes of confusional states, almost never give rise to unilateral neglect.

In those cases in which unilateral neglect is the outcome of stroke, most patients recover. Life table analysis indicated that the median time required for 50 percent of the patients to recover from neglect behavior varied from 9 to 43 weeks.[63] In our experience, the most persistent cases of neglect occur with subcortical lesions in the basal ganglia, thalamus, or white matter. In patients who do recover, the mechanism for the restitution of function is not entirely clear. It is conceivable that other areas of the brain assume some of the lost function. Experimental evidence in rhesus monkeys that develop neglect after prefrontal cortex lesions shows (1) that the period of neglect is associated with metabolic depression in additional cerebral areas that may be synaptically related to the area of destruction and (2) that restitution of function may correspond to the normalization of metabolism in these distant regions.[31] The nature of the behavioral deficits may give the impression that unilateral neglect would be amenable to behavioral modification. In individual cases, a number of behavioral strategies may indeed improve performance.[33] However, it is the impression of many clinicians that these behavioral gains are rarely generalized to everyday performance.

Left unilateral neglect commonly occurs in conjunction with other behavioral manifestations that are commonly attributed to right-hemisphere injury. These include denial of illness (anosognosia), constructional deficits, and dressing difficulty (apraxia). The correlation coefficients are 0.64 between extinction and apraxia, 0.46 between extinction and anosognosia, and 0.4 between extinction and defective block designs.[62]

ANATOMY OF UNILATERAL NEGLECT

The Parietal Lobe and Neglect

Unilateral neglect has been described in patients with lesions of posterior parietal cortex, lateral prefrontal cortex, the cingulate gyrus, the striatum, and thalamus. Although this multiplicity of responsible lesions may suggest the absence of topographic specificity, close examination suggests that each of these areas belongs to a complex but well-delineated neural network.[90]

Influential papers on unilateral neglect by Brain,[11] Patterson and Zangwill,[107] and McFie and co-workers[87] designated the parietal lobe as a principal site of damage in cases of left neglect. However, more precise localization was hampered because these patients had either multilobar infarcts, head trauma, or large neoplasms. Hécaen and associates[50] subsequently described the emergence of unilateral neglect in patients who underwent excision of the right inferior parietal lobule for the control of focal epilepsy. This localization acquired even greater certitude when unilateral neglect was described following infarctions in the region of the inferior parietal lobule in individuals without a history of prior neurologic impairment.[22,27,55] Recent evidence, however, suggests that the neglect that results from parietal lobe damage alone may be quite subtle and that lesions in the *superior* rather than in the inferior parietal lobule may be more closely associated with neglect behavior.[111] With respect to clinical features, patients with parietal lobe damage tend to have prominent multimodal extinction when presented with bilateral simultaneous stimulation;[55] they neglect the internal sensory representation of the left extrapersonal space;[8] and they have difficulty with leftward shifts of covert attention.[111]

Considerable insight has been acquired on the physiology of parietal neglect with the help of an animal model for this syndrome. It is important to realize, however, that homologies between cerebral areas of different species are always difficult to establish. For example, in the monkey brain, the entire *inferior parietal lobule* carries the Brodmann designation of area 7. The caudal part of area 7 in the monkey is also known as area PG of the von Economo nomenclature, while the rostral part of area 7 is known as area PF. In the human brain, however, area 7 is used to designate the *superior* parietal lobule and is used synonymously with the von Economo designation of PE. The inferior parietal lobule in the human contains area PG (angular gyrus) caudally and area PF (supramarginal gyrus) anteriorly. In the human brain, PG and PF correspond to areas 39 and 40 in the Brodmann nomenclature, not to area 7. It is interesting that Brodmann's map of the monkey brain does not contain any regions designated with the number 39 or 40. In the following discussion, common usage will be followed in designating the caudal inferior parietal lobule of the monkey as area PG or, alternatively, as area 7. It should be realized, however, that the homologous part of the human brain could correspond to area 7 in the *superior* parietal lobule, area PG in the *inferior* parietal lobule, or both (see Chapter 1). Whatever the exact cytoarchitectonic homologies may be, it is known that rhesus monkeys also develop neglect in the form of multimodal contralateral extinction following unilateral lesions that involve the posterior parietal cortex and immediately surrounding regions.[28,53]

It has been possible to record the activity of single neurons in this part of the brain in awake and behaving animals. Neurons in the inferior parietal lobule, especially in its PG sector (caudal part of area 7), have been shown to increase their firing rates when the animal detects, looks at, or reaches toward a motivationally relevant object such as food when hungry or liquid when thirsty.[13,68,81,103,117] Most of these neurons are responsive to stimuli in the contralateral hemifield even though some ipsilateral representation is also present. These neurons are not as responsive if the visual event has no motivational significance or if equivalent eye and limb movements are performed passively or spontaneously rather than being directed toward the stimulus.[81] Additional observations show that the sensitivity of these neurons to behavioral relevance is not contingent on the presence of movements directed toward the stimulus. This was demonstrated by training monkeys to maintain central fixation while a spot of light appeared in the peripheral fields. In some trials, detection of subsequent dimming of the peripheral spot was rewarded while in other trials, identically placed spots had no such behavioral relevance. Even in the absence of any head and eye movements, the response of these neurons to the onset of the light spot was much more vigorous when reward was made contingent on the detection of subsequent dimming.[13] Thus, neurons in this part of the brain are sensitive to shifts in behavioral relevance even when all other variables remain constant. This body of evidence based on single-unit recordings strongly suggests that the inferior parietal lobule of the monkey, especially its more caudal PG sector (area 7), contains neurons that are responsive not only to sensory aspects of events in the extrapersonal space but also to their motivational relevance and perhaps even to the likelihood that they will become targets for subsequent manual or visual grasp and exploration. If similar neurons exist in the human brain, an increase in their activity may well correspond to the psychologic state of heightened attentiveness. Unilateral damage to these neurons could give rise to contralateral neglect, since the organism would have lost a neural mechanism for registering the motivational impact of sensory events and also for increasing the likelihood that they will become the targets of subsequent behavior.

These experiments based on single-unit recordings also indicated that neurons in the PG sector of the inferior parietal lobule obey three realms of contingencies. First, there is a sensory contingency since these units do recognize complex objects. Second, there is a motivational or limbic contingency in that these units are sensitive to the behavioral rel-

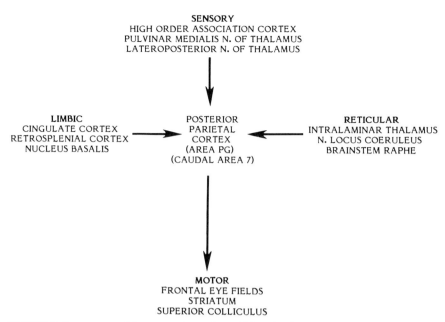

SENSORY
HIGH ORDER ASSOCIATION CORTEX
PULVINAR MEDIALIS N. OF THALAMUS
LATEROPOSTERIOR N. OF THALAMUS

LIMBIC
CINGULATE CORTEX
RETROSPLENIAL CORTEX
NUCLEUS BASALIS

POSTERIOR
PARIETAL
CORTEX
(AREA PG)
(CAUDAL AREA 7)

RETICULAR
INTRALAMINAR THALAMUS
N. LOCUS COERULEUS
BRAINSTEM RAPHE

MOTOR
FRONTAL EYE FIELDS
STRIATUM
SUPERIOR COLLICULUS

FIGURE 7. A summary of the connections of area PG in the rhesus monkey.

evance of the sensory events. Third, there is a motor contingency in that neural activity is increased when the sensory event becomes the target of looking or reaching behavior. These three contingencies parallel the three behavioral components that can be seen in patients with unilateral neglect. Neuroanatomic investigations have shown that the overall connectivity pattern of the PG region is quite consistent with these three behavioral and physiologic components.[95]

Area PG has remarkably selective connections, which can be subdivided into the four major categories shown in Figure 7. With respect to sensory information, this area receives its input exclusively from cortical regions that are best described as high-order heteromodal association areas.[95] Thus, area PG could be considered an association area for high-order association areas. The major thalamic input originates from two association nuclei of the thalamus: the medial pulvinar nucleus and the lateroposterior nucleus. Before it can have access to area PG, incoming sensory input must therefore be processed first in primary sensory cortex, then in unimodal association cortex, and then in polymodal cortex. Few, if any, other cortical areas depend exclusively on such extensively preprocessed information.[91] This may allow neurons in PG to have an abstract overview of the entire extrapersonal space. A distortion in the dynamic aspects of this representation could lead to some of the manifestations in unilateral neglect. For example, Denny-Brown and co-workers[28] attributed many components of parietal neglect to a phenomenon that was designated as amorphosynthesis. The connectivity pattern of PG suggests that this area could well be a major site for the complex multimodal integration that the concept of morphosynthesis appears to entail.

The Limbic Connection in Neglect

The observation that many PG units were sensitive to motivational relevance was somewhat puzzling until it was shown that area PG receives massive neural input form limbic areas including the cingulate gyrus, the retrosplenial area, and the nucleus basalis.[90,91,93,95] Projections from the nucleus basalis are also distributed to virtually all other cortical areas and could influence the state of the entire cortical mantle. However, the cingulate and retrosplenial projections are much more selective and may be related to more complex and learned aspects of motivation. The convergence of this limbic input form the cingulate-retrosplenial cortex with the extensively preprocessed sensory information may allow PG neurons to recognize motivational relevance in complex sensory events.

In cats, stimulation of several brain regions causes a cessation of spontaneous activity and the onset of searching head and eye movements toward the contralateral side.[71] The areas from which these "attention" responses are most consistently obtained include the cingulate region. In monkeys, unilateral lesions of the cingulum bundle and adjacent cingulate cortex result in contralateral somatosensory extinction.[149] In the human, medial frontal infarcts focused around the cingulate region of the right hemisphere have given rise to typical unilateral neglect syndromes for the left hemispace.[59] Such lesions may yield neglect, because they interfere with the sensory-limbic convergence in area PG.

The Reticular Component

The intralaminar thalamic nuclei, the brainstem raphe nuclei, the nucleus locus coeruleus, and probably also the pontomesencephalic cholinergic nuclei of the reticular formation project to area PG as well as to virtually all other parts of cortex (see Chapter 1). The role of these structures in the ascending reticular activating system were reviewed in the section on the biology of the attentional matrix. It is conceivable that this input regulates the overall attentional tone in area PG as well as in other cortical regions. As discussed in the introductory section of this chapter, a vector function such as directed attention operates on a background provided by the underlying matrix of attentional tone. Unilateral lesions in the intralaminar nuclei and even in the mesencephalic reticular formation may produce contralateral neglect in the cat and in the rhesus monkey.[104,150,151] Perhaps such lesions lead to unilateral neglect because they interfere with reticular activation in one hemisphere.

The Motor Connection and Frontal Neglect

The pattern of neural inputs shown in Figure 7 provides considerable insight into the type of information processing that occurs in area PG. However, in order to understand how this information may be transformed into action, it is also necessary to examine the pattern of neural outputs that emanate from area PG. Many outputs of this area are reciprocally directed to sources of input and are undoubtedly important for feedback regulation. Three of the neural outputs from area PG—those to the striatum, to the superior colliculus, and to the frontal eye fields—are worth emphasizing. These three regions, especially the superior colliculus and frontal eye fields, are crucial for the modulation of visual scanning, for the motor aspects of orienting, for exploratory behavior with head and eyes, and perhaps also for reaching movement with the limbs.[32,45,121,135,139]

In both man and monkey there is a cortical area just rostral to premotor cortex in which stimulation results in contralateral head and eye deviation. This area has been called the frontal eye fields. The frontal eye fields and the superior colliculus contain units that could be considered command neurons for eye movements.[45,154] Combined lesions in both of these areas yield a severe depression of contralaterally directed saccades.[121] It has been suggested that the superior colliculus may mediate the foveation of peripheral stimuli, whereas the frontal eye fields may mediate the internal planning and spatial organization of exploration with the head and eyes and perhaps also with the limbs.[20,90] The region of the inferior parietal lobule is one of the few cortical areas that has direct projections to both the superior colliculus and the frontal eye fields.[75,89]

Single-unit studies in monkeys show that frontal eye field neurons give a burst of activity just before a visually guided saccade. The mere appearance of the stimulus without eye movements or spontaneous saccades not directed toward a relevant object do not elicit such bursts. These neurons have relatively large and mostly contralateral visual fields. Furthermore, the coordinates of eye movements are spatially organized within the frontal eye fields, so that the direction of a saccade that will occur upon microstimulation of a particular neuron can be predicted by mapping its visual field.[45] It appears, therefore, that neurons in the frontal eye fields may coordinate the spatial distribution of scanning eye movements toward relevant sensory events in the visual space. Indeed, bifrontal lesions, which probably include the frontal eye fields in humans, severely disturb the ability to systematically scan complex visual scenes even in the absence of any obvious gaze paresis.[90]

Unilateral lesions that include the frontal eye fields can cause profound contralateral neglect in humans as well as in monkeys.[7,127] It has been our experience, in fact, that such lesions result in more dramatic neglect behavior than comparable lesions confined to posterior parietal cortex. There may be a tendency to attribute this neglect behavior to head and eye deviation that results form a primary weakness of neck or eye muscles. However, clinical observation would suggest that in most cases the head and eye deviation is a consequence of the inattention rather than a manifestation of weakness.[6] Furthermore, gaze paresis and head deviation are usually confined to the acute period (several days) following cerebral damage, whereas neglect behavior may last much longer. Sometime the visual neglect that follows such frontal lesions may be so pronounced that it may clinically look like a hemianopia. The clinical picture of frontal neglect is not yet fully worked out. It appears that these patients share many features with patients who have parietal neglect, even though some systematic differences are also beginning to emerge.[56,111] One might have expected that frontal neglect would have more of the motor components of neglect behavior, whereas parietal neglect would be characterized by a predominance of sensory features. We do not have enough information to know if such distinctions can be made in the human.

In monkeys, lesions in the area of the frontal eye fields have been known to result in marked contralateral neglect.[7,73,151,152] These animals do not orient toward the contralateral hemispace. They also fail to retrieve motivationally relevant objects from the neglected side even with the intact arm ipsilateral to the lesion. This shows that these animals have a hemispatial attentional deficit rather than just a hemiparesis. As in humans, the lack of response to events in the neglected hemispace may be so profound that it may be difficult to make a distinction from hemianopia.[73] In contrast to monkeys with parietal lesions, which show neglect mostly under conditions of bilateral simultaneous stimulation, monkeys with frontal eye field lesions consistently show poor orientation and exploration within the hemispace contralateral to the lesion even in the absence of competing stimuli from the other side. In this respect, there is some support for a sensory-motor dichotomy between the parietal and frontal neglect syndromes in monkeys. However, this dichotomy is not absolute, because there are motor components in parietal neglect[145] and because monkeys with frontal neglect may act as if they were blind in the field contralateral to the lesion.[78]

The anatomic connectivity of the frontal eye fields is quite consistent with its role in directed attention. This region has efferent projections to premotor cortex as well as to the superior colliculus and subthalamic nucleus.[3,89] These projections provide direct access to pathways that control head, eye, and limb movements necessary for scanning and exploratory activities. Up to 51 percent of all the afferent input into the caudal portion of the frontal eye fields originates in unimodal visual association areas in the peristriate and inferotemporal regions.[5] This pattern suggests that the frontal eye field region may be profoundly influenced by (and probably also profoundly influences) visual information at a relatively early stage of analysis. This may explain why a lesion so far removed from traditional visual pathways may appear to result in a major disturbance of visually guided behavior. Parts of the frontal eye fields also receive auditory input, and this connection may mediate orientation to auditory stimuli.[5] Furthermore, the frontal eye fields receive extensive limbic inputs from the cingulate cortex. This connection may be important in directing exploratory movements toward motivationally relevant segments of the extrapersonal space. It is interesting that area PG and the frontal eye fields receive inputs form overlapping groups of cingulate neurons.[5,95] This arrangement would ensure that the frontal and parietal regions that are important for attention receive similar information about the distribution of motivational relevance.

Subcortical Neglect and Phylogenetic Corticalization of Attention

Unilateral neglect in the human has also been observed after lesions of the thalamus.[14,123,148] It is not yet clear which of the many thalamic nuclei are responsible for the resultant neglect. Patients have been described in whom the medial dorsal and medial pulvinar nuclei were at the focus of the thalamic involvement. The intralaminar thalamic nuclei were probably also involved in these cases. It is interesting to note that the medial pulvinar is the major thalamic nucleus for area PG while the medialis dorsalis provides the major thalamic input to the frontal eye fields.[5,95] It remains to be seen if the thalamic nuclei that give rise to neglect when damaged are predominantly those that have direct connections with area PG and the frontal eye fields.

Unilateral striatal damage has also been associated with contralateral neglect in humans[80,49] and in cats.[48] Interrupting the dopaminergic nigrostriatal pathway has also been reported to yield contralateral neglect in cats and rats.[40,79,84] The motor aspects of the nigrastriatal neglect has been stressed by Ljundberg and Ungerstedt,[79] who point out that the animals have a deficit in orienting to the contralateral side even if significant events are presented unilaterally. On the other hand, hemispatial neglect has not yet been described in humans with unilateral involvement of the nigrostriatal system, such as patients with unilateral parkinsonism.

Although subcortical lesions can cause deficits of directed attention in humans, there is also evidence for a phylogenetic trend toward corticalization of this function. In the cat, for example, prefrontal lesions do not cause neglect, whereas they do in monkeys and humans. Furthermore, the parietal neglect in the monkey is mild compared to the neglect that is seen after equivalent lesions of the human brain. There may also be a trend toward a decrease in the role of subcortical structures for directed attention. For example, unilateral superior colliculus ablations in the cat result in marked contralateral neglect, whereas equivalent lesions in the monkey yield only subtle delays of contralateral saccades but no neglect.[134,155] Perhaps the severe neglect syndromes observed in patients with subcortical lesions largely reflect the destruction of fiber bundles that supply and interconnect the cortical areas involved in spatial attention.

A Neural Network for the Distribution of Attention

The information reviewed in this section shows that the areas that cause neglect when damaged are not randomly distributed, and that a direct connection with posterior parietal cortex and perhaps also with the frontal eye fields is a shared characteristic of all such regions (see Fig. 7). In the monkey brain, the part of the parietal lobe most closely associated with directed attention is within area PG (caudal part of area 7) of the inferior parietal lobule. Although similar patterns of connectivity with analogous behavioral implications are likely to exist in the human brain, it is not yet clear whether the comparable posterior parietal component is in the superior or in the inferior parietal lobule.

None of the connections shown in Figure 7 are unique to area PG. However, the overall pattern is remarkably well suited for the process of directed attention. First, the convergence of polymodal sensory input with limbic information may allow neurons in PG to shift motivational valence across the sensory space. Second, the reticular input may provide the necessary arousal tone. Third, the motor output may provide access to the mechanisms that are necessary for orientation, search, and exploration.

It is interesting that only three regions in the entire cortical mantle of the primate brain consistently lead to unilateral neglect when damaged: the frontal eye fields, the inferior parietal lobule, and cingulate cortex. These three cortical areas are also reciprocally and monosynaptically interconnected with one another.[5,95,106] These observations have led to a hypothetical model according to which the distribution of directed attention is coordinated by a neural network that contains three independent but interacting representations of the extrapersonal world (Fig. 8) One such representation, centered on posterior parietal cortex, may contain a sensory template of the extrapersonal world. The frontal eye fields and the adjacent polymodal cortex may contain a motor map for the distribution of orienting and exploratory movements within the extrapersonal space. A third representation, perhaps centered on cingulate cortex, may contain a map for the distribution of expectancy and relevance. Each of these three representations receives a set of reticular inputs that provides a matrix for the attentional tone. Furthermore, each of the three representations have specific connections with the striatum and thalamus.

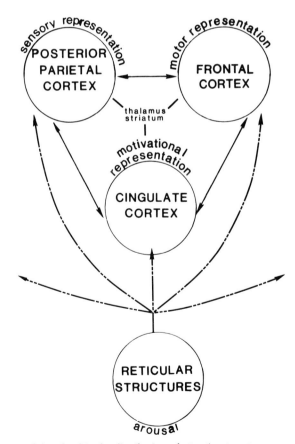

FIGURE 8. A network involved in the distribution of attention to extrapersonal targets.

Individual segments of the extrapersonal space and events within them might have a neural code in all three of these maps, and this could determine the distribution of directed sensory attention. In the monkey brain, it appears that the three representations in each hemisphere are responsive mostly to the contralateral hemispace even though considerable ipsilateral representation is also present. In the human, however, there are marked hemispheric asymmetries, and these will be discussed in the next section.

According to this hypothetical mechanism, the effective distribution of attention within the extrapersonal space requires a flexible interaction among the three representations shown in Figure 8. Damage to any one of these, to their interconnections, or to the thalamic and striatal regions with which they are connected gives rise to unilateral neglect. It is conceivable that the clinical flavor of the resulting neglect syndrome in each case could reflect the anatomic specialization that characterizes the site of involvement. However, it is unreasonable to expect rigid distinctions, since the components of this network are tightly interconnected.

It is also important to stress that the three cortical regions in Figure 8 have other behavioral affiliations in addition to directed attention. This is consistent with the network approach to cortical localization, which states that each cortical area is a component of several independent but intersecting neural networks, each network subserving a different complex behavior (see Chapter 1).

Neglect of Body Surface and Motor Neglect

The preceding discussion has concentrated on directed attention to sensory targets within the extrapersonal space. However, many patients with unilateral neglect, especially after right hemisphere lesions, also neglect the left side of the body. Sometimes this leads to denial of left-sided hemiparesis and, on occasion, even to the delusion that the paralyzed limb belongs to somebody else. This is usually known as anosognosia.[4] In addition to a map of the extrapersonal space, the parietal lobe probably also contains a representation of the body surface (body schema). Lesions that include the site of this representation may be responsible for neglect of the body and anosognosia.

Another phenomenon often associated with right hemisphere lesions is motor neglect (negligence motrice), which is used to describe a reluctance to use the contralateral limbs even in the absence of discernible weakness. Specific commands to use that limb usually overcome this resistance. This type of motor neglect, also known as hypokinesia, has been described with lesions of the striatum, of the lateral thalamus, and of frontoparietal cortex.[59,76,77]

It is important to stress that this motor neglect is confined to the limb contralateral to the lesion, regardless of the hemispace within which the movement is to be discharged. Furthermore, some patients with this kind of hypokinesia may have no neglect for the extrapersonal space.[144] Thus, whereas hypokinesia may represent another aspect of neglect for the body, it is not a necessary or even essential feature of neglect for the extrapersonal space.

HEMISPHERIC DOMINANCE FOR DIRECTED ATTENTION

It has been suggested that the right hemisphere in many animal species is more specialized for spatial functions, including attention.[25] In subhuman species this difference is subtle at best. In the human, on the other hand, there are marked hemispheric asymmetries with respect to directed attention. Definitive evidence now leads to the conclusion that left-sided neglect after right-hemisphere lesions is more common, more severe, and more lasting than right-sided neglect following lesions in the left hemisphere.[15,26,29,43,105] This has led to the assumption that the right hemisphere is dominant for the process of directed sensory attention. A simple model based on three assumptions can account for this specialization: (1) The intact right hemisphere may contain the neural apparatus for attending to both sides of space; (2) In contrast, the left hemisphere is almost exclusively concerned with attending to the contralateral right hemispace; and (3) More synaptic activity is devoted to all attentional tasks in the right hemisphere than in the left. According to this model, unilateral lesions of the left hemisphere are unlikely to yield neglect, since the intact right hemisphere may take over the task of attending to the right side. Right-hemisphere lesions, on the other hand, will result in left unilateral neglect, since the intact left hemisphere lacks the neural mechanisms for ipsilateral attention (Fig. 9).

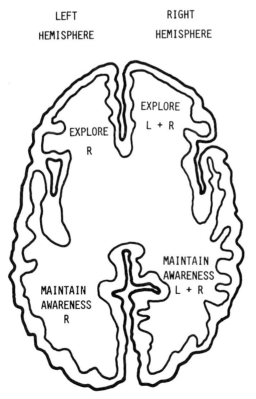

LEFT HEMISPHERE   RIGHT HEMISPHERE

EXPLORE L + R

EXPLORE R

MAINTAIN AWARENESS L + R

MAINTAIN AWARENESS R

FIGURE 9. A model for the right-hemisphere specialization in directed attention.

Indirect evidence in favor of this model has been gathered in normal subjects form various lines of investigation. In favor of the first two assumptions, it has been shown that event-related potentials and EEG desynchronizations are recorded in the left hemisphere only after stimulation of the right hand, whereas the right hemisphere shows these changes after stimulation of either hand.[30,58] In another line of investigation, positron emission tomography with 18F labeled 2-deoxyglucose indicated that the right side of the brain showed more activation than the left in response to behaviorally relevant visual stimulation, even though the stimulus was centrally placed (see Chapter 10).

It is also interesting to note that unilateral posterior parietal or prefrontal lesions on the right side of the brain (but not on the left) give rise to confusional states.[39,47,96,110] There are at least two possible explanations for this. One explanation is that the left hemisphere in some individuals has a negligible participation even in attentional tasks directed to the right hemispace. In such individuals, unilateral lesions in the right hemisphere may lead to deficits of directed attention in the entire extrapersonal space, leading to a clinical picture that looks like a confusional state. Alternatively, it is conceivable that the right hemisphere has greater overall influence on the reticular activating system and that deficits of attentional tone are therefore more substantial after right-sided hemispheric lesions. In that case, the right hemisphere could have a dominant function not only in directed attention but also in the modulation of the overall attentional matrix. It is not known whether lesions in similar parts of the brain give rise to unilateral neglect in some patients and to confusional states in others or whether there are systematic differences in the location of the lesion giving rise to each of these two clinical conditions.

Indirect evidence for a right-hemisphere specialization in regulating the overall attentional tone has been gathered from several sources. First, simple reaction times to ipsilateral visual stimuli are faster with the left hand.[2] Second, patients with right-hemisphere lesions are more likely to have bilateral deficits in reaction time.[65] Third, in split-brain patients, there is evidence that vigilance performance is better when the task is being performed by the right hemisphere.[34] Fourth, in patients with unilateral brain injury, those with right-hemisphere damage show smaller galvanic skin responses in reaction to stimulation than patients with equivalent damage to the left hemisphere.[54,101] If the galvanic skin response is accepted as an index of arousal, this last set of observations suggests that right-hemisphere damage has a greater influence on this aspect of attentional tone. These lines of evidence are consistent with the suggestions that there may be a relative right-hemisphere dominance for vigilance and arousal as well as for directed attention.

There is very little information on the hemispheric specialization pattern for directed attention in left-handed individuals. In the case of language, most left-handed individuals appear to have left-hemisphere dominance for language-related functions (see Chapter 5). An analogous relationship may occur in attentional functions, so that most left-handers may still have a right-hemisphere specialization for directed attention. In fact, the hemispheric asymmetry for attention may be conserved even in

minance for language. For example,
ls with crossed aphasias also show
ting that the right-hemisphere spe-
served even when the same hemi-
e.

tic instances of unilateral neglect
y, marked right-sided neglect may
here lesions in dextrals. The most
mineglect that we have seen, how-
ilateral injury to the brain. Severe
ould therefore raise the possibility

TIONS WITH
DEFICITS

components of the clinical picture
dromes of unknown etiology: the
le autism, and schizophrenia.
sites for advanced mental activity
nment is the capacity for the flex-
d distractibility. This is undoubt-
master, and the directive to "pay
ion of children. Adequate atten-
n have short attention spans and
reasonable to expect that devel-
among the most common distur-
uch attentional disturbances are
s known as the hyperactive child
nergic drugs frequently alleviate
l interruption of ascending nor-
in abnormal distractibility.[85] The
a background of so-called mini-
ndicate the presence of relatively
clinical neurologic examination.
een specific findings in the neu-
y syndrome. This is not surpris-
levels in the neuraxis that are
tional matrix.
unds with speculation on caus-
in this literature is the descrip-
t.[9,128] This deficit is sometimes
al filtering mechanism that nor-
nt stimuli. It remains to be seen
e primary or secondary to the
pathogenetic mechanism that underlies schizophrenia.

Whereas the attentional filter seems leaky or erratic in schizophrenia, it appears that this same hypothetical filter may be too rigid and
impermeable in children with the classic form of infantile autism. In a
classic monograph, Bernard Rimland[116] has suggested that the reticular

activating system regulates the ratio between channel width and fidelity in attentional functions and that both autism and schizophrenia may share a fundamental disturbance in this cerebral mechanism. While the excessive channel width in schizophrenia may lead to abnormal distractibility and a flooding by irrelevant or threatening percepts, the autistic child may suffer the consequences of an abnormally high fidelity, which leads to a pathologically narrowed band width and therefore to extreme restrictions in the ability to interact with the environment. This is an attractive hypothesis that deserves to be tested.

OVERVIEW

Attention permeates all aspects of behavior. For heuristic purposes, it is possible to identify two physiologic components. First, there is a state function that determines attentional tone and capacity. The pacemaker for this attentional tone is often associated with the reticular activating system—a system that itself contains complex interactions among structures in the brainstem, thalamus, and cortex. The attentional tone provides a matrix upon which the vector (goal-directed) aspects of attention operate. All attentional behaviors constitute mixtures of both of these physiologic components, some tasks relying more heavily on the state function and others on the vector aspects. In clinical practice, the confusional states provide a group of disorders in which the attention tone is disturbed, giving rise to perturbations in all other functions, including the vector aspects of attention. Unilateral neglect, on the other hand, represents a condition in which one vector aspect, directed sensory attention, is severely impaired, even though the overall attentional tone may remain relatively intact.

Although the maintenance of attentional tone is itself vastly complex, it is the vector aspects of attention that raise the most challenging questions in brain behavior relations. Even directed sensory attention, which is a relatively simple vector function, requires the harmonious regulation of a complicated neural network. How much more complex must be the physiology of other vector aspects of attention—the ability to concentrate on single trends of thought, the ability to covertly and voluntarily shift attention from one stimulus to another or from one modality to the other. In the future, it may be possible to address these issues directly in a way that will contribute not only to the understanding of normal brain function but also to the management of neuropsychiatric conditions with prominent attentional disturbances.

REFERENCES

1. ADAMS, RD, AND VICTOR, M: *Delirium and other confusional states.* In WINTROBE, MM, THORN, GW, ADAMS, RD, BRAUNWALD, E, ISSELBACHER, KJ, AND PETERSDORF, RG (EDS): *Principles of Internal Medicine.* McGraw-Hill, New York, 1974, pp 149–156.

2. ANZOLA, GP, BERTOLONI, G, BUCHTEL, HA, AND RIZZOLATTI, G: *Spatial compatibility and anatomical factors in simple and choice reaction times.* Neuropsychologia 15:295–302, 1977.

3. ARIKUNI, T, SAKAI, M, HAMADA, I, AND KUBOTA, K: *Topographical projections from the prefrontal cortex to the post-arcuate area in the rhesus monkey, studied by retrograde axonal transport of horseradish peroxidase.* Neurosci Lett 19:155–160, 1980.

4. BABINSKI, J: *Contribution a l'étude des troubles mentaux dans l'hémiplégie organique cerebrale (anosognosie).* Rev Neurol (Paris) 27:845–848, 1914.

5. BARBAS, H, AND MESULAM, M-M: *Organization of afferent input to subdivisions of area 8 in rhesus monkeys.* J Comp Neurol 200:407–431, 1981.

6. BARD, L: *De l'origine sensorielle de la déviation conjuguée des yeux avec rotation de la tête chez les hémiplégiques.* Sem Med 24:9–13, 1904.

7. BIANCHI, L: *The functions of the frontal lobes.* Brain 18:497–522, 1895.

8. BISIACH, E, LUZZATTI, C, AND PERANI, D: *Unilateral neglect, representational schema and consciousness.* Brain 102:609–618, 1979.

9. BLEULER, E: *Dementia Praecox oder Gruppe der Schizophrenien.* Deuticke, Leipzig. 1911.

10. BOYD, EH, BOYD, ES, AND BROWN, LE: *Precentral cortex unit activity during the M-Wave and contingent negative variation in behaving squirrel monkeys.* Exp Neurol 75:535–554, 1982.

11. BRAIN, WR: *Visual disorientation with special reference to lesions of the right cerebral hemisphere.* Brain 64:244–272, 1941.

12. BROADBENT, DE: *Perception and Communication.* Oxford, 1958.

13. BUSHNELL, MC, GOLDBERG, ME, AND ROBINSON, DL: *Behavioral enhancement of visual responses in monkey cerebral cortex: 1. Modulation in posterior parietal cortex related to selective visual attention.* J Neurophysiol 46:755–771, 1981.

14. CAMBIER, J, ELGHOZI, D, AND STRUBE, E: *Lésions du thalamus droit avec syndrome de l'hémisphère mineur. Discussion du concept de négligence thalamique.* Rev Neurol 136:105–116, 1980.

15. CHAIN, F, LEBLANC, M, CHEDRU, F, AND LHERMITTE, F: *Négligence visuelle dans les lésions posterieures de l'hémisphere gauche.* Rev Neurol 135:105–126, 1979.

16. CHEDRU, F, AND GESCHWIND, N: *Disorders of higher cortical functions in acute confusional states.* Cortex 8:395–411, 1972.

17. CHU, N, AND BLOOM, FE: *Norepinephrine-containing neurons: Changes in spontaneous discharge patterns during sleeping and waking.* Science 179:907–910, 1973.

18. COHEN, J: *The contingent negative variation in visual attention.* Electroencephalogr Clin Neurophysiol 31:287–305, 1971.

19. COHEN, S: *The toxic psychoses and allied states.* Am J Med 15:813–828, 1953.

20. COLLIN, NG, COWEY, A, LATTO, R, AND MARZI, C: *The role of frontal eye-fields and superior colliculi in visual search and non-visual search in rhesus monkeys.* Behav Brain Res 4:177–193, 1982.

21. COLLINS, RC, AL-MONDDHIRY, H, CHERNIK, NL, AND POSNER, JB: *Neurological manifestations of intravascular coagulation in patients with cancer.* Neurology 25:795–806, 1975.

22. CRITCHLEY, M: *The Parietal Lobes.* Edward Arnold, London, 1953.

23. DAMASIO, AR, DAMASIO, H, AND CHUI, HC: *Neglect following damage to frontal lobe or basal ganglia.* Neuropsychologia 18:123–132, 1980.

24. DEMPSEY, EW, AND MORISON, RS: *The production of rhythmically recurrent cortical potentials after localized thalamic stimulation.* Am J Physiol 135:293–300, 1942.

25. DENENBERG, VH: *Hemispheric laterality in animals and the effects of early experience.* Behav Brain Sci 4:1–49, 1981.

26. DENES, G, SEMENZA, C, STOPPA, E, AND LIS, A: *Unilateral spatial neglect and recovery from hemiplegia.* Brain 105:543–552, 1982.

27. DENNY-BROWN, D, AND CHAMBERS, RA: *The parietal lobe and behavior.* A Res Nerv Ment Dis 36:35–117, 1958.

28. DENNY-BROWN, D, MEYER, JS, AND HORENSTEIN, S: *The significance of perceptual rivalry resulting from parietal lobe lesion.* Brain 75:434–471, 1952.

29. DERENZI, E, FAGLIONI, P, AND SCOTTI, G: *Hemispheric contribution to exploration of space through the visual and tactile modality.* Cortex 6:191–203, 1970.

30. DESMEDT, JE: *Active touch exploration of extrapersonal space elicits specific electrogenesis in the right cerebral hemisphere of intact right handed man.* Proc Natl Acad Sci USA 74:4037–4040, 1977.

31. DEUEL, RK, AND COLLINS, RC: *Recovery from unilateral neglect.* Exp Neurol 81:733–748, 1983.

32. DEUEL, RK, AND DUNLOP, NL: *Role of frontal polysensory cortex in guidance of limb movements.* Brain Res 169:183–188, 1979.

33. DILLER, L, AND WEINBERG, J: *Hemi-inattention in rehabilitation: The evolution of a rational remediation program.* Adv Neurol 18:63–80, 1977.

34. Dimond, SJ: *Depletion of attentional capacity after total commissurotomy in man.* Brain 99:347–356, 1976.

35. Dines, DE, Louis, W, Burgher, and Okazaki, H: *The clinical and pathological correlation of fat embolism syndrome.* Mayo Clin Proc 50:407–411, 1975.

36. Eidelberg, E, and Schwartz, AS: *Experimental analysis of the extinction phenomenon in monkeys.* Brain 94:91–108, 1971.

37. Engel, GL, and Romano, J: *Delirium. A syndrome of cerebral insufficiency.* J Chron Dis 9:260–277, 1959.

38. Faraj, BA, Bowen, PA, Isaacs, JW, and Rudman, D: *Hypertyraminemia in cirrhotic patients.* N Engl J Med 294:1360–1364, 1976.

39. Feeley, MP, O'Harre, J, Veale, D, and Calloghan, M: *Episodes of acute confusion or psychosis in familial hemiplegic migraine.* Acta Neurol Scand 65:369–375, 1982.

40. Feeney, DM, and Wier, CS: *Sensory neglect after lesions of substantia nigra or lateral hypothalamus: Differential severity and recovery of function.* Brain Res 178:329–346, 1979.

41. Ferrier, D: *Functions of the Brain.* GP Putnam's Sons, New York, 1880.

42. Fuster, JM, and Jervey, JP: *Inferotemporal neurons distinguish and retain behaviorally relevant features of visual stimuli.* Science 212:952–955, 1981.

43. Gainotti, G, Messerli, P, and Tissot, R: *Qualitative analysis of unilateral spatial neglect in relation to laterality of cerebral lesions.* J Neurol Neurosurg Psychiatry 35:545–550, 1972.

44. Glenn, LL, and Steriade, M: *Discharge rate and excitability of cortically projecting intralaminar thalamic neurons during waking and sleep states.* J Neurosci 2:1387–1404, 1982.

45. Goldberg, ME, and Bushnell, MC: *Behavioral enhancement of visual responses in monkey cerebral cortex: II. Modulation in frontal eye fields specifically related to saccades.* J Neurophysiol 46:773–787, 1981.

46. Goodman, SJ: *Visuo-motor reaction times and brain stem multiple-unit activity.* Exp Neurol 22:367–378, 1968.

47. Guard, O, Delpy, C, Richard, D, and Dumas, R: *Une cause mal connue de confusion mentale: le ramollissement temporal droit.* Rev Med 40:2115–2121, 1979.

48. Hagamen, TC, Greeley, HP, Hagamen, WD, and Reeves, AG: *Behavioral asymmetries following olfactory tubercle lesions in cats.* Brain Behav Evol 14:241–250, 1977.

49. Healton, EB, Navarro, C, Bressman, S, and Brust, JCM: *Subcortical neglect.* Neurology 32:776–778, 1982.

50. Hécaen, H, Penfield, W, Bertrand, C, and Malmo, R: *The syndrome of apractognosia due to lesions of the minor cerebral hemisphere.* Arch Neurol Psychiatry 75:400–434, 1956.

51. Heilman, KM, Bowers, D, and Watson, RT: *Performance on hemispatial pointing task by patients with neglect syndrome.* Neurology 33:661–664, 1983.

52. Heilman, KM, and Howell, GJ: *Seizure-induced neglect.* J Neurol Neurosurg Psychiatry 43:1035–1040, 1980.

53. Heilman, KM, Pandya, DN, and Geschwind, N: *Trimodal inattention following parietal lobe ablations.* Trans Am Neurol Assoc 95:259–261, 1970.

54. Heilman, KM, Schwartz, HD, and Watson, RT: *Hypoarousal in patients with the neglect syndrome and emotional indifference.* Neurology 28:229–232, 1978.

55. Heilman, KM, and Valenstein, E: *Auditory neglect in man.* Arch Neurol 26:32–35, 1972.

56. Heilman, KM, and Valenstein, E: *Frontal lobe neglect in man.* Neurology 22:660–664, 1972.

57. Heilman, KM, and Valenstein, E: *Mechanisms underlying hemispatial neglect.* Ann Neurol 5:166–170, 1979.

58. Heilman, KM, and Van Den Abell, T: *Right hemisphere dominance for attention: The mechanism underlying hemispheric asymmetries of inattention (neglect).* Neurology 30:327–330, 1980.

59. Heilman, KM, Watson, RT, Valenstein, E, Damasio, AR: *Localization of lesions in neglect.* In Kertesz, A (ed): *Localization in Neuropsychology.* Academic Press, New York, 1983, pp 455–470.

60. Hernandez-Peón, R: *Neurophysiologic aspects of attention.* In Vinken, PJ, and Bruyn, GW (eds): *Handbook of Clinical Neurology,* Vol 3. Elsevier, New York, 1969, pp 155–186.

61. Hernandez-Peón, R, Scherrer, H, and Jouvet, M: *Modification of electric activity in cochlear nucleus during "attention" in unanesthetized cats.* Science 123:331–332, 1956.

62. Hier, DB, Mondlock, J, and Caplan, LR: *Behavioral abnormalities after right hemisphere stroke.* Neurology 33:337–344, 1983.

63. Hier, DB, Mondlock, J, and Caplan, LR: *Recovery of behavioral abnormalities after right hemisphere stroke.* Neurology 33:345–350, 1983.

64. HORENSTEIN, S, CHAMBERLIN, W, AND CONOMY, J: *Infarction of the fusiform and calcarine regions: Agitated delirium and hemianopia.* Trans Am Neurol Assoc 92:85–89, 1967.

65. HOWES, D, AND BOLLER, F: *Simple reaction time: Evidence for focal impairment from lesions of the right hemisphere.* Brain 98:317–332, 1975.

66. HUBEL, DH, HENSON, CO, RUPERT, A, AND GALAMBOS, R: *"Attention" units in the auditory cortex.* Science 129:1279–1280, 1959.

67. HYVARINEN, J, LAAKSO, M, ROINE, R, LEINONEN, L, AND SIPPEL, H: *Effect of ethanol on neuronal activity in the parietal association cortex of alert monkeys.* Brain 101:701–715, 1978.

68. HYVARINEN, J, AND PORANEN, A: *Function of the parietal associative area 7 as revealed from cellular discharges in alert monkeys.* Brain 97:673–692, 1974.

69. HYVARINEN, J, PORANEN, A, AND JOKINEN, Y: *Influence of attentive behavior on neuronal responses to vibration in primary somatosensory cortex of the monkey.* J Neurophysiol 43:870–882, 1980.

70. JAMES, W: *The Principles of Psychology.* New York, Holt, 1890.

71. JANSEN, J, ANDERSEN, P, AND KAADA, BP: *Subcortical mechanisms in the "searching" or "attention" response elicited by prefrontal cortical stimulation in unanesthetized cats.* Yale J Biol Med 28:331–341, 1955.

72. JONES, EG: *Some aspects of the organization of the thalamic reticular complex.* J Comp Neurol 162:285–308, 1975.

73. KENNARD, MA: *Alterations in response to visual stimuli following lesions of frontal lobe in monkeys.* Arch Neurol Psychiatry 41:1153–1165, 1939.

74. KRNJEVIC, K: *Cellular mechanisms of cholinergic arousal.* Behav Brain Sci 4:484–485, 1981.

75. KUYPERS, HGJM, AND LAWRENCE, DG: *Cortical projections to the red nucleus and the brain stem in the rhesus monkey.* Brain Res 4:151–188, 1967.

76. LAPLANE, D, AND DEGOS, JD: *Motor neglect.* J Neurol Neurosurg Psychiatry 46:152–158, 1983.

77. LAPLANE, D, ESCOUROLLE, R, DEGOS, JD, SAURON, B, AND MASSIOU, H: *La négligence motrice d'origine thalamique.* Rev Neurol (Paris) 138:201–211, 1982.

78. LATTO, R, AND COWEY, A: *Visual field defects after frontal eye-field lesions in monkeys.* Brain Res 30:1–24, 1971.

79. LJUNGBERG, T, AND UNGERSTEDT, U: *Sensory inattention produced by 6-hydroxydopamine-induced degeneration of ascending dopamine neurons in the brain.* Exp Neurol 53:585–600, 1976.

80. LURIA, AR, KARPOV, BA, AND YARBUSS, AL: *Disturbances of active visual perception with lesions of the frontal lobes.* Cortex 2:202–212, 1966.

81. LYNCH, JC: *The functional organization of posterior parietal association cortex.* Behav Brain Sci 3:485–499, 1980.

82. LYNN, R: *Attention, Arousal and the Orientation Reaction.* Pergamon Press, Oxford, 1966.

83. MARKAND, ON, WHEELER, GL, AND POLLACK, SL: *Complex partial status epilepticus (psychomotor status).* Neurology 28:189–196, 1978.

84. MARSHALL, JF: *Somatosensory inattention after dopamine-depleting intracerebral 6-OHDA injections: Spontaneous recovery and pharmacological control.* Brain Res 177:311–324, 1979.

85. MASON, ST, AND FIBIGER, HC: *Noradrenaline and selective attention.* Life Sci 25:1949–1956, 1979.

86. MCCANCE, I, PHILLIS, JW, AND WESTERMAN, RA: *Acetylcholine-sensitivity of thalamic neurones: Its relationship to synaptic transmission.* Br J Pharmacol 32:635–651, 1968.

87. MCFIE, J, PIERCY, MF, AND ZANGWILL, OL: *Visual-spatial agnosia associated with lesions of the right cerebral hemisphere.* Brain 73:167–190, 1950.

88. MEDINA, JL, RUBINO, FA, AND ROSS, A: *Agitated delirium caused by infarction of the hippocampal formation and fusiform and lingual gyri: A case report.* Neurology 24:1181–1183, 1974.

89. MESULAM, M-M: *Tetramethyl benzidine for horseradish peroxidase neurohistochemistry: A non-carcinogenic blue reaction-product with superior sensitivity for visualizing neural afferents and efferents.* J Histochem Cytochem 26:106–117, 1978.

90. MESULAM, M-M: *A cortical network for directed attention and unilateral neglect.* Ann Neurol 10:309–325, 1981.

91. MESULAM, M-M: *The functional anatomy and hemispheric specialization for directed attention—the role of the parietal lobe and its connectivity.* Trends in Neuroscience 6:384–387, 1983.

92. MESULAM, M-M, AND GESCHWIND, N: *Disordered mental states in the post-operative period.* Urol Clin North Am 3:199–216, 1976.

93. MESULAM, M-M, MUFSON, EJ, LEVEY, AI, AND WAINER, BH: *Cholinergic innervation of cortex by the basal forebrain: Cytochemistry and cortical connections of the septal area, diagonal band nuclei, nucleus basalis (substantia innominata) and hypothalamus in the rhesus monkey.* J Comp Neurol 214:170–197, 1983.

94. MESULAM, M-M, MUFSON, EJ, WAINER, BH, AND LEVEY, AI: *Central cholinergic pathways in the rat: An overview based on an alternative nomenclature (Ch1–Ch6).* Neuroscience 10:1185–1201, 1983.

95. MESULAM, M-M, VAN HOESEN, GW, PANDYA, DN, AND GESCHWIND, N: *Limbic and sensory connections of the inferior parietal lobule (area PG) in the rhesus monkey: A study with a new method for horseradish peroxidase histochemistry.* Brain Res 136:393–414, 1977.

96. MESULAM, M-M, WAXMAN, SG, GESCHWIND, N, AND SABIN, TD: *Acute confusional states with right middle cerebral artery infarctions.* J Neurol Neurosurg Psychiatry 39:84–89, 1976.

97. MEYER, A: *Historical Aspects of Cerebral Anatomy.* Oxford University Press, London, 1971.

98. MORAY, N: *Attention in dichotic listening: Affective cues and the influence of instructions.* QJ Exp Psychol 11:56–60, 1959.

99. MORISON, RS, AND DEMPSEY, EW: *A study of thalamo-cortical relations.* Am J Physiol 135:281–292, 1942.

100. MORRISON, JH, AND MAGISTRETTI, PJ: *Monamines and peptides in cerebral cortex.* Trends in Neuroscience 6;146–151, 1983.

101. MORROW, L, VRTUNSKI, PB, KIM, Y, AND BOLLER, F; *Arousal responses to emotional stimuli and laterality of lesion.* Neuropsychologia 19:65–71, 1981.

102. MORUZZI, G, AND MAGOUN, HW: *Brain stem reticular formation and activation of the EEG.* Electroencephalogr Clin Neurophysiol 1:459–473, 1949.

103. MOUNTCASTLE, VB, LYNCH, JC, GEORGOPOULOUS, A, SAKATA, H, AND ACUNA, A: *Posterior parietal association cortex of the monkey: Command functions for operations within extrapersonal space.* J Neurophysiol 38:871–908, 1975.

104. OREM, J, SCHLAG-REY, M, AND SCHLAG, J: *Unilateral visual neglect and thalamic intralaminar lesions in the cat.* Exp neurol 40:784–797, 1973.

105. OXBURY, JM, CAMPBELL, DC, AND OXBURY, SM: *Unilateral spatial neglect and impairments of spatial analysis and visual perception.* Brain (Part III) 97:551–564, 1974.

106. PANDYA, DN, VAN HOESEN, GW, AND MESULAM, M-M: *Efferent connections of the cingulate gyrus in the rhesus monkey.* Exp Brain Res 42:319–330, 1981.

107. PATTERSON, A, AND ZANGWILL, OL: *Disorders of visual space perception associated with lesions of the right cerebral hemisphere.* Brain 67:331–358, 1944.

108. PENFIELD, W, AND JASPER, H: *Epilepsy and the Functional Anatomy of the Human Brain.* Little, Brown & Co, Boston, 1954.

109. PERRIN, RG, HOCKMAN, CH, KALANT, H, AND LIVINGSTON, KE: *Acute effects of ethanol on spontaneous and auditory evoked electrical activity in cat brain.* Electroencephalogr Clin Neurophysiol 36:19–31, 1974.

110. PEROUTKA, SJ, SOHMER, BH, KUMAR, AJ, FOLSTEIN, M, AND ROBINSON, RG: *Hallucinations and delusions following a right temporoparieto-occipital infarction.* Johns Hopkins Med J 151:181–185, 1982.

111. POSNER, MI, WALKER, JA, FRIEDRICH, FJ, AND RAFAL RD: *Effects of parietal injury on covert orienting of visual attention.* J Neurosci, 1984, in press.

112. PLUM, F, AND POSNER, JB: *The Diagnosis of Stupor and Coma.* FA Davis, Philadelphia, 1972.

113. RAPOPORT, JL, BUCHSBAUM, MS, ZAHN, TP, WEINGARTNER, H, LUDLOW, C, AND MIKKELSEN, EJ: *Dextroamphetamine: Cognitive and behavioral effects in normal prepubertal boys.* Science 199:560–563, 1978.

114. RAY, CL, MIRSKY, AF, AND PRAGAY, EB: *Functional analysis of attention-related unit activity in the reticular formation of the monkey.* Exp Neurol 77:544–562, 1982.

115. RISBERG, J, AND INGVAR, DH: *Patterns of activation in the grey matter of the dominant hemisphere during memorizing and reasoning—a study of regional cerebral blood flow changes during psychological testing in a group of neurologically normal patients.* Brain 96:737–756, 1973.

116. RIMLAND, B: *Infantile Autism.* Appleton-Century-Crofts, East Norwalk, CT, 1962.

117. ROBINSON, DL, GOLDBERT, ME, AND STANTON, GB: *Parietal association cortex in the primate: Sensory mechanisms and behavioral modulations.* J Neurophysiol 41:910–932, 1978.

118. ROHRBAUGH, JW, SYNDULKO, K, AND LINDSLEY, DB: *Brain wave components of the contingent negative variation in humans.* Science 191:1055–1057, 1976.

119. ROLAND, PE: *Cortical regulation of selective attention in man. A regional cerebral blood flow study.* J Neurophysiol 48:1059–1078, 1982.

120. SCHEIBEL, ME, AND SCHEIBEL, AB: *Anatomical basis of attention mechanisms in vertebrate brains.* In QUARTON, GC, MELNECHUK, T, AND, SCHMITT, FO (EDS): *The Neurosciences.* Rockefeller University Press, New York, 1967, pp 577–602.

121. SCHILLER, PH, TRUE, SD, AND CONWAY, JL: *Effects of frontal eye field and superior colliculus ablations on eye movements.* Science 206:590–592, 1979.

122. SCHLAG, J, AND WASZAK, M: *Electrophysiological properties of units of the thalamic reticular complex.* Exp Neurol 32:79–97, 1971.

123. SCHOTT, B, LAURENT, B, MAUGUIERE, F, AND CHAZOT, G: *Négligence motrice par hematome thalamique droit.* Rev Neurol 137:447–455, 1981.

124. SELTZER, B, AND PANDYA, DN: *Some cortical projections to the hippocampal area in the rhesus monkey.* Exp Neurol 50:146–160, 1976.

125. SHARPLESS, S, AND JASPER, H: *Habituation of the arousal reaction.* Brain 79:655–680, 1956.

126. SHERRINGTON, CS: *Man on His Nature.* Cambridge University Press, 1951.

127. SILBERPFENNIG, J: *Contribution to the problem of eye movements.* Confin Neurol 4:1–13, 1941.

128. SILVERMAN, J: *Problem of attention in research and theory in schizophrenia.* Psychol Rev 71:352–379, 1964.

129. SINGER, W: *Control of thalamic transmission by corticofugal and ascending reticular pathways in the visual system.* Physiol Rev 57:386–420, 1977.

130. SOKOLOV, EN: *Neuronal models and the orienting reflex.* In BRAZIER, MAB (ED): *The Central Nervous System and Behavior.* Madison Printing, Madison, NJ, 1960, pp 187–276.

131. SOPER, HV, DIAMOND, IT, AND WILSON, M: *Visual attention and inferotemporal cortex in rhesus monkeys.* Neuropsychologia 13:409–419, 1975.

132. SPARKS, R, AND GESCHWIND, N: *Dichotic listening in man after section of neocortical commissures.* Cortex 4:3–16, 1968.

133. SPEHLMANN, AND SMOTHERS, CC: *The effects of acetylcholine and of synaptic stimulation in the sensorimotor cortex of cats. II. Comparison of the neuronal responses to reticular and other stimuli.* Brain Res 74:243–253, 1977.

134. SPRAGUE, JM, AND MEIKLE, TH, JR: *The role of the superior colliculus in visually guided behavior.* Exp Neurol 11:115–146, 1965.

135. STAMM, JS: *Functional dissociation between the inferior and arcuate segments of dorsolateral prefrontal cortex in the monkey.* Neuropsychologia 11:181–190, 1973.

136. STERIADE, M: *State-dependent changes in the activity of rostral reticular and thalamocortical elements.* Neurosci Res Prog Bull 18:83–91, 1980.

137. STERIADE, M: *EEG desynchronization is associated with cellular events that are prerequisites for active behavioral states.* Behav Brain Sci 4:489–492, 1981a.

138. STERIADE, M: *Mechanisms underlying cortical activation: Neuronal organization and properties of the midbrain reticular core and intralaminar thalamic nuclei.* In POMPEIANO, O, AND AJMONE MARSON, C (EDS): *Brain Mechanisms and Perceptual Awareness.* Raven Press, New York, 1981b, pp 327–377.

139. STEVENS, JR, KIM, C, AND MACLEAN, PD: *Stimulation of caudate nucleus: behavioral effects of chemical and electrical excitation.* Arch Neurol 4:47–54, 1961.

140. SZERB, JC: *Cortical acetylcholine release and electroencephalographic arousal.* J Physiol 192:329, 1967.

141. TEBECIS, AK: *Studies on cholinergic transmission in the medial geniculate nucleus.* Br J Pharmacol 38:138–147, 1970.

142. TREISMAN, AM: *Verbal cues, language and meaning in selective attention.* Am J Psychol 77:206–219, 1967.

143. TUNE, LE, DAMLOUJI, NF, HOLLAND, A, GARDNER, TJ, FOLSTEIN, MF, AND COYLE, JT: *Association of postoperative delirium with raised serum levels of anticholinergic drugs.* Lancet 2:651–653, 1981.

144. VALENSTEIN, E, AND HEILMAN, KM: *Unilateral hypokinesia and motor extinction.* Neurology 31:445–448, 1981.

145. VALENSTEIN, E, HEILMAN, KM, WATSON, RT, AND VAN DEN ABELL, T: *Nonsensory neglect from parietotemporal lesions in monkeys.* Neurology 32:1198–1201, 1982.

146. VANDERWOLF, CH, AND ROBINSON, TE: *Reticulo-cortical activity and behavior: A critique of the arousal theory and a new synthesis.* Behav Brain Sci 4:459–514, 1981.

147. VOLPE, BT, LEDOUX, JE, AND GAZZANIGA, MS: *Information processing of visual stimuli in an 'extinguished' field.* Nature 282:722–724, 1979.

148. WATSON, RT, AND HEILMAN, KM: *Thalamic neglect.* Neurology 29:690–694, 1979.

ATTENTION,
CONFUSIONAL
STATES, AND
NEGLECT

167

149. WATSON, RT, HEILMAN, KM, CAUTHEN, JC, AND KING, FA: *Neglect after cingulectomy.* Neurology 23:1003–1007, 1973.

150. WATSON, RT, HEILMAN, KM, MILLER, D, AND KING, FA: *Neglect after mesencephalic reticular formation lesions.* Neurology 24:294–298, 1974.

151. WATSON, RT, MILLER, BD, AND HEILMAN, KM: *Nonsensory neglect.* Ann Neurol 3:505–508, 1978.

152. WELCH, K, AND STUTEVILLE, P: *Experimental production of unilateral neglect in monkeys.* Brain 81:341–347, 1958.

153. WILLANGER, R, DANIELSEN, UT, AND ANKERHUS, J: *Denial and neglect of hemiparesis in right-sided apopleptic lesions.* Acta Neurol Scand 64:310–326, 1981.

154. WURTZ, RH, AND GOLDBERG, ME: *Activity of superior colliculus in behaving monkey. III. Cells discharging before eye movements.* J Neurophysiol 35:575–585, 1972a.

155. WURTZ, RH, AND GOLDBERG, ME: *Activity of superior colliculus in behaving monkey. IV. Effects of lesions on eye movements.* J Neurophysiol 35:587–596, 1972b.

156. YINGLING, CD, AND SKINNER, JE: *Regulation of unit activity in nucleus reticularis thalami by the mesencephalic reticular formation and the frontal granular cortex.* Electroencephalogr Clin Neurophysiol 39:635–642, 1975.

Chapter 4

MEMORY AND AMNESIAS*

Jean-Louis Signoret, M.D.

Our experience of the past allows us at each moment to adapt ourselves to the present and to look into the future. This capacity for an awareness of the past implies that experience can leave durable traces in our organism, almost certainly in the form of structural or biochemical modifications within cerebral neuronal networks. Since memory is such a pervasive component of all mental life, one might have expected that only the largest lesions in the brain would impair it and that isolated memory disorders would be rare. Nothing could be further from the truth. In fact, isolated memory disorders (or amnesias) are frequently seen in behavioral neurology and can be caused by relatively small lesions, when they occur within certain parts of the brain. In such cases, the lesions result in amnesia not because they destroy individual memory traces (wherever these may be) but because they interfere with the mechanisms for learning and retrieving information. The clinical pathologic study of patients with amnesia allows us to gain some insight into the structural organization of memory. This is the purpose of this chapter. The following two clinical sketches illustrate the clinical diversity of the amnesias.

Alcoholic Korsakoff's Syndrome. At 55 years of age, this office worker has been an alcoholic for many years and is overly thin. He has been hospitalized for a month, in the same room.

*We would like to thank the amnesic patients that we have examined—many have forgotten us—and also the colleagues who referred these patients to us.

We would like to express our gratitude to Mr. Parth Bhatt, who helped us with the English text of this manuscript; to Dr. Mortimer Mishkin, who read the first draft; and to Dr. M-Marsel Mesulam, who provided us with the clarity of his vision.

—Why were you hospitalized?

—I haven't been hospitalized. I'm in here for an examination; it's my ankle that's hurting (the patient points to a plaster in his foot).

—What happened to you?

—It's an old accident; I was in the country last weekend, I climbed up a tree to gather some fruit, and I fell (the patient had suffered a fracture of this foot a week earlier; he had in fact incurred a twisted ankle after a fall from a tree, but this occurred 5 years ago).

—Did you have a good day yesterday?

—Oh yes. I had dinner with my aunt after leaving work.

The patient is told a short story; after 1 minute of conversation he is unable to remember the story; the patient even denies that a story was told to him. Arithmetic problems are rapidly solved without error; proverbs are correctly defined and commented on.

Transient Global Amnesia. The wife of a retired 65-year-old-man accompanies her husband to the doctor's office.

—I have been worried; 2 days ago my husband completely lost his memory for a few hours. He is alright now, but he does not remember what happened.

—Please continue.

—We were at our house in the country; in the afternoon my husband mowed the lawn for 2 hours. He was sweating and took a cold shower. When he came out of the shower, he came to see me. His words upset me—he kept saying "what's happening to me?" He didn't know that we had been in the country for 2 days or that our children had spent a day with us and had offered him a tie for his birthday. He would forget everything I said to him, but he spoke correctly. I thought he was having some sort of attack. The next morning everything was alright, but he still has a hole in his memory. He does not know that a doctor came to see him and he does not remember having mowed the lawn.

CLINICAL FEATURES OF THE AMNESTIC SYNDROME

Memory becomes disturbed in a great number of clinical conditions, ranging from metabolic encephalopathy, to head injury, tumor, epilepsy, and so on. However, in many of these conditions, memory is either secondarily impaired as a consequence of attentional disorders (as in confusional states) or it emerges merely as one of the many other cognitive deficits (as in certain tumors or dementias). In certain conditions, however, a relatively isolated amnesia is seen. It is these cases that offer the best opportunity for studying memory and therefore constitute the basis of this chapter.

The clinical and neurologic data lead to the formulation of a general rule: *almost all severe amnesias occur after bilateral involvement of limbic structures*. The behavioral and clinical similarities of severe amnesias justify the description of a core amnestic syndrome. It is important to realize that this is perhaps a mere abstraction, but it is useful to keep this reference in mind. In essence, patients with the amnestic syndrome have a global anterograde and retrograde memory deficit and occasionally confabulate. Attention, motivation, and other cognitive functions are essentially intact.

Anterograde amnesia refers to the inability to learn new material. The existence of an anterograde memory impairment can be confirmed with test batteries. Identical overall scores in these batteries do not always signify identical amnesias. There is as yet no universally applicable set of tests for the assessment of memory impairment (for review, see Erickson and Scott[28]). The Wechsler Memory Scale,[108] which consists of seven subtests of which only three are in fact memory tasks, is still the standard battery. It allows the establishment of a memory quotient (MQ), which can then be compared with the intelligence quotient (IQ). The experimental paradigm proposed by Peterson and Peterson[80] provides another means of demonstrating the presence of anterograde amnesia. This test requires subjects to recall information following a period of time during which they are occupied by an intervening task. Sometimes the amnesic patient, instead of remembering the event that has just been experienced, recalls a previous unrelated event from a previous testing session. This phenomenon is known as proactive interference (or as intrusion error) and may be experimentally investigated.

Retrograde amnesia indicates the inability to remember events that occurred in the premorbid period. This is almost always present in the amnestic syndrome and affects a limited period of time (from a few months to few years). This period may decrease during the evolution of the disease.[4] Many investigations have confirmed Ribot's Law,[84] which states that the oldest memories are the most resistant to amnesia.[2] Retrograde amnesia with this type of temporal gradient must be differentiated from amnesia for remote events, in which the deficit may affect a very large period of time with no temporal gradient.[17] It is certainly possible to confuse the two disorders in the clinic; these two types of impairment may imply two different mechanisms.

Confabulation can be a most spectacular clinical manifestation, even to the point of masking the memory loss. Confabulation takes the form of fabricated verbal responses concerning the patient's memory of recent experiences. It is sometimes spontaneous but more often induced by the examiner's questions. Confabulation can appear as a kind of filling in of "memory gaps" and can be interpreted as the result of a compensatory mechanism. However, only a minority of patients with the amnestic syndrome confabulate. Confabulation and denial of memory loss are linked. Patients may vary greatly in their awareness of their memory defect during the course of the disease. In fact, increased awareness and reactive depression sometimes herald the onset of improvement.

Certain learning tasks—especially those that depend on the acquisition of motor skills (habits)—may be preserved in patients with the amnestic syndrome.[76] These tasks include mirror drawing, rotary pursuit, bimanual tracking, tapping, and maze learning.[19,66,69]

Amnesic patients are also able to retain certain perceptual information[69,105] and are able to acquire conditioned responses.[110] Furthermore, it is said that experiences with strong emotional content are retained much better. However, in sharp contrast to their ability to acquire these skills, amnesic patients may deny familiarity with the tasks that they have nonetheless learned.

In the pure amnestic syndrome, attention span (unfortunately, referred to as immediate memory by some) is intact. In fact, when attention is impaired, memory function cannot be assessed adequately. Other cognitive abilities, such as language, construction, reasoning, and calculation, are also relatively preserved in patients with the amnestic syndrome. Chapter 2 contains further discussion on the assessment of memory and related functions.

SYNDROMES OF DISORDERED MEMORY

The clinical varieties and etiologic diversity of the amnesias are such that all classifications are subject to criticism. The classification proposed here is simply pragmatic. We shall thus treat the amnesias in the following order: amnesias with bilateral limbic involvement (the amnestic syndrome), selective amnesias, amnesias of unknown anatomy, transient global amnesias, benign senescent forgetfulness, and psychogenic amnesias.

Amnesias with Bilateral Limbic Involvement (the Amnestic Syndrome)

Medial Temporal Lesions

Neurosurgical Intervention. Bilateral medial temporal lobectomies were performed on an epileptic patient, HM, who subsequently and unexpectedly became amnesic.[91] This type of surgery is no longer being done, because of this devastating complication. Following surgery, the anterograde amnesia of HM was global for verbal and nonverbal material, even though certain motor skills could be acquired. This anterograde amnesia has remained essentially unchanged for several decades. Retrograde amnesia was initially extensive but started to decrease after a period of a year. HM has always shown good reasoning abilities with no confabulation and has presented no cognitive deficit (IQ = 117). Since attention span is preserved, HM does well if he is allowed silent rehearsal. However, forgetting occurs as soon as the verbal rehearsal is prevented by some interference task, even if this is brief. HM is able to retain some complex perceptual and motor tasks even though he denies being exposed to them.[69] The amnesia in HM was initially attributed to hippocampal involvement.[77,78] Most investigators would now interpret HM's case as being an example of amnesia resulting from combined bilateral lesions of hippocampus, amygdala, and adjacent cortex.

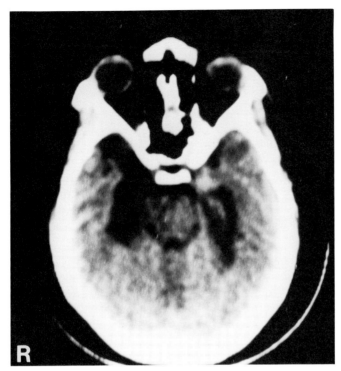

FIGURE 1. CT of a patient with an amnestic syndrome (herpes encephalitis, 7 months after onset). The photo shows bilateral lesions of the mesial temporal lobe, more extensive on the right side. (From Charcot Neuroradiological Department, Hopital de la Salpetriere, Pr. J. Bories, with permission.)

Herpes Simplex Encephalitis. This disease has a predilection for limbic and paralimbic parts of the brain and can result in very pure amnestic syndromes, especially in instances when the lesion is limited to the two medial temporal regions including the amygdala, the hippocampus, and the parahippocampal gyrus (Fig. 1). The amnestic syndrome in such cases is identical to that observed in HM.[85] When the herpetic encephalitis is more widespread, additional dramatic deficits of comportment are also seen. These can include severe disturbances of sexual and ingestive behavior and can lead to a picture reminiscent of the Klüver-Bucy syndrome (see Chapter 1). Judgment, insight, and affect are also severely disrupted in such patients with more widespread encephalitis.[32]

Infarcts and Tumors. Bilateral posterior cerebral infarcts may cause a severe amnestic syndrome, initially accompanied by a confusional state (see Chapter 3) but without confabulation.[23] The amnesia is frequently associated with visual field deficits. With the exception of the amygdala, the entire medial temporal lobe is supplied by the posterior cerebral artery. Occlusions of this artery can result in infarction within this territory. Since both posterior cerebral arteries attach to a common basilar trunk, bilateral occlusions are not rare. Severe memory defects are most commonly seen after bilateral infarcts.[104] Whereas left unilateral infarcts can also give rise to an amnestic syndrome, this is almost always transient.[5,73] Patients with this kind of unilateral left posterior cerebral artery infarcts may also have pure alexia without agraphia (see Chapters 1, 5, and 7).

Tumors in the medial temporal lobe also give rise to similar memory disorders. Some tumors that arise on one side can spread to the other side and then give rise to an amnestic syndrome.

Anoxia. Patients surviving carbon monoxide poisoning, cardiac arrest, or acute anoxia may also come to medical attention with an amnestic syndrome. Additional personality changes including indifference and apathy are common. Cognitive deficits and attentional difficulties are also common, so it is rare to find an isolated amnestic syndrome in this setting. The hippocampus is particularly vulnerable to anoxia, and this may be the pathologic substrate for the amnesia.

Primary Degenerative Diseases. Memory disorders are a nearly constant feature of Alzheimer's disease. The presence of other cognitive disorders is characteristic of the disease and cause a complex form of memory failure.[109] The first signs of the disease are usually forgetfulness for everyday events, especially personal experiences. Alzheimer's disease can, although only rarely, take the form of an isolated amnestic syndrome for a number of years. It is difficult to establish precise anatomic correlations in Alzheimer's disease. However, it is important to note the prevalence of lesions in the hippocampal and parahippocampal regions[11] and particularly the impairment of cholinergic innervation in limbic structures.[22]

In Pick's disease, neuropathologic data have shown a correlation between memory defects and medial temporal lesions.[100]

Bilateral Limbic Diencephalic Lesions

Korsakoff's Syndrome. In 1887, Korsakoff described a new disease with both peripheral and central nervous manifestations and gave it the name "cerebropathia toxaemia psychica." The causes of the disease are varied.[48] The nutritional origin caused by a thiamine deficit has been established; alcoholism is the major but not the exclusive setting for the thiamine deficit. The onset of difficulties may either be progressive or may occur following a global confusional state; it is often accompanied by signs characteristic of Wernicke's encephalopathy such as ataxia, nystagmus, and ophthalmoplegia. After the acute period passes, the patient is left in a relatively pure amnestic state. The severity of the deficit in acquiring new information is variable, confabulation is frequent, and retrograde amnesia that follows a fixed temporal gradient is a constant factor.[2] According to our experience it seems legitimate to distinguish three major clinical types of Korsakoff's syndrome: (1) those with anosagnosia, confabulation, and a certain euphoria; (2) those with severe disorientation and difficulty in maintaining a coordinated goal-directed mental activity; and (3) those in whom the memory impairment seems isolated and pure. It is possible to observe each of these three clinical types during the temporal evolution of the syndrome in a single patient.

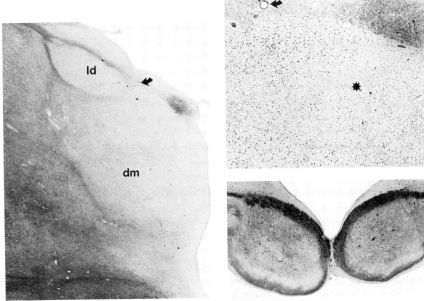

FIGURE 2. Alcoholic Korsakoff's syndrome. *Left:* Coronal section through the thalamus (Nissl Luxol stain) showing the lateral dorsal (ld) and dorsomedial (dm) nuclei. There is cell loss and gliosis in the lateral dorsal and dorsomedial nuclei. *Upper right:* Dorsomedial nucleus: neuronal depopulation, cellular gliosis. The arrowhead points to the same vascular structure as in the photomicrograph on the left. The area surrounding the asterisk is almost completely depopulated of neurons. *Lower right:* Coronal section through mammillary bodies showing pseudonecrosis; myelinated fibers have disappeared in the central region (Loyez myelin stain). (From University Neuropathological Laboratory, Hopital Sainte-Anne, Pr. S. Brion, with permission.)

This diversity may account for the difficulties and the contradictions that arise in the interpretation of neuropsychologic investigations of Korsakoff's syndrome.[40]

The lesions are always bilateral and symmetrical and always affect subcortical structures (Fig. 2).[24] Lesions of the mammillary bodies principally affect the medial nuclei; more rarely, the lesions can extend to the anterior columns of the fornix and even—although in a much less severe fashion—to other nuclei of the hypothalamus. Lesions of the thalamus include in particular the dorsomedial nuclei, the medial pulvinar, and the lateral dorsal nuclei. Lesions of the midbrain are also reported. The clinicopathologic significance of these lesions is the subject of debate. Lesions of the mammillary bodies are universally present in Korsakoff's syndrome, whereas lesions of the thalamic dorsomedial nuclei are seen in a large number of—but not in all—cases. For Victor, Adams, and Collins,[103] it is the thalamic lesions that account for the memory impairment; they have observed five patients with lesions to the mammillary bodies but with intact thalamus who had no memory disorders. Brion and coworkers,[9] on the other hand, reported a series of 14 amnesic patients with Korsakoff's syndrome only eleven of whom had thalamic lesions in addition to the mammillary lesions; all 11 patients confabulated. The three patients with mammillary body lesions but without thalamic lesions also had amnestic syndromes but did not confabulate. This point is confirmed by two anatomoclinical observations reported by Mair and associates.[56]

It is possible that the biochemical approach will provide a new understanding of this complex syndrome. For example, there seems to be a decreased level of the primary brain metabolite of norepinephrine in patients with Korsakoff's syndrome.[61]

Hypothalamic Lesions. Amnestic syndromes have been reported in cases of hypothalamic tumors. In the evolution of such tumors, additional disorders of attention and vigilance make it difficult to interpret the memory impairment. Cerebral tumors developing in the critical area surrounding the floor and walls of the third ventricle are most likely to cause an amnestic syndrome.[114]

Thalamic Lesions. The limbic and paralimbic nuclei of the thalamus are shown in Figure 8 of Chapter 1. Lesions affecting these nuclei are often associated with memory disorders.

Bilateral dorsomedial thalamotomies have been performed to relieve intractable psychiatric disorders. Spiegel and co-workers[94] have described, after this type of intervention, disorders affecting the temporal organization and chronology of the patient's experiences. This has been termed "chronotaraxis." This disorder is not a regular feature and is transitory in nature; perhaps this is because the thalamotomies are always partial.

Bilateral thalamic infarcts affecting the paramedian and more particularly the dorsomedial nuclei can give rise to severe memory defects, both anterograde and retrograde in nature.[57,60,81] This defect may be associated with additional behavioral disturbances and especially with hypokinetic inertia and oculomotor disorders.[65] The study of the sites of the lesion shows that the establishment of correlations between a memory defect and a thalamic lesion is highly complex, especially since the mammillothalamic tract may also be affected.[14] Unilateral left thalamic lesions in the dorsomedial or anterior thalamic nuclei can also result in an incomplete anterograde amnesia, principally (but not exclusively) affecting verbal material[36] (Fig. 3).

An incomplete anterograde amnesia in which verbal material was more affected than nonverbal material resulted from a stab wound to the left dorsomedial thalamic region of the extensively studied patient, NA.[16,96,99] A fencing foil entered the brain through the right nostril, taking a slightly oblique course to the left. A paralysis of upward gaze was associated with the amnesia, so that additional extrathalamic lesions are almost certainly present in this patient.

Basal Forebrain Lesions

Basal forebrain tumors may involve the septum, hypothalamus, and caudal orbitofrontal regions. They can lead to amnesia. Additional impairments of consciousness are frequent, so these cases might be difficult to examine.[24] Anterior communicating artery aneurysms can also cause an amnesia that may be similar to Korsakoff's syndrome. This amnesia can be associated with elements of a frontal syndrome that makes the interpretation of these disorders somewhat problematic.[54] Such aneurysms can cause amnesia either because they bleed directly into the basal fore-

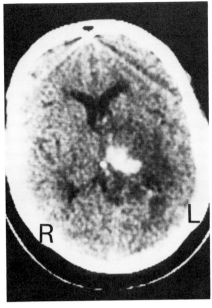 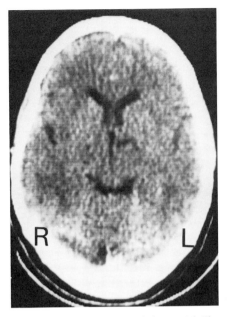

FIGURE 3. *Left:* CT of patient with memory impairment, more marked for verbal material. The photo shows a left thalamic hemorrhage (dorsomedial nucleus and pulvinar). *Right:* CT of a patient with, initially "transient global amnesia," later "incomplete anterograde amnesia." The photo shows a left thalamic infarct (anterior and probably dorsomedial nuclei). (From Charcot Neuroradiological Department, Hopital de la Salpetriere, Pr. J. Bories, with permission.)

brain or because of distal spasm or embolization in arteries that supply cingulate and orbitofrontal areas. In some patients, the amnesia does not appear until after surgical intervention, probably because of orbitofrontal and medial frontal damage during the procedure. One of the several mechanisms that may account for the amnesia in these cases is the damage to septal and nucleus basalis cell bodies, which provide virtually all of the cholinergic innervation for distant limbic and cortical areas.[63]

The extraordinary observation reported in 1913 by Mabille and Pitres[55] still remains a puzzle. Following a patient's amnesia that had lasted 23 years, postmortem examination showed a bilateral and symmetrical lacunar cavity located anterior to the head of each of the caudate nuclei.

Bilateral Fornix Lesions

Some patients with tumors or bilateral section of the fornix have developed an amnestic syndrome,[39,98] but there are also other reports that deny this.[116]

Cingulate Lesions

Bilateral infarcts of the anterior cingulate region give rise to akinetic mutism, so that memory cannot be tested in these patients. Bilateral cingulotomies usually lead to transitory memory disorders that have been compared to the confabulation observed in the course of Korsakoff's syndrome.[111] Another disorder in these patients is the impossibility of recalling recent experiences in their proper temporal order. It is likely that the severe amnestic syndromes after callosal commissurotomy[117] and with posterior corpus callosum tumors[46] are due to bilateral posterior cingulate involvement.

Selective Amnesias

Unilateral lesions in the limbic system or those that interfere with its connections give rise to more selective amnesias than the amnestic syndrome.

Unilateral Temporal Lesions

A material specific (nonverbal versus verbal) memory loss can be the result of a unilateral temporal lobectomy. For example, the recognition of recurring nonsense figures, visual maze learning,[66] and the recognition of faces are disturbed in the case of right temporal lobectomy.[67] The recall of verbal alphabetic material, on the other hand, is disturbed in the case of left temporal lobectomy.[68] These deficits are proportional to the posterior extension the hippocampal excision. These impairments appear to be independent of the sensory modality in which the material is presented; thus, the disorder persists if a verbal stimulus is presented orally or visually. The difference between the effects of right and left medial temporal removals is thus according to the *nature of the material* presented but not according to the *modality of presentation.* Furthermore, patients with left medial temporal lesions can improve their performance on verbal paired-associate learning by using visual imagery.[47]

The existence of hemispheric differences in memory for a given type of material is confirmed by the superiority of the right hemisphere in recognizing nonverbalizable tactile patterns after cerebral commissurotomy.[70]

Limbic Disconnections

A disconnection between the cortical structures, where information is received and analyzed, and the structures of the medial temporal lobe could account for some modality-specific amnesias. Such modality-specific amnesias have been described in the visual and tactile modalities by Ross.[87] Object agnosia, as well as prosopagnosia or even anomia, could also be considered disconnection amnesias. See Chapters 1 and 7 for more detail on this topic.

Amnesias of Unknown Anatomy

Closed Head Trauma

Traumatic amnesias are frequent and varied. Interpretation of these amnesias is difficult because of many factors: the possible multiplicity of anatomic lesions that are difficult to locate, the associated cognitive disorders, possible changes in personality, and the possibility of associated litigation and secondary gain.[51,89]

The areas that are most susceptible to contusion are the temporal poles and the orbitofrontal surface.[20] Second, long fiber bundles including the fornix are very much susceptible to sheering forces.[97] Therefore, there is reason to believe that limbic and paralimbic areas may have a special vulnerability to head trauma. But the lesions in head trauma are multifocal and no two head traumas are exactly identical. Generally, but not always, there is a correlation between the severity of the anterograde amnesia and the span of time covered by the retrograde amnesia in patients who sustain head trauma.[88]

Electroconvulsive Therapy

Memory disorders associated with electroconvulsive therapy (ECT) are, to a certain extent, comparable with those observed in head injuries.[113] Limited knowledge of the neurobiologic consequences of ECT make interpretation difficult. Unilateral ECT gives rise to material-specific amnestic syndromes, depending on the side that receives the treatment.[30]

Epilepsy

Stimulation in the white matter of the temporal lobe can produce amnesia.[7] The affects of bilateral stimulation of the hippocampus or the amygdala or both are less clear;[15] a memory impairment may occur, but as a rule this is in conjunction with other complex disorders. Fugues and amnesias may occur in patients with complex partial seizures.[58] Many epileptic patients complain of memory impairments in daily life. The interpretation of these disorders must take into account many factors, including the type of seizures and the medication.[53] Delay and co-workers[25] reported an interesting patient who developed an intense amnestic syndrome in association with isolated lesions of pyramidal cells of the hippocampus after repetitive epileptic seizures. The amygdala in this patient was spared. Thus, hippocampal involvement may be one mechanism that contributes to the amnesia of chronic epilepsy.

Drugs

Benzodiazepines, especially when given intravenously, may cause transient amnesia lasting a few hours, which is very similar to transient global amnesia.[10] Alcohol can also have analogous effects, often called "blackouts."[37] Scopolamine, a cholinergic antagonist, causes a memory loss with many features of the amnestic syndrome.[26]

Huntington's Disease

Memory deficits can occur early in Huntington's disease. The comparison of the memory impairment with those observed in Korsakoff's syndrome shows consistent differences.[13] The existence of this memory deficit raises questions about the possible role of the striatum in memory processes (see section on the striatum in Chapter 1).

Depression

Depression is commonly associated with partial and relatively mild memory disorders. The major problem appears to be at the stage of encoding.

Transient Global Amnesias

In a typical case, the patient, more often a man than a woman, aged between 50 and 70 years, is suddenly struck by global anterograde amnesia, which regresses rapidly, usually within 3 to 24 hours. The acute onset, similar to a stroke, provided justification for the first designation of this clinical picture in 1956 as "amnesic ictus" by Guyotat and Courjon.[38] In 1964, this syndrome was designated as "transient global amnesia" by Fisher and Adams.[29] Anterograde amnesia is total, and awareness of the disorder is acute and is accompanied by considerable anxiety and perplexity; frequently, a constantly repeated phrase is emitted ("What's happening to me?"). Temporal disorientation is always severe; spatial disorientation is variable, but frequently leads to constantly repeated questions ("What am I doing here?"). Retrograde amnesia affects the last few hours preceding the onset of the disorder; certain biographic data—although in no particular order—are also forgotten, whereas others are not. General knowledge is intact, and there are no other accompanying intellectual or neurologic disorders. Recovery is rapid but not sudden; it is quite frequent that, for a few days or weeks, the patient complains of memory disorders.[59] A permanent memory gap remains for the time period in which the disorder occurred and also for an interval of up to an hour before the onset of the impairment. The clinical diagnosis is often retrospective, through accounts provided by family or friends. The authenticity of the memory impairment is no longer questioned.

Transient global amnesia is a syndrome with various possible etiologies.[101] It is conceivable that a common denominator is injury to the hippocampus and associated medial temporal structures. Ischemia in the territory of the posterior cerebral arteries is possible but difficult to prove; the presence of visual troubles in amnesic episodes is an important argument in its favor. Furthermore, the occurrence of transient global amnesia after cerebral angiography supports this etiology. A thalamic infarct can initially resemble a transient global amnesia but does not lead to recovery.[36] Migraine headaches also can cause transient global amnesia. An epileptic mechanism has been suggested as well, but this must be rare and has to be differentiated from epileptic fugue states. Transient global amnesias often occur during a period of high stress: emotional tension, physical effort, a cold bath, physical pain, sexual activity, and so on. This relationship of stress to the syndrome of transient global amnesia remains to be elucidated.

Benign Senescent Forgetfulness

In 1962, Kral described a condition of benign senescent forgetfulness in elderly individuals.[49] This is an essentially indolent condition that does not progress to dementia. The greatest difficulty is for names and details, although events are easily recalled. An item that cannot be recalled at one point can be recalled with ease the next moment. Cueing improves performance, so the major deficit may be at the stage of retrieval. It is difficult to differentiate this from the memory difficulties that appear intrinsic to normal aging.

Psychogenic Amnesias

The presentation of these amnesias is radically different from that of the organic amnesias. Psychogenic amnesias are often interpreted as manifestations of hysteria or as the expression of "forgetting the disagreeable."[82] They are also known as functional retrograde amnesia.[90] Medicolegal issues are often involved. It is perhaps important to note that onset of psychogenic amnesias occurs after strong emotion or psychologic trauma. The loss of personal identity results in patients forgetting their names and their past, whereas their memory for recent events is normal. This syndrome has provided great inspiration to literary writers. The onset is often abrupt and, curiously enough, patients will themselves often report their disorder to the legal authorities or the police, who may then refer them to a hospital. Anxiety in such patients is not very great, although fugue often accompanies this type of amnesia. The disorder ceases after a few days or weeks. It has been said that persuasion, or hypnosis, may have a therapeutic effect.

PATHOPHYSIOLOGY

Anatomic Data

A severe amnestic syndrome occurs only with bilateral lesions in the limbic system. The relevant parts of the brain are closely associated with a circuit described in 1937 by Papez.[75] The components of this circuit and their interconnectivity are as follows: there are projections from the subiculum of the hippocampus to the mammillary body (via the fornix), from the mammillary body to the anterior thalamic nucleus, from the anterior thalamus to the cingulate, from the cingulate gyrus to the presubiculum, from the presubiculum to the entorhinal area, and from the entorhinal area back the the hippocampus.[102] Then there are two additional extensions of this circuit that are relevant to memory: (1) Reciprocal connections exist between the entire hippocampal formation and the septum.[83] (2) There are reciprocal connections between the amygdala and the hippocampus and also projections from the amygdala to the dorsomedial nucleus of thalamus.[86] All of these structures are parts of the limbic system (see Chapter 1). The lesions responsible for the amnestic syndrome affect, in bilateral fashion, one or more of these structures (Fig. 4).

Animal Behavior Data

Mishkin,[71,72] working with monkeys, has studied visual recognition memory in a "delayed nonmatching to sample" task after experimental

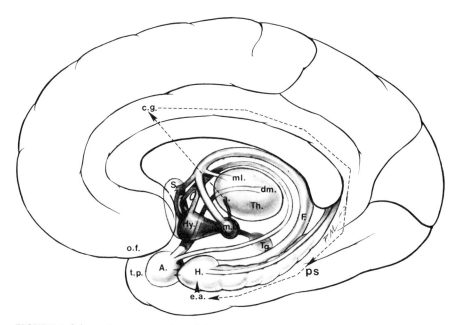

FIGURE 4. Schematic representation of the limbic structures that are relevant to memory. *A.*: amygdala; *F.*: fornix; *H.*: hippocampus; *Hy.*: hypothalamus; *m.b.*: mammillary body; *S.*: septum; *Th.*: thalamus; *a.*: anterior nucleus; *d.m.*: dorsomedial nucleus; *m.l.*: midline nuclei; *Tg.*: midbrain tegmentum; *c.g.*: cingulate gyrus; *e.a.*: entorhinal area; *o.f.*: orbitofrontal area; *ps.*: presubiculum; *t.p.*: temporal pole. The *dotted lines* show (1) connection between anterior thalamic nucleus and cingulate gyrus; (2) the cingulate–presubiculum–entorhinal cortex–hippocampus pathway. (This representation of the limbic system was adapted from another diagram, proposed by J. Bach-evalier, whom we would like to thank for permitting us to use it. This drawing was made by P. McDonnell.)

lesions. A bilateral lesion only of the hippocampus or only of the amygdala leads to a moderate impairment in this task. A bilateral lesion of both the hippocampus and the amygdala is required in order to produce a dramatic impairment of visual recognition memory. Bilateral lesions of the visual association cortex in the inferior temporal cortex (Bonin and Baily's area TE) also cause a dramatic impairment of visual recognition memory, probably by preventing the interaction between visual information and the limbic system. A lesion in the TE area of one side, and the hippocampus and amygdala on the opposite side, causes only a moderate impairment; but sectioning the anterior white commissure, thus preventing visual information from reaching the intact contralateral limbic system, causes a dramatic impairment. This animal model may have some relevance to the case of HM, who had bilateral surgical resection of the hippocampus, the amygdala, and the adjacent areas. In 1982, Duyckaert and co-workers[27] reported a human neuropathologic case of a total anterograde amnesia with bilateral amygdalohippocampal lesions (Fig. 5).

In monkeys, experiments also show the emergence of amnesias after medial thalamic lesion[1] and after mammillary body lesion.[41] As in humans, fornix lesions in monkeys do not always cause amnesia.[31,74] Horel[42] criticized the role of the hippocampus in memory and proposed that temporal stem damage is responsible for the amnesia. However, in

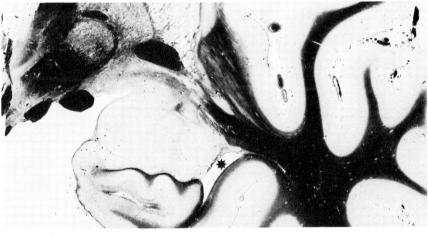

FIGURE 5. A patient with global and complete anterograde amnesia; unknown etiology.[27] Lesions were symmetrical on the two sides. *Top:* Coronal section through the uncus. A reduction in the size of the amygdala results in an enlargement of the temporal horn *(asterisk).* Atrophy predominates in the laterobasal part of the amygdala. Note that the temporal stem is preserved, while the hippocampus is reduced in size (Loyez myelin stain). *Bottom:* Normal amygdala from another patient at about the same coronal level. The amygdala and temporal horn *(asterisk)* are of normal size. (From Charles Foix Laboratory, Hopital de la Salpetriere, Pr. R. Escourolle, with permission.)

monkeys, transection of the temporal stem has not affected recognition memory.[118]

Pharmacologic Data

All limbic and neocortical structures receive cholinergic innervation. This originates from cell bodies in the basal forebrain.[63] In 1976, a decreased level of the enzyme choline acetyl transferase (ChAT) in postmortem cerebral cortex and hippocampal formation was reported in patients with Alzheimer's disease.[22] Since it had been known that memory disorders could be induced in normal volunteers by the injection of the anticholinergic agent scopolamine,[26] the hypothesis was advanced that the amnesia of Alzheimer's disease may reflect a deficiency of central cholinergic transmission.

Other neurotransmitters may play a role, directly or indirectly, in memory processes as well (for review, see Squire and Davis[95]). A deficit in norepinephine, was reported in alcoholic Korsakoff's syndrome.[61] Neuropeptides may also modulate memory storage. In rats, vasopressin enhances retention; oxytocin impairs retention.[8]

ORGANIZATION OF MEMORY

The problem that we face, as far as memory is concerned, is the explanation of how certain units of information will lead to long-lasting cerebral modifications—that is to say, memory traces. Solving the memory problem also requires an explanation of how these memory traces, after a certain timespan, are retrievable, or accessible. We propose to use the following terms: (1) "memorizing process," to refer to the set of processes that leads to the formation and the construction of memory traces; (2) "storing process," to refer to the set of processes that contributes to the continued retention of these traces; and (3) "remembering process," to refer to the set of processes that allows the use of memory traces. These three types of processes can be disrupted differentially or jointly in the course of the amnesias.

The Memorizing Process

The memorizing process is composed of two kinds of processes: the holding process and the acquiring process.

The Holding Process

The information in a given event may exceed the attention span. It is thus necessary that there be a process that holds the information. If not, two possibilities exist: either the information cannot be entirely absorbed, or each successive unit of information will erase the preceding one. The attention span in amnesic patients is normal. However, a defect of the holding process is a factor that is constant but with varying degrees of severity in all patients with anterograde amnesia.[44] This may be the explanation of their vulnerability to interference.

The Acquiring Process

At least three operations can be described in the acquiring process:

1. Encoding capacity and "levels of processing" refers to the level of analysis performed by a subject on the information; the level can range from a "superficial" (or automatic?) analysis limited only to its sensory attributes, to a "deep" analysis examining its semantic or conceptual attributes.[21] This analysis may also include a change in modality (verbal information becomes an image and vice versa).
2. "Chunking" permits a reduction in the quantity of information and obeys certain rules of organization.[112]
3. Linking or binding of information to the context in which it is situated; this includes all the cues that surround the event, as well as their temporal and spatial attributes.[93]

The acquiring process depends necessarily on the holding process; an inadequate holding process leads to a failure to complete the acquiring process.

It is a defect in the acquiring process that Butters and Cermak[12] give importance to in Korsakoff's syndrome and call a "defect of information processing." This defect leads to a deficit for remembering contextual cues[52,115] and temporal organization.[43]

The Storing Process

The storing process could be subdivided into two kinds of mechanisms: the consolidating process and the reconstructing process.

The Consolidating Process

Memory traces are made up of information that has been integrated and that has become long lasting, if not permanent. It is thus legitimate to infer a process of consolidation. This process allows the transfer of a transient memory into permanent storage. An impairment in consolidation or in the acquiring process could account for amnesia due to bitemporal lesion.[69] The acquiring and the consolidating processes are linked. The consolidating process seems to continue operating even after the acquisition of information. One argument in favor of this is that recent but not distant memory traces are selectively vulnerable to retrograde amnesia, perhaps because of insufficient consolidation. The consolidating process is one of the factors at the origin of the "strength" of memory traces; the weakening of these traces seems to be common to all amnestic syndromes and leads to forgetting.[64] The speed of decay or forgetting is increased in amnestic syndromes due to bitemporal lesions but not in alcoholic Korsakoff's syndrome.[45]

The Reconstructing Process

Memory traces do not remain fixed once and for all. They are constantly used to acquire new traces which organize themselves among the old ones. The old traces are thus, so to speak, reassembled. This process of reassembly may be crucial for the upkeep of memory for remote events.

The Remembering Process

Remembering is composed of the retrieving process and the scanning process.

The Retrieving Process

In order to be used, memory traces must be activated so that they are available to the individual according to the requirements of the current situation. The absence of this activation, which is the basis of retrieval, leads to a "false" forgetting. For example, the *transient* retrograde amnesia seen after head injury is likely to reflect a sudden transitory disruption of the retrieving process rather than a destruction of the memory traces.[4,34]

The Scanning Process

Memory traces must be chosen and selected according to the circumstances.[18] This selective choice, which is based on the contextual cues, can be disrupted as shown by the confabulations occurring in Korsakoff's syndrome, in which erroneous verbal responses appear to be beyond the patient's control. If the amnesic patient is provided with specific cues, this leads to an improvement in his performance, perhaps because it assists in the scanning process.[115] Disturbance in the scanning process is all the more evident when there is a weakness in the memory traces. It is probably the association of these two factors, that accounts for the frequency of proactive interference in certain amnestic syndromes. In proactive interference, one or more bits of information that have previously been acquired—rather than the relevant information—are recalled. The permanent and uncontrolled existence of such interference has been proposed as an explanation of amnestic syndromes by Warrington and Weiskrantz.[106] However, the same authors later rejected this interpretation.[107] This change of position is probably because of the fact that proactive interference is observed—not only in patients with the amnestic syndrome but also in those with frontal lesions.[54]

Some Pathophysiologic Speculations

Patients suffering from an amnestic syndrome are capable of learning and retaining certain tasks that are generally perceptuomotor in nature and that do not require language.[76] Perhaps it is necessary to distinguish between habits and memory, as Bergson did.[6] Habits refer to behaviors that are learned without being necessarily integrated into conscious biographic experiences. Habits are relatively well preserved in patients with the amnestic syndrome.

Memory is obviously based on several component processes. We also know from the anatomic data that lesions in many different parts of the limbic system result in memory loss. It is conceivable that each of these limbic structures has a different contribution to make to each one of the psychologic processes. Clinical data gathered from humans lead one to suggest certain tentative correlations between anatomic structures and functional process.[92]

One of the many remaining puzzles is the fact that memory disorders are so much more severe after left posterior cerebral artery infarcts, thalamic lesions, and electroconvulsive therapy than after equivalent right-hemisphere involvement. This leads one to ask whether there might be cerebral dominance within the limbic system. This apparent left-hemispheric dominance could also reflect the dependency on verbal mediation of many memory processes.

TREATMENT, MANAGEMENT, AND FUTURE DIRECTIONS

The link between memory impairment and deficits in cerebral cholinergic activity has led to many therapeutic experiments based on a simple principle: to increase the amount of intracerebral acetylcholine, either by providing a precursor of acetylcholine (choline salt, lecithin), by administering cholinergic agonists (pilocarpine, arecoline), or by preventing the hydrolysis of acetylcholine with anticholinesterase agents such as physostigmine. These trials mainly involved patients with Alzheimer's disease (for review, see Bartus and associates[3]), as well as patients with postencephalitic amnesia[79] and post-traumatic amnesia.[35] Mixed results of the experiments using these patients have been reported and these observations await further confirmation.

Other chemical agents have also been used. After administration of clonidine, which presumably enhances norepinephrine activity, an improvement of the Wechsler Memory score was observed in eight patients with alcoholic Korsakoff's syndrome.[62] A neuropeptide, vasopressin, was also used as a therapeutic agent in amnesia, based on data from experiments on animals.[50] This has also led to mixed success. One should also mention that behavioral strategies have been proposed to rehabilitate memory, particularly in traumatic amnesias.[33]

Although amnestic syndromes have led to a considerable development of knowledge, many questions on amnesias still remain to be answered. To help classify the different varieties of amnesias, it would be desirable to develop a standardized method of assessment that would take into account both the patient's daily activities and the results of neuropsychologic tests. It would be equally desirable to avoid superfluous psychologic analyses devoid of pathologic context. Work on the limbic system and on animal models should be intensified. These developments may all lead to new therapeutic approaches.

REFERENCES

1. AGGLETON, JP, AND MISHKIN, M: *Visual recognition impairment following medial thalamic lesions in monkeys* Neuropsychologia 21:189–197, 1983.
2. ALBERT, MS, BUTTERS, N, AND LEVIN, J: *Temporal gradients in the retrograde amnesia of patients with alcoholic Korsakoff's disease.* Arch Neurol 36:211–216, 1979.

3. BARTUS, RT, DEAN III, RL, BEER, B, AND LIPPA, AS: *The cholinergic hypothesis of geriatric memory dysfunction.* Science 217:408–417, 1982.

4. BENSON, DF, AND GESCHWIND, N: *Shrinking retrograde amnesia.* J Neurol Neurosurg Psychiatry 30:539–544, 1967.

5. BENSON, DF, MARSDEN, CD, AND MEADOWS, JC: *The amnesic syndrome of posterior cerebral artery occlusion.* Acta Neurol Scand 50:133–145, 1974.

6. BERGSON, H: *Matière et Mémoire.* Felix Alcan, Paris, 1896.

7. BICKFORD, RG, MULDER, DW, DODGE, HW, SVIEN, HJ, AND ROME, HP: *Changes in memory function produced by electrical stimulation of the temporal lobe in man.* Res Publ Assoc Res Nerv Ment Dis 36:227–243, 1958.

8. BOHUS, B, CONTI, L, KOVACS, GL, AND VERSTEEG, DHG: *Modulation of memory processes by neuropeptides: Interaction with neurotransmitter systems.* In AJMONE MARSAN, C, AND MATHIES, H (EDS): *Neuronal Plasticity and Memory Formation.* Raven Press, New York, 1981 pp 75–87.

9. BRION, S, MIKOL, J, AND PLAS, J: *Mémoire et specialisation fonctionnelle hémisphérique. Rapport anatomoclinique.* Rev Neurol 139:39–43, 1983.

10. BROWN, J, LEWIS, V, BROWN, M, AND HORN, G: *A comparison between transient amnesias induced by two drugs (diazepam or lorazepam) and amnesia of organic origin.* Neuropsychologia 20:55–70, 1982.

11. BRUN, A, AND ENGLUND, E: *Regional pattern of degeneration in Alzheimer's disease: Neuronal loss and histopathological grading.* Histopathology 5:549–564. 1981.

12. BUTTERS, N, AND CERMAK, LS: *Alcoholic Korsakoff's Syndrome. An Information Processing Approach to Amnesia.* Academic Press, New York, 1980.

13. BUTTERS, N, TARLOW, S, CERMAK, AND SAX, D: *A comparison of the information processing deficits of patients with Huntington's chorea and Korsakoff's syndrome.* Cortex 12:134–144, 1976.

14. CASTAIGNE, P, LHERMITTE F, BUGE, A, ESCOUROLLE, R, HAUW, JJ, AND LYON-CAEN, O: *Paramedian thalamic and midbrain infarcts: clinical and neuropathological study.* Ann Neurol 10:127–148, 1981.

15. CHAPMAN, LF, WALTER, RD, MARKHAM, CH, RAND, RW, AND CRANDALL, PH: *Memory changes induced by stimulation of hippocampus or amygdala in epilepsy patients with implanted electrodes.* Trans Am Neurol Assoc 92:50–56, 1967.

16. COHEN, NJ, AND SQUIRE, LR: *Preserved learning and retention of pattern analyzing skill in amnesia; dissociation of knowing how and knowing what.* Science 210:207–209, 1981.

17. COHEN, NJ, AND SQUIRE, LR: *Retrograde amnesia and remote memory impairment.* Neuropsychologia 19:337–356, 1981.

18. CORBALLIS, MC: *Memory retrieval and the problem of scanning.* Psychol Rev 86:157–160, 1979.

19. CORKIN, S: *Acquisition of motor skill after bilateral medial-temporal lobe excision.* Neuropsychologia 6:255–265, 1968.

20. COURVILLE, CB: *Trauma of the Central Nervous System.* Williams & Williams, Baltimore, 1945.

21. CRAIK, FIM, AND LOCKHART, RS: *Levels of processing: A framework for memory research.* J Verb Learn Verb Behav 11:671–684, 1972.

22. DAVIES, P, AND MALONEY, AJF: *Selective loss of central cholinergic neurons in Alzheimer's disease.* Lancet 2:1403, 1976.

23. DEJONG, RN, ITABASHI, HH, AND OLSON, JR: *Memory loss due to hippocampal lesions.* Neurology 20:339–348, 1969.

24. DELAY, J, AND BRION, S: *Le syndrome de Korsakoff.* Masson, Paris, 1969.

25. DELAY, J, BRION, S, LEMPERIERE, TH, AND LECHEVALLIER, B: *Cas anatomo-clinique de syndrome de Korsakoff post-comitial après corticothérapie pour asthme subintrant.* Rev Neurol 113:583–594, 1965.

26. DRACHMAN, DA, AND LEAVITT, JL: *Human memory and the cholinergic system: A relationship to aging?* Arch Neurol 30:113–121, 1974.

27. DUYCKAERTS, CH, GRAY, F, SIGNORET, JL, AND ESCOUROLLE, R: *Amnesic syndrome with bilateral amygdalohippocampal lesions.* In *IXth International Congress of Neuropathology* (abstract). Vienna, 1982, p 216.

28. ERICKSON, RC, AND SCOTT, ML: *Clinical memory testing: A review.* Psychol Bull 84:1130–1149, 1977.

29. FISHER, CM, AND ADAMS, RD: *Transient global amnesia.* Acta Neurol Scand 40 (Suppl 9):7–83, 1964.

30. FLEMINGER, JJ, HORNE, DJ, HORNE, L, AND NOTT, PN: *Unilateral electroconvulsive therapy and cerebral dominance: Effect of right and left sided electrode placement on verbal memory.* J Neurol Neurosurg Psychiatry 33:408–411, 1970.

31. GAFFAN, D: *Monkey's recognition memory for complex pictures and the effect of the fornix transection.* QJ Exp Psychol 29:505–514, 1977.

32. GASCON, GG, AND GILLES, F: *Limbic dementia.* J Neurol Neurosurg Psychiatry 36:421–430, 1973.

33. GIANUTSOS, R, AND GIANUTSOS, J: *Rehabilitating the verbal recall of brain-injured patients by mnemonic training: An experimental demonstration using single-case methodology.* J Clin Neuropsychol 1:117–135, 1979.

34. GOLDBERG, E, ANTIN, SP, BILDER, RM, GERSTMAN, LJ, HUGHES, JEO, AND MATTIS, S: *Retrograde amnesia: Possible role of mesencephalic reticular activation in long-term memory.* Science 213:1392–1394, 1981.

35. GOLDBERG, E, GERSTMAN, LJ, MATTIS, SJ, HUGUES, JEO, BILDER, RM JR, AND SIRIO, CA: *Effects of cholinergic treatment on post-traumatic anterograde amnesia.* Arch Neurol 39:581, 1982.

36. GOLDENBERG, G, WIMMER, A, AND MALY, J: *Amnesic syndrome with a unilateral thalamic lesion: A case report.* J Neurol 229:79–86, 1983.

37. GOODWIN, DC, BRUCE CRANE, J, AND GUZE, SB: *Phenomenological aspects of the alcoholic "black-out."* Br J Psychiatry 115:1033–1038, 1969.

38. GUYOTAT, J, AND COURJON, J: *Les ictus amnésiques.* J Med Lyon 37:697–701, 1956.

39. HEILMAN, KM, AND SYPERT, GW: *Korsakoff's syndrome resulting from bilateral lesions of fornix.* Neurology 27:490–493, 1977.

40. HIRST, W: *The amnesic syndrome: Descriptions and explanations.* Psychol Bull 91:435–460, 1982.

41. HOLMES, EJ, JACOBSON, S, STEIN, BM, AND BUTTERS, N: *Ablations of the mammillary nuclei in monkeys: Effects on post-operative memory.* Exp Neurol 81:97–113, 1983.

42. HOREL, JA: *The neuroanatomy of amnesia. A critique of the hippocampal memory hypothesis.* Brain 101:403–445, 1978.

43. HUPPERT, FA, AND PIERCY, M: *Temporal context and familiarity of material.* Cortex 13:3–20, 1976.

44. HUPPERT, FA, AND PIERCY, M: *Recognition memory in amnesic patients: A defect of acquisition?* Neuropsychologia 15:643–652, 1977.

45. HUPPERT, FA, AND PIERCY, M: *Normal and abnormal forgetting in organic amnesia: Effect of locus of lesion.* Cortex 15:385–390, 1979.

46. IRONSIDE, R, AND GUTTMACHER, R: *The corpus callosum and its tumours.* Brain 52:442–483, 1929.

47. JONES, MK: *Imagery as a mnemonic aid after left temporal lobectomy: Contrast between material-specific and generalized memory disorders.* Neuropsychologia 12:21–30, 1974.

48. KORSAKOFF, SS: *Sur une forme de maladie mentale combinée avec la neurite multiple dégénérative.* In *Congrès International de Médecine Mentale.* Masson, Paris, 1889, pp 75–94. (Translation by Victor, M, and Yakovlev, PI) Neurology 5:395–406, 1955.

49. KRAL, VA: *Senescent forgetfulness: Benign and malignant.* Can Med Assoc J 86:257–260, 1962.

50. LE BOEUF, A, LODGE, J, AND EAMES, PG: *Vasopressin and memory in Korsakoff syndrome.* Lancet 21:1370, 1978.

51. LEVIN, HS, BENTON, AL, AND GROSSMAN, RG: *Neurobehavioral consequences of closed head injury.* Oxford University Press, New York, 1982.

52. LHERMITTE, F, AND SIGNORET, JL: *Analyse neuropsychologique et différentiation des syndromes amnésiques.* Rev Neurol 126:161–178, 1972.

53. LOISEAU, P, SIGNORET, JL, STRUBE, E, BROUSTET, D, AND DARTIGUES, JF: *Nouveaux procédés d'appréciation des troubles de la mémoire chez les épileptiques.* Rev Neurol 138: 387–400, 1982.

54. LURIA, AR: *The neuropsychology of memory.* John Wiley & Sons, New York, 1976.

55. MABILLE, H, AND PITRES, A: *Sur un cas d'amnésie de fixation post-apoplectique ayant persisté pendant vingt-trois ans.* Revue de Médecine 33:257–279, 1913.

56. MAIR, WGP, WARRINGTON, EK, AND WEISKRANTZ, L: *Memory disorder in Korsakoff's psychosis. A neuropathological and neuropsychological investigation of two cases.* Brain 102:749–783, 1979.

57. MARKOWITSCH, HJ: *Thalamic mediodorsal nucleus and memory: a critical evaluation of studies in animals and man.* Neurosci Biobehav Rev 6:351–380, 1982.

58. Mayeux, R, Alexander, MP, Benson, DF, Brandt, J, and Rosen, J: *Poriomania*. Neurology 29:1616–1619, 1979.

59. Mazzuchi, A, Moretti, G, Caffara, P, and Parma, M: *Neuropsychological functions in the follow-up of transient global amensia*. Brain 103:161–178, 1980.

60. Mc Entee, WJ, Biber, MP, Perl, DP, and Benson, DF: *Diencephalic amnesia: A reappraisal*. J Neurol Neurosurg Psychiatry 39:436–441, 1976.

61. Mc Entee, WJ, and Mair, RG: *Memory impairment in Korsakoff's psychosis: A correlation with brain noradrenergic activity*. Science 202:905–907. 1978.

62. Mc Entee, WJ, and Mair, RG: *Memory enhancement in Korsakoff's psychosis by clonidine: Further evidence for a noradrenergic deficit*. Ann Neurol 7:466–470, 1980.

63. Mesulam, MM, Mufson, EJ, Levey, AI, and Wainer, BH: *Cholinergic innervation of cortex by the basal forebrain: Cytochemistry and cortical connections of the septal area, diagonal band nuclei, nucleus basalis (substantia innominata), and hypothalamus in the Rhesus Monkey*. J Comp Neurol 214:170–197, 1983.

64. Meudell, P, and Mayes, A: *Normal and abnormal forgetting: Some comments on the human amnesic syndrome*. In Ellis, AW (ed): *Normality and Pathology in Cognitive Functions*. Academic Press, New York, 1982, pp 203–237.

65. Mills, RP, and Swanson, PD: *Vertical oculomotor apraxia and memory loss*. Ann Neurol 4:149–153, 1978.

66. Milner, B: *Visually-guided maze learning in man: Effects of bilateral hippocampal, bilateral frontal and unilateral cerebral lesions*. Neuropsychologia 3:317–338, 1965.

67. Milner, B: *Visual recognition and recall after right temporal-lobe excision in man*. Neuropsychologia 6:191–209, 1968.

68. Milner, B: *Hemispheric specialization: Scope and limits*. In Schimtt, FO, and Worden, FG (eds): *The Neurosciences: Third Study Program*. MIT Press, Boston, 1973, pp 75–89.

69. Milner, B, Corkin, S, and Teuber, HL: *Further analysis of the hippocampal amnesic syndrome: 14 years follow-up study of H.M.* Neuropsychologia 6:215–234, 1968.

70. Milner, B, and Taylor, L: *Right-hemisphere superiority in tactile pattern-recognition after cerebral commissurotomy: Evidence for non verbal memory*. Neuropsychologia 10:1–15, 1972.

71. Mishkin, M: *Memory in monkeys severely impaired by combined but not separate removal of amygdala and hippocampus*. Nature 273:297–298, 1978.

72. Mishkin, M: *A memory system in the Monkey*. Trans R Soc London B 298:85–95, 1982.

73. Mohr, JP, Leicester, J, Stoddard, LT, and Sidman, M: *Right hemianopia with memory and color deficits in circumscribed left posterior cerebral artery territory infarction*. Neurology 21:1104–1113, 1971.

74. Moss, M, Mahut, H, and Zola-Morgan, SM: *Concurrent discrimination learning of monkeys after hippocampal, entorhinal or fornix lesions*. J Neurosci 1:227–240, 1981.

75. Papez, JW: *A proposed mechanism of emotion*. Arch Neurol Psychiatry 38: 725–743, 1937.

76. Parkin, AJ: *Residual learning capability in organic amnesia*. Cortex 18:417–440, 1982.

77. Penfield, W, and Mathieson, G: *Memory. Autopsy findings and comments on the role of hippocampus in experiential recall*. Arch Neurol 31:145–154, 1974.

78. Penfield, W, and, Milner, B: *Memory deficit produced by bilateral lesions in the hippocampal zone*. Arch Neurol Psychiatry 79:475–497, 1958.

79. Peters, BH, and Levin, HS: *Memory enhancement after physostigmine treatment in the amnesic syndrome*. Arch Neurol 34:215–219, 1977.

80. Peterson, LR, and Peterson, MJ: *Short-term retention of individual verbal items*. J Exp Psychol 58:193–198, 1959.

81. Poirier, J, Barbizet, J, Gaston, A, and Meyrignac, C: *Démence thalamique. Lacunes expansives du territoire thalamo-mésencéphalique paramédian. Hydrocéphalie par sténose de l'aqueduc de Sylvius*. Rev Neurol 139:349–358, 1983.

82. Pratt, RTC: *Psychogenic loss of memory*. In Witty, CWM, and Zangwill, OL (eds): *Amnesia*, ed 2. Butterworths, London, 1977, pp 224–232.

83. Raisman, G: *The connexions of the septum*. Brain 89:317–348, 1966.

84. Ribot, Th: *Les Maladies de la Mémoire*. Félix Alcan, Paris, 1881.

85. Rose, FC, and Symonds, CP: *Persistant memory defect following encephalitis*. Brain 83:195–212, 1960.

86. Rosene, DL, and Van Hoesen, GW: *Hippocampal efferents reach widespread areas of cerebral cortex and amygdala in the rhesus monkey*. Science 198:315–317, 1977.

87. Ross, ED: *Sensory-specific and fractional disorders of recent memory in man. I. Isolated loss of visual recent memory.* Arch Neurol 37:193–200, 1980. *II. Unilateral loss of tactile recent memory.* Arch Neurol 37:267–272, 1980.

88. Russel, WR: *The Traumatic Amnesia.* Oxford University Press, London, 1971.

89. Schacter, DL, and Crovitz, HF: *Memory function after closed head injury: A review of the quantitative research.* Cortex 13:150–176, 1977.

90. Schacter, DL, Wang, PL, Tulving, E, and Freedman, M: *Functional retrograde amnesia: A quantitative case study.* Neuropsychologia 20:523–532, 1982.

91. Scoville, WB, and Milner, B: *Loss of recent memory after bilateral hippocampal lesions.* J Neurol Neurosurg Psychiatry 20:11–21, 1957.

92. Signoret, JL: *Memory and functional hemispheric specialization: Neuropsychological analysis.* In *Proceedings of the 12th World Congress of Neurology*, International Congress Series 568. Excerpta Medica, Amsterdam, 1981, pp 36–44.

93. Spear, NE: *The Processing of Memories: Forgetting and Retention.* Lawrence Erlbaum, Hillsdale, 1978.

94. Spiegel, EA, Wycis, HT, Orchinik, C, and Freed, H: *Thalamic chronotaraxis.* Am J Psychiatry 113:97–105, 1956.

95. Squire, LR, and Davis, HP: *The pharmacology of memory: A neurobiological perspective.* Ann Rev Pharmacol Toxicol 21:323–356, 1981.

96. Squire, LR, and Moore, RY: *Dorsal thalamic lesion in a noted case of chronic memory dysfunction.* Ann Neurol 6:503–506, 1979.

97. Strich, SJ: *The pathology of brain damage due to blunt head injuries.* In Walker, AE, Caveness, WF, and Critchley, M (eds): *The Pathology of Trauma.* Charles C Thomas, Springfield, 1969, pp 501–526.

98. Sweet, WH, Talland, GA, and Ervin, FR: *Loss of recent memory following section of fornix.* Trans Am Neurol Assoc 84:76–82, 1959.

99. Teuber, HL, Milner, B, and Vaughan, HG Jr: *Persistent anterograde amnesia after stab wound of the basal brain.* Neuropsychologia 6:267–282, 1968.

100. Tissot, R, Constantinidis, J, and Richard, J: *La maladie de Pick.* Masson, Paris, 1975.

101. Trillet, M, Laurent, B, and Fisher, C: *Les troubles transitoires de mémoire. Congrès de Psychiatrie et de Neurologie de Langue Française.* Masson, Paris, 1983.

102. Van Hoesen, GW, Rosene, DL, and Mesulam, M-M: *Subicular input from temporal cortex in the rhesus monkey.* Science 205:608–610, 1979.

103. Victor, M, Adams, RD, and Collins, GH: *The Wernicke-Korsakoff syndrome.* Blackwell Scientific, Oxford, 1971.

104. Victor, M, Angevine, JB, Mancall, EL, and Fisher, CM: *Memory loss with lesions of the hippocampal formation.* Arch Neurol 5:244–263, 1961.

105. Warrington, EK, and Weiskrantz, L: *New method of testing long-term retention with special reference to amnesic patients.* Nature 217:972–974, 1968.

106. Warrington, EK, and Weiskrantz, L: *Further analysis of the prior learning effect in amnesic patients.* Neuropsychologia 16:169–177, 1978.

107. Warrington, EK, and Weiskrantz, L: *Amnesia: A disconnection syndrome?* Neuropsychologia 20:233–248, 1982.

108. Wechsler, DA: *A standardized memory scale for clinical use.* J Psychol 19: 87–95, 1945.

109. Weingartner, H, Grafman, J, Boutelle, W, Kaye, W, and Martin, PR: *Forms of memory failure.* Science 221:380–382, 1983.

110. Weiskrantz, L, and Warrington, EK: *Conditioning in amnesic patients.* Neuropsychologia 17:187–194, 1979.

111. Whitty, CWM, and Lewin, W: *A Korsakoff syndrome in the post cingulectomy confusional state.* Brain 83:648–653, 1960.

112. Wickelgreen, WA: *Chunking and consolidation: A theoretical synthesis of semantic networks, configuring in conditioning S-R versus cognitive learning, normal forgetting, the amnesic syndrome, and the hippocampal arousal system.* Psychol Rev 86:44–60, 1979.

113. Williams, M: *Memory disorders associated with electroconvulsive therapy.* In Whitty, CWM, and Zangwill, OL (eds): *Amnesia,* ed 2. Butterworths, London, 1977, pp 183–198.

114. Williams, M, and Pennybaker, J: *Memory disturbances in third ventricle tumours.* J Neurol Neurosurg Psychiatry 17:115–123, 1954.

115. Winocur, G, and Kinsbourne, M: *Contextual cueing as an aid to Korsakoff amnesics.* Neuropsychologia 16:671–682, 1978.

116. Woolsey, RM, and Nelson, JS: *Asymptomatic destruction of the fornix in man.* Arch Neurol 32:566–568, 1975.

117. Zaidel, D, and Sperry, R: *Memory impairment after commissurotomy in man.* Brain 97:263–279, 1974.

118. Zola-Morgan, SM, Squire, LR, and Mishkin, M: *The neuroanatomy of amnesia: Amygdala-hippocampus versus temporal stem.* Science 218:1337–1339, 1982.

Chapter 5

APHASIA AND RELATED DISORDERS: A CLINICAL APPROACH*

D. Frank Benson, M.D., and Norman Geschwind, M.D.†

Clinical studies of aphasia dating from the last half of the 19th century have contributed significantly to the understanding of the cerebral substrates of language and have provided a model for the elucidation of structure-function relationships in other areas of neurology. The clinical aspects of aphasia are the focus of the present chapter.

Aphasia is defined as a loss or impairment of language function caused by brain damage. It is used by some authors to signify only disturbances of spoken language, but in more common usage it also includes disturbances of other aspects of language function, such as alexia or agraphia—impairments of reading and writing, respectively. The term "related disorders" is used for impairments often associated with or caused by aphasia such as acalculia, apraxia, Gerstmann's syndrome, and so on. The three terms *language, speech,* and *thought* are commonly used in the literature. In neurologic usage *speech* is the term applied to the coordinated muscle activity of oral communication and to the neural control of this activity; *language* is the signal system used by one individual to communicate with another; *thought* designates all forms of mental activity in both linguistic and nonlinguistic form. Although precise non-overlapping definitions are difficult to formulate, one can find relatively pure clinical disorders of each of these functions. Bulbar poliomyelitis and progressive bulbar paralysis cause disordered speech (i.e., impaired articulatory function) without disturbing language or thought. In some types of aphasia, abnormal language and thought occur without disturbance of speech; other types disturb language and

*Some of the work reported here was supported in part by grants from the National Institutes of Health (NINCDS, NS 14018),The Orton Research Fund, and the Essel Foundation.
†Work on this chapter was completed through the stage of galley proofs before the untimely death of Dr. Geschwind.

speech but not thought; while still others manifest only disordered language. Finally, patients suffering from schizophrenia, depression, and some types of dementia can show seriously disturbed thought without significant speech or language abnormality.

As a general rule, one can say that in aphasia the disturbance of language function is manifested by either incorrect grammar or incorrect choice of words. In principle, disturbances of phonology (i.e., the learned rules for producing the sounds characteristic of any given language) would have to be regarded as aphasic. In some cases it is easy to recognize a phonologic disorder—for example, when one well-articulated sound is substituted incorrectly for another (so-called phonemic or literal paraphasias). On the other hand, when the individual sounds are incorrectly produced, it is difficult to separate a disturbance of phonology from a disorder of neuromuscular control. This chapter focuses on disorders of language, but two conditions frequently misinterpreted as aphasia deserve comment first.

Mutism

Of conditions misdiagnosed as aphasia, the most dramatic is the total cessation of verbal output, called mutism. Many causes of mutism are recognized,[30] including such diverse entities as severe laryngitis, damage to the appropriate peripheral nerves, mesencephalic or frontal lesions producing akinetic mutism,[142] left thalamic disturbance (thalamotomy or thalamic infarct),[129] lesions of the supplementary motor area,[124] and acute damage to the left posterior inferior frontal region (Broca's area).[12] Mutism may or may not be associated with aphasia, but in the presence of mutism the existence of aphasia in spoken language cannot be determined until there is sufficient return of verbalization to permit assessment of language function (i.e., correctness of grammar and word choice). One important clue may be available in some cases of mutism. If the patients writes normal sentences, one can predict, with high accuracy, that when speech returns it will be either only very mildly aphasic or not at all so, although articulatory problems may persist.

Dysarthria

Abnormal utterances may result from several varieties of motor disorder and can lead to errors in diagnosis. Motor speech disturbances, which may be spastic, hyper- or hypokinetic, ataxic, and so on,[55] are frequently misinterpreted as aphasia. The rule already stated is that aphasia can be diagnosed only when a disturbance of language, not just a motor speech problem, can be demonstrated. Aphasia and dysarthria often coexist, but one is often observed without the other and they must be recognized as separate disorders.

HISTORIC BACKGROUND

The history of aphasia has been well reviewed elsewhere,[14,26] and we present only a synopsis. Whereas some cases of language loss had been recognized even in antiquity,[28] the history of modern aphasia dates from Paul Broca's description of the brains of patients who have lost language following a vascular accident. Broca[32] suggested that the loss of language observed clinically was secondary to a focal left posterior frontal lesion. This observation not only marked the birth of aphasiology but also helped to usher in the era of clinicoanatomic correlations that became a fundamental aspect of modern neurology. It was soon demonstrated that not all aphasics had frontal pathology. Wernicke[161] described aphasics with a copious but abnormal verbal output (i.e., fluent aphasia with significant problems in language comprehension and in the choice of individual words). He pointed out that aphasics with diminished verbal output (nonfluent) usually had the frontal pathology described by Broca, whereas the fluent aphasics were more likely to suffer from temporal lesions. Many other clinicopathologic correlations were added to this simple dichotomy over the years, on the basis of which increasingly complex classifications of language disturbance were devised. In the hands of some investigators, the classifications became excessively detailed with the inclusion of clinicoanatomic correlations based on inadequate evidence.

A number of investigators challenged the strict localizationist approach.[60,112] Marie[112] argued that there was only one aphasia, a general disturbance of language produced by damage in Wernicke's area. The many clinical variations described by other investigators were regarded by him as mere variations of this basic language disorder produced by extension of the lesion into primary motor or sensory regions. Aphasia itself was thus a single, basic disturbance, a disorder of the central language processing function. This holistic viewpoint gained ascendancy after World War I.[74,87,141,159]

To some extent the holistic approach was a response to the excesses of the localizationist approach,[87] but it foundered on its inability to account for the existence of significantly different language syndromes with a single simplified theoretical structure. There has been a return by most, although not all, investigators, to a localizationist approach, particularly since the mid-1960s. This trend has been furthered by the development of new techniques for imaging the brain such as isotope brain scans, x-ray, and positron CT scans.[21,101,117,121] In addition, psychologists, linguists, speech pathologists, and educators have brought new concepts and methods to the growing body of knowledge.

Definition of the Syndrome Concept

Of paramount importance for discussion of the clinical approach to aphasia is an understanding of the term *syndrome*. Each type of aphasia discussed in this chapter constitutes a syndrome, but this term is easily misinterpreted. A syndrome can be defined as a collection of clinical findings, the simultaneous occurrence of which points with high probability to a common pathogenetic mechanism or a common localization. Even in cases of obscure origin and localization, a syndrome can lead to an accurate prediction of either the future clinical course or of associated clinical findings. The ideal syndrome would be a perfect predictor. In clinical practice, however, there are always individual cases with signs that overlap different syndromes. Syndromes are thus idealized groupings of signs and symptoms. Whereas the aphasia syndromes to be discussed are based on clinical findings, they are neither totally consistent nor immutably stable, and almost no patient exactly fits the descriptions outlined. These descriptions of syndromes have proved highly useful, however, since a large number of aphasic patients do have a cluster of findings that more closely resembles one clinical type than any of the others.[101] Above all, the predictions of localization, clinical course, and associated findings that are based on these findings have a high degree of accuracy.

It should be stressed that the syndromes are useful in two ways. First, they are important clinically. Even in this era of powerful imaging techniques, clinical localization is important; there are many cases of aphasia in which the imaging methods fail to demonstrate the lesion at all or show only lesions irrelevant to the clinical picture. Furthermore, knowledge of the syndromes is very important for prognosis and guidance of therapy. Second, the method of careful examination of patients combined with anatomic study has been the major source of our knowledge concerning the cerebral organization of language. The importance of the syndromes at the theoretical level cannot be overemphasized.

TESTING FOR APHASIA

Correct diagnosis of the syndromes of aphasia depends on the proper use of tests for the major language disturbances. In recent years, tests for aphasia have grown steadily in sophistication; this is true of formal tests (e.g., aphasia batteries[78,128,149]) and of the many research tools designed to probe specific theoretical questions.[76,114,165] Whereas both the formal batteries and the research techniques have proved valuable in advancing the fundamental understanding of aphasia, clinical assessment techniques have advanced considerably and remain the major tool for the clinician.

The clinician, using a fairly simple approach, can reliably diagnose most aphasias rapidly and correctly. Six major language functions should be tested; in addition, three portions of the basic neurologic examination frequently contribute important information. Tables 1 and 2 show the relevance of these basic clinical determinations for the diagnosis of each major syndrome of aphasia (also see Chapter 2 for testing of language).

SPONTANEOUS SPEECH. All examinations for aphasia should start with careful consideration of the patient's verbal production during conversation; most aphasic verbal output can be classed as fluent or nonfluent.[12,97,161] Nonfluent aphasia is characterized by sparse verbal output (less than 50 words per minute), considerable effort to produce words, poor articulation of speech sounds, a tendency for short phrase length (often a single word), loss of some melodic and inflectional qualities of language (dysprosody),[80,119] and a tendency to use only meaningful content words. A salient feature is agrammatism—that is, a tendency to omit the grammatical, syntactically significant function ("closed class") words and the syntactic endings such as verb tense and plurals.[76] In contrast, fluent aphasia is characterized by a normal to excessive, effortlessly produced verbal output (100 to 200 words per minute); absence of dysarthria; normal phrase length (5 to 8 words between pauses); and normal prosodic quality. The language content, on the other hand, lacks substantive, lexically significant words (leading to "empty" speech); and there is a tendency to substitute incorrect, although well-articulated, words (paraphasia). In extreme cases the patient with a posterior lesion produces a very rapid sequence of well-articulated but paraphasic words, to which the term *jargon* is applied. Some patients with posterior lesions may give an appearance of slow speech production because of long pauses before specific content words. An important clue in these cases is preservation of grammatical structure. On the other hand, some patients with anterior lesions will produce a rapid output that is nevertheless easy to distinguish from fluent aphasia since the output will consist of a series of indistinct mumbled sounds from which individual words cannot be made out.

The fluent/nonfluent dichotomy has strong anatomic correlates. Almost without exception patients with fluent, paraphasic verbal output have lesions located posterior to the fissure of Rolando; in contrast, most adult patients with a persistent nonfluent verbal output have an anterior locus of pathology.[12,126,157] Children with acquired aphasia, even those suffering posterior lesions, almost invariably produce an apparently nonfluent verbal output, and the same is true of many young adults. Certain features, particularly dysarthria, phrase length, and agrammatism will only be present in those with an anterior lesion, however. Also note that we have stressed that most patients with *persistent* nonfluent language disorders have anterior lesions, since, in the acute stage, many patients with posterior lesions have nonfluent output that is usually replaced within a few days by a fluently abnormal pattern of language production. It is also important to be aware that the patient with a persistent nonfluent language disorder from an anterior lesion almost invariably suffers from persistent hemiparesis or brachial monoparesis. Aphasias resulting from anterior lesions without hemiparesis or with transient paralysis have an excellent prognosis for recovery.

REPETITION. The ability to repeat spoken language is relatively easy to test. The examiner says single words, phrases, multisyllabic words, and sentences, including some that are rich in function (closed class) words. The patient is then graded on the ability to repeat accurately. Repetition of closed class words is a particularly important test, since errors may be more frequent in repetition of a series of words of this type than in repetition of long words with an equal or even greater number of syllables. Thus, "No ifs, ands, or buts" or "Is he also there now?" may be repeated more poorly than "Hospital administration." Failure in repetition also has reliable anatomic correlates. Aphasics who cannot repeat well usually have lesions involving the perisylvian language structures, either anterior or posterior. In contrast, patients with moderate or severe aphasia who repeat in a normal or nearly normal manner usually have lesions either in the vascular border zone that surrounds the areas of the perisylvian region or in regions even more distant from the perisylvian area (e.g., the frontal parasagittal area). In *mild* aphasia, intactness of repetition may not be a reliable anatomic guide.

COMPREHENSION OF SPOKEN LANGUAGE. The ability to understand is much more difficult to test, at times demanding considerable ingenuity in examination. Protocols for the examination of aphasia often suggest assessment of comprehension by the response of the patient to verbal commands (e.g., "raise your arm," "shut your eyes," "stand up," "walk to the door," and so on), but this approach may be treacherous. Some aphasics are apraxic to verbal command—that is, they fail to carry out verbal commands despite adequate comprehension and intact motor capability. On the other hand, other aphasics may carry out verbal commands for movements involving the trunk muscles or the eyes (e.g., "stand up," "walk," "close your eyes") and yet fail all other tests of comprehension. Questions requiring yes/no responses are probably the most useful single method, but some aphasic patients cannot produce "yes" or "no" reliably. Another method is to ask the patient to point to specific objects or body parts. Once the idea of pointing is established, the patient simply responds to the named object. The language demands of the pointing task can be made complex by use of elaborate descriptions (e.g., "point to the receptacle for the residuals of something that has burned"). Comprehension is rarely an all-or-none phenomenon; most aphasics show some understanding, and most augment their limited comprehension of verbal language by careful monitoring of nonverbal cues such as gesture and prosody. It is easy to overestimate the language comprehension capabilities of an aphasic, if the examiner is not careful to avoid changes in prosody, facial expression, or the use of gestures that accompany the utterances.

NAMING. Testing the ability of the patient to produce the names of objects on confrontation is a standard part of the aphasia examination. The examiner presents an object, the picture of an object, a body part, a color, or a movement and asks the patient to give the name. It is the patient's production of the name that is tested; if the patient fails to name the object correctly but replies accurately to questions such as "Is this chalk?" the patient has failed the naming test but passed a test of comprehension. In addition to the test of confrontation naming, it is often useful to ask the patient to produce lists of words in a given category (animals, automobiles, words beginning with the letter "b," and so on), but failures on this type of testing may also occur in nonaphasic patients, e.g., those with bilateral frontal lesions. Normal individuals routinely produce more than 12 category names in a minute. Some degree of naming disturbance is present in almost every aphasic patient,[77] and naming tests may therefore be used as a screening test for aphasia. A failure to name well is not specifically localizing in itself and should be evaluated in conjunction with other findings. On the other hand, the *types* of errors in naming tend to vary with the localization of the lesion.

READING. Evaluation of reading can be performed in a simple manner, usually by presenting a few words or sentences to the patient. Both reading aloud and reading comprehension are important, but they must be carefully distinguished. The failure only to *comprehend* written material is called *alexia*.[20] Whereas failure to comprehend written material is usually accompanied by incorrect reading aloud, there are many aphasic patients who cannot read aloud correctly but have good comprehension. If the patient successfully comprehends words and sentences, the capacity to comprehend short paragraphs can also be tested.

WRITING. Few individuals write as well as they speak, and the ability to write often deteriorates with brain damage (agraphia). Some causes of agraphia are purely mechanical but others are true language disturbances (i.e., forms of aphasia). It is best to start by asking the patients to sign their name, the most commonly performed writing task. A successful signature does not, however, exclude agraphia as many aphasic patients can sign their name but write nothing else. Patients should be asked to write words to dictation and also to formulate descriptions of the weather, their job, and so on. Disturbances in writing are less useful in localization than are disturbances of spoken language. When spoken production is aphasic in character written language is invariably involved (exceptions must be very rare). On the other hand, abnormal writing can be accompanied by normal spoken language.

NEIGHBORHOOD NEUROLOGIC FINDINGS. Other neurologic abnormalities often accompany many of the aphasic syndromes, including disturbances of cognition or emotion, and also elementary impairments. The accompanying disorders are important both in analyses of mechanism and in localization. Thus, the presence of motor paralysis is an important adjunct finding in aphasia, indicating anatomic localizations that are more precisely defined when correlated with the language disorder findings. Similarly, both sensory loss and visual field disorder can provide useful correlations to the aphasia examination. These basic neurologic functions are sufficiently defined that testing measures need not be outlined here.

APRAXIA. This is the inability to carry out a motor task, which cannot be explained by a basic neurologic disorder. It is another related disturbance, often associated with aphasia, that may provide valuable adjunct information. Tests for apraxia may be subdivided several ways. The physical area of requested movement—e.g., buccofacial (whistle, smile, cough), limb (make a fist, salute, wave goodbye), or whole body (stand, walk, dance)—is of importance. The command for the act, whether verbally requested or demonstrated by the examiner for imitation, is of consequence. Finally, differences between pretended use of an object and actual handling of the real object (i.e., use of a comb or toothbrush) may be indicative, particularly since brain-damaged subjects have a tendency to substitute a body part for a real object when pretending. Although sometimes difficult to interpret, apraxia may prove useful in the understanding of the mechanisms and the localization of language disorders.

THE ANATOMY OF THE LANGUAGE SYSTEMS

We do not discuss here the voluminous literature on the anatomic organization of the language systems that has been built up over the past 120 years, including many very important advances in the past 30 years. What is presented in this section is merely a very brief sketch of the major outlines of the anatomy of language and aphasia that will facilitate understanding the material on the aphasic syndromes.

Figure 1 is a simple diagram of the major structures of the language system. As in other neural systems, there are regions of gray matter in which arriving information is processed. In addition, there are fiber pathways connecting these regions with each other as well as with other parts of the brain. The diagram includes only some of the major cortical centers and pathways. Subcortical regions are omitted as are some of the alternative routes that may be operative after lesions. It should be kept in mind, however, that such alternative routes are probably not fully equivalent functionally to the normal routes.

For the overwhelming majority of humans who have ever lived, language was learned only in auditory verbal form, and the auditory association cortex of the posterior superior temporal lobe (part of Wernicke's area) has an overwhelming importance. One may, as a very simple approximation, consider this area as one that processes the sound patterns of the language, including the patterns of individual words.

Consider how a child learns the name of a seen object. The child must link the name spoken by another individual to the visual image of the object. The angular gyrus region at the temporoparietal occipital junction appears to play an important role in this linkage. The visual impression of the object is transmitted to the primary visual cortex and from this region is transmitted to visual association regions (both in the occipital and temporal lobes), which probably establish neural patterns of seen objects. When the object is seen and the name is heard simulta-

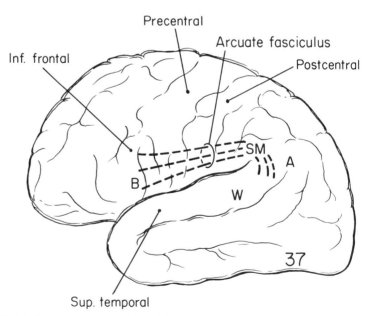

FIGURE 1. A diagrammatic view of the left hemisphere indicating major cortical language areas. A = angular gyrus; B = Broca's area; SM = supramarginal gyrus; W = Wernicke's area; and 37 = Brodmann area 37.

neously, the angular gyrus region probably acts to link the neural pattern of the visualized object to that of the heard word.

In the future when the name is heard it is transmitted to Wernicke's area for recognition of the sound pattern of the word; this in turn arouses the angular gyrus, and a visual memory of the seen object is evoked. Wernicke's area and the angular gyrus are essential way-stations in the comprehension of certain aspects of spoken language.

How does the child learn to say the name? Broca's area lies anterior to the lower end of the rolandic (precentral) cortex, the so-called face area, which controls all of the musculature involved in speech. Broca's area can be thought of as containing the programs for the control of the precentral face area, including the learned rules for converting auditory patterns into spoken form.

When the child first hears a name, the appropriate neural pattern is transmitted from Wernicke's area to Broca's area via the arcuate fasciculus. Broca's area eventually acquires the rules for turning the heard word into spoken form. Thus, three structures, Wernicke's area, the arcuate fasciculus, and Broca's area are essential parts of the pathway for repetition.

Now consider how the child names an object. The sight of the object arouses the angular gyrus and then Wernicke's area (as outlined earlier), which then transmits the neural pattern to Broca's area, whose programs lead the precentral face region to produce the correct movements for the utterance of the name. Thus, Wernicke's area, the arcuate fasciculus, and Broca's area all play crucial roles in the production of language.

How does the child learn to read? For most children, the acquisition of reading skills depends on forming links between the seen and heard word. Thus, the angular gyrus region again plays a major role and is of central importance for reading comprehension. As writing depends on having learned to read, the angular gyrus is also essential for writing. In addition, when reading, the seen word typically arouses the auditory patterns in Wernicke's area, so this region also plays a major role in this language performance. Other mechanisms of reading comprehension probably exist but are not discussed at this time.

A brief note is appropriate on the anatomy of cerebral dominance. Although it had been argued that there were anatomic asymmetries in the language regions, these were neglected or overlooked, until Geschwind and Levitsky[70a] showed that the planum temporale in the posterior part of the upper surface of the temporal lobe (which contains a portion of Wernicke's area) is larger on the left in most brains. The anatomy of asymmetry has been discussed in several papers.[61]

SYNDROMES OF SPOKEN LANGUAGE IMPAIRMENT (APHASIA)

An important outcome of the 19th century investigations of aphasia was delineation of a number of syndromes—that is, clusters of clinical findings that were found to be associated with specific loci of brain damage. "Syndromology" was discouraged during the heyday of the holistic

TABLE 1. Clinical Features of Aphasias With Disturbance of Repetition

| | Language Function | | | | | | | Elementary Findings | | |
	Spontaneous Speech	Auditory Comprehension	Repetition	Naming	Reading	Writing	Motor	Sensory	Visual Field
Broca's	NF	+	–	∓	–*	–	hemiparesis	± hemianesthesia	normal
Wernicke's	F,P	–	–	–	–	–	normal	± hemisensory	∓ quadrantanopsia
Conduction	F,P	+	–	∓	+	–	normal	± hemisensory	normal
Global	NF	–	–	–	–	–	hemiparesis	hemisensory	hemianopsia
Aphemia†	NF	+	–	–	+	+	± hemiparesis	normal	normal
Pure word deafness†	F	–	–	+	+	+	normal	normal	normal

F = Fluent
NF = Nonfluent
P = Paraphasic
+ = Relatively preserved function
*See discussion in text.
†Although not strictly an aphasia, included for convenience.

∓ = Often but not always impaired
– = Almost always impaired
± = Occasionally impaired

approach but has resurfaced in recent years. As we have already noted, despite its shortcomings this approach provides an invaluable framework for investigating the neural basis of language. Recent experience with newer laboratory methods has demonstrated that this century-old system of clinical/anatomic correlation does in fact supply markers of high reliability for localization[101] and prognosis. Some of these new developments in the localization of aphasic syndromes are reviewed in Chapter 11.

The syndromes are discussed in two major groupings: (1) those aphasias in which there is a significant impairment of the ability to repeat spoken language, and (2) those aphasias in which repetition is relatively normal. Table 1 presents the basic clinical findings of the language disorders with disturbed repetition.

Broca's Aphasia

The clinical picture of the type of aphasia named for Paul Broca is the most widely recognized. The verbal output is invariably nonfluent so that a small number of words (usually 10 to 12 per minute) are produced with great effort. The words produced are often unclear, because of poor production of the sounds of language. It is, in fact, almost impossible for the examiner to reproduce the incorrect sounds made by a patient with Broca's aphasia. These patients produce very short phrases, sometimes containing only a single word between pauses. The speech output is characterized by a grossly abnormal melody, inflection, and timbre—that is, by *dysprosody*.[119] In addition, the linguistic content of the verbal output is distinct. While uttering relatively few words, the patient with Broca's aphasia often conveys ideas by producing meaningful substantive words. In sharp contrast, there is agrammatism—a notable decrease of syntactically significant function (closed class) words (e.g., prepositions, articles, and other grammatic modifiers) and of endings (e.g., those marking plurals, possessives, and tenses).[76] The lack of syntactic language combined with the sparse output, loss of melody, and short phrase length compose a complex strongly diagnostic of Broca's aphasia.[80]

Although agrammatism has long been described as a feature of Broca's aphasia,[100,125] only recently has there been proof that this represents a specific, anatomically derived linguistic disturbance. For years it was popularly believed that the patient with anterior aphasia produced almost exclusively substantive words merely as an economy measure, designed to overcome the enormous articulatory difficulties. Marie's[112] remark that Broca's original patient was not truly aphasic emphasized this interpretation, and additional support came from a number of 20th-century investigators.[44,141] The features of agrammatism were precisely described[76] and proved consistent, but that it was a true disorder of syntactic language and not a semiconscious correction for articulatory disorder was not conclusively proved until the past two decades. First, the inordinate problems that patients with anterior aphasia have repeating the function words was noted,[67] and later their inability to comprehend these same words when presented either aurally or written was outlined.[17,62,138,165] Anterior aphasics appear to have an excessive, and apparently specific, disturbance in handling the syntactic aspects of language.

While repetition is almost invariably abnormal, it does vary from case to case and is often superior to spontaneous verbal output. Characteristically, the patient with Broca's aphasia finds it most difficult to repeat the seemingly simple syntactic (closed class) words and endings. Comprehension of spoken language also varies but is considerably better than comprehension of expressive language; rarely, however, is comprehension fully normal. In particular, an inability to comprehend the syntactic function words (e.g., the difference between *in* and *on*) may constitute the greatest comprehension problem. Naming on confrontation is usually abnormal in Broca's aphasia but with distinctive characteristics. Although failing to name on confrontation, the patient often produces the correct word when a *phonetic* cue (the beginning sound of the word) or an open-ended *contextual* cue ("you pound a nail with a _____") is presented. While patients with other types of aphasia may also respond to a cue, this trait is seen most clearly in the anterior aphasias.

Reading comprehension is often impaired in Broca's aphasia, most often as the result of inability to comprehend the grammatically significant words.[138] Thus, the patient with Broca aphasia often understands written substantive words but cannot comprehend their relationships. Writing, even with the unaffected left hand, is almost invariably abnormal; if exceptions exist, they must be extremely rare. The written production is usually large, scrawling, and poorly formed but is also aphasic, since grammatical words are routinely omitted and substantive words are often misspelled.

Broca's aphasia is usually accompanied by a right hemiplegia in the early stages; when there is a persistent Broca's aphasia, a hemiplegia is very nearly invariable. Characteristically the patient has little use of the right upper extremity and limited use of the lower extremity. Ideomotor apraxia of the left arm to verbal command is commonly present; that is, when given a verbal command to use the left hand, aphasics may fail to make the desired movement, even though they can correctly indicate whether a movement made by the examiner is the correct one. The patient thus shows comprehension of the spoken command. Although about 35 percent of those apraxic to verbal command will imitate correctly the movement carried out by the examiner, about 65 percent will fail. Thus, apraxia on imitation is present in most cases. A smaller number will fail even in handling the actual object (i.e., they are apraxic in the handling of objects). Thus, ideomotor apraxia affecting the "normal" left arm and leg is a common feature of Broca's aphasia. Facial apraxia—failure to make correct movements of the cranial musculature (excluding those of the eyes)—is even more common.

Wernicke's Aphasia

The second major type of aphasia, also named for the neurologist who originally described the findings, is Wernicke's aphasia. The language characteristics differ considerably from those of Broca's aphasia (see Table 1): the verbal output is fluent with normal or even above-normal rate, ease of production, normal articulation, normal melodic and inflectional qualities, and normal phrase length. Grammatical structure is usually normal or nearly so. The content is, however, abnormal, because of a lack of words of specific meaning. Substitution of nonspecific words (it, thing, us, and so on) is common and, thus, despite a copious verbal output, little information is conveyed. This emptiness is further enhanced by a tendency for paraphasic substitutions. In Wernicke's aphasia the most common form is substitution of one word for another (*verbal* or *semantic* paraphasia), but phonemic substitutions (*literal* or *phonemic* paraphasia) or even nonexistent words (*neologisms*) may be produced. An excess of nonspecific words, multiple paraphasic substitutions, and a lack of significant content words often destroys or markedly reduces the ability of the patient with Wernicke's aphasia to communicate except by gesture and inflection.

Patients with Wernicke's aphasia show a serious impairment of the ability to comprehend spoken language and a similar disturbance of repetition. If able to understand a few words, they will also be able to repeat a few words. One aspect of auditory comprehension may be spared: the performance of axial movements (those involving movements of the trunk or eyes) to verbal command (close your eyes, turn around, stand at attention, and so on) is usually relatively preserved—in some cases dramatically so. By contrast, these patients usually fail to carry out commands involving the face or a single limb, and they also fail on other forms of comprehension testing. The ability to name is almost always seriously impaired. Reading comprehension and writing are impaired, usually severely, but the degree of impairment of written and of spoken language comprehension may not be equal. Some patients with Wernicke's aphasia are more "word deaf" than "word blind," whereas others show the opposite tendency, comprehending spoken language somewhat better than written.[88,95] Apraxia to verbal command is difficult to demonstrate in Wernicke's aphasia, because the spoken command is not understood. Some of the patients will, however, be apraxic on imitation and object handling. Hemiparesis is rare, except in the first few days after onset when it is sometimes seen; sensory loss is sometimes demonstrable, but is not helpful diagnostically. Some patients with Wernicke's aphasia may have a visual field defect, classically a superior quadrantanopia indicating involvement of the geniculocalcarine pathway as it courses through the temporal lobe.

In some patients with Wernicke's aphasia, there is striking lack of concern (a finding not seen in those with Broca's aphasia), which may be replaced by paranoid behavior. The combination of agitated paranoid behavior and abnormal language easily leads to a mistaken diagnosis of acute psychosis. The clinician should remember that acute psychoses in patients who come to medical attention with grossly abnormal speech are very rare compared with those in patients with Wernicke's aphasia, especially in middle age.

Conduction Aphasia

Wernicke postulated the existence of another language syndrome, resulting from separation of the auditory language comprehension center from the verbal motor center, and suggested the term *Leitungsaphasie* (conduction aphasia). His theoretical description differs somewhat from that of the syndrome now recognized, and for years the entity was considered either nonexistent or extremely rare. Recent investigations have shown that conduction aphasia not only exists but is relatively common.[14,81] It can be demonstrated, however, only if the language evaluation includes tests of repetition. Table 1 outlines the primary characteristics of conduction aphasia.

The verbal output in conduction aphasia is fluent and paraphasic; the paraphasias are characteristically literal, with verbal paraphasias and neologisms less common. The ability to comprehend spoken language is relatively well preserved, but repetition is dramatically difficult. The sharp difference between preserved comprehension and impaired repetition characterizes conduction aphasia. While unable to repeat the words of the examiner, the patient often demonstrates clear understanding and memory of the word (e.g., although failing to repeat the name of the President of the United States, the patient may correctly describe him). Naming is usually impaired, most often because of multiple literal paraphasic substitutions. Thus, when asked to name an object the patient may present a word with the right number of syllables and correct inflection but with one or more incorrect phonemes, so that the response is incorrect. Despite this failure, the patient often insists that he knows the correct name and can select it from a group offered by the examiner.

Reading aloud is severely disturbed in conduction aphasia, but reading for comprehension is relatively preserved. Writing is disturbed but sometimes to a milder degree than is spoken language. The patient produces real letters and real words but makes spelling errors, omits words, and alters the sequence of letters and words.

Conduction aphasia may be accompanied by paresis or visual field loss, but neither is a necessary feature. Sensory loss is more common than in Broca's aphasia but is also not universal; it is often a true cortical sensory loss, involving position sense and stereognosis but not pain. Apraxia on verbal command, imitation, and object handling is common in conduction aphasia, but it too may be absent.[22]

The traditional location of pathology in cases of conduction aphasia was in the supramarginal gyrus; that is, the gyrus lying above and around the posterior end of the sylvian fissure, and in the subjacent white matter tracts (the arcuate fasciculus). This lesion separates the temporal from the frontal language area.[53,67,107] Although this is the most common lesion, many exceptions have been reported.[22,105,116] In some cases only the cortex of the supramarginal gyrus is involved. This may reflect the fact that many of the fibers of the arcuate fasciculus do not run directly between the two language areas but have synapses in the parietal operculum. In other reported cases the dominant Wernicke's area has been destroyed. The preservation of comprehension suggests that in these cases the right temporal area is capable of comprehension of spoken language.[105]

Aphemia

This disorder has been discussed under many names, including cortical dumbness, anarthria, and subcortical motor aphasia. It is likely that aphemia is not a true aphasia, because these patients can express them-selves normally in written language and fully understand both written and spoken language.[14] When spoken language returns, although it may be poorly articulated, slow and effortful, both word choice and syntax are normal or nearly so. Aphemia thus appears to be a disturbance of motor verbal output alone, to be classed as a speech disturbance and not a language disorder.[140] Aphemia must, however, be considered in the dif-ferential diagnosis of aphasia.

In the acute stage, patients with aphemia are often mute. Although unable to converse, they can write full sentences correctly (or nearly so) that express their thoughts and desires. Similarly, comprehension of both spoken and written language is intact or nearly intact, but repetition is usually as impaired as spontaneous verbal output. Naming cannot be performed in a verbal mode, although the patient can write the names of objects. A right hemiparesis is often seen but is usually transient.

The location of pathology in cases of aphemia has remained unset-tled, but two closely connected sites are most often described. There may be a lesion within Broca's area itself—an almost purely cortical lesion involving a small part of the left frontal opercular region.[118,147] Alterna-tively, there is a subcortical lesion that undercuts Broca's area, possibly disconnecting its outflow channels.[11]

With recovery, an abnormal articulatory output appears, often hypophonic, breathy, and poorly articulated. In contrast to that of Broca's aphasia, the output is grammatically intact or very nearly so, even in the early stages of recovery. With full recovery a prominent alteration in the prosodic quality of output,[119] the so-called foreign-accent syndrome, is often noted; the patient produces a linguistically normal output with an altered inflection.

A syndrome that can resemble aphemia occurs when pathology affects the left medial frontal cortex, the supplementary motor area, and/or the cingulate gyrus. Mutism or severely sparse verbal output with nor-mal language characteristics is noted originally, usually accompanied by either akinesia[54] or paresis of the contralateral lower extremity and prox-imal upper extremity.[45] Unlike in aphemia, writing is usually limited and recovery most often leads to a transcortical motor aphasia (see subse-quent discussion).

Pure Word Deafness, Auditory Agnosia, and Cortical Deafness

Pure word deafness is characterized by an inability to comprehend spo-ken language, even though the patient is not truly deaf (nonverbal sounds are correctly identified). In early stages some degree of paraphasic verbal output may be present but this usually disappears. The patient with pure word deafness has a normally fluent verbal output, severe dis-turbance of spoken language comprehension and repetition but adequate naming and, most significantly, no problems with reading or writing. Normal verbal output and retention of the ability to comprehend written material in the face of severe spoken language comprehension distur-bance characterize pure word deafness and separate it from Wernicke's aphasia. As can be anticipated, there are cases in which the two syndromes overlap, which produces a "word deaf" variation of Wernicke's aphasia, a fluent aphasia in which comprehension of spoken language is far more impaired than that of written language.[95]

Classically, no additional neurologic abnormalities are present in the individual with pure word deafness, although the presence of a superior quadrantanopic visual field defect may be present.

Two totally different sites of pathology have been described in pure word deafness.[159] In the classic form, a single lesion involves the region of the dominant hemisphere Heschl's gyrus and the white matter pathways coming into this area. The white matter lesion is thought to destroy both the auditory projection from the medial geniculate nucleus and the callosal fibers from the opposite superior temporal region. This lesion thus effectively isolates the left-hemisphere auditory association cortex from auditory input. A second, different localization is also well documented; in this localization, temporal lobe damage, most often affecting middle portions of the first temporal gyrus, is present bilaterally. Auerbach and associates[8] have argued that this variation produces somewhat different linguistic disturbances, but the number of cases studied so far is far too small to permit the use of these features for clinical diagnosis.

Pure word deafness must be separated from two other conditions. The first is *auditory agnosia*, a much misused term given several different definitions. For many, auditory agnosia indicates a loss of higher function dependent on auditory input, which can be divided into the two main groups, verbal and nonverbal. *Verbal auditory agnosia*, an inability to interpret spoken language despite intact hearing, is thus synonymous with pure word deafness. In contrast, *nonverbal auditory agnosia* describes an inability to interpret (recognize) nonverbal sounds, such as the ringing of a telephone or the bark of a dog, despite adequate hearing. The two disorders can occur independently.[156] Reports in the literature suggest that the two types of auditory agnosia are dependent on damage to different temporal areas, nonverbal auditory agnosia occurring predominantly with right-hemisphere involvement.[8,156]

A differential diagnostic consideration in cases of pure word deafness concerns *cortical deafness*. It has been stated that bilateral involvement of Heschl's gyrus or its connections can lead to failure to interpret either verbal or nonverbal sounds, although an awareness of the occurrence of the sound can be demonstrated. Cortical deafness thus shows the lack of spoken language comprehension that occurs in pure word deafness plus an inability to name or otherwise interpret nonverbal sounds. While it has generally been assumed that bitemporal damage underlies this condition, there are few actual case reports and many of them do not adequately document examination for dissociation of verbal and nonverbal hearing. It is conceivable that in some cases selected brainstem or diencephalic lesions, if bilateral, could also be responsible. This issue requires further study.

Global Aphasia

Global or total aphasia is a common language disturbance in which most language functions are impaired (see Table 1). The verbal output is nonfluent and comprehension of spoken language, repetition, and naming are all severely compromised. In addition, these patients neither read nor write and cannot be tested for apraxia to verbal command because of severe comprehension disturbance. Global aphasia is usually accompanied by a right hemiplegia but can occur with little or no motor or sensory disturbance; this is most common when the global nature of aphasia is transient. In general, the lesion producing persistent global aphasia is of sufficient size to involve both frontal and parietotemporal language regions. Conceivably, small, correctly placed lesions could produce global aphasia without motor or sensory loss, but in most instances the area damaged is large, encompassing much of the language cortex. Very large anterior lesions sometimes produce this clinical picture in the acute stage, but the clinical picture alters to that of severe Broca's aphasia in most such cases.

Transcortical Motor (TCM) Aphasia

Some patients with true aphasia retain considerable competency in repeating spoken language. Wernicke[161] theorized that in these cases, for which he used the term *transcortical aphasia,* there was a second, uninvolved pathway. Several varieties of transcortical aphasia were subsequently described and the data were summarized in the book *Die transkortikalen Aphasien.*[73] The Wernicke theory was found faulty and abandoned, but the term *transcortical* has persisted as a purely descriptive term to indicate retention of repetition in aphasia. The term *transcortical* is applied to three syndromes, but several other language disturbances, most prominently anomic aphasia, also show relatively normal repetition (Table 2).

Transcortical motor (TCM) aphasia is characterized by considerable difficulty in spontaneous language production (Table 2), often a halting, dysarthric, incomplete verbalization. In sharp contrast, when the patient is requested to repeat a sentence spoken by the examiner, the verbal output is almost flawless, an attribute that clearly separates TCM from Broca's aphasia. Associated with the dramatic discrepancy between spontaneous and repeated verbal output is a relatively well-preserved ability to comprehend spoken language. Difficulty in naming is often present. In another contrast to Broca's aphasia, patients with TCM aphasia often read adequately (both aloud and for comprehension). They do, however, show a significant degree of agraphia.

TCM aphasia is often accompanied by hemiplegia. Hemisensory loss and/or a visual field disturbance may occur, but neither is consistent. Apraxia to verbal command involving limb movements of the nonparalyzed side may be present but is again inconsistent.

Lesions underlying TCM aphasia were originally found in the frontal lobe, either low (anterior to Broca's opercular area) or high on the lateral convexity (above Broca's area). In recent years the characteristic findings of TCM aphasia have also been described during the recovery stage from dominant (left) anterior cerebral infarction.[5,135] The lesion in these cases involves the supplementary motor area (SMA) in the medial

TABLE 2. Clinical Features of Aphasias Without Disturbance of Repetition

	Language Function						Elementary Findings		
	Spontaneous Speech	Auditory Comprehension	Repetition	Naming	Reading	Writing	Motor	Sensory	Visual Field
Transcortical Motor	NF	+	+	−	+	−	∓ hemiparesis	normal	normal
Transcortical Sensory	F,P	−	+	−	−	−	± hemiparesis	hemisensory	hemianopsia
Mixed Transcortical	NF	−	+	−	−	−	hemiparesis	hemisensory	hemianopsia
Anomic	F	+	+	−	±	∓*	normal	normal	normal

F = Fluent
NF = Nonfluent
P = Paraphasic
+ = Relatively preserved function

− = Relatively impaired function
∓ = Often but not always present
± = Occasionally present
*See text

parasagittal region of the frontal lobe, usually the left. In a recently reported investigation of CT lesions in cases with TCM aphasia, Freedman and co-workers[59] demonstrated pathology involving a wide area of dominant frontal lobe but always in a position to separate the medially located SMA from Broca's area. Thus, TCM aphasia appears in some cases to be the result of SMA damage, whereas in other cases to be the result of disconnection of the SMA from Broca's region. Other mechanisms may yet be elucidated.

Transcortical Sensory (TCS) Aphasia

A second, quite different transcortical aphasia resembles Wernicke's aphasia, except that patients with this disorder can repeat spoken language (see Table 2). The verbal output in patients with transcortical sensory (TCS) aphasia is fluent, usually contaminated with paraphasia, and its content is often unrelated to the questions or conversation of the examiner. Although the patient does not appear to comprehend spoken language, repetition is relatively and often dramatically preserved. The patient often repeats (or even paraphrases) the question of the examiner but then may discuss a totally unrelated topic. Because this disorder is frequently accompanied by dramatic agitation, the patient with TCS aphasia is often misdiagnosed as a psychotic with "schizophrenic word salad," and many such patients have been admitted to psychiatric units. It should be kept in mind, however, that schizophrenic word salad is extremely rare and that it is seen not so much in patients suffering an acute schizophrenic break as in chronically neglected psychotic patients.

The remaining clinical findings of TCS aphasia are similar to those of Wernicke's aphasia—that is, the patients neither comprehend written language nor write it, and they also fail in naming. Some degree of paresis and significant hemisensory disturbance and/or visual field defect are not uncommon.

The minimum extent of the pathology underlying TCS aphasia remains unsettled. Damage in the border-zone region of the parietotemporal junction is found most often. Whereas the immediate perisylvian temporal and parietal structures are preserved, the cortex around them—particularly the angular gyrus and the second and third temporal gyrus areas—are often damaged.[102] Few well-described clinical cases have come to detailed postmortem examination.

Mixed Transcortical (MTC) Aphasia

An unusual disturbance, sometimes called isolation of the speech area,[71,74] can be considered analogous to global aphasia except for the ability to repeat. The syndrome is both rare and clinically unique (see Table 2). These patients do not speak unless spoken to; they then repeat (echo) the remarks of the examiner, but no conversational language follows. Careful testing shows no competence in the usual tests of comprehension, naming, reading, or writing. In some cases, repetition of spoken language is only relatively preserved but even this lesser degree of sparing is dramatic in comparison to failure in all other language modes. In other cases, repetition may be nearly faultless.

The few cases of mixed transcortical (MTC) aphasia studied at postmortem have had a consistent localization of pathology (i.e., damage in both the posterior and anterior vascular/border-zone areas). The perisylvian language area structures are preserved, but there is widespread destruction of the surrounding cortex of the language-dominant hemisphere. The etiology of several of the originally described cases was hypoxia, most often from suicide attempts with gas, but both acute carotid occlusion and head trauma with severe cerebral edema have also caused the MTC picture.[14]

Anomic Aphasia

This is the most common of all aphasias, and, in the purest examples, the inability to find the correct word, both in spontaneous speech and writing, and in naming on confrontation by speech or writing are the only observed defects. In such patients, the verbal output is fluent, repetition and comprehension are preserved, and the patient can comprehend written language. The ability to find words in writing, although never fully intact, may be conspicuously better than in speaking (see Table 2). A more common clinical picture is less pure; in particular, some degree of impairment in comprehending spoken and written language and a more severe impairment in writing are frequent. Anomic aphasia, not infrequently persistent, is often the most notable residual defect following recovery from one of the other types of aphasia. In this situation, a patient with the characteristics of anomic aphasia may, particularly when tired, show remnants of prior language impairments.

As a syndrome, anomic aphasia is often difficult to localize. When there is an acute onset of a severe anomic aphasia, the most commonly involved area is the left temporoparietal junction. On the other hand, when there is a mild anomic aphasia, localization is much less reliable, In this case, dominant-hemisphere frontal, parietal, and temporal pathology may be found, especially in the recovery phase from a more readily localizable syndrome. Even right-hemisphere and subcortical damage can cause a comparable clinical picture. It should also be stressed that when a space-taking lesion is present in either hemisphere, the patient may manifest an anomic aphasia. In the presence of this aphasia, therefore, one must be very cautious about localization. Furthermore, brain tumor must rank high in the differential diagnosis of a slowly progressive anomic aphasia. Anomic aphasia is also the most common form seen after head injury and in the dementia of the Alzheimer type. Anomia is also common in metabolic or toxic disorders, although in this situation certain other clinical features are helpful. The patient often exhibits a much more severe impairment of writing than of speaking, and features of a confusional state are usually present as described in Chapter 3.

Variations in the type of word-finding impairment can provide valuable evidence for more exact localization (see the section on syndromes of written language impairment), especially when the aphasic syndrome is not a pure anomic aphasia. Pure anomic aphasia remains difficult to localize.

Another clinical point that deserves emphasis is the common error of applying the term *anomic aphasia* to any patient who has trouble naming on confrontation. Since this difficulty is almost invariably present in all patients with aphasia, it is both unnecessary and misleading to describe a patient as suffering from, for example, both Broca's aphasia and anomic aphasia. The term *anomic aphasia* is appropriately confined to those patients with fluently abnormal language output whose repetition and comprehension are much better preserved or intact.

Subcortical Aphasia

Although some cases of aphasia from pathologically proven subcortical lesions were described in the past, the extensive use of computerized tomography (CT) as a diagnostic tool has demonstrated that a number of patients who formerly would have been diagnosed as suffering from aphasia due to cortical infarction have subcortical damage—either hemorrhage or infarction. As a result another, clinically separable, aphasic syndrome has been outlined. Table 3 presents an outline of the major clinical characteristics of the subcortical aphasias but must be considered tentative as the number of cases studied in detail remains small. Furthermore, some cortical lesions can produce findings that resemble the clinical picture of subcortical aphasia. Infarction or hemorrhage in the caudate,[52] putamen,[120] and thalamus,[4,35] as well as in the supplementary motor area and other portions of the anterior cerebral artery territory,[135] have produced the clinical picture outlined in Table 3. One major problem in separating different syndromes of subcortical aphasia is that many different gray matter structures and fiber tracts are tightly packed together in the basal regions. One must be cautious about attributing the clinical syndrome found in any particular case to the most severely involved region of subcortical gray matter. Subcortical aphasia usually has an acute onset with mutism or notably hypophonic (soft) voicing. In the early recovery stages, hypophonia is consistent. The verbal output tends to be slow and poorly articulated (slurred), but other articulatory qualities vary considerably. Some patients show a significant spastic dysarthria (i.e., high pitched slow and effortful speech), while others produce either hyper- or hypokinetic speech (i.e., rapid or slow speech, respectively, with very slurred articulation). The dysarthria is characteristically complicated by the presence of paraphasia. Although characteristically present in the subcortical aphasias, paraphasias may be difficult to determine, because of the poor articulation. One striking finding is dis-

TABLE 3. Clinical Features of Subcortical Aphasia

1. Acute onset:	Usually mutism, hemiparesis, and/or hemisensory loss
2. Improve to:	a. Hypophonic, paraphasic output (fluent or nonfluent)
	b. Repetition without paraphasia
	c. Comprehension, naming and reading variable but often relatively good
3. Recovery to either:	a. Nonaphasic state
	b. Residual aphasia

appearance of paraphasia during repetition. During verbal output, repetition, although not always normal, often improves, primarily because the quantity of paraphasia decreases. Another important characteristic of subcortical aphasia is a strong tendency for improvement; pure subcortical aphasia is frequently a transient condition.

To date, most studies of subcortical aphasia have used either neuropathologic or x-ray CT data for localization. An apparently paradoxic finding becomes apparent: whereas a specific aphasia syndrome (subcortical aphasia, Table 3) can be correlated with a specific neuroanatomic locus, total clearing of the language impairment often occurs despite the continued presence of structural damage to the subcortical structures. Thus, several months following onset, many individuals with subcortical aphasia have a return to normal language production despite persistent evidence of significant damage to the subcortical structures.[14] The paradox has been clarified somewhat by positron CT studies. In the few subcortical aphasia cases studied by this method,[117] a striking metabolic disturbance was found not only in the area of subcortical damage but also in portions of the dominant hemisphere cortex. In the few cases restudied later, the cortical hypometabolism had disappeared in those whose language had recovered but remained in those with residual aphasia. This suggests but does not prove that the aphasia that follows subcortical damage may be based on altered cortical dysfunction. Either an acute loss of essential activation from subcortical structures or an actual change in cortical function resulting from vascular insufficiency is a possible mechanism. Subcortical aphasias are not rare, but much more study will be required to elucidate their clinical pictures and mechanisms.

SYNDROMES OF WRITTEN LANGUAGE IMPAIRMENT (ALEXIA AND AGRAPHIA)

From the early days of the study of aphasia, it was recognized that brain lesions could lead to impairments of the production and comprehension of written language. As the majority of the population was then illiterate, these lesions did not produce functional deficits in many patients. As literacy expanded, however, the acquired inability to read (alexia) and write (agraphia) became significant problems. Interest in alexia and agraphia grew against a background of organized knowledge of aphasia. It was stimulated in 1891 and 1892 when Dejerine published two cases in which an acquired inability to read was the major problem, one associated with agraphia, the other not, and with different anatomic locations of the underlying pathology; these findings have been confirmed repeatedly. Alexia with agraphia and alexia without agraphia were almost immediately recognized as specific language disorders.[163] Despite the powerful confirmatory evidence, these differentiations were generally neglected in the English-speaking world, until the appearance of a number of studies in the 1960s.[1,20,39,66] It should be stressed, however, that these syndromes were continuously recognized by French-language neurologists. Despite renewed interest during the past two decades, no complete consensus exists as to terminology. This section presents a simple

TABLE 4. Language Changes in Alexia

	Posterior Alexia	Central Alexia	Anterior Alexia
1. Reading	very poor	very poor	partially impaired
2. Writing, dictation, copying	no agraphia; slavish	severe agraphia; slavish	severe agraphia; poor, clumsy
3. Comprehension of spelled words	good	very poor	some success
4. Spelling aloud	good	very poor	poor
5. Verbal output	normal	+	nonfluent aphasia
6. Letter naming	*	severe letter anomia	severe letter anomia
7. Paralexia	semantic paralexia occasional	semantic paralexia frequent	semantic paralexia rare

* = Varies from severely impaired to unimpaired; usually relatively good.
\+ = May be normal; anomic aphasia of some degree frequent.

clinical classification of alexia; writing impairment is discussed in another section.

Table 4 offers a classification of the alexias. Included are the two varieties whose pathology was described by Dejerine, which in this classification are called *posterior alexia* (alexia without agraphia) and *central alexia* (alexia with agraphia). We will also discuss two other syndromes of reading impairment that have been reported recently.

Posterior Alexia (Alexia without Agraphia)

One of the most dramatic entities in the language impairment literature is the acute loss of the ability to read despite full retention of the ability to write. This syndrome was already known before the first description of the causative lesion by Dejerine,[56b] and many cases have been described since that time.[50,69] The major clinical finding is an acute inability to comprehend written material. By contrast, the patient readily recognizes words spelled aloud, words spelled on the palm, or words formed with anagram blocks that can be palpated. Only when spelled words are presented visually is there a problem in recognition. Some patients cannot recognize any written letters but, more often, some individual letters can be identified. With considerable practice these patients learn to read most letters out loud. When this happens, words can be spelled aloud by the patient and then recognized auditorily.

Some associated neurobehavioral findings are frequent. A complete right homonymous hemianopsia is present in many cases, but there are exceptions.[84,150] In some cases, a partial visual field defect may be observed.[85] Impaired naming and understanding of color names (sometimes called "color agnosia") in the presence of intact or nearly intact color vision is common. On the other hand, some patients suffer from an actual impairment of color vision *(achromatopsia)*, and some have no problem with color recognition.[50] A mild anomia is common but not always present. In most cases, there are no other aphasic disturbances or abnormalities of the primary motor or sensory systems. The related deficits that these patients may have in complex visual processing are described further in Chapter 7.

The most commonly reported pathology is occlusion of the dominant (left) posterior cerebral artery, which leads to infarction of both the left occipital lobe (causing partial or complete right homonymous hemianopsia) and the splenium of the corpus callosum. There is considerable evidence for the mechanism of this syndrome. The clearest cases are those in which there is a complete right hemianopsia, so that the patient cannot read in the right visual field. Words can be seen only in the left visual field and are therefore projected only to the right hemisphere. The lesion in the splenium prevents transfer of the visual information from the right hemisphere to the left, and, as Damasio[51] has pointed out, even when the right hemianopsia is only partial, the lesion of the left occipital lobe is so placed as to prevent transfer of the visual image from the intact portion of the left visual cortex to the angular gyrus area. Thus, visual-verbal disconnection primarily affects the written word and colors.

The patients can see (they copy written language as though it were a foreign language), can name objects, can recognize people, and otherwise show intact visual function, but cannot identify words in visual form, a task involving pure visual-verbal association. In Table 4 the language features of the various alexias are outlined.

Central Alexia

Dejerine[56a] described the postmortem findings of another type of alexia, one that may be best characterized as a return to illiteracy; that is, an acquired inability to read, write, or carry out any other activity that can be performed only by those who have learned to read. In Dejerine's case, there were only minimal language and neurologic residua other than alexia and agraphia. Many published cases are closely similar to Dejerine's; others manifest a greater degree of neurologic and behavioral abnormality. The primary clinical findings are readily characterized as an acquired impairment in the ability to recognize and produce written language. These patients do not recognize words spelled aloud, written in the palm, or composed of anagram blocks that are palpated. They do retain some ability to copy written language but perform this slavishly and without comprehension.

Although the purest cases have little in the way of other neurologic findings, such findings are present in a high proportion. This is not surprising, since a common cause of central alexia is an occlusion of the middle cerebral artery or its distal branches. As a result, many different neurologic signs may be present. Some degree of hemiparesis can be seen, at least early in the course and, even more frequently, a hemisensory disturbance. Visual field defects vary from full to partial to none;[84] absence of a field defect is common, in contrast to the case in posterior alexia. Some degree of aphasia—most commonly, anomic aphasia—is often present. Wernicke's or transcortical sensory aphasia may also be present, but in these cases the language disturbance is often of sufficient severity that the alexia and agraphia may be regarded as mere indicators of the underlying language problem,[57] and the dominance of the reading and writing problems may be overlooked. When alexia and agraphia are both present and dominate the clinical picture, their characteristics remain the same, regardless of whether other language difficulties are present or not. When a lesion in Wernicke's area extends into the adjacent angular gyrus, a Wernicke's aphasia with more severe comprehension difficulty in written than in spoken language may be seen.

Dejerine's original patient was found to have suffered a dominant-hemisphere parietal infarction, and pathology involving this area appears to be the critical lesion underlying central alexia.[20,151]

The "Third Alexia" (Anterior Alexia)

Disorders of reading comprehension with anterior lesions were mentioned in the older literature. It has been suggested that the reading impairment that occurs with anterior aphasia, particularly Broca's aphasia, is distinctive and can be called *anterior alexia*.[13] One of the most striking features of this alexia is impaired ability to comprehend syntactic structures, in particular the function words and endings. As was pointed out earlier, many patients with anterior aphasia comprehend substantive words but fail to recognize many relational words.[138,165] This affects both spoken and written language and the term *syntactic alexia* has been suggested for the effect of this specific linguistic defect on written language. In addition, although patients with anterior reading disturbances demonstrate competence in handling single, semantically significant words, they often cannot deal with a sequence of such words,[2] a defect that further disturbs their ability to comprehend sentence-length material. The neurologic findings associated with this form of alexia are, of course, also distinctive. As noted, Broca's aphasia is common (most patients with TCM aphasia comprehend written language considerably better than do those with Broca's aphasia). A nonfluent output, relatively good auditory comprehension, poor repetition, and some degree of naming difficulty are common; hemiparesis and some degree of hemisensory loss are usual, but visual field defect is comparatively uncommon.

The location of the lesions in anterior alexia is the same as for Broca's aphasia. The posterior inferior frontal cortex of the dominant hemisphere is involved, and in most cases there is extension to subcortical structures.

Deep Dyslexia

Marshall and Newcombe[115] studied the varieties of paralexia (substitutions in reading aloud) and reported a syndrome that features synonym substitutions *(semantic paralexia)*. Thus, *automobile* might be read aloud as *car*, *carpet* as *rug*, and so on. Each patient studied had a severe (almost total) reading disturbance with a notable tendency to substitute a synonym word for the actual written word. The disorder was called "deep dyslexia."

Additional studies have been reported,[43] including attempts to correlate the syndrome with neuroanatomic abnormalities.[113] Other features of the syndrome include a tendency to turn adjectives into the corresponding noun (e.g., *rain* for *rainy*), a special difficulty in reading function (closed class) words, and inability to read nonsense words. Deep dyslexia almost always follows severe damage to the dominant hemisphere language area. In fact, semantic paralexia often appears to be a sign of partial recovery from global (total) alexia. Several theories have been advanced to account for this condition. Many of the original suggestions were attempts to account for the syndrome on the basis of selective impairment of different hypothesized methods of reading. Thus, normal individuals may read phonetically—that is, by using rules for "sounding out" words. This method enables one to read regularly written words, whether they are familiar real words (e.g., "administrative"), unfamiliar ("reflame"), or nonsense ("disbrate") words. Other words cannot be read by this route, particularly words with an irregular spelling that is unique or that has components pronounced in a different way in different words (yacht, slough, bough, thought). In this case, correct reading comprehension might depend on a direct connection between the word and its meaning. Thus, it is argued that the patient with deep dyslexia has lost the route for sounding out words (phonemic) but retains parts of the other normal route (semantic).

In alternative theories, however, it is argued that the clinical picture in deep dyslexia results from use of anatomic pathways that are not normally used in reading. Thus, Benson[16] suggested that although the pathways normally used for reading aloud are cut off by damage in the dominant left hemisphere, a visual image may be elicited in the right hemisphere from an imageable written word and this image can sometimes be named. It is understandable that a synonym would often be given (e.g., *infant* read aloud as *baby*) and that nonsense words and function words are not read correctly. Space does not permit us to discuss in detail the arguments advanced for the different theories of this intriguing phenomenon.

Agraphia

Clinicoanatomic correlations of isolated writing disturbances are much less clear than for the other forms of language disorders. It has been suggested that because writing is either a poorly mastered language competence or one that depends on many systems, it is to be expected that many types of impairment will interfere with writing. Chédru and Geschwind[40] noted that severe agraphia was often present in the acute confusional state, even in cases without tissue destruction or other language impairment. The confusional state is the most common cause of an isolated disorder of written language. The other signs of a confusional state make this an easy disorder to diagnose (see Chapter 3). Interference with the motor activities of the preferred hand (e.g., writer's cramp, choreiform movements, tremor, and so on) can produce significant writing impairment, but unless the production is totally illegible it should be clear that in these cases the grammar and word choice are intact. There are, however, also varieties of writing disturbance clearly associated with focal brain lesions, especially when aphasia is present in spoken language. In fact, when spoken language is aphasic, the written output is also abnormal; if exceptions exist, they are extremely rare.

Two distinctly different types of agraphia are clearly related to aphasic disorders and can be called anterior (frontal) or posterior (parietotemporal) agraphia. Writing in cases of dominant-hemisphere, anterior aphasia-producing lesions is usually performed by the nondominant or left hand because of right upper extremity weakness. The output is characteristically large, crude, poorly constructed, and scrawled. In addition to these mechanical abnormalities, misspelling is frequent and there is a strong tendency to agrammatism (i.e., the omission of function words and endings). The agraphia of anterior aphasia thus resembles the verbal output but is even more limited in quantity.

In contrast, the agraphia seen with posterior, dominant-hemisphere lesions often consists of well-formed letters, joined together in actual words and sentences. Spelling is usually incorrect,[103,104,132] word order is abnormal, and omissions are frequent. Thus, the written output is often unintelligible and nonsensical, even though individual letters are well formed. On the other hand, although posterior lesions may lead to a very copious production of fluently abnormal language (logorrhea), abnormal written language of this type (graphorrhea) is never (or extremely rarely) observed. On the other hand, graphorrhea is occasionally reported in patients with schizophrenia or those with the interictal personality syndrome of complex-partial epilepsy.[158] Furthermore, in some cases of posterior lesions, the writing output resembles that of anterior lesions (although the reverse is rarely seen). Thus, writing is much less useful in differential diagnosis of the type or localization of aphasia than is spoken language output.

A third distinct type of agraphia is caused by nondominant hemisphere damage and is called *visual-spatial agraphia*. The patient writes real letters and real words but has considerable difficulty orienting them on the page, tending to make increasingly large margins at one side of the page, sometimes actually writing off the page, and having considerable difficulty progressing from one line to the next.[90] The final product is unacceptable, even though individual words and even phrases are clearly legible.

Isolated agraphia (without significant aphasia or alexia) occasionally results from lesions in the dominant angular gyrus, often with the other components of *Gerstmann's syndrome* (finger identification problems, right-left difficulties, and acalculia).[7a,46] An aphasic agraphia limited to the left hand (or the right hand in some strongly left-handed individuals) may be seen following damage to the corpus callosum.[70]

A relatively isolated agraphia can also be observed on recovery from aphasia in left-handed individuals. The mechanism of this is still not fully clear and is another instance in which isolated agraphia provides little localizing information.

In recent years, distinctions in writing disorders have been demonstrated in selected patients by linguistically derived tests. Shallice[143] noted that some brain-damaged patients could write real words but could not write nonsense words to dictation. A corollary, an inability to write nonphonetic real words (e.g., although) with retention of the ability to write phonetically true nonsense words has also been demonstrated.[131]

The former is thought to represent a disturbance of phoneme-grapheme transformations (surface agraphia), whereas the latter is considered a disorder of lexical-grapheme transformation (deep agraphia).[34] Roeltgen and Heilman,[131] on the basis of a few carefully studied cases, suggested an anatomic localization for the two disorders: the surface agraphia was seen with posterior superior temporal pathology, whereas the deep agraphia was more indicative of posterior inferior parietal damage.

SYNDROMES OF WORD-FINDING IMPAIRMENT

Difficulty in finding specific words, either in spontaneous speech or in naming on confrontation, is the most common and frequently noted disturbance in aphasia. It has been stated that almost no patient with discernible aphasia in spoken language has fully normal word-finding abilities.[77] Word-finding impairment may result from lesions or dysfunction in many different parts of the brain and, by itself, has little localizing value. Increased intracranial pressure, certain types of progressive dementia,[7,49] acute confusional states,[40] psychogenic disturbances,[68] and even subcortical lesions[41,123] can cause word-finding difficulties. If one examines different syndromes in which this problem occurs, however, distinct differences become apparent. Such differences were noted by Geschwind[68] and Luria[110] and have been reviewed recently.[15] Table 5 presents a classification of clinically differentiable word-finding problems. Word-finding difficulty is usually noted as poor performance on confrontation naming tests, emptiness (decreased number of substantive words) in conversational speech, and decreased production of category word lists; in selected circumstances one test may be abnormal without the others.

Word-Production Anomia

In this disturbance, the patient appears to possess knowledge of the desired name but cannot produce it correctly. The existence of two types has been suggested: an articulatory initiation problem and a paraphasic disturbance.[15] One should also add word-evocation anomia. In the artic-

TABLE 5. Aphasic Anomia

Varieties of Aphasic Anomia	Location of Pathology
1. Word production anomia	
	Dominant Hemisphere
a. articulatory initiation anomia	Broca's area
b. word-evocation anomia	Many locations
c. paraphasic anomia	Wernicke's area and arcuate fasciculus
2. Word selection anomia	Brodmann area 37
3. Semantic anomia	Angular gyrus
4. Disconnection anomia	
	Either or Both Hemispheres
a. modality specific anomia	Posterior white matter tracts
b. category specific anomia	Posterior white matter tracts
c. callosal anomia	Corpus callosum

ulatory initiation type the patient characteristically fails to name an object on confrontation but, if given a clue (either a phonetic cue such as the initial phoneme or a contextual cue such as an open-ended sentence using the desired word), is able to produce the name. The ease of production with minimal prompting suggests that in at least some cases the patient is actually *aware* of the appropriate word, failing to name it because of difficulty in initiating articulation. It may also be true in other cases that the clue evokes the correct word, of which the patient was not consciously aware, just as often occurs in normal individuals (word-evocation anomia). It is not always easy to distinguish between these two mechanisms.

Paraphasic word-production anomia is different: the patient easily initiates a word, but the verbal production is so contaminated with literal paraphasic substitutions that the response is wrong. Patients often insist that they know the correct name and, not infrequently, the number of syllables produced and the inflections are correct. Occasionally, true but incorrect names are substituted, but more often the incorrect response is a neologism. The articulatory initiation problem occurs primarily in the anterior language disturbances, Broca's and transcortical motor aphasias, in which prompting is often helpful. Paraphasic anomia is most characteristic of conduction aphasia, in which literal paraphasic substitution is a major feature of verbal output, and in Wernicke's aphasia, in which verbal and/or neologistic substitutions are common. Word-evocation anomia may occur with both anterior and posterior lesions.

Word-Selection Anomia

This is a pure word-finding problem. The patient fails to name an object on presentation in any sensory modality but can describe the use of the object and can select the appropriate object when the examiner presents the name. The difficulty appears to be isolated to thinking of the word for the presented object. This is an unusual disorder in pure form and almost invariably indicates damage in Brodmann area 37, the temporal-occipital junction region of the dominant hemisphere.[15,122]

Semantic Anomia

This disturbance is characterized by a combined inability to name an object or point to the correct object when the name is presented (i.e., a two-way defect). Both the comprehension and use of the name are disturbed. Semantic anomia has been described as a situation in which the word has lost its symbolic meaning.[74,110] In some instances, the patient can correctly repeat the name when offered by the examiner even though failing to recognize the object mentioned or to name the object. The latter phenomenon is classically seen in TCS aphasia, but some degree of semantic anomia is often present in Wernicke's aphasia also. A variation of semantic anomia is a curious phenomenon that French authors call "intoxication by the word." A patient who can usually recognize the correct name of an object may fail to do so just after producing an incorrect name.

Disconnection Anomia

Finally, at least three different types of anomia based on disconnection of posterior cortical areas have been described. In *modality-specific anomia*[33] the patient can name normally in all sensory modalities except one. Thus, a patient with associative visual agnosia cannot name an object on visual confrontation, even though the same object is rapidly and correctly named when palpated.[137] Auditory agnosia (see earlier) produces a similar disturbance—for example, the patient cannot name the source of a sound, such as a bell, but can name it when the object is seen. The distinction between modality-specific anomia and agnosia is difficult and often incomplete. Further discussion of this topic can be found in Chapter 7. *Category specific anomia* is somewhat different. In this situation, the patient can name adequately except for a given stimulus category. Thus, an inability to name seen colors or to select a seen color in response to the spoken name may occur (e.g., in posterior alexia) even though the patient competently matches colors and correctly uses and comprehends color names in purely auditory verbal tasks (see Chapter 7 for additional descriptions of this syndrome). Patients may thus fail to name the color of an object shown to them or to pick out an object of a color requested by the examiner. They may, however, respond correctly to questions such as "What color is the sky?" and "Give me the name of a yellow fruit."[69] Some patients show greater difficulty in body-part naming, others in object naming. Category naming problems may be present although not easy to define in the presence of other naming disturbances. It is likely that there are other category-specific anomias with different mechanisms. Nielsen[122] discussed patients who could not name inanimate objects.

Callosal anomia occurs when a patient whose corpus callosum has been lesioned is asked to name an unseen object placed in the left hand. The patient may show recognition of the object (demonstrated by correct use or by after-selection) but can neither give a correct name for the object[63,70] nor recognize the correct name when it is offered by the examiner. The key feature of all disconnection anomias is pathology that effectively separates a primary sensory or sensory association area from the dominant-hemisphere language area.[67]

APRAXIA

Apraxia is the inability to carry out a motor act despite intact motor and sensory systems, good comprehension, and full cooperation. It was clearly described by Hughlings Jackson,[98] who noted that certain patients with aphasia could not protrude their tongue on command. The term came into widespread use following the writings of Liepmann,[108–109] but, unfortunately, apraxia is now used to describe so many faulty motor acts that it lacks specificity. Many disorders that use this term such as dressing apraxia, constructional apraxia, ocular apraxia, oral apraxia, and verbal apraxia deserve more accurate titles. Three varieties of apraxia originally described by Liepmann deserve discussion here.

Limb-Kinetic Apraxia

This term is rarely used at present; it appears to identify a focal (sometimes unilateral or even confined to a single limb) clumsiness in performance of fine motor acts. Many regard limb-kinetic apraxia as a mild pyramidal or extrapyramidal disorder and thus not a true apraxia.

Ideomotor Apraxia

This disturbance, which is also called *motor apraxia*, and not infrequently just *apraxia*, appears to have a close relationship to language function. *Ideomotor apraxia* is best defined as an inability to carry out, in response to verbal command, a motor activity that is easily performed spontaneously. Thus, a patient with ideomotor apraxia may fail to protrude the tongue on command but easily licks the lips spontaneously. As noted in the section on anatomy of the language system, ideomotor apraxia commonly occurs with certain types of aphasia. In particular, buccofacial and bilateral limb apraxia is common in conduction aphasia, and apraxia of the nonparalyzed limb is frequent in cases of Broca's aphasia. Apraxia was demonstrated in 40 percent of aphasic subjects studied at the Boston Veterans Administration Aphasia Research Center.[14] Comprehension problems (e.g., Wernicke's aphasia) can also interfere with the ability to carry out verbal commands, but, by definition, this problem is not routinely considered apraxia.

Several mechanisms have been suggested for ideomotor apraxia. One, a pure disconnection theory, posits that damage to selected neural pathways or cortical areas is necessary to carry out a spoken command.[67] Specifically, damage to the arcuate fasciculus is said to cause *parietal apraxia*, damage to Broca's area underlies *sympathetic apraxia* (the inability of the nonpathologic left hand to carry out commanded movements), and anterior callosal damage can cause *callosal apraxia* (the inability of one limb—usually the left—to perform on command, even though the other limb performs easily). A second theoretical mechanism uses the concept of an engram, a cortically stored movement pattern. In this presentation, apraxia would result from damage or interference to the neural structures that store the learned movement pattern.[91] In either situation, the presence of ideomotor apraxia, particularly when elicited as described in the section on testing for aphasia, offers valuable, language-oriented, brain-localization information.

Ideational Apraxia

The third type of apraxia often given importance in brain function has proved very difficult to define or localize. Originally *ideational apraxia* was considered an inability to carry out a sequence of related activities (such as filling and lighting a pipe) although each separate activity can be successfully performed.[109] Later works use the term to refer to an inability to handle real objects, even though mimed use of the object is successful.[58] Neither type is seen often in pure state but the originally described disturbance can be demonstrated in dementias of the Alzheimer type[49] and in confusional states (see Chapter 3).

SYNDROMES OF CALCULATING IMPAIRMENT ACALCULIA

A number of different classifications of acalculia have been suggested in the past,[86] but a consistent agreement on terminology has not been established. The approach suggested by Hecaen and co-workers[89] is used here.

Aphasic Acalculia

One form can be called *aphasic acalculia*, a disturbance in which the patient is unable to handle numbers as language entities. The patient either cannot comprehend or write numbers correctly or substitutes one number for another.[19] These language problems produce a severe acalculia, and some degree of aphasic acalculia occurs in many (in fact, most) types of aphasia. The pathology affects the dominant-hemisphere language area. The fact that some of these patients fail to comprehend the written arithmetical signs is attributed by some to the aphasic disorder itself, although others attribute this failure to accompanying anarithmetria (see later discussion).

Visual-Spatial Acalculia

Visual-spatial acalculia is quite different. Here the patient understands numbers and computational signs and can compute, being able to add, subtract, multiply, and so on. The problem lies in the placement of the numbers in the correct position in space. The patient cannot align numbers in columns for addition or multiplication, cannot place decimal signs, and so on.[42] The visual-spatial disturbance interferes with complex calculations, even though the patient is competent at individual number manipulations. The lesion almost invariably affects the nondominant hemisphere, usually the parietal-occipital junction region; visual-spatial acalculia represents a true visual-spatial discrimination problem.

Anarithmetria

Anarithmetria, a third type of calculation disorder is characterized by impairment of the ability to perform number manipulation, a true loss of computational competency. Patients correctly recognize and reproduce individual numbers, know their value, and can handle numbers in space, but they cannot perform computations. They often correctly perform elementary rote addition and multiplication, but fail to combine these to solve more complex problems. Anarithmetria is not common in pure state; Grewel[86] suggested that the most common cause was dementia (of the Alzheimer type). Benson and Weir[23] presented a single case in which anarithmetria was the major residual of a dominant hemisphere posterior parietal traumatic lesion. It seems probable that anarithmetria is not uncommon but is difficult to delineate as it is so often associated with aphasic acalculia and/or severe dementia.

GERSTMANN'S SYNDROME

Over a period of years Josef Gerstmann[64-65] described the combination of findings that now carry his name. The four major findings included agraphia, acalculia, right-left disorientation, and finger agnosia. Gerstmann suggested that this combination of findings indicated pathologic involvement of the dominant parietal lobe. Additional cases were subsequently recorded and the material reviewed by Critchley,[46] who noted that most reported cases had pathology involving the parietal-occipital junction area, almost without exception in the left or dominant hemisphere. Of the few cases of Gerstmann's syndrome demonstrated to have pathology elsewhere, either the patient did not have standard hemispheric dominance for language or multiple lesions were present.[46] Thus, Gerstmann's syndrome came to be accepted as a strong neurologic localizing syndrome.

This position was challenged; Benton,[25] Heimberger and co-workers,[92] and, in later years, Critchley[47] himself argued that the syndrome consisted of four entities with little or no relationship to each other and that these findings had neither specific neuropsychologic meaning nor consistent localization. Forceful arguments in favor of this view have been presented by several authors.[27,127] Others, however, have given equally strongly their reasons for rejecting these arguments.[133,153] The interested reader may review the papers cited, but the individual components of Gerstmann's syndrome deserve discussion.

Finger Agnosia

First recognized by Gerstmann,[64] finger agnosia is the most difficult of the tetrad to define or demonstrate. Classically, the patient cannot name the fingers or point or move a finger when the name is given; some cannot recognize which finger has been touched or moved by the examiner. Aphasia may interfere with tests for finger agnosia; either a severe naming disturbance or a severe comprehension problem can cloud the picture. Similarly, sensory loss can interfere with recognition of the finger touched. Nevertheless, appropriate testing can demonstrate that finger recognition and naming are far more difficult for some patients than are tasks featuring other body parts and also are more difficult than naming other items.

Right-Left Disorientation

This disturbance is demonstrated by asking the patient to show his right or left hand or point to right or left body parts on himself or the examiner. If successful, crossed-pointing can be requested (i.e., a named digit of one side is to touch a named body part of the other).[24,104] Again, aphasia, particularly comprehension disturbance, interferes with this assessment, and unilateral inattention can cause failure. As with finger agnosia, however, a significant incompetency can be demonstrated in right-left orientation in comparison to other mental tasks. Nonverbal tests of right-left difficulty have not, however, received fully adequate attention.

Acalculia

As noted in the section on syndromes of calculating impairment, acalculia is not a localizing finding by itself. In a study of the acalculia of Gerstmann's syndrome, Benson and Denckla[19] reported that the calculation errors made by two individuals with full Gerstmann's syndrome were based on verbal paraphasic substitutions of numbers. The patients could perform the computational tasks (as demonstrated by multiple-choice responses) but either read or wrote the numbers incorrectly and produced wrong answers. On the other hand, in some cases with Gerstmann's syndrome acalculia has been reported in the absence of aphasia.[71a]

Agraphia

Writing impairment, the fourth component of Gerstmann's syndrome, is common following brain damage, particularly if the dominant hemisphere is involved and, when considered by itself, is not a specific finding (see the section on anatomy of the language system).

Each individual component of Gerstmann's syndrome can result from lesions in several different locations; an individual component is, therefore, a weak localizing feature. When all four occur together in a single individual, however, dominant parietal involvement is overwhelmingly probable and the combination of findings composing Gerstmann's syndrome has strong localizing value. Certain other findings are commonly associated with Gerstmann's syndrome. Stengel[152] noted that individuals showing all four components of Gerstmann's syndrome almost invariably suffer constructional disturbance and proposed this as a fifth component. Further, involvement of the dominant angular gyrus often causes central alexia and anomic aphasia. The angular gyrus syndrome[14,18] is a combination of central alexia and anomic aphasia with Gerstmann's syndrome. Aphasia is common in patients with Gerstmann's syndrome[127]—usually a variety of posterior aphasia. It must be noted, however, that many severely aphasic individuals do not have all four components, and Gerstmann's syndrome can exist in the total absence of aphasia.[153]

LOCALIZATION OF SELECTED LANGUAGE DEFICITS AND DOMINANCE

The original clinical/postmortem correlations on which the syndrome approach to aphasia is based have been confirmed by the new techniques that study brain structure and function in life. Both approaches support the view that different language functions (e.g., reading) require the integrity of selected areas of the brain. Although many questions remain, a large body of knowledge supports the concept that specific neural structures underlie specific aspects of language function. The brain is a collection of components selectively integrated on a functional basis among which are the neural structures that underlie language.

A key aspect of neural localization, long accepted, is dominance for language function within the left hemisphere in most humans. Left-hemisphere dominance for language is the rule for right-handed individuals; it is sometimes stated that as many as 99 percent of those who prefer the right hand for writing and other motor activities have dominance for language in their left hemisphere, although some[6] argue that this figure may be too high. Among left-handed individuals some investigators suggest a 60/40 left/right split.[79,130,154] Other studies, however, suggest another interpretation of these figures—that many individuals who prefer the left hand actually carry out language functions with both hemispheres.[72] It is likely that many right-handed individuals also have some bilaterality of language function. Thus, some language function may be carried out by the right hemisphere in some individuals with either right- or left-hand preference. Crossed aphasia, the occurrence of aphasia in a right-handed individual following right-hemisphere damage is relatively rare,[38] and an unusually high percentage of the reported cases are based on trauma or tumor, suggesting that many, but not all, cases actually represent bilateral damage.[29] Hemispheric dominance for language is a strongly localized neural function.

Differences in the pattern of spontaneous speech of aphasic patients also have strong implications for localization. For over a century, different forms of aphasic verbal output, now often called fluent and nonfluent (described in the first section of this chapter) have been noted.[97,161] Both early reports[161] and recent work[12,126,157] demonstrate relatively consistent anatomic correlations, which have already been presented. Despite some problems (e.g., aphasia in non–right-handed individuals is often difficult to classify), the fluent-nonfluent dichotomy permits anterior or posterior localization with high accuracy.

Repetition is another useful aid for localization of language disorders. Some patients with aphasia have great problems in repetition, whereas others have relatively preserved ability to repeat (see sections on historic background, testing for aphasia, and anatomy of the language system). The anatomic correlations have already been discussed. Determination of competence in repetition provides localizing information different from and complementary to that of the evaluation of fluency.

Comprehension of spoken language is often considered a single entity, but this is incorrect, since definite qualitative differences in comprehension can be observed. Thus, some individuals who cannot understand spoken language will comprehend written language (pure word deafness). Other patients have problems comprehending both auditory and written material (e.g., in Wernicke's aphasia). Patients with both of these types fail at repetition tasks, but those with a third disturbance (transcortical sensory aphasia) show similarly impaired comprehension of both spoken and written language even though repetition is relatively normal. Comprehension of spoken language is abnormal in each of these conditions, but the neighborhood signs indicate different causes of comprehension failure. Thus, Wernicke's aphasia is usually thought to include damage to the dominant auditory association cortex, whereas in pure word deafness there is a disconnection of auditory input from these regions. In transcortical sensory aphasia (angular gyrus damage) both the auditory primary and association cortices of the auditory system are intact; the patient can hear and repeat spoken language but its meaning cannot be understood. A fourth, distinctly different comprehension disturbance, impairment of syntactic comprehension, has been studied very extensively in recent years.[37,165] When this disorder exists in relative isolation from other comprehension disorders, damage in the anterior language area is probable. Thus, four distinct variations of comprehension disturbance can be noted clinically, each typically indicating a different area of brain involvement (see Chapter 1 for related anatomy).[14–15]

Naming disturbance is another disorder of language that is often incorrectly discussed as a single entity. A number of variations of naming disturbances can be separated clinically, indicative of lesions in different neuroanatomic areas, as has already been discussed (see Syndromes of Word-Finding Impairment).

Similar correlations have been presented for reading and writing disorders in the section on syndromes of written language impairment. When considered together, these localizations of varieties of language disorders provide considerable insight into the role played by different portions of the brain in language. Future investigation of the neural substrates necessary for individual language functions will remain an important source of knowledge of the biologic study of language.

EMOTIONAL ACCOMPANIMENTS OF APHASIA

Aphasia is a consequence of damage to brain structures, and only rarely is the area of damage so precise or limited that the patient has no other neurobehavioral problems of significance.[75] In addition, emotional disturbances further complicate the clinical picture. The emotional disturbances are complex and, for simplicity, are discussed here under two subsections—psychosocial and neurobehavioral. In addition, the risk of suicide is discussed.

The psychosocial complications of aphasia are disturbing in the acute stage and become crucial during rehabilitation. Most patients who are suddenly bereft of language understandably suffer from a feeling of loss, and as a consequence, a depressive or grief reaction appears in many aphasic patients. This reactive depression, which often does not appear in severe form for weeks or even months after onset, can be a serious complication. The patient becomes less and less communicative; withdraws from the nursing staff, family members, and therapists; cries easily; and sleeps and eats poorly. The condition is upsetting to all but is usually self-limited. After 7 to 10 days of an increasing grief reaction, most patients return to a more normal state and often perform better in rehabilitative measures. During the period of deep depression, however, they are seriously ill and demand considerable attention.

The factors underlying the grief reaction appear obvious. Not only has the patient lost a significant function, language, but there is often a hemiparesis limiting physical activity and producing further loss of self-image. In addition, alterations occur in the patient's social structure; for example, a lessening of the patient's importance in the family, a total change in work patterns, altered recreational activities, and a change in friends and acquaintances. Furthermore, many aphasic patients are secretly concerned about their sexual competency. An alteration of life-style is necessary, a difficult process that requires considerable support. Although some families can help a patient with aphasia return to optimal activity, others may be sufficiently disrupted to be of little or no help.

In some cases these reactive depressive responses are unusually severe. Some patients with aphasia, who are fully aware of their own problems and become frustrated by the inability to express their thoughts and desires in language, become depressed and may even suffer what has been called a catastrophic reaction[74]—an unusually severe state of agitated depression. Fortunately, the latter is rare.

The depressive reactions we have so far discussed are similar to those that might occur with any similar loss (e.g., an amputation or spinal injury). For this type of expected reaction to occur, it is reasonable to assume that the parts of the brain that control emotional responses are intact. On the other hand, one must be aware that in other cases the patient may exhibit an unexpected pattern of emotional reaction, which appears to be associated with specific localizations of the aphasia-producing lesions. Thus, there are aphasic patients who appear unaware of their own problems, act relatively unconcerned or even euphoric, and tend to blame others for their communication problems. They can, however, suddenly shift from unconcern to suspicion, sometimes developing paranoid reactions. Patients with aphasia who exhibit unconcern, euphoria, or paranoia almost invariably have posterior pathology, especially Wernicke's aphasia in which early euphoria is not at all uncommon. Serious paranoid reactions are most frequent in patients with temporal lobe pathology; many pure word deaf patients will eventually show this response, while only some with Wernicke's aphasia and few with transcortical sensory aphasia manifest this change.

Finally, it must be recognized that suicidal ideation can be a significant problem in the management of the aphasic. Many individuals showing the frustration/depression tendencies ruminate about their own worthlessness, questioning whether continued life is worthwhile, worrying about being a burden to their family, and so on. They can become severely depressed, and the risk of suicide must be kept in mind. Fortunately, few act on their suicidal ideation. On the other hand, patients with the unawareness/unconcern syndrome may suddenly shift to suspicion and act impulsively. Suicide is rare among individuals with aphasia and does not demand constant attention in most, but even as an uncommon event it deserves consideration.

RECOVERY FROM APHASIA

It has long been recognized that there is a strong tendency for some degree of spontaneous recovery from aphasia. At times recovery is early, rapid, and extensive. Recovery in the very early stages (i.e., within the first week or two) probably reflects improvement of the anoxia, edema, cellular infiltration, and increased intracranial pressure caused by the acute insult.[101] Language recovery, however, routinely continues over a much longer period and the later mechanisms are not fully understood.

To deal with the problem of recovery, both early and late, a number of factors that have been thought by different investigators to affect the process can be listed (Table 6).[56]

Some of the factors listed in Table 6 are obvious. The patient's age at the onset of aphasia is important. The young child almost always makes a good recovery from a unilateral brain lesion that has produced aphasia. With advancing age, recovery becomes slower and less certain, but many believe that the younger the adult, the better the prognosis for recovery from an aphasia-producing lesion. Other investigators, however, believe that once adult maturation is attained, age is no longer a major factor in recovery. The age at which the brain reaches maturity is still unsettled, however.[14−15,106,164] Handedness is another factor of importance; Luria,[111] Subirana,[154] Gloning and associates,[72] and many others have demonstrated that left-handed individuals and their first-degree relatives, even if they favor the right hand, recover from aphasia better than do right-handed individuals.

Many authors[55,101] recognize etiology as a significant factor in recovery. The degree of recovery also varies with site and extent of lesion. The type of aphasia is significant; Vignolo[155] demonstrated that spontaneous recovery was, on the average, better for comprehension defects than for output problems. Aphasic syndromes caused by smaller lesions (such as conduction aphasia) characteristically show a better overall improvement.

Some investigators suggest that the educational level, intelligence, current milieu, and prior language competence are significant factors, but proof that they influence recovery is less compelling. The presence of nonlanguage behavioral complications is an important factor that has only recently been emphasized.[55,75] Amnesia, dementia, unilateral inattention, agnosia, apraxia, and many other disorders are not uncommon in patients with aphasia and may complicate the recovery process.

It has been suggested that the time elapsed between the onset of aphasia and the beginning of therapy, and the type (and quality) of aphasia therapy, may influence recovery. While many therapists encourage commencement of therapy at an early stage, the data supporting this stand are not overwhelming. Vignolo[155] reported that the number of subjects who improved spontaneously decreased with the passage of time

TABLE 6. Factors Affecting Recovery from Aphasia

1. Age of onset of aphasia
2. Handedness
3. Drive and motivation
4. Educational level/complications
5. Social milieu
6. Type of aphasia
7. Etiology of aphasia
8. Neurobehavioral complications
9. Time lapsed before therapy started
10. Type of therapy

from the onset. Wepman[160] claimed that individuals given language training in the first year following onset attained better recovery than did patients who began later. Sarno and co-workers,[139] however, could not demonstrate any significant difference in recovery when treatment was started immediately after stroke or 4 months after stroke. While early treatment is often recommended, there is evidence suggesting that treatment begun late can also be successful.[9,145]

Finally, the type of language therapy offered may well be of significance (see the section on therapy for aphasia). It is clear that aphasic patients suffer from many different types of language disturbance, and therapy designed to aid a given problem may be more successful than a broader, less-directed approach. Spontaneous recovery occurs in many individuals with aphasia but the ultimate degree of recovery is variable and dependent on a number of factors, one of which may be the type and amount of language therapy.

THERAPY FOR APHASIA

In recent decades, aphasia therapy has become increasingly common, but spontaneous recovery makes demonstration of the efficacy of language therapy difficult. Current evidence, both direct and indirect,[9,162] suggests that patients given active rehabilitation regain on the average more language competence than do untreated patients. Only in recent years have special forms of language therapy been devised for specific language disturbances and these efforts are still in their early stages.

In general, speech pathologists prefer to manage therapeutic efforts on an individual basis, tailoring the program either to the patient's major problem or to the therapy technique best known to the therapist. "Traditional" language therapy has concentrated on attempts to improve spoken language. Some common techniques have included those used by speech therapists to improve dysarthria and by educators to teach language to the young. These techniques have met with considerable success and remain a basic structure of most current language therapy programs.

In the past decade, a number of completely different therapeutic techniques have been introduced, some of which have proved useful for a limited type of aphasia only. One of these, Melodic Intonation Therapy (MIT),[3,148] utilizes the ability of many aphasic patients to sing better than they speak. Short sentences are put into an exaggerated melodic form, the patient and therapist intoning and beating the rhythm of the sentence in unison until the patient can produce the sentence independently. In selected patients this technique has led to considerable increase in adequacy of verbal output over relatively short periods of time. Helm[93] noted four major characteristics of the patient who will respond to MIT: (1) nonfluent verbal output; (2) relatively good comprehension; (3) poor repetition; and (4) poor oral agility. Conversely, MIT has generally proved ineffective in treating patients with more posterior lesions or global aphasia.

Another relatively new technique for the treatment of aphasia, American Indian Sign (AMERIND), uses a gestural pantomime communication system. This has proved effective for patients with aphasia of relatively recent onset, particularly for those with either severe comprehension deficit or a total language disturbance.[144] AMERIND is most effective when given in a group setting, and in some rehabilitation centers AMERIND groups have become a regular facet of the language therapy.

Another specific technique that has received attention recently is called Visual Action Therapy (VAT).[94] With this technique simple gesture-pantomime activity is combined with pictures of an implement and intonation of its name or action. Through considerable practice, the patient becomes adept at recognizing the specific object from any of the three presentations—that is, the pantomime, the picture, or the verbalization. VAT has proved most successful for individuals with serious comprehension problems, including both Wernicke's aphasia and global aphasia.

Additional therapeutic methods have been developed in recent years, some of which show promise for the future. These include approaches designed to improve naming disturbances, to overcome agrammatism, or to increase attention to auditory cues.

Therapeutic approaches are also available for alexia and agraphia. Most remain in the traditional fold, consisting of classwork under the therapist's supervision, abetted by homework as the patient improves. Some individuals, particularly those with associative alexia or mild degrees of aphasic alexia, may show considerable improvement in reading and writing.

A number of mechanical aids have been introduced to compensate for language loss, most with limited success. Patients with minimal anterior aphasia, fully capable of monitoring their own verbalizations, can sometimes be helped by a Language-Master®, a miniaturized taperecorder that allows the patient to hear a word or sentence and then repeat it. This technique is successful only if the patient retains a good ability to monitor his own output. Other machine approaches have attempted to replace human verbalization with electronic output. Unfortunately, most aphasic patients cannot operate the devices, as they demand language skills; effective operation of the machine comes about only as a result of improved language competency. For individuals with a severe dysarthric component to their aphasic problem, some of the more recent mechanical voice machines can be of help; all too often, however, these also prove too difficult for the aphasic patient to master.

In summary, aphasia therapy often appears to be effective and should be considered for every aphasic. Innovations are being made in aphasia therapy, some of which are effective for a specific type of aphasia only. Future improvements in language rehabilitation appear likely so that even more patients with aphasia will regain increased language competence.

REFERENCES

1. AJAX, ET: *Acquired dyslexia.* Arch Neurol 11:66–72, 1964.
2. ALBERT, ML: *Auditory sequencing and left cerebral dominance for language.* Neuropsychologia 10:245–248, 1972.

3. Albert, ML, Sparks, R, and Helm, N: *Melodic intonation therapy for aphasia*. Arch Neurol 29:130–131, 1973.

4. Alexander, MP, and LoVerme, SR, jr: *Aphasia after left hemispheric intracerebral hemorrhage*. Neurology 30:1193–1202, 1980.

5. Alexander, MP, and Schmidt, MA: *The aphasia syndrome of stroke in the left anterior cerebral artery territory*. Arch Neurol 37:97–100, 1980.

6. Annett, M: *Hand preference and the laterality of cerebral speech*. Cortex 11:305–328, 1975.

7. Appell, J, Kertesz, A, and Fisman, M: *A study of language functioning in Alzheimer patients*. Brain Lang 17(1):73–91, 1982.

7a. Auerbach, SH, and Alexander, MP: *Pure agraphia and unilateral optic ataxia associated with a left superior parietal lobule lesion*. J Neurol Neurosurg Psychiatry 44:430–432, 1981.

8. Auerbach, SH, Alland, T, Naeser, M, Alexander, MP, and Albert, ML: *Pure word deafness: Analysis of a case with bilateral lesions and a defect at the prephonemic level*. 105:271–300, 1982.

9. Basso A, Capitani, E, and Vignolo, LA: *Influence of rehabilitation on language skills in aphasia patients*. Arch Neurol 36:190–195, 1979.

10. Basso, A, Faglioni, P, and Vignolo, LA: *Etude controlee de la reeducation du language dans l'aphasie: Comparison entre aphasiques traites et nontraites*. Rev Neurol 131:607–614, 1975.

11. Bastian HC: *Aphasia and Other Speech Defects*. HK Lewis, London, 1898.

12. Benson, DF: *Fluency in aphasia: Correlation with radioactive scan localization*. Cortex 3:373–394, 1967.

13. Benson, DF: *The third alexia*. Arch Neurol 34:327–331, 1977.

14. Benson, DF: *Aphasia, Alexia, and Agraphia*. Churchill Livingstone, New York, 1979.

15. Benson, DF: *Neurologic correlates of anomia*. In Whitaker, H, and Whitaker, HA (eds): *Studies in Neurolinguistics*. Vol 4. Academic Press, New York, 1979, pp 293–328.

16. Benson, DF: *Alexia and the neuroanatomical basis of reading*. In Pirozzolo, FJ, and Wittrock, MC (eds): *Neuropsychological and Cognitive Processes in Reading*. Academic Press, New York, 1981, pp 69–92.

17. Benson, DF, Brown, J, and Tomlinson, EB: *Varieties of alexia*. Neurology 21:951–957, 1971.

18. Benson, DF, Cummings, JL, and Tsai, SY: *Angular gyrus syndrome simulating Alzheimer disease*. Arch Neurol 39:616–620, 1982.

19. Benson, DF, and Denckla, MB. *Verbal paraphasia as a source of calculation disturbance*. Arch Neurol 21:96–102, 1969.

20. Benson, DF, and Geschwind, N. *The alexias*. In Vinken, PJ, and Bruyn, GW (eds): *Handbook of Clinical Neurology*. Vol. 4. North Holland Publishing, Amsterdam, 1969, pp 112–140.

21. Benson, DF, and Patten, DH: *The use of radioactive isotopes in the localization of aphasia-producing lesions*. Cortex 3:258–271, 1967.

22. Benson, DF, Sheremata, WA, Buchard, R, Segarra, J, Price, D, and Geschwind, N: *Conduction aphasia*. Arch Neurol 28:339–346, 1973.

23. Benson, DF, and Weir, W: *Acalculia: Acquired anarithmetria*. Cortex 8:465–474, 1972.

24. Benton, AL: *Right-Left Discrimination and Finger Localization: Development and Pathology*. Hoeber, New York, 1959.

25. Benton, AL: *The fiction of the "Gerstmann syndrome."* J Neurol Neurosurg Psychiatry 24:176–181, 1961.

26. Benton, AL: *Contributions to aphasia before Broca*. Cortex 1:314–327, 1964.

27. Benton, AL: *Reflection on the Gerstmann syndrome*. Brain Lang 4:45–62, 1977.

28. Benton, AL, and Joynt, RJ: *Early descriptions of aphasia*. Arch Neurol 3:205–222, 1960.

29. Boller, F: *Destruction of Wernicke's area without language disturbance. A fresh look at crossed aphasia*. Neuropsychologia 11:243–246, 1973.

30. Botez, MI, and Barbeau, A: *Role of subcortical structures, and particularly of the thalamus, in the mechanism of speech and language*. Int J Neurol 8:300–320, 1971.

31. Broca, P: *Remarques sur le siège de la faculté du langage articulé, suivies d'une observation d'aphemie*. Bull Soc Anatomique (Paris) 2:330–357, 1861.

32. Broca, P: *Sur la faculté du langage articulé*. Bull Soc Anthropol (Paris) 6:337–393, 1865.

33. Brown, JW: *Aphasia, Apraxia, and Agnosia*. Charles C Thomas, Springfield, IL, 1972.

34. Bub, D, and Kertesz, A: *Deep agraphia*. Brain Lang 17:146–165, 1982.

35. Cappa, SF, and Vignolo, LA: *"Transcortical" features of aphasia following left thalamic hemorrhage*. Cortex 15:121–130, 1979.

APHASIA AND
RELATED
DISORDERS: A
CLINICAL APPROACH

233

36. CARAMAZZA A, AND BERNDT, RS: *Semantic and syntactic processes in aphasia: A review of the literature.* Psychol Bull 85:898–918, 1978.

37. CARAMAZZA, A, BERNDT, RS, BASLI, AG, AND KOLLER, JJ: *Syntactic processing deficits in aphasia.* Cortex 17:333–348, 1981.

38. CARR, MS, JACOBSON, T, AND BOLLER, F: *Crossed aphasia: Analysis of four cases.* Brain Lang 14:190–202, 1981.

39. CASEY, T, AND ETTLINGER, G: *The occasional "independence" of dyslexia and dysgraphia from dysphasia.* J Neurol Neurosurg Psychiatry 23:228–236, 1960.

40. CHÉDRU, F, AND GESCHWIND, N: *Disorders of higher cortical functions in acute confusional states.* Cortex 8(4):395–411, 1972.

41. CIEMENS, VA: *Localized thalamic hemorrhage: A cause of aphasia.* Neurology 20:776–782, 1970.

42. COHN, R: *Dyscalculia.* Arch Neurol 4:301–307, 1961.

43. COLTHEART, M, PATTERSON, K, AND MARSHALL, JC (EDS): *Deep Dyslexia.* Routledge & Kegan Paul, London, 1980.

44. CONRAD, K: *New problems in aphasia.* Brain 77:491–509, 1954.

45. CRITCHLEY, M: *The anterior cerebral artery and its syndromes.* Brain 53:120–165, 1930.

46. CRITCHLEY, M: *The Parietal Lobes.* Edward Arnold and Co, London, 1953.

47. CRITCHLEY, M: *The enigma of the Gerstmann's syndrome.* Brain 89:183–198, 1966.

48. CROSBIE, EC, HUMPHREY, E, AND LAUER, EW: *Correlative Anatomy of the Nervous System.* Macmillan, New York, 1962.

49. CUMMINGS, JL, AND BENSON, DF: *Dementia: A Clinical Approach.* Butterworths, Boston, 1983.

50. DAMASIO, A, YAMADA, T, DAMASIO, H, CORBETT, J, AND McKEE, J: *Central achromatopsia: Behavioral, anatomic and physiologic aspects.* Neurology 30:1064–1071, 1980.

51. DAMASIO, AR, AND DAMASIO, H. *The anatomic basis of pure alexia.* Neurology 33:1573–1583, 1983.

52. DAMASIO, AR, DAMASIO, H, RIZZO, M, VARNEY, N, AND GERSH, F: *Aphasia with non-hemorrhagic lesions in the basal ganglia and internal capsule.* Arch Neurol 39:15–20, 1982.

53. DAMASIO, H, AND DAMASIO, A: *The anatomical basis of conduction aphasia.* Brain 103:337–350, 1980.

54. DAMASIO, AR, AND VAN HOESEN, GW: *Emotional disturbances associated with focal lesions of the limbic frontal lobe.* In HEILMAN, KM, AND SATZ, P (EDS): *Neuropsychology of Human Emotion.* Guilford, New York, 1983, pp 85–110.

55. DARLEY, FL: *Treatment of acquired aphasia.* In FRIEDLANDER, WJ (ED): *Advances in Neurology,* Vol. 7. Raven Press, New York, 1975, pp 111–145.

56. DARLEY, FL, ARONSON, AE, AND BROWN, JR: *Motor Speech Disorders.* WB Saunders, Philadelphia, 1975.

56a. DEJERINE, J: *Sur un cas de cecite verbale avec agraphie, suivi d'autopsie.* Mem Soc Biol 3:197–201, 1891.

56b. DEJERINE, J: *Contribution a l'etude anatomoclinique et clinique des differentes varietes de cecite verbale.* Mem Soc Biol 4:61–90, 1892.

57. DEMASSARY, J: *L'alexie.* Encephale 27:134–164, 1932.

58. DERENZI, E, PIECZURO, A, AND VIGNOLO, L: *Ideational apraxia: A quantitative study.* Neuropsychologia 6:41–52, 1968.

59. FREEDMAN, M, ALEXANDER, MP, AND NAESER, MA: *Anatomical basis of transcortical motor aphasia.* Neurology 34:409–417, 1984.

60. FREUD, S: *On Aphasia.* (Trans, STENGEL, E.) International University Press, New York, 1953.

61. GALABURDA, AM, LEMAY, M, KEMPER, TL, AND GESCHWIND, N: *Right-left asymmetries in the brain.* Science 199:852–856, 1978.

62. GARDNER, H, AND ZURIF, E: *BEE but not be: Oral reading of single words in aphasia and alexia.* Neuropsychologia 13:181–190, 1975.

63. GAZZANIGA, MS, AND SPERRY, RW: *Language after section of the cerebral commissures.* Brain 90:131–148, 1967.

64. GERSTMANN, J: *Fingeragnosie: Eine (umschriebene) Storung der Orienterung am eigenen Korper.* Wien Klin Wschr 31:1010–1012, 1924.

65. GERSTMANN, J: *Zur symptomatologie der Hirnlasionen im Uebergangsgebiet der unteren Parietal und mittleren Occipitalwindung.* Nervenarzt 3:691–695, 1931.

66. Geschwind, N: *The anatomy of acquired disorders of reading.* In Money, J (ED): *Reading Disability.* Johns Hopkins, Baltimore, 1962.

67. Geschwind, N: *Disconnection syndromes in animals and man.* Brain 88:237–294, 1965.

68. Geschwind, N: *The varieties of naming errors.* Cortex 3:97–112, 1967.

69. Geschwind, N, and Fusillo, M: *Color naming defects in association with alexia.* Arch Neurol 15:137–146, 1966.

70. Geschwind, N, and Kaplan, E: *A human cerebral disconnection syndrome.* Neurology 12:675–685, 1962.

70a. Geschwind, N, and Levitsky, W: *Human brain: Left-right asymmetries in temporal speech region.* Science 161:186–187, 1968.

71. Geschwind, N, Quadfasel, FA, and Segarra, J: *Isolation of the speech area.* Neuropsychologia 6:327–340, 1968.

71a. Geschwind, N, and Strub, R: *Gerstmann syndrome without aphasia: A reply to Poeck and Orgass.* Cortex 11:296–298, 1975.

72. Gloning, I, Gloning, K, Haub, C, and Quatember, R: *Comparison of verbal behavior in right-handed and non right-handed patients with anatomically verified lesion of one hemisphere.* Cortex 5:43–52, 1969.

73. Goldstein, K: *Die transkortikalen Aphasien.* Gustav Fischer, Jena, 1917.

74. Goldstein, K: *Language and Language Disturbances: Aphasic Symptom Complexes and their Significance for Medicine and Theory of Language.* Grune & Stratton, New York, 1948.

75. Goodglass, H, Benson, DF, and Helm, N: *Aphasia and Related Disorders; Assessment and Therapy.* Cerebrovascular Survey Report, National Institute of Neurological and Communicative Disorders and Stroke. Whiting Press, Rochester, MN, 1980.

76. Goodglass, H, and Berko, J: *Agrammatism and inflectional morphology in English.* J Speech Hear Res 3:257–267, 1960.

77. Goodglass, H, and Geschwind, N: *Language disturbance (aphasia).* In Carterette, EC, and Friedman, MP (EDS): *Handbook of Perception,* Vol 7. Academic Press, New York, 1976, pp 389–428.

78. Goodglass, H, and Kaplan, E: *The Assessment of Aphasia and Related Disorders.* Lea & Febiger, Philadelphia, 1972.

79. Goodglass, H, and Quadfasel, F: *Language laterality in left handed aphasics.* Brain 77:521–548, 1954.

80. Goodglass, H, Quadfasel, F, and Timberlake, W: *Phrase length and the type and severity of aphasia.* Cortex 1:133–153, 1964.

81. Green, H, and Howes, D: *Conduction aphasia.* In Whitaker, H, and Whitaker, HA (EDS): *Studies in Neurolinguistics,* Vol 3. Academic Press, New York, 1977, pp 123–156.

82. Greenblatt, S: *Neurosurgery and the anatomy of reading: A practical review.* Neurosurgery 1:6–15, 1977.

83. Greenblatt, SH: *Localization of lesions in alexia.* In Kertesz, A (ED): *Localization in Neuropsychology.* Academic Press, New York, 1977, pp 123–156.

84. Greenblatt, SH: *Alexia without agraphia or hemianopsia.* Brain 96:307–316, 1973.

85. Greenblatt, SH: *Subangular alexia without agraphia or hemianopsia.* Brain Lang 3:229–245, 1976.

86. Grewel, F: *Acalculia.* Brain 75:397–407, 1952.

87. Head, H: *Aphasia and Kindred Disorders* (2 vols). Cambridge University Press, London, 1926.

88. Hécaen, H, and Albert, ML: *Human Neuropsychology.* John Wiley & Sons, New York, 1978.

89. Hécaen, H, Angelergues, R, and Houillier, S. *Les varietes cliniques des acalculies au cours des lesion retrorolandiques: Approche statistique du probleme.* Rev Neurol 105:85–103, 1961.

90. Hécaen, H, and Marcie, P: *Disorders of written language following right hemisphere lesions: Spatial dysgraphia.* In Dimond, S, and Beaumont, L (EDS): *Hemispheric Function in the Human Brain.* Paul Elek, London, 1974, pp 345–366.

91. Heilman, K: *Apraxia.* In Heilman, K, and Valenstein, E (EDS): *Clinical Neuropsychology.* Oxford, New York, 1979, pp 159–185.

92. Heimberger, RF, Demyer, W, and Reitan, RM: *Implications of Gerstmann's syndrome.* J Neurol Neurosurg Psychiatry 27:52–57, 1964.

93. Helm, NA: *Criteria for selecting aphasia patients for melodic intonation therapy.* Presentation, Annual meeting, American Association for the Advancement of Science, Washington, DC, 1978.

94. HELM-ESTABROOKS, N, FITZPATRICK, PM, AND BARRESI, B: *Visual action therapy for global aphasia.* J Speech Hear Dis 47:385–389, 1982.

95. HIER, DB, AND MOHR, JP: *Incongruous oral and written naming: Evidence for a subdivision of the syndromes of Wernicke's aphasia.* Brain Lang 4:115–126, 1977.

96. HINSHELWOOD, J: *Letter-, Word- and Mind-Blindness.* HK Lewis, London, 1900.

97. JACKSON, JH: *Clinical remarks on cases of defects of expression (by words, writing, signs, etc) in diseases of the nervous system.* Lancet 1:604–605, 1864.

98. JACKSON, JH: *Remarks on non-protrusion of the tongue in some cases of aphasia.* Lancet 1:716, 1878; Brain 38:104, 1915 (reprint).

100. JAKOBSON, R: *Towards a linguistic typology of aphasic impairments.* In DEREUCK, AVS, AND O'CONNOR, M: *Disorders of Language.* Little, Brown, Boston, 1964, pp 21–46.

101. KERTESZ, A: *Aphasia and Associated Disorders.* Grune & Stratton, New York, 1979.

102. KERTESZ, A, SHEPPARD, A, AND MACKENZIE, R: *Localization in transcortical sensory aphasia.* Arch Neurol 39:475–478, 1982.

103. KINSBOURNE, M, AND ROSENFIELD, DB: *Agraphia selective for written spelling.* Brain Lang 1:215–226, 1974.

104. KINSBOURNE, M, AND WARRINGTON, EK: *The developmental Gerstmann syndrome.* Arch Neurol 8:490–501, 1963.

105. KLEIST, K: *Leitungsaphasie (Nachsprechaphasie).* In BONHOEFFER, K (ED): *Handbuch der artzlichen Erfahrungen im Weltkriege 1914/1918.* Barth, Leipzig, 1934, pp 725–737.

106. LENNENBERG, E: *Biological Foundations of Language.* John Wiley & Sons, New York, 1967.

107. LICHTHEIM, L: *On aphasia.* Brain 7:434–484, 1885.

108. LIEPMANN, H: *Das Krankheitsbild der Apraxie ('motorischen Asymbolie').* Karger, Berlin, 1900.

109. LIEPMANN, H: *Der weitere Krankheitsverlauf bei dem einseitig Apraktischen und der Gehirnbefund auf Grund von Serienschnitten.* Monatschr Psychiatr Neurol 17:289–311, 1905.

110. LURIA, AR: *Higher Cortical Functions in Man.* Basic Books, New York, 1966.

111. LURIA, AR: *Traumatic Aphasia.* Mouton, The Hague, 1947, 1970.

112. MARIE, P: *Revision de la question de l'aphasie: La troisieme circonvolution frontale gauche ne joue aucun role special dans la fonction du langage.* Sem Med 26:241–247, 1906.

113. MARIN, OSM: *CAT scans of five deep dyslexic patients.* In COLTHEART, M, PATTERSON, K, AND MARSHALL, JC (EDS): *Deep Dyslexia.* Routledge & Kegan Paul, London, 1980, pp 407–411.

114. MARSHALL, JC, AND NEWCOMBE, F: *Syntactic and semantic errors in paralexia.* Neuropsychologia 4:169–176, 1966.

115. MARSHALL, JC, AND NEWCOMBE, F: *Patterns of paralexia—a psycholinguistic approach.* J Psychol Res 2:175–199, 1973.

116. MENDEZ, M, AND BENSON, DF: *Atypical conduction aphasia (a disconnection syndrome.* Arch Neurol (in press).

117. METTER, EJ, WASTERLAIN, CG, KUHL, DE, HANSON, WR, AND PHELPS, ME: *[18]FDG positron emission computed tomography in a study on aphasia.* Ann Neurol 10:173–183, 1981.

118. MOHR, JP: *Rapid amelioration of motor aphasia.* Arch Neurol 28:77–82, 1973.

119. MONRAD-KROHN, GH: *Dysprosody or altered melody of language.* Brain 70:405–415, 1947.

120. NAESER, MA, ALEXANDER, MP, HELM-ESTABROOKS, N, LEVINE, H, LAUGHLIN, SA, AND GESCHWIND, N: *Aphasia with predominantly subcortical lesion sites.* Arch Neurol 39:2–14, 1982.

121. NAESER, MA, AND HAYWARD, RW: *Lesion localization in aphasia with cranial computed tomography and the Boston Diagnostic Aphasia exam.* Neurology 28:545–551, 1978.

122. NIELSEN, JM: *Agnosia, Apraxia and Aphasia. Their Value in Cerebral Localization,* ed 2. Hafner Publishing, New York, 1946.

123. OJEMANN, GA, FEDIO, P, AND VANBUREN, JM: *Anomia from pulvinar and subcortical parietal stimulation.* Brain 91:99–116, 1968.

124. PENFIELD, W, AND JASPER, HH: *Epilepsy and the Functional Anatomy of the Human Brain.* Little, Brown, Boston, 1954.

125. PICK, A: *Die Agrammatischen Sprachstorungen.* Springer, Berlin, 1913.

126. POECK, K, KERSCHENSTEINER, M, AND HARTJE, W: *A quantitative study on language understanding in fluent and nonfluent aphasia.* Cortex 8(3):299–305, 1972.

127. POECK, K, AND ORGASS, B: *Gerstmann's syndrome and aphasia.* Cortex 2:421–437, 1966.

128. PORCH, B: *Porch Index of Communicative Ability.* Consulting Psychologists, Palo Alto, CA, 1967.

129. RIKLAN, M, LEVITA, E, ZIMMERMAN, J, AND COOPER, IS: *Thalamic correlates of language and speech.* J Neurol Sci 8:307, 1969.

130. ROBERTS, L: *Aphasia, apraxia and agnosia in abnormal states of cerebral dominance.* In VINKEN, PJ, AND BRUYN, GW (EDS): *Handbook of Clinical Neurology,* Vol 4. North Holland Publishing, Amsterdam, 1969, pp 312–326.

131. ROELTGEN, D, AND, HEILMAN, K: *A neurological and anatomical model of writing.* Presentation to Annual Meeting, Academy of Aphasia, Minneapolis, Minnesota, October 23, 1983.

132. ROELTGEN, DP, SEVUSH, S, AND HEILMAN, KM: *Phonological agraphia: Writing by the lexical-semantic route.* Neurology 33:755–765, 1983.

133. ROELTGEN, DP, SEVUSH, S, AND HEILMAN, KM: *Pure Gerstmann's syndrome from a focal lesion.* Arch Neurol 40:46–47, 1983.

134. RUBENS, AB: *Aphasia with infarction in the territory of the anterior cerebral artery.* Cortex 11:239–250, 1975.

135. RUBENS, AB: *Transcortical motor aphasia.* In WHITAKER, H, AND WHITAKER, HA (EDS): *Studies in Neurolinguistics,* Vol 1. Academic Press, New York, 1976, pp 293–304.

136. RUBENS, AB. *Agnosia.* In HEILMAN, KM, AND VALENSTEIN, E (EDS): *Clinical Neuropsychology.* Oxford University Press, New York, 1979, pp 233–267.

137. RUBENS, AB, AND BENSON, DF: *Associative visual agnosia.* Arch Neurol 24:305–315, 1971.

138. SAMUELS, JA, AND BENSON, DF: *Some aspects of language comprehension in anterior aphasia.* Brain Lang 8:275–286, 1979.

139. SARNO, MT, SILVERMAN, M, AND SANDS, E: *Speech therapy and language recovery in severe aphasia.* J Speech Hear Res 13:607–623, 1970.

140. SCHIFF, HB, ALEXANDER, MP, NAESER, MA, AND GALABURDA, AM: *Aphemia.* Arch Neurol 40:720–727, 1983.

141. SCHUELL, H, JENKINS, JJ, AND JIMENEZ-PABON, E: *Aphasia in Adults—Diagnosis, Prognosis and Treatment.* Harper & Row, New York, 1964.

142. SEGARRA, JM: *Cerebral vascular disease and behavior.* Arch Neurol 22:408–418, 1970.

143. SHALLICE T. *Phonological agraphia and the lexical route in writing.* Brain 104:412–429, 1981.

144. SKELLY, M, SCHINSKY, L, SMITH, R, AND FUST, R: *American Indian sign (Amerind) as a facilitator of verbalization for the oral verbal apraxic.* J Speech Hear Dis 39:445–456, 1974.

145. SMITH, A: *Diagnosis, Intelligence and Rehabilitation of Chronic Aphasics: Final Report.* University of Michigan, Ann Arbor, 1972.

147. SOQUES, A: *Quelques cas d'anarthrie de Pierre Marie.* Rev Neurol 2:319–368, 1928.

148. SPARKS, R, HELM, N, AND ALBERT, M: *Aphasia rehabilitation resulting from melodic intonation therapy.* Cortex 10:303–316, 1974.

149. SPREEN, O, AND BENTON, A: *Neurosensory Center Comprehensive Examination for Aphasia.* Neuropsychology Laboratory, University of Victoria, Victoria, British Columbia, Canada, 1969.

150. STACHOWIAK, FJ, AND POECK, K: *Functional disconnection in pure alexia and color naming deficit demonstrated by deblocking methods.* Brain Lang 3:135–143.

151. STARR, A: *The pathology of sensory aphasia, with an analysis of fifty cases in which Broca's centre was not diseased.* Brain 12:82–99, 1889.

152. STENGEL, E: *Loss of spatial orientation, constructional apraxia and Gerstmann's syndrome.* J Ment Sci 90:753–760, 1944.

153. STRUB, R, AND GESCHWIND, N: *Gerstmann syndrome without aphasia.* Cortex 10:378–387, 1974.

154. SUBIRANA, A: *The relationship between handedness and language function.* Int J Neurol 5:215–234, 1964.

155. VIGNOLO, L: *Evolution of aphasia and language rehabilitation: A retrospective exploratory study.* Cortex 1:344–367, 1964.

156. VIGNOLO, LA: *Auditory agnosia: A review and report of recent evidence.* In BENTON, AL (ED): *Contributions to Clinical Neuropsychology.* Aldine, Chicago, 1969, pp 172–208.

157. WAGENAAR, E, SNOW, C, AND PRINS, R: *Spontaneous speech of aphasic patients: A psycholinguistic analysis.* Brain Lang 3:281–303, 1975.

158. WAXMAN, SG, AND GESCHWIND, N: *Hypergraphia in temporal lobe epilepsy.* Neurology 24:629–636, 1974.

159. WEISENBURG, TS, AND McBRIDE, KL: *Aphasia.* Hafner Publishing, New York, 1964.

160. WEPMAN, JM: *Recovery from Aphasia.* Ronald, New York, 1951.

161. WERNICKE, K: *Der aphasische Symptomencomplex*. Cohn & Weigert, Breslau, West Germany, 1874. (Engl trans: COHEN, RS, AND WARTOFSKY, MW (EDS): *The Symptom-Complex of Aphasia. In Boston Studies in the Philosophy of Science,* Vol IV. 1969, Reidel Dordrecht-Hollands, pp 34–97.

162. WERTZ, RT, COLLINS, MJ, WEISS, D, KURTZKE, JF, FRIDEN, T, BROOKSHIRE, RH, PIERCE, J, HOLZAPPLE, P, HUBBARD, DJ, PORCH, BE, WEST, JA, DAVIS, L, MATOVICH, V, MORLEY, GK, AND RESSURECTION, E: *Veterans Administration cooperative study on aphasia: A comparison of individual and group treatment.* J Speech Hear Res 24:580–594, 1981.

163. WYLIE, J: *The Disorders of Speech.* Oliver & Boyd, Edinburgh, 1894.

164. ZANGWILL, OL: *Cerebral Dominance and its Relation to Psychological Function.* Charles C Thomas, Springfield, IL, 1960.

165. ZURIF, EB, CARAMAZZA, A, AND MYERSON, R: *Grammatical judgements of agrammatic aphasics.* Neuropsychologia 10:405–417, 1972.

Chapter 6

MODULATION OF AFFECT AND NONVERBAL COMMUNICATION BY THE RIGHT HEMISPHERE

Elliott D. Ross, M.D.

The neurologic basis for affective behaviors, emotions, and nonverbal communication is not well understood. This state of affairs probably reflects our relative ignorance concerning the extensive contributions of the right hemisphere to behavior. Until recently, most clinical inquiries into human language and behavior have been directed toward understanding left-hemisphere functions. This particular focus of interest can be traced to the fundamental discoveries of Broca[5,6] and Wernicke[68] that lesions in the left hemisphere cause aphasias (see Chapter 5). Subsequently, vast numbers of studies have been done to delineate the linguistic and related functions of the left hemisphere; this has led to the widely held opinion that the "dominant" or "major" hemisphere is the left and that the right is relegated to a "minor" or "nondominant" role. Nevertheless, neurologic and neuropsychologic evidence has been gathered in the last decade to show that this particular point of view is incorrect[56] and that the true richness and complexity of human communication and behavior is only partially represented in the left hemisphere (see also Chapter 1). This chapter shows that the right hemisphere has a major role in the prosodic, attitudinal, emotional, and gestural aspects of language and behavior.

PROSODY

The smallest segment of language is the phoneme. It forms the building blocks for words that compose the *lexicon*. Words are temporally strung together into *syntactic* relationships to form phrases, sentences, and paragraphs, which then convey meanings and concepts beyond the single word.[65] These particular features of language (phonemes, words, the syntactic relationships among words, and the formation of phrases, sentences, and paragraphs) are considered to be the linguistic or *propositional* components of language that give rise to the *articulatory line*. However, other features are also present in the articulatory line that, in addition to enhancing the linguistic aspects of speech by modifying meaning, also convey attitudes and emotions.[10,11,48,65] These particular features are best embraced under the term *prosody*, which refers to the melody, pauses, intonation, stresses, and accents applied to the articulatory line. The prosodic features of language are a graded phenomenon, making them more difficult to study than the discretely organized linguistic features.[10,11,65] Nevertheless, since prosody adds such complexity and richness to language, one must consider it a very crucial part of communication. In fact, studies of infant communication have demonstrated that the fundamental building blocks for language are the prosodic, not the linguistic, elements.[10,11,41,48]

The first systematic inquiry into the neurology of prosody was initiated in the late 1940s by Monrad-Krohn.[45] During World War II, he observed a female patient who had received a shrapnel wound to the left frontal area, causing a Broca's aphasia. The woman made a rather remarkable recovery except that she was left with a Germanic accent even though she was a native Norwegian. This caused her great distress during the Nazi occupation of Norway, because she was consistently mistaken for a German and socially ostracized. She was reported to have preserved overall speech melody, as evidenced by her ability to sing, intonate, and speak with emotion. The shift in accent reflected an inappropriate distribution of stresses and pauses during articulation. On the basis of this patient and others, Monrad-Krohn began developing his ideas concerning prosody.[46,48]

He felt that prosody could be divided into four components: intrinsic, intellectual, emotional, and inarticulate. *Intrinsic* prosody serves specific linguistic purposes and gives rise to dialectical and idiosyncratic differences in speech quality. The case of the patient just described illustrates a disturbance of this particular component of prosody. Examples of linguistic uses of intrinsic prosody are: raising the voice at the end of a statement to indicate a question; changing the stress on certain segments of a word to clarify its grammatical class[4] (i.e., cónvict [noun] versus convíct [verb]; or changing the stress on certain words and altering the pause structure in a sentence to clarify potentially ambiguous statements[11] (i.e., "The mán . . . and the womán dressed in black . . . came to see us," implying that only the woman was dressed in black, versus "The man and the womán . . . dressed in black . . . came to see us," implying that both were dressed in black and distinguishing them from other couples that were not dressed in black). *Intellectual* prosody adds attitudinal components to language. For example, if the sentence, "He is clever," is emphatically stressed on the *"is,"* it becomes a resounding acknowledgment of the person's ability. If the emphatic stress resides on *"clever,"* with a slight terminal rise in intonation, sarcasm becomes apparent. If *"he"* is emphatically stressed, then the sentence again acknowledges the person's ability but also implies that perhaps his associates are not very clever. *Emotional* prosody is used for imparting emotions into speech, and *inarticulate* prosody is the use of paralinguistic elements, such as grunts and sighs.

Monrad-Krohn then went on to describe various clinical disorders of prosody. He felt that *dysprosody* was occasionally observed in patients who had fairly good recovery from lesions causing Broca's or motor types of aphasia. These patients may experience a change in their voice quality, giving them a different accent because of the inability to properly stress segments and words. Dysprosody, therefore, is a disorder of intrinsic prosody. *Aprosody* is the general lack of prosody encountered in Parkinson's disease as part of the akinesia and masked facies. *Hyperprosody* is the excessive use of prosody that is often observed in manic patients or in Broca's aphasics who have very few words at their disposal but use them to their utmost to convey attitudes and emotions. Although Monrad-Krohn did not describe disorders of prosody from focal right brain damage, he was the first neurologist to observe in detail this severely neglected aspect of human communication.

Recent studies in patients with focal brain lesions have shown that right brain damage may seriously impair the affective (intellectual and emotional) components of prosody and gestures without disrupting the linguistic features of language.[32,53-56,64] These prosodic components, coupled with gestures (see further on), impart vitality to discourse and communicate important social, pragmatic, and emotional messages that in many instances are far more important than the actual words chosen.[16] This affective aspect of language and behavior is the focus of the remainder of this chapter.

KINESICS

Gestural behavior can be considered a nonverbal or paralinguistic component of communication that is closely aligned with the prosodic features of speech.[10,11] The study of limb, body, and facial movements associated with nonverbal communication is called *kinesics*.[9] Movements that are used for semiotic or referential purposes should be classified as pantomime, according to Hughlings-Jackson[35] and Critchley,[8] since they convey specific semantic information, while movements used to color, emphasize, and embellish speech are best classified as gestures. It should be noted, however, that this classification is not universally adhered to in the literature, which may help explain some of the "contradictory" observations and conclusions that have been published. Since most kinesic activity blends gestures and pantomime into a single movement (e.g., the "V" used for the victory sign may be displayed matter of factly, angrily, or jokingly), any analysis of kinesic activity should always pay attention both to its referential (pantomimal) content and to its emotional and attitudinal (gestural) content.

Despite an extensive literature on kinesic behavior, beginning with Darwin,[15] only a handful of contemporary studies have addressed the neurology of pantomime and gestures. Disturbances in the performance and comprehension of pantomime have been firmly linked to aphasia and left brain damage.[19,24] The pantomime decoding disorder in aphasics with comprehension difficulties has been attributed to their general inability to comprehend symbols. On the output side, Goodglass and Kaplan[28] have demonstrated that difficulty with the execution of pantomime in aphasics who do not have comprehension problems is best correlated with the presence of ideomotor apraxia. Cicone and co-workers[7] found that patients with Broca's aphasia used more referential (pantomimal) kinesic activity to embellish their telegraphic speech, whereas those with Wernicke's aphasia utilized more nonreferential (gestural) kinesic activity, similar to behavior of normal individuals.[18] Other studies, however, have not shown such correlations between the specific kinesic disturbance and the linguistic disturbance,[21] although all studies to date have found that disorders of pantomime are almost always the result of left-hemisphere damage causing aphasic deficits.[21,28,61]

Gestural kinesics, on the other hand, has not been well studied neurologically, although it has been noted periodically in the literature that gestural activity is often preserved in aphasic patients.[8,34] The first clinical publication to address the relationship of gestures and right brain damage was published in 1979 by Ross and Mesulam.[56] They observed that lesions of the right frontal operculum can cause a complete loss of spontaneous gestural activity without disturbances in praxis. The suggestion was made, therefore, that gestural behavior was a dominant function of the right hemisphere, in keeping with its putative role in the modulation of affective behavior. Since then, a number of studies have lent further support to this hypothesis.

Using a bedside examination technique, Ross[53] found that patients with right frontal damage who acutely lost the ability to gesture retained their ability to comprehend gestures. In contrast, patients with right parietotemporal opercular lesions were unable to comprehend the meaning of gestures even though they could gesture spontaneously. DeKosky and associates[17] found that right brain–damaged patients were severely impaired when asked to recognize the emotional content of facial expressions, whereas left brain–damaged patients performed almost as well as did normal individuals. Benowitz and co-workers[1] studied nonverbal communication in brain-damaged patients, by using the Profile of Nonverbal Sensitivity (PONS) test, which evaluates the subject's ability to comprehend facial expressions, limb and body movements, and intonational qualities of the voice. They found that right-hemisphere damage severely impaired the patients' overall ability to comprehend gestural kinesics—in particular, those involving facial expressions. Left brain damage, even of considerable extent, led to only very mild deficits. Thus, the right hemisphere appears specialized not only for emitting gestures but also for comprehending their meaning.

RIGHT HEMISPHERE AND AFFECTIVE BEHAVIORS

Until very recently, our knowledge concerning the right hemisphere's contribution to both language and behavior has been very rudimentary, even though Hughlings Jackson[34] pointed out a potential role for the right hemisphere as early as 1879. He made the observation that aphasic patients with left brain lesions, who have great difficulty with the propositional aspects of communication, seem to be quite capable of communicating emotions. Hughlings Jackson even went so far as to hypothesize that the emotional aspects of communication might be a function of the right hemisphere. Although he made this perceptive observation over 100 years ago, the first neurologic study to corroborate his hypothesis was not published until 1975.

In the 1970s, a series of neurolinguistic studies by Zuriff,[71] Blumstein and Cooper,[3] and others[65] used dichotic auditory testing techniques to show that the left ear was better (right-hemisphere advantage) at discerning the intonational aspects of speech, whereas the right ear was better (left-hemisphere advantage) at the linguistic aspects of speech. The first clinical study to formally test the hypothesis that right-hemisphere damage severely disrupts the decoding or comprehension of the affective components of language was published by Heilman and co-workers[32] in 1975. They tested brain-damaged patients for their ability to recognize emotions inserted into linguistically neutral statements. They found that right brain–damaged patients were markedly impaired on the recognition tasks when compared with normal individuals and aphasic left brain–damaged patients. In a followup study by Tucker and associates,[64] they also showed that right brain–damaged patients had great difficulty inserting affective intonation into linguistically neutral statements on request and by imitation. Although the patients did not have CT scans, it was felt clinically and by isotope scans that most of the lesions in the right brain–damaged patients involved the superior-posterior temporal and inferior parietal lobes.

In 1979, Ross and Mesulam[56] published a descriptive study that examined the performance (encoding) aspect of affective behavior. They reported two patients who suffered ischemic infarctions, verified by CT, involving the right inferior frontal and anterior-inferior parietal regions. Neither patient was aphasic or apraxic, but both complained bitterly of their almost total inability to insert affective and attitudinal variation into their speech and gestural behavior. One patient was a school teacher who, despite her persistent left arm monoplegia, was able to return to the classroom. Her main difficulty was that she could no longer control the classroom because (said in a monotone voice), "I cannot put any emotion into my voice or actions, and the pupils do not know when I am angry and mean business." The other patient was a surgeon who had marked problems interacting with his wife and family because of his inability to modulate his tone of voice to make it socially acceptable. Thus, when he asked his wife to do something, it always came out as a flat commanding statement rather than a pleasant request for a favor. Neither of these patients seemed to have difficulty perceiving the display of affect in others, and they both insisted that they could feel and experience emotions inwardly. However, their monotone voices and lack of gestures gave the impression that they had a flat affect or that they might be depressed.

Based on the study of these patients and on the previous publications by Heilman and associates,[32,64] Ross and Mesulam[56] hypothesized that (1) the right hemisphere was dominant for organizing the affective-prosodic components of language and gestural behavior and (2) the functional/anatomic organization of affective language in the right hemisphere mirrored the organization of propositional language in the left hemisphere. Additional indirect evidence to support the second hypothesis was also provided in the 1978 study of Larsen and co-workers.[39] These investigators measured focal changes in cerebral blood flow over the cortex during an automatic speech task in which patients repeatedly counted aloud from one to 20. They used the technique of direct unilateral intracarotid injection of radioactive 133-xenon, measuring the washout of xenon with 256 surface detectors, in order to calculate focal cerebral bloodflow.[40] To their surprise, the investigators found focal and homologous increases in bloodflow in both hemispheres involving the frontal operculum, temporal operculum, and supplementary motor areas. Their observation suggests that whatever the right hemisphere's contribution is to language, it is probably organized anatomically and functionally in a parallel fashion to propositional language in the left hemisphere.[53]

An issue not resolved by the Ross and Mesulam[56] paper, however, was whether the prosodic deficits from right brain damage were just relegated to the attitudinal and emotional aspects of prosody or if the deficits also involved the more linguistic (intrinsic) aspects of prosody. The studies by Weintraub,[67] Heilman,[30] and their co-workers as well as publications by Danly and associates[13,14] and Blumstein and Goodglass[4] have looked more carefully at this issue in both right and left brain–damaged patients. The composite data from these publications indicates that the linguistic components of prosody may be impaired by either right[30,67] or left[4,13,14] brain damage but that the affective and perhaps attitudinal components are disrupted almost exclusively by right brain damage.[30]

In 1981, Ross[53] approached the issue of whether the anatomic organization of the affective components of language in the right hemisphere was, in fact, similar to the organization of the linguistic components in the left hemisphere. He published a series of 10 patients with focal right brain damage from infarction, localized by CT scan, who were assessed at the bedside in a manner similar to the method used for propositional language. Thus, the patients were examined for (1) their spontaneous use of affective prosody and gesturing during conversation; (2) their ability to repeat, through imitation, linguistically neutral sentences with affective prosody; (3) their ability to auditorily comprehend affective prosody; and (4) their ability to visually comprehend the gestural components of kinesic behavior. The patients were collected consecutively, and it was found that all patients who had lesions bordering the right sylvian fissure had some disorder of affective language. Because the clustering of affective-prosodic deficits appeared similar to the clustering of linguistic deficits observed in patients with left brain damage,[2] these particular syndromes were called *aprosodias* (ā·pro·sō·di·a·s) and the same modifiers as those used in the aphasias were applied for classification purposes (Fig. 1, Table 1). Aprosodias, therefore, should be viewed as encoding/decoding disorders of affective behavior.

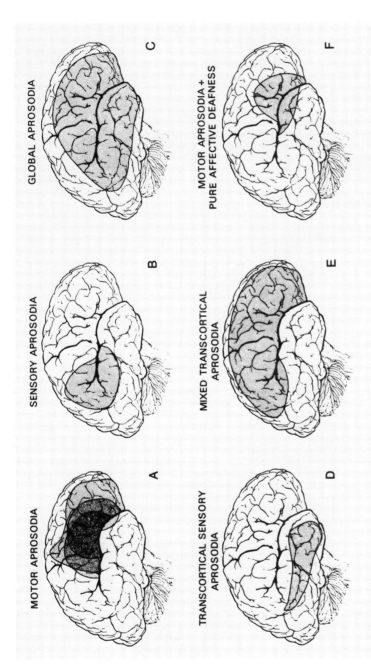

FIGURE 1. Right lateral brain templates showing the distribution of infarctions seen on CT in eight of 10 patients reported by Ross[26] with various aprosodias.

MODULATION OF
AFFECT AND
NONVERBAL
COMMUNICATION BY
THE RIGHT
HEMISPHERE

245

TABLE 1. The Aprosodias*

	Spontaneous Affective Prosody and Gesturing	Affective Prosodic Repetition	Affective Prosodic Comprehension	Comprehension of Emotional Gesturing
Motor	poor	poor	good	good
Sensory	good	poor	poor	poor
Global	poor	poor	poor	poor
(Conduction)	(good)	(poor)	(good)	(good)
Transcortical Motor	poor	good	good	good
Transcortical Sensory	good	good	poor	poor
Mixed Transcortical	poor	good	poor	poor
(Anomic)	(good)	(good)	(good)	(poor)

*The existence of conduction and anomic aprosodias has been hypothesized. Motor, sensory, global, and transcortical sensory aprosodias have good anatomic correlation with lesions in the left hemisphere known to cause homologous aphasias.

APROSODIAS

Bedside Evaluation

In order to assess the affective components of language and behavior at the bedside, some practice is required by the examiner. However, once the clinician becomes familiar with the analysis, it can readily be incorporated into the routine neurologic examination in much the same way that one conventionally tests for propositional language. The tasks described below are easily and flawlessly done by normal individuals, regardless of educational level (see also Chapter 2).

Spontaneous Affective Prosody and Gesturing

During the interview, it should be observed whether or not there is affective prosody in the patient's voice—especially when the patient is asked emotionally loaded questions (for example, "How do you feel about your neurologic deficits?" "Have you experienced the loss of a loved one?" or "Have you had any close calls with death or serious injury?"). One should ignore the overall loudness or softness of speech and pay strict attention to the finer aspects of prosodic variation and gesturing, to see whether they convey emotional or attitudinal information and whether or not they are appropriate to the situation under discussion.

Repetition of Affective Prosody

In this evaluation, a declarative sentence, void of emotional words, should be used. The patient is then asked to repeat the sentence with exactly the same affective tone used by the examiner. Thus, the statement, "I am going to the other movies," is said in a happy, sad, tearful, disinterested, angry, or surprised voice. Repetition ability should be judged by how well the patient imitates the affective prosody of the examiner. Merely raising or lowering the overall loudness of voice without other prosodic variation, or raising the voice at the end of a statement to indicate a question, should not be interpreted as constituting good affective-prosodic repetition.

Comprehension of Affective Prosody

To examine comprehension of affective prosody, a declarative statement, void of emotional words, should be produced by the examiner with differing affective tones. Standing behind the patient during this assessment will avoid giving the patient visual clues (see further on). The patient is then asked to identify what kind of affect was injected into the statements. Sometimes a multiple-choice format is necessary initially to orient the patient to the requested task.

Comprehension of Gestural Kinesics

This is done with the examiner facing the patient and conveying a particular affective state using gestural activity involving the face and limbs. As with affective-prosodic comprehension, the patient is requested to identify the emotion by name or description. Occasionally, a multiple-choice format is needed during initial testing to help orient the patient to the task.

Functional-Anatomic Correlates

Using the bedside evaluation just outlined, one may classify disorders of affective behavior following brain damage into the subtypes shown in Table 1. The mapping of lesions onto a right hemisphere template based on CT scans of published (Fig. 1)[53] and unpublished[55] data leads to a number of distinct functional-anatomic correlates.

Motor Aprosodia (Fig. 1A)

Patients with motor aprosodia usually come to medical attention with a moderate to severe left hemiplegia and variable left-sided sensory loss. On CT, the lesions have involved the right frontal and anterior parietal opercula. This distribution is analogous to that of lesions in the left hemisphere known to cause persistent Broca's (motor) aphasia.[2,36,44] Patients have also been encountered with motor aprosodia from subcortical lesions, two with right basal ganglia intracerebral hemorrhages (unpublished), and one with an infarction of the right internal capsule confirmed at autopsy.[54] Most patients with motor aprosodia will have transient anosognosia and dysarthria early in their course. Their persistent language deficit, however, is characterized by flat monotone speech with loss of spontaneous gesturing. Repetition of affective prosody is severely compromised, but comprehension of affective prosody and visual comprehension of emotional gesturing remains intact. Some patients who have been followed up to 1 year have shown remarkable improvement in their gestural ability but only incomplete improvement of spontaneous affective prosody. This observation suggests that the neurology of gestures and affective prosody share a common anatomic substrate but that the overlap is not total. Two patients became severely depressed during their illness but never displayed a depressive affect in speech or gestures.[53,57] Their motor aprosodia remained unchanged even after the depression was treated. Thus, the flattening of spontaneous affective behavior in motor aprosodia is a true neurologic deficit that cannot be accounted for by either psychiatric or psychologic explanations. Another interesting finding in these patients is that under extreme emotional conditions, patients with motor aprosodia are often able to laugh or cry in a fleeting all-or-none fashion reminiscent of the pathologic affect encountered in pseudobulbar palsy. Thus, the ability to display extremes of emotional motor behavior appears to be available to these patients despite their flattened affect, which strongly suggests that these displays are organized by motor systems that are separate from the right frontal neocortex.

Sensory Aprosodia (Fig. 1*B*)

Patients with sensory aprosodia are characterized by having excellent affective prosody in speech and active spontaneous gesturing. However, in these patients auditory comprehension of affective prosody, visual comprehension of emotional gesturing, and repetition of affective prosody are severely impaired. Some of the patients will have moderate deficits in vibratory sense, position sense and stereognosis on the left side, and a dense left hemianopsia. Hemiplegia in these patients has not been encountered. The CT lesions, so far, have involved the right posterior temporal and posterior parietal opercula. This particular distribution is analogous to left-hemisphere lesions known to cause Wernicke's (sensory) aphasia.[2,36,37] During the interview, the patients often appear somewhat euphoric and overly happy, even when they talk about their strokes and the possibility of losing their jobs.

Global Aprosodia (Fig. 1*C*)

Patients with global aprosodia usually demonstrate large right perisylvian lesions involving the frontal, parietal, and temporal lobes on CT or occasionally a deep right intracerebral hemorrhage. A severe left hemiplegia with hemisensory loss and hemianopsia is usually present. Since the patients lack the ability to display affect through prosody and gestures, they give the impression of a flattened affect. Comprehension and repetition of affective prosody and visual comprehension of emotional gesturing are severely compromised. Over time, affective-prosodic comprehension and gesturing may improve, but spontaneous affective prosody usually remains severely curtailed. The lesions encountered so far have been consistent with the left perisylvian distribution of brain damage known to produce global aphasia.[2,36] The improvement in affective-prosodic comprehension is also consistent with Kertesz's[36] observations that the most likely parameter to improve in global aphasics is comprehension.

Transcortical Motor Aprosodia

Patients with transcortical motor aprosodia have aprosodic-agestural speech with preserved repetition and comprehension of affective prosody and emotional gesturing. The two reported patients with this condition had left hemipareses without sensory loss. One of the patients had a metastatic tumor with tracking edema involving the right frontal, parietal, and temporal lobes, thus making it impossible to reach precise functional-anatomic correlations. The second patient had a right striatal infarction confirmed by CT. Interestingly, her transcortical motor aprosodia rapidly resolved over a 2-week period, which would be consistent with some of the transient motor aphasias that have been observed following left striatal lesions.[2,33]

Transcortical Sensory Aprosodia (Fig. 1D)

Transcortical sensory aprosodia has been described in one patient. Spontaneous affective prosody and its repetition, as well as emotional gesturing, were preserved; however, comprehension of affective prosody was severely impaired. Comprehension of emotional gesturing could not be tested in this patient, because he was blind secondary to severe cataracts. The CT scan showed a large intracerebral hemorrhage, involving the right anterior-inferior temporal lobe with sparing of its posterior-superior aspect. This particular distribution, with sparing of the posterior-superior temporal lobe, is consistent with left hemisphere lesions associated with transcortical sensory aphasia.[2,36] Except for transient coma and left hemiplegia, which completely resolved within 6 hours, the patient did not sustain any elementary neurologic deficits from the hemorrhage.

Mixed Transcortical Aprosodia (Fig. 1E)

One patient has been reported to have a mixed transcortical aprosodia. The lesion was shown by CT scan to involve the right suprasylvian region and a small portion of the posterior temporal operculum. This patient had a severe left hemiplegia with hemisensory loss. Her affective language was first evaluated 6 months after the stroke, at which time, the patient had markedly attenuated gesturing and spontaneous affective prosody. Repetition of affective prosody, however, was present but not fully normal, whereas comprehension of affective prosody and emotional gesturing was very poor. Although repetition of affective prosody was not perfectly normal, it was so much better than her spontaneous affective prosody that she was classified as having a mixed transcortical aprosodia, even though her lesion seemed to be most consistent with a global aprosodia. Perhaps if she had been evaluated closer to the time of her ictus, she would have demonstrated a global aprosodia. Interestingly, Kertesz[36] has reported that patients with large suprasylvian lesions with minimal involvement of the superior temporal lobe may initially exhibit a global aphasia that subsequently evolves into a mixed transcortical or even a transcortical motor aphasia. These observations are pertinent to this patient, since her mixed transcortical aprosodia eventually evolved into a transcortical motor aprosodia.

Pure Affective Deafness + Motor Aprosodia (Fig. 1F)

One patient has been encountered with a syndrome best described as *pure affective deafness* (in the original paper[53] the term *pure prosodic deafness* was used) and motor aprosodia. The patient was admitted to the hospital with a severe left hemiplegia without sensory loss or aphasia. His voice was flat and devoid of affective variation and his gesturing was very blunted. Repetition and comprehension of affective prosody were poor. However, comprehension of emotional gesturing was flawless. The CT scan showed an enhancing lesion consistent with an acute infarction, involving the right frontal operculum, anterior insula, and the anterior temporal operculum, with sparing of the posterior temporal and parietal opercula. With the exception of his ability to visually decode affective expression, this patient had all the other components of global aprosodia. This is similar to left brain–damaged patients who have been reported to have "global aphasia without alexia."[31] The right frontal opercular lesion easily accounted for the patient's aprosodia-agestural speech and poor affective-prosodic repetition. The auditory comprehension difficulty for affective prosody could be attributed to the anterior temporal lesion, assuming it isolated the right posterior-superior temporal lobe from ipsilateral and contralateral auditory inputs—very similar to mechanism postulated to explain pure word deafness arising from a unilateral lesion in the left temporal lobe.[25] If the visual connections to the right posterior temporal regions remained intact, as suggested by the CT scan, then it is not surprising that this patient was able visually to comprehend emotional gesturing, even though he was unable to auditorily comprehend affective prosody.

Summary

The existence of various aprosodias following focal right brain damage suggests that the right hemisphere dominantly modulates the affective components of language and behavior. This modulation is organized in much the same way as is the left hemisphere's contribution to the phonologic, lexical, and grammatical aspects of language. In addition, some of the recovery patterns observed in patients with aprosodia are similar to those described with aphasias.[2,36] Subcortical lesions in the right hemisphere may produce aprosodias,[53,54] just as subcortical lesions in the left hemisphere may produce aphasic deficits.[2,12,49] It has also been postulated that appropriate lesions in the right hemisphere, involving the parietal operculum or the angular gyrus, might produce aprosodias that are homologous to conduction and anomic aphasia, but to date these syndromes have not yet been encountered (see Table 1).[53]

EMOTIONS

The previous sections address only the encoding and decoding of affective behaviors by the right hemisphere. However, there are a number of complex interactions between the experience of emotions[51] and their expression through affective behavior that need clarification. The relevant issues include extreme emotional displays, changes in vegetative functions associated with mood disturbances, the verbal description of internal emotional states by patients, and the actual internal experience of emotions.

Patients who have motor or global aprosodia (producing a loss of spontaneous affective behaviors) have been reported to be able to display the extremes of emotions during very sad, happy, or angry situations.[53,57] These displays tend to be all or none, uncontrollable, and socially embarrassing, giving them the quality of pathologic affect.[2,50,70] Unlike true pathologic affect, however, the displays are generally mood congruent. Thus, one must conclude that the motoric organization of extreme emotional displays are modulated by areas of the brain outside the right neocortex. Most likely, these areas reside in the temporal limbic system, basal forebrain, and diencephalon, since epileptic discharges in these regions may induce extreme outbursts of emotion that usually take the form of pathologic affect display.[42, 50-52] In contrast, the role of the right neocortex seems confined to the more graded and subtle aspects of affective expression.

Although patients with aprosodia lose their ability to encode and decode affective behavior, they may continue to experience the entire range of emotional feeling states.[53,56,57] This suggests, therefore, that emotional experience and affective behaviors are dissociable from each other and that they may each have a different neuroanatomic substrate. A dissociation between affective display and emotional experience is also well known to occur in patients who have pathologic regulation of affect either in conjunction with bilateral lesions involving the descending neocortical motor pathways (pseudobulbar palsy) or as a result of unilateral lesions involving diencephalic or limbic structures.[2,50,70] These patients are characterized as having uncontrollable bursts of laughing and crying that are usually precipitated by trivial environmental stimuli. Their emotional display generally occurs in the absence of a corresponding emotional experience or is out of proportion to it.

Furthermore, in a study of the diagnostic and neuroanatomic correlates of depression in brain-damaged patients, Ross and Rush[57] reported two cases that demonstrate the dissociability of verbal behaviors from both internal mood states and affective behaviors. One patient with mixed transcortical aprosodia verbally denied being depressed for months on end; yet she was observed to have severe anorexia, insomnia, psychomotor retardation, irritability, and abnormal cortisol secretions—signs and laboratory findings that are consistent with endogenous depression. Appropriate treatment with a tricyclic antidepressant completely resolved her abnormal behaviors and vegetative signs and also normalized her cortisol secretions. Her spontaneous affective behavior, however, remained flat as a result of her aprosodia. This uncoupling of verbal reports from the patient's presumed experience of depression is similar to the striking verbal-behavioral dissociations that have been described in split brain patients by Gazzaniga and co-workers.[26] A second patient with depression was reported who had Wernicke's aphasia, which severely impaired her ability to communicate through propositional language. Nevertheless, this patient displayed the full range of vegetative signs, depressive behaviors, suicidal acts, and abnormal cortisol suppression indicative of endogenous depression. Following treatment with a tricyclic antidepressant, her depressive behaviors and abnormal cortisol secretions completely resolved. This patient provides further evidence that verbal behavior can be uncoupled from affective behavior, vegetative regulation, and emotional experience by appropriate brain lesions.

From these observations, it is clear that clinicians must be exceedingly careful about assessing the internal emotional state of patients who suffer brain damage. Merely observing overt affective behaviors or relying solely on the patient's verbal reports can lead to faulty impressions and conclusions. For instance, although many right brain–damaged patients are traditionally described as being indifferent,[22] they may actually harbor a full component of concern[53,56] even to the extent of a suicidal depression.[53,57] Conversely, even in the absence of underlying sadness and dysphoria, the flat affect caused by right brain damage may give the erroneous impression that the patient is depressed.[57]

CONCLUSION

The pathophysiologic correlations in patients with aprosodia have indicated that the right hemisphere may play a dominant role in the decoding and encoding of affective expression. Additional evidence suggests that this may reflect a more general specialization of the right hemisphere for all nonverbal and paralinguistic aspects of communication.[1] However, we do not yet know if the right hemisphere is also specialized for modulating the entire range of emotional experience. Although several authors believe in this type of right-hemisphere specialization,[20,23,60,62,63,69] others stress that both hemispheres may participate in the experience of emotion with each hemisphere having its greatest impact on a different type of emotion.[57–59] For example, it appears that the left hemisphere may play a greater role in modulating happiness and laughter, whereas the right hemisphere may modulate sadness and crying.[58,59] This may help to explain why right-hemisphere damage often leads to a state of euphoria or indifference (see also Chapter 1).

Many lines of evidence show that the most important brain area involved in emotional experience is the limbic system.[27,29] Although powerful reciprocal connections occur between the neocortex and the limbic system,[38,43] most details about the functional interaction between these areas in regulating emotions and emotional behavior are unknown (see Chapters 1 and 8 for further discussion). Figures 2 and 3 attempt to summarize schematically the information presented in this chapter regarding the known and hypothetical mechanisms that may regulate the experi-

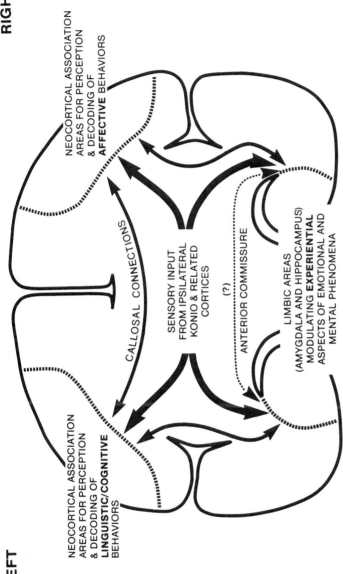

RIGHT

LEFT

NEOCORTICAL ASSOCIATION AREAS FOR PERCEPTION & DECODING OF **AFFECTIVE** BEHAVIORS

NEOCORTICAL ASSOCIATION AREAS FOR PERCEPTION & DECODING OF **LINGUISTIC/COGNITIVE** BEHAVIORS

CALLOSAL CONNECTIONS

SENSORY INPUT FROM IPSILATERAL KONIO & RELATED CORTICES

(?)

ANTERIOR COMMISSURE

LIMBIC AREAS (AMYGDALA AND HIPPOCAMPUS) MODULATING **EXPERIENTIAL** ASPECTS OF EMOTIONAL AND MENTAL PHENOMENA

FIGURE 2. Decoding by the forebrain of linguistic, cognitive, and affective behaviors. The essential neocortical areas for these various functions reside around the temporal-parietal-occipital confluence. The limbic areas modulate the experiential aspects of emotions and mental phenomena. Current evidence suggests that certain emotional experiences may be lateralized in the temporal limbic system.

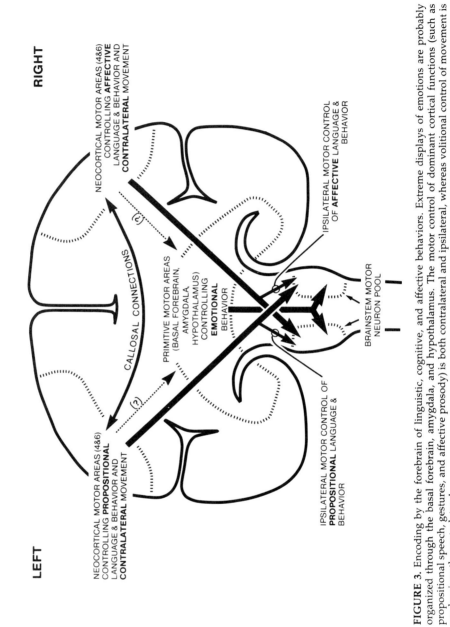

FIGURE 3. Encoding by the forebrain of linguistic, cognitive, and affective behaviors. Extreme displays of emotions are probably organized through the basal forebrain, amygdala, and hypothalamus. The motor control of dominant cortical functions (such as propositional speech, gestures, and affective prosody) is both contralateral and ipsilateral, whereas volitional control of movement is predominantly contralateral.

ence, expression, and decoding of affect-related behaviors and emotions by the human brain.

REFERENCES

1. BENOWITZ, LI, BEAR, DM, ROSENTHAL, R, MESULAM, M-M, ZAIDEL, E, AND SPERRY, RW: *Hemispheric specialization in nonverbal communication.* Cortex 19:5, 1983.

2. BENSON, DF: *Aphasia, Alexia, and Agraphia.* Churchill Livingstone, New York, 1979.

3. BLUMSTEIN, S, AND COOPER, W: *Hemispheric processing of intonation contours.* Cortex 10:146, 1974.

4. BLUMSTEIN, S, AND GOODGLASS, H: *The perception of stress as a semantic cue in aphasia.* J Speech Hear Res 15:800, 1972.

5. BROCA, P: *Remarques sur le siège de la faculté du langage articulé, suivies d'une observation d'aphemie.* In VON BONIN, G (TRANS): *The Cerebral Cortex.* Charles C Thomas, Springfield, IL, 1960.

6. BROCA, P: *Sur le siège de la faculté du langage articulé.* Bulletin d'Anthropologie 6:377, 1865.

7. CICONE, M, WAPNER, W, FOLDI, N, ZURIF, E, AND GARDNER, H: *The relationship between gesture and language in aphasic communication.* Brain Lang 8:324, 1979.

8. CRITCHLEY, M: *The Language of Gesture.* Edward Arnold, London, 1939.

9. CRITCHLEY, M: *Aphasiology and Other Aspects of Language* (Chapter 15). Edward Arnold, London, 1970.

10. CRYSTAL, D: *Prosodic Systems and Intonation in English.* University Press, Cambridge, England, 1969.

11. CRYSTAL, D: *The English Tone of Voice.* St. Martin's Press, New York, 1975.

12. DAMASIO, AR, DAMASIO, H, RIZZO, M, VARNEY, N, AND GERSCH, F: *Aphasia with nonhemorrhagic lesions in the basal ganglia and internal capsule.* Arch Neurol 39:15, 1982.

13. DANLY, M, COOPER, WE, AND SHAPIRO, B: *Fundamental frequency, language processing, and linguistic structure in Wernicke's aphasia.* Brain Lang 19:1, 1983.

14. DANLY, M, AND SHAPIRO, B: *Speech prosody in Broca's aphasia.* Brain Lang 16:171, 1982.

15. DARWIN, C: *The Expression of the Emotions in Man and Animals.* Philosophical Library, New York, 1955.

16. DEGROOT, A: *Structural linguistics and syntactic laws.* Word 5:1, 1949.

17. DEKOSKY, ST, HEILMAN, KM, BOWERS, D, AND VALENSTEIN, E: *Recognition and discrimination of emotional faces and pictures.* Brain Lang 9:206, 1980.

18. DELIS, D, FOLDI, NS, HAMBE, S, GARDNER, H, AND ZURIF, E: *A note on temporal relations between language and gestures.* Brain Lang 8:350, 1979.

19. DE RENZI, E, MOTTI, F, AND NICHELLI, P: *Imitating gestures: A quantitative approach to ideomotor aproxia.* Arch Neurol 37:6, 1980.

20. DIMOND, SJ, FARRINGTON, L, AND JOHNSON, P: *Differing emotional responses from right and left hemispheres.* Nature 261:690, 1976.

21. FEYEREISEN, P, AND SERON, X: *Nonverbal communication and aphasia: A review.* I. *Comprehension,* II. *Expression.* Brain Lang 16:191, 213, 1982.

22. GAINOTTI, G: *Emotional behavior and hemispheric side of the lesion.* Cortex 8:41, 1972.

23. GAINOTTI, G: *The relationships between emotions and cerebral dominance: A review of clinical and experimental evidence.* In GRUZELIER, J, AND FLOR-HENRY, P (EDS): *Hemisphere Asymmetries of Function in Psychopathology.* Elsevier/North-Holland Biomedical Press, Amsterdam, 1979.

24. GAINOTTI, G, AND LEMMO, M: *Comprehension of symbolic gestures in aphasia.* Brain Lang 3:451, 1976.

25. GAZZANIGA, MS, GLASS, AV, SARNO, MT, ET AL: *Pure word deafness and hemispheric dynamics: A case history.* Cortex 9:136, 1973.

26. GAZZANIGA, MS, AND LEDOUX, JE: *The Integrated Mind* (Chapters 5–7). Plenum Press, New York, 1978.

27. GLOOR, P, OLIVIER, A, QUESNEY, LF, ANDERMANN, F, AND HOROWITZ, S: *The role of the limbic system in experimental phenomena of temporal lobe epilepsy.* Ann Neurol 12:129, 1982.

28. GOODGLASS, H, AND KAPLAN, E: *Disturbance of gesture and pantomine in aphasia.* Brain 86:703, 1963.

MODULATION OF
AFFECT AND
NONVERBAL
COMMUNICATION BY
THE RIGHT
HEMISPHERE

255

29. HALGREN, E, WALTER, RD, CHERLOW, DG, AND CRANDALL, PE: *Mental phenomena evoked by electrical stimulation of the human hippocampal formation and amygdala.* Brain 101:83, 1978.

30. HEILMAN, KM, BOWERS, D, SPEEDIE, L, AND COSLETT, HB: *The comprehension of emotional and nonemotional prosody.* Neurology (Suppl 2)33:241, 1983.

31. HEILMAN, KM, ROTHI, L, CAMPANELLA, D, ET AL: *Wernicke's and global aphasia without alexia.* Arch Neurol 36:129, 1979.

32. HEILMAN, KM, SCHOLES, R, AND WATSON, RT: *Auditory affective agnosia: Disturbed comprehension of affective speech.* J Neurol Neurosurg Psychiatry 38:69, 1975.

33. HERMANN, K, TURNER, JW, GILLINGHAM, FJ, AND GAZE, RM: *The effects of destructive lesions and stimulation of the basal ganglia on speech mechanisms.* Confin Neurol 27:197, 1966.

34. HUGHLINGS JACKSON, J: *On affections of speech from disease of the brain.* Brain 2:202, 1879.

35. HUGHLINGS JACKSON, J: *Words and other symbols.* In TAYLOR, J (ED): *Selected Writings of John Hughlings Jackson.* Hodder & Stoughton, London, 1932, p 205.

36. KERTESZ, A: *Aphasia and Associated Disorders.* Grune & Stratton Inc, New York, 1979.

37. KERTESZ, A, AND BENSON, DF: *Neologistic jargon: A clinicopathological study.* Cortex 6:362, 1970.

38. LAMENDELLA, JT: *The limbic system in human communication.* In WHITAKER, H, AND WHITAKER, HA (EDS): *Studies in Neurolinguistics,* Vol 3. Academic Press, New York, 1977, p 157.

39. LARSEN, B, SKINHØJ, E, AND LASSEN, NA: *Variations in regional cortical blood flow in the right and left hemispheres during automatic speech.* Brain 101:193, 1978.

40. LASSEN, NA, INGVAR, DH, AND SKINHØJ, E: *Brain function and blood flow.* Sci Am 239:62, 1978.

41. LEWIS, A: *Infant Speech: A Study of the Beginnings of Language.* Harcourt, Brace, & World, New York, 1936.

42. MALAMUD, N: *Psychiatric disorder with intracranial tumors of limbic system.* Arch Neurol 17:113, 1967.

43. MESULAM, M-M, VAN HOESEN, GW, PANDYA, DN, ET AL: *Limbic and sensory connections of the inferior parietal lobule (area PG) in the rhesus monkey: A study with a new method for horseradish peroxidase histochemistry.* Brain Res 136:393, 1977.

44. MOHR, JP: *Broca's area and Broca's aphasia.* In WHITAKER, H, AND WHITAKER, HA (EDS): *Studies in Neurolinguistics,* Vol 1. Academic Press, New York, 1976, p 201.

45. MONRAD-KROHN, GH: *Dysprosody or altered 'melody of language.'* Brain 70:405, 1947.

46. MONRAD-KROHN, GH: *The prosodic quality of speech and its disorders.* Acta Psychiatr Neurol 22:255, 1947.

47. MONRAD-KROHN, GH: *Altered melody of language ('dysprosody') as an element of aphasia.* Acta Psychiatr Neurol (Suppl) 46:204, 1947.

48. MONRAD-KROHN, GH: *The third element of speech: Prosody and its disorders.* In HALPERN, L (ED): *Problems of Dynamic Neurology.* Hebrew University Press, Jerusalem, 1963, p 101.

49. NAESER, MA, ALEXANDER, MP, HELM-ESTABROOKS, N, LEVINE, HL, LAUGHLIN, SA, AND GESCHWIND, N: *Aphasia with predominantly subcortical lesion sites: Description of three capsular/putaminal aphasia syndromes.* Arch Neurol 39:2, 1982.

50. POECK, K: *Pathophysiology of emotional disorders associated with brain damage.* In VINKEN, PJ, AND BRUYN, GW (EDS): *Handbook of Clinical Neurology,* Vol 3. North-Holland, Amsterdam, 1969, p 343.

51. PRIBRAM, KH, AND MELGES, FT: *Psychophysiological basis of emotion.* In VINKEN, PJ, AND BRUYN, GW (EDS): *Handbook of Clinical Neurology,* Vol 3. North-Holland, Amsterdam, 1969, p 316.

52. REEVES, AG, AND PLUM, F: *Hyperphagia, rage, and dementia accompanying a ventromedial hypothalamic neoplasm.* Arch Neurol 20:616, 1969.

53. ROSS, ED: *The aprosodias: Functional-anatomic organization of the affective components of language in the right hemisphere.* Arch Neurol 38:561, 1981.

54. ROSS, ED, HARNEY, JH, DELACOSTE, C, AND PURDY, P: *How the brain integrates affective and propositional language into a unified brain function. Hypotheses based on clinicopathological correlations.* Arch Neurol 38:745, 1981.

55. ROSS, ED, HOLZAPFEL, D, AND FREEMAN, F: *Assessment of affective behavior in brain damaged patients using quantitative acoustical-phonetic and gestural measurements.* Neurology (Suppl 2) 33:219, 1983.

56. ROSS, ED, AND MESULAM, M-M: *Dominant language functions of the right hemisphere? Prosody and emotional gesturing.* Arch Neurol 36:144, 1979.

57.   Ross, ED, and Rush, AJ: *Diagnosis and neuroanatomical correlates of depression in brain-damaged patients: Implications for a neurology of depression.* Arch Gen Psychiatry 38:1344, 1981.

58.   Sackeim, HA, Greenberg, MS, Weiman, AL, Gur, RC, Hungerbuhler, JP, and Geschwind, N: *Hemispheric asymmetry in the expression of positive and negative emotions: Neurologic evidence.* Arch Neurol 39:210, 1982.

59.   Sackeim, HA, and Gur, RC: *Lateral asymmetry in intensity of emotional expression.* Neuropsychologia 16:473, 1978.

60.   Schwartz, GE, Davidson, RJ, and Maer, F: *Right hemisphere lateralization for emotion in the human brain: Interactions with cognition.* Science 190:286, 1975.

61.   Seron, X, Van Der Kaa, MA, Remitz, A, and Van Der Linden, M: *Pantomime interpretation and aphasia.* Neuropsychologia 17:661, 1979.

62.   Suberi, M, and McKeever, WF: *Differential right hemispheric memory storage of emotional and non-emotional faces.* Neuropsychologia 15:757, 1977.

63.   Terzian, H: *Behavioral and EEG effects of intracarotid sodium amytal injection.* Acta Neurochirurgica 12:230, 1964.

64.   Tucker, DM, Watson, RT, and Heilman, KM: *Discrimination and evocation of affectively intoned speech in patients with right parietal disease.* Neurology 27:947, 1977.

65.   Van Lancker, D: *Cerebral lateralization of pitch cues in the linguistic signal.* Int J Hum Comm 13:201, 1980.

66.   Van Lancker, D, Canter, GJ, and Terbeek, D: *Disambiguation of distropic sentences: Acoustic and phonetic cues.* J Speech Hear Res 24:330, 1981.

67.   Weintraub, S, Mesulam, M-M, and Kramer, L: *Disturbances in prosody.* Arch Neurol 38:742, 1981.

68.   Wernicke, C: *Der aphasische Symptomencomplex. Eine psychologische Studie auf anatomischer Basis.* In Eggert, GH (trans): *Wernicke's Works on Aphasia. Sourcebook and Review.* Mouton, Paris, 1977.

69.   Wexler, BE: *Cerebral laterality and psychiatry: A review of the literature.* Am J Psychiatry 137:279, 1980.

70.   Wilson, SAK: *Pathological laughing and crying.* J Neurol Psychopathol 4:299, 1924.

71.   Zurif, EB: *Auditory lateralization: Prosodic and syntactical factors.* Brain Lang 1:391, 1974.

MODULATION OF
AFFECT AND
NONVERBAL
COMMUNICATION BY
THE RIGHT
HEMISPHERE

257

DISORDERS OF COMPLEX VISUAL PROCESSING: AGNOSIAS, ACHROMATOPSIA, BALINT'S SYNDROME, AND RELATED DIFFICULTIES OF ORIENTATION AND CONSTRUCTION*

Antonio R. Damasio, M.D., Ph.D.†

Disorders of complex visual processing are frequently encountered in neurologic practice. These disorders can be evaluated at the bedside, on the basis of history and neurologic observation, but the comprehensive appraisal of the defects requires neuropsychologic and neuro-ophthalmologic laboratory techniques. In this chapter, those disorders that are important either because of their frequency or because of their scientific implications are discussed.

DISORDERS OF PATTERN RECOGNITION

Visual Agnosia

Teuber's definition of agnosia as "a normal percept stripped of its meanings"[107] remains lapidary and applies to what was designated by Lissauer as "associative" agnosia.[81] (The definition does not apply to Lissauer's visual "apperceptive" agnosia, in which visual perception is manifestly abnormal; we have reservations about the use of the term *agnosia* in such cases.) An operational definition of visual agnosia could read as follows: a disorder of higher behavior *confined to the visual realm*, in which an *alert, attentive, intelligent*, and *nonaphasic* patient with *normal visual perception* gives evidence of *not knowing the meaning* of those stimuli—that is, of *not recognizing* them. We shall begin by examining the components of this definition.

DISORDERS OF COMPLEX VISUAL PROCESSING: AGNOSIAS, ACHROMATOPSIA, BALINT'S SYNDROME, AND RELATED DIFFICULTIES OF ORIENTATION AND CONSTRUCTION

259

*Supported by NINCDS Grant PO1 NS 19632-01.
†The author thanks Dr. Hanna Damasio for permission to use CT, MRI, and SPET images from her collection.

The statement that the deficit pertains to the visual realm reminds us that agnosias usually occur in relation to one sensory modality, leaving recognition intact in other channels. Visual and auditory agnosias are clinically the most dramatic and, from the standpoint of behavioral neurology, the most revealing. Alertness, intelligence, and intact command of language are prerequisites for the diagnosis of agnosia. A disturbance of attention precludes both the normal perception and the normal cognitive search process that are necessary for the attribution of meaning and recognition. Likewise, a severe disturbance of intellect (e.g., in a syndrome of dementia) impairs this process. It is especially important to note the latter requisite, because in the past (see the work of German neurologist Bay[9]) there was an attempt to explain agnosia as the consequence of dementia combined with decayed perception. Needless to say, a patient with visual agnosia may appear demented, because the magnitude of the defect impairs a normal appraisal of the environment; conversely, a patient with primary dementia may develop a superimposed visual agnosia. But the point is that the two processes are generally independent and neither leads to the other.

Let us now turn to the fundamental criterion of normal perception. The issue is complex, because the practical distinction between perception and recognition may or may not have a neurophysiologic counterpart and because it is difficult to determine whether a patient perceives stimuli as well as the examiner does. We believe that although perception and recognition are part of a continuum, it is possible to identify, both behaviorally and neurophysiologically, some components of the process that are mainly related to perception and some that are mainly related to recognition.

The painstaking search for supporting evidence of effective perception should proceed in the following ways. First, because most patients with visual agnosia are not aphasic, the examiner should try to obtain a workable verbal description of what the patient sees. A fairly accurate description of stimuli in terms of general shape, number, position, and distribution in space should help establish whether the basics of visual perception are impaired or normal. Such features as definition of contour and sense of depth should also be investigated and described verbally. A statement by the patient to the effect that vision is "blurred" or "foggy" is generally *not* compatible with the diagnosis of visual associative agnosia. Further confirmation of the patient's ability to perceive normally should come from procedures such as copying a drawing of an object or matching an actual object with the appropriate drawing or photograph of a similar object.

Patients affected by true visual (associative) agnosia will perform normally in all of these procedures. They will also be able to draw or describe the major details of the stimuli, with the exception of color, in that visual field. Most patients with visual agnosia also have achromatopsia (impairments of color perception), as discussed in the next section of this chapter. This is due to the contiguity of lesions that cause both impairments. In contrast, patients with so-called visual apperceptive agnosia fail in their ability both to copy a drawing and, generally, to match visual stimuli; usually, they complain of unclear vision. They may or may not have achromatopsia.

The observations of a behavioral neurologist or neuropsychologist are thus limited to establishing whether the patient's perceptual state is compatible with *seeing* an object and claiming not to know the meaning of it. For patients who offer a detailed description of the visual structure of the object, who can copy it and match it with another—that is, who behave in relation to the object in much the same manner as the observer—the answer is affirmative. But from a theoretical standpoint, the observer should be aware of the following: (1) the patient may have fine disturbances of perception not detectable by this form of testing, and such potential disturbances may actually contribute to the recognition defect; and (2) the fact that the patient behaves as if he *sees* the object does not guarantee that *all* cerebral structures involved in the recognition process "see" the object too (i.e., by a mechanism of disconnection, the patient could see with one part of the brain and yet not have another crucial brain area access that perception).

The final criterion in the definition of agnosia pertains to the inability to arrive at the meaning of (to recognize) the stimuli. Once again, the first step to establish that the patient no longer recognizes a stimulus is verbal inquiry. Most patients with agnosia will actually state that they do not know what a given object is; but the examiner should make sure that the statement does not reflect an inability to *name* the object, as opposed to an inability to *recognize* the object. Naturally, a patient who cannot recognize an object will not be able to name it either, but the converse is not true. When the disturbance affects naming only, patients will still be able to know what the object is and what it is used for, and they often have an appropriate affective reaction to the object. That can be established by inquiring about the use and functional category of the object or by having the patient produce responses pertinent to the object both by manipulating it or by choosing logically connected items from a multiple-choice display. In other words, once the examiner suspects or is told by the patient that a stimulus cannot be recognized, both verbal and nonverbal analyses are necessary in order to establish whether the defect is one of naming or of recognition.

We must emphasize that the distinction between knowing and naming is crucial and that it is often missed in the clinical evaluation of patients. The error goes both ways—that is, anomic patients classified as agnosic and agnosic patients classified as anomic. This common error can result from incomplete observation, but more often than not it occurs because the distinction between naming and knowing is improperly drawn in the examiner's mind. It is clear that full knowledge of the universe, as possessed by normal humans, includes both nonverbal and verbal components. However, it is certainly true that most stimuli in the environment can be adequately known by nonverbal means alone—that is, they can be properly recognized without appeal to a verbal memory. Examples of this are readily available in everyday life. One is often introduced to new human faces, socially or through the media, that one can later recognize at a strictly nonverbal level but that one is entirely unable

DISORDERS OF COMPLEX VISUAL PROCESSING: AGNOSIAS, ACHROMATOPSIA, BALINT'S SYNDROME, AND RELATED DIFFICULTIES OF ORIENTATION AND CONSTRUCTION

261

to name, either because the verbal tag of that face was never learned or because it was learned but quickly forgotten. In the animal world, it is clear that high-level recognition of individual stimuli and of complex configurations of visual stimuli can be achieved, for example, by a dog or by a monkey, without any intervening verbal mediation. The clinical analysis of visual agnosia, then, is aimed at establishing the presence of a *nonverbal recognition defect* and at eliminating the possibility that what poses as impaired recognition is impaired verbal communication. Finally, it should be noted that visual recognition defects are generally accompanied by visual learning defects.

Prosopagnosia

The word "prosopagnosia" derives from the Greek *prosopon* (face) and *gnosis* (knowledge) and was introduced by Bodamer.[23] As the term implies, prosopagnosia is a visual agnosia hallmarked by an inability to recognize previously known human faces (the retrograde defect) and to learn new ones (the anterograde defect). The onset almost always involves the sudden realization that the well-known face of a relative can no longer be recognized visually and appears entirely unfamiliar. Yet the voice easily gives away the identity of the unrecognized face, as do a variety of clues such as body build, attire, or posture. The use of such strategies highlights the remarkable preservation of recognition through other channels (generally auditory) as well as the intactness of intellectual means to formulate intelligent guesses about the environment. The notion that the deficit in prosopagnosia is limited to human faces is erroneous; the defect is of a far greater magnitude. For example, a farmer will no longer be able to recognize his cows individually and a bird watcher will no longer identify different species of birds; in other words, patients with prosopagnosia also fail to recognize stimuli that, as faces, belong to a group containing numerous, visually "ambiguous" stimuli. We define visually "ambiguous" stimuli operationally, as stimuli that share the same type of subcomponents, arranged in the same way, and that can be distinguished only by relatively minor differences of shape or size of subcomponents. Human faces form groups of visually "ambiguous" stimuli, with numerous members. The specific recognition of those members is generally a necessity, and the survival value of this seemingly trivial ability is incalculable. On a different level of importance, automobiles or a collection of dresses in a closet also constitute groups of visually "ambiguous" stimuli, in the sense that recognition of each separate one depends on minor distinguishing features.

Patients with prosopagnosia can recognize any object in the environment, provided the examiner does not require a recognition of the specific object within the group (e.g., provided the examiner does not inquire into the historic relation between that specific object and the patient). Thus, patients are able to recognize a pencil, or an article of furniture, or a car, as, respectively, pencils, furniture, and cars; but they cannot decide whether such an article belongs to them or not, or what the specific manufacturer of a given car is. In other words, these patients can perform a *generic recognition*, which shows that they still possess the knowledge of the class to which a stimulus belongs. But the identification of a *specific member* within the generic class eludes these patients. Nowhere is this dissociation more clear than with human faces themselves. Patients with prosopagnosia are unable to recognize any previously known face, including their own. Yet they do know that a face *is* a face, and they can generally point to or name, without difficulty, the eyes, ears, nose, or mouth of the examiner or of their own reflection in a mirror. Testing for such an ability will immediately assure the examiner that the patient's perception of both the whole and the parts of a facial stimulus is intact and that the real defect lies in the appropriate evocation of associated information (i.e., that it lies in a failure to activate "contextual" memory). Prosopagnosic patients are generally able to perform complex perceptual tasks (such as the Benton and Van Allen's test of unfamiliar facial discrimination).[19] On the other hand, many patients with severe disturbances of visuospatial performance do not have prosopagnosia.[86,91] In most instances, prosopagnosia is a defect of visually triggered memory, one in which the memories pertinent to a given visual stimulus fail to be evoked, thus blocking the activation of the multimodal memory traces on which the matrix of recognition is built. Patients with prosopagnosia can evoke those multimodal traces without difficulty when a different sensory channel is used. Thus, the patient who fails to recognize his wife visually will easily do so when she speaks and makes herself known acoustically. Such an experiment indicates the intactness of the patient's multimodal memory banks. Resorting to an arbitrary and necessarily crude distinction between perception and recognition, we would say that the defect in most instances of associative agnosias is placed beyond the early stages of perception but before the stage of multimodal memory activation on which recognition depends. In the model of facial recognition we have proposed,[39] we presume that the dysfunction responsible for prosopagnosia may (1) interfere with the arrival of information at the "facial template system," or (2) destroy the template system, or (3) block the anatomic pathways that permit the template system to activate the multimodal memory banks. Depending on the exact placement of the lesion in each case, it is possible that any of the three mechanisms can be responsible for the defect.

The studies of Hecaen and Angelergues[60] and of Meadows,[84] based on clinical data alone, raised the possibility that prosopagnosia was preferentially linked to right-hemisphere lesions. But analysis of the pathologic material published in the literature (Table 1), and seen more recently with CT scans, reveals that all cases of prosopagnosia in which the defect endured beyond the acute period proved to have bilateral lesions (Fig. 1).[6,10,27,30,32,39,53,60,62,80,95,115] The lesions compromise either the inferior and mesial visual association cortices (in the lingual and fusiform gyri) or their subjacent white matter. We have noted that those lesions are approximately symmetric from the functional point of view; that is, they involve equivalent portions of the central visual pathways in the left and right hemispheres (Fig 2). Cases of prosopagnosia with bilateral lesions located exclusively in the superior visual association cortices have not been reported, although the inferiorly located lesions can extend into parts of the superior visual association cortex in some patients. This may

DISORDERS OF COMPLEX VISUAL PROCESSING: AGNOSIAS, ACHROMATOPSIA, BALINT'S SYNDROME, AND RELATED DIFFICULTIES OF ORIENTATION AND CONSTRUCTION

263

TABLE 1. Pathologic Correlates of Prosopagnosia

	Left Hemisphere	Right Hemisphere
Wilbrand (1982)	The white matter of the entire occipital lobe is destroyed. The cortical component is located between 1st and 2nd occipital gyri.	Damage to fusiform gyrus, cuneus (posterior portion) and cortex of calcarine fissure.
Heidenhain (1927)	Damage to striate cortex. Lesion extends anteriorly into mesial and inferior portions of the occipital lobe. Third occipital gyrus is destroyed too. Lingual gyrus as well as inferior lip of calcarine cortex are involved.	Design of lesion is identical to the left side, but damage is more at the occipital pole.
Pevzner et al (1962)	The core of the lesion is in angular gyrus, but damage extends into occipitoparietal sulcus.	Lower lip of calcarine fissure.
Gloning and Gloning (1970)	Damage to lingual and fusiform gyri. There is involvement of optic radiations and extension into occipital pole.	Fusiform and lingual gyri. Lesion extends into supramarginal gyrus and involves optic radiations.
Lhermitte et al (1972)	Cortex of fusiform gyrus. Lesion extends into occipital horn.	Fusiform and lingual gyri, predominantly the latter. Lesion extends into inferior lip of calcarine fissure and into hippocampus.
Benson et al (1974)	Brunt of damage is in parahippocampal cortex and anterior third of fusiform gyrus. Lesion extends to posterior periventricular white matter and to cingulate gyrus (lesion is much larger on the left).	Cyst in fusiform gyrus.
Cohn et al	Case 1: Upper and lower lip of calcarine fissure (spares polar portion of area). Caudal portion of hippocampal gyrus. Fusiform and lingual gyri. Case 2: All of lingual and fusiform gyrus (the lesion is much larger than on the right).	Upper and lower lip of calcarine fissure all the way to occipital pole. Lower portion of cingulate gyrus and precuneus. Caudal portion of hippocampal gyrus. Fusiform gyrus. Lingual and fusiform gyri close to the pole of the occipital lobe.
Hecaen and Angelergues (1962)	Glioblastoma extending to mesial occipital region through the splenium of the corpus callosum.	Invasion in parietotemporo-occipital region.
Arseni and Botez (1965)	Spongioblastoma invading the posterior half of fusiform and lingual gyri, and splenium of the corpus callosum.	The tumor occupied the posterior portion of the centrum ovale, in the temporo-occipital junction, level of the splenium.
Bornstein (1965)	*Note:* Reviews on prosopagnosia often include one postmortem case of neoplasm by Bornstein (Lhermitte, 1972; Benton, 1979). However, the autopsy description of that particular case is not published. To complicate matters Bornstein has commented elsewhere (1965) on an additional tumor case about which no postmortem details are available. It is known, however, that both cases had primary involvement of the left hemisphere.	

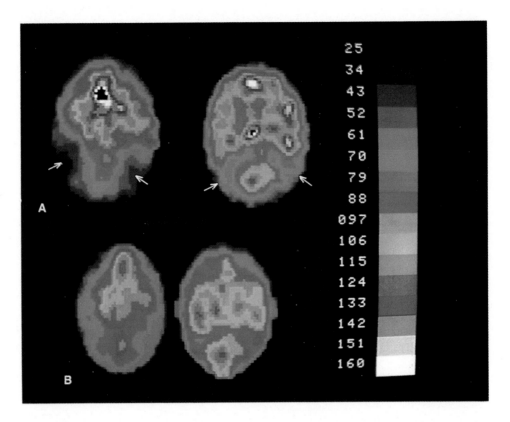

FIGURE 1. *A.*, Xenon[133] SPECT imaging of a prosopagnosic patient, obtained with eyes open. Note the decreased cerebral bloodflow level in both occipitotemporal regions (corresponding to visual association cortices), and contrast it with the intact cerebral bloodflow level in the paramedian occipital regions (primary visual cortices).

    B., Comparable images in a normal control, also with eyes open, showing normal levels of cerebral bloodflow. (From Hanna Damasio and Peter Kirchner, University of Iowa, with permission.)

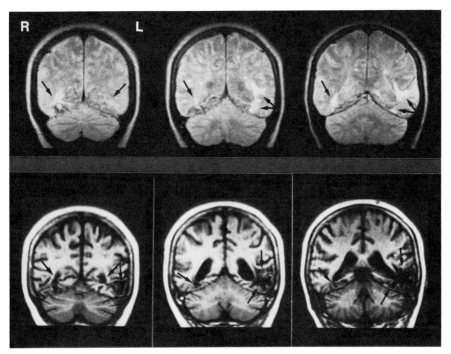

FIGURE 2. Magnetic resonance imaging (MRI) of a patient with prosopagnosia. The coronal slices in the top row were obtained with a spin echo technique. There are two areas of infarction, one in each hemisphere, indicated by the bright white signal (*black arrows*). Both involve inferior visual association cortex, in the occipitotemporal region. The corresponding coronal slices in the lower row were obtained with an inversion recovery technique.

cause visuospatial disturbances, in addition to the agnosia. There is additional evidence showing that right-hemisphere lesions alone are not likely to cause prosopagnosia. For example, patients with right hemispherectomy recognize faces normally[34] and split-brain patients have no difficulty recognizing faces presented tachistoscopically to each of their hemifields.[78] The value of human facial recognition is such that the process appears to be represented in both hemispheres. But we have postulated that the mechanism for facial learning and recognition should be different in each hemisphere,[39] and there is some evidence to support that concept.[50,106]

From the pathologic standpoint, most cases are caused by cerebrovascular accidents, generally the occlusion of branches of the posterior cerebral arteries by emboli. In some cases, the lesions occur at different times and may be separated by months or years. Patients may have a mere upper quadrantic visual field defect, hemiachromatopsia, or alexia, on the basis of a unilateral occipitotemporal lesion, and may suddenly develop the second contralateral lesion that renders them agnosic and severely handicapped. In a small number of cases, prosopagnosia is due to cerebral tumors (especially gliomas, which can originate in one hemisphere and traverse to the opposite side through the splenium of the corpus callosum). Prosopagnosia can occur as an ictal phenomenon related to a bilateral epileptic focus or to a unilateral focus with momentary spreading to the opposite hemisphere.[1]

DISORDERS OF COMPLEX VISUAL PROCESSING: AGNOSIAS, ACHROMATOPSIA, BALINT'S SYNDROME, AND RELATED DIFFICULTIES OF ORIENTATION AND CONSTRUCTION

265

Patients with defective visual perception, such as those described subsequently in the section on Balint's syndrome, often appear not to recognize objects and faces and can thus be misdiagnosed as agnosic. Their disorder is primarily perceptual.

Visual Object Agnosia

Patients with visual object agnosia are behaviorally and anatomically distinct from those with prosopagnosia. In addition to prosopagnosia, these patients have an inability to recognize even the generic class to which an object belongs. Those patients will be unable to know that a face is a face or that a car is a car. They may have also a defect of visual naming, which, in the presence of intact auditory and tactile naming, is traditionally known as "optic aphasia" (Freund's[48] rarely used term). Such patients are almost invariably alexic. Unlike patients with prosopagnosia, some patients with object agnosia may suddenly "unblock" their agnosia and either name or describe the use of an object when it is moved or rotated slowly. Because of the effect of movement upon performance, this defect has been called "static visual agnosia."[28]

Some patients with visual object agnosia complain of "unclear," "blurred" vision, and they probably have a selective defect of low spatial frequency vision that especially impairs appreciation of static, low-contrast stimuli but that may leave intact the vision of moving, high-contrast stimuli. Those patients certainly do not fit the diagnosis of visual agnosia in the associative sense, as we defined it here. In some patients, provided they have stable lesions, the perceptual impairment may diminish and a truly associative type of agnosia may thus appear.

Visual object agnosia is associated with comparable but more extensive damage than prosopagnosia (bilateral lesions in the ventral and mesial parts of occipitotemporal visual areas that often extend dorsally and laterally). It is possible that polar and lateral occipital lesions may actually be sufficient in the absence of mesial occipitotemporal damage, but there is insufficient evidence to be conclusive at this point.

Cortical Blindness and Anton's Syndrome

Severe bilateral damage to visual cortices and to optic radiations may obliterate vision entirely. This dramatic condition is commonly accompanied by Anton's syndrome—the denial of blindness. (The condition known as Anton's syndrome was not described in Anton's 1899 paper that dealt with denial of somatosensory defects,[3] but the eponym is traditionally used in connection with denial of blindness.) In many cases, both the blindness and its denial are transient. The reasons for the denial of blindness are as unclear as those for the denial of left hemiplegia or for anosognosia in general. The right hemisphere component of the dysfunction is crucial perhaps, but there is no evidence to help us decide whether the parietal or the temporal structures are more closely correlated with the denial reaction.

Pure Alexia

Pure alexia, otherwise known as alexia-without-agraphia or "word-blindness," is also a disorder of visual pattern recognition. The reading of most words and of sentences is severely impaired. In many patients even the reading of single letters is defective. Patients with pure alexia are able to *see* the sentences, words, or letters that they are unable to read. This is easily demonstrated by showing that patients can copy what they cannot read. In most instances in spite of quadrantanopic or hemianopic field defects, the patient has normal visual acuity and has normal recognition of nonverbal visual stimuli. Patients with pure alexia cannot recognize words, probably because they are unable to evoke the appropriate associated material when confronted with those words. Thus, a phenomenon of agnosia can probably explain many cases of pure alexia.

Any lesion capable of disconnecting *both* visual association cortices from the dominant, language-related, temporoparietal cortices will cause pure alexia. In most instances a lesion in the corpus callosum or in its outflow, on either hemisphere, achieves an *inter*hemispheric disconnection (of right-to-left visual information), whereas an additional lesion in the left occipital lobe achieves an *intra*hemispheric disconnection (from left visual association cortex to left language cortex). In some cases, the lesion in the left occipital lobe actually prevents the arrival of visual information in the visual cortex. (Such cases have a right homonymous hemianopia, a common but not necessary accompaniment of pure alexia). In many cases, a single lesion strategically located in what we have termed the left paraventricular area (the region behind, beneath, and under the occipital horn) can cause both types of disconnection by damaging pathways en route from the callosum and pathways en route from the left visual association cortex. Dejerine's original case of a patient with pure alexia[47] had the prototype of that lesion (see Damasio and Damasio[40] for additional anatomic details and Chapter 5 for additional discussion on alexia). It should be noted that patients with other forms of alexia (e.g., alexia-with-agraphia) may fail to read owing to different mechanisms that are related to aphasia rather than agnosia. Some patients with alexia can also have visual agnosia when an additional lesion is present in the right hemisphere.[10,40]

Disorders of Topographic (Spatial) Orientation

A variety of disorders fall under this designation. The acquired inability to locate a public building in a city, to find one's room at home, or to describe either verbally or by means of a map how to get to a specific room or place have been classified under this heading. Some patients develop such deficits on the basis of unilateral hemispatial neglect or on the basis of a global amnestic syndrome. In this chapter, we do not deal with these causes, but we do focus on disorders of topographic orientation, which are based on the mechanisms of agnosia.

Such defects are likely to reflect an impairment of visuospatial memory, and the kinship to the visual agnosias is apparent. When con-

DISORDERS OF COMPLEX VISUAL PROCESSING: AGNOSIAS, ACHROMATOPSIA, BALINT'S SYNDROME, AND RELATED DIFFICULTIES OF ORIENTATION AND CONSTRUCTION

267

fronted by a given architectural or topographic detail, these patients can no longer conjure previously stored memories that would help them establish their bearings and plan their route to the desired destination. The complex operations involved in topographic orientation are likely to involve a variety of structures in both inferior and superior visual association cortices, with a special contribution from the right occipitoparietal region. The limited literature on the subject supports this view. Hecaen and Angelergues[60] noted that the defect is strongly associated with either bilateral posterior lesions or with right posterior lesions and also remarked that, in either case, the disturbance is infrequent.

The ability to learn *new* topographic routes after focal brain lesions certainly depends on yet another type of process. It has been our experience that patients with bilateral lesions of the visual system, both superiorly and inferiorly, have difficulty in learning new topographic tasks. (The inability to learn how to get back to the room in the novel hospital environment is often seen.) Patients with prosopagnosia are certainly disturbed in this regard.[99] Studies based on patients with well-localized lesions are not available.

Semmes and associates[105] developed a laboratory technique to study route learning disturbances in which patients were asked to follow a route depicted in a map. The patients could not read the map visually or tactually (with either hand). The impairment was strongly associated with parietal lesions and clearly independent from the mode in which the map was presented. The results highlight the supramodal nature of route learning and its functional association with a brain structure in which visual and somatosensory information can interact. Ratcliff and Newcombe[99] using a similar task noted that only patients with bilateral posterior lesions performed defectively. Bowen, Hoehn, and Yahr[29] found that patients with parkinsonism, who presumably had active bilateral dysfunction of the neostriatum, also were defective in this task. Patients with parkinsonism whose signs predominated in the left side (whose dysfunction was more marked in the right striatum) performed more poorly.

A different but perhaps related impairment, that of geographic orientation, denotes the inability to identify cities in a map or to construct a map of a country or a state. This appears to be a far more frequent defect and is seen in patients with both bilateral or unilateral posterior cerebral lesions.[18,60] In Benton and Van Allen's test, a map of the United States is presented and the subject asked to point to the places where major cities or states are supposed to be. Patients with neglect characteristically fail this test, either by not supplying the location of cities in one of the halves of the map or by displaying them eastward or westward of their true locations. The performance of these patients may reflect the presence of hemispatial neglect (see Chapter 3).

DISORDERS OF COLOR PERCEPTION

In this section, we concentrate on acquired deficits of perception, the central achromatopsias. This group of deficits can be defined as an *acquired disorder of color perception involving all or part of the visual field*, with *preservation of the vision of form,* caused by *focal damage of the visual association cortex* or of its subjacent white matter. The existence of these defects has been known since the last quarter of the 19th century. For historic reasons to be discussed later the concept all but vanished from clinical neurology and was only reinstated in the past decade. The disorder occurs frequently in the setting of cerebrovascular disease and is important both as a clinical sign and as an indicator of disordered physiology.

Patients with achromatopsia experience a loss of color vision in a quadrant, a hemifield, or the whole visual field. The loss of color may be complete to the point that a patient experiences only black and white or it may be partial, generally described as a "washing out" of the colors as if they were "dirty" or "dulled" yet still distinguishable as individual colors. In general, the precipitous loss of color vision is a noticeable phenomenon. One of our patients described it as if the "color of my TV screen had gone out of tune," giving way to a black and white image. Depending on the ambient lighting conditions, patients describe the affected field as "whitish" or "blackish" or merely as a collection of shades of gray. But although color is lost, the perception of forms is maintained, so that patients can still see accurately in the colorless region of their visual field. Furthermore, definition of contour and the sense of depth of the colorless field are generally not compromised. In short, achromatopsia amounts to a loss of color vision with preservation of form vision. In patients in whom only a hemifield or a quadrant is affected, the disorder can be tested by moving stimuli in and out of the achromatopsic area and crossing into the area of intact color perception.[37]

Full-field achromatopsia is usually associated with visual agnosia, especially prosopagnosia. Most of the patients also have some impairment of the visual field; for example, a bilateral or unilateral upper quadrantic defect or a combination of a hemifield defect with a quadrant defect on the opposite side. In other words, patients with full-field achromatopsia come to medical attention with (1) visual agnosia, (2) achromatopsia in the portion of the field with intact form vision, and (3) blindness for both form and color vision in the remainder of the field. Some of those patients may have pure alexia as well, and all have bilateral occipitotemporal lesions, as described for prosopagnosia.

Some patients exhibit hemiachromatopsia only. Patients with left hemiachromatopsia have a unilateral lesion in the right occipitotemporal region and usually have no other neuropsychologic defects. This is the purest form of achromatopsia and can be associated with a single lesion located inferiorly in the occipitotemporal cortex. Patients with right hemiachromatopsia have unilateral lesions of the left occipitotemporal region and may or may not have associated alexia.[40] Some of these patients have both pure alexia and blindness in the right upper quadrant. In these patients, the upper quadrant blindness results from damage to the inferior optic radiations or, on occasion, to the lower calcarine cortex; the lower quadrant achromatopsia results from damage to color processing regions located medially and ventrally in occipitotemporal cortex; and the alexia follows damage to intra- and interhemispheric visual pathways within the left occipital cortex, which connect visual association

DISORDERS OF COMPLEX VISUAL PROCESSING: AGNOSIAS, ACHROMATOPSIA, BALINT'S SYNDROME, AND RELATED DIFFICULTIES OF ORIENTATION AND CONSTRUCTION

269

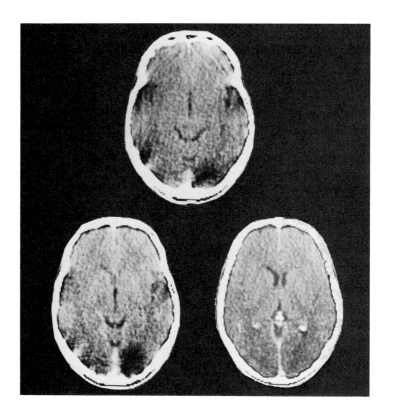

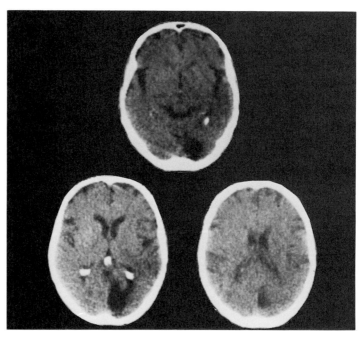

FIGURE 3. *Top,* CT scan of a patient with left hemiachromatopsia. The lesion involves the polar and mesial visual cortex, located below the calcarine fissure in the right hemisphere.

Bottom, This CT scan is that of a patient with left hemianopia, i.e., with an entire blind left field. Note that the lesion encompasses an additional cut which involves the primary visual cortex and, of necessity, the optic radiations traveling toward it.

cortex with language areas. Form vision is preserved in the colorless right lower quadrant because the superior calcarine cortex and the superior contingent of optic radiations *are* intact.

In most cases of pure achromatopsia, the lesion is located in the visual association cortex of the occipitotemporal region. The lesion is usually an infarct in the territory of one of the branches of the posterior cerebral artery, often of embolic origin. The lesion is positioned in such a way that both optic radiations and primary visual cortex are mostly spared (Fig. 3, *top*). Because it is far more common for the primary visual cortex or the attending optic radiations to be damaged than spared in occlusions of the posterior cerebral arteries, the occurrence of hemianopia is far more common than is that of hemiachromatopsia (Fig. 3, *bottom*). In addition, it is possible that the defect may not be brought to the attention of physicians as often or as quickly as with hemianopia.

Our current view regarding the position of the lesion that causes achromatopsia is that, as mentioned earlier, damage has to be located in the ventral part of visual association cortex in an occipitotemporal area that encompasses the fusiform and lingual gyri. This was the location first described by Verrey[111] in a patient who developed a right hemiachromatopsia with no other defect of reading or form vision. A similar location can be surmised in the renowned patient of Dejerine[47] in whom right hemiachromatopsia was associated with alexia but not with blindness. As noted in the detailed neuropathologic study of this case by Dejerine's disciple Vialet, the primary visual cortex in the calcarine region was almost entirely spared, and the lesion was located in the inferior visual association cortex as well as in the subjacent white matter, extending into the paraventricular region.[40,113] In the cases we have studied to date, our impression is that lesions in this vicinity will cause achromatopsia provided they are located fairly posteriorly in the visual association cortex; that is, immediately below the occipital pole, under the caudal half of the calcarine cortex. By the same token, we have seen patients with lesions located more anteriorly in the visual association cortex (i.e., immediately below the rostral half or one third of the calcarine cortex) who did not have achromatopsia. Those patients had form vision defects in the periphery of the opposite upper quadrant, but they did not complain of color loss in the remainder of their intact field nor did they exhibit color loss when tested in the color perimeter. On the other hand, we have studied patients with lesions in the white matter of the occipitotemporal region who also developed achromatopsia and we have seen rare instances of patients with achromatopsia related to superior occipital lesions. It should be clear that the exact location and range of the lesions that can cause achromatopsia is a matter of current research.

Several neuropsychologic studies published during the 1960s had indicated an apparent preponderance of color perception defects in patients with right-hemisphere lesions.[4,44,79,103] All of these studies were based on large series of patients with damage to either the right or the left hemisphere. But exact anatomic definition was lacking, and the groups were heterogenous from the pathologic standpoint. Furthermore, the defect in color perception (color "imperception," according to some of the authors) was defined on a statistical basis, as a performance below

DISORDERS OF COMPLEX VISUAL PROCESSING: AGNOSIAS, ACHROMATOPSIA, BALINT'S SYNDROME, AND RELATED DIFFICULTIES OF ORIENTATION AND CONSTRUCTION

271

that of normal subjects. In fact, few if any of these patients had real achromatopsia as defined here. Evidence from the literature on achromatopsia has not supported the notion that color processing is the specialty of the right hemisphere. Nonetheless, the right hemisphere might process color differently than the left, even though unilateral right hemisphere damage may not lead to clinical complaints characteristic of achromatopsia.

We have studied patients with achromatopsia who recover markedly; but we have also seen patients in whom the disturbance remains unchanged. It is reasonable to suspect that the position and extent of the lesion will play a major role in the patient's recovery. This is also a matter of current study, and it is not yet possible to make reliable predictions of outcome on the basis of available information. We suspect that the most posterior lesions in visual association cortex are the ones that cause the most permanent forms of achromatopsia.

The reasons why the concept of central achromatopsia vanished from clinical neurology and is rarely mentioned in textbooks of neurology or ophthalmology deserve a special mention. The two fundamental descriptions of the phenomenon, complete with appropriate postmortem correlation, were available between 1888 and 1893 (see the articles of the ophthalmologist Verrey;[111,112] the notes by Landolt, which were included in Dejerine's paper of 1892;[47] and Vialet's postmortem study of Dejerine's patient[113]). Other cases of achromatopsia were described, with autopsy.[83] None of these articles gained a foothold in the German and English literatures of the time. Furthermore, after World War I, when Gordon Holmes made his contribution to the understanding of the visual system based on patients with gunshot wounds, he did not cite these articles and unwittingly denied their substance. Most of Holmes' patients had dorsal occipitoparietal lesions, and none exhibited achromatopsia.[63,64] Holmes saw few if any pure occipitotemporal lesions and, not realizing that his sample of patients was highly skewed, concluded that disorders of color perception could not be caused by focal damage of the visual cortices. His authority and the importance of his other superb observations cast a shadow on the descriptions of acquired achromatopsia. The existence of acquired achromatopsia secondary to lesions of the central nervous system was only acknowledged by Critchley, in 1965,[33] even though no pathologic correlate was proposed then. Another decade would pass before a handful of papers would revive and expand the observations of Verrey and Landolt.[2,37,56,83,94]

The finding of patients with color loss due to focal lesions has been clarified by developments in animal physiology. Zeki[117,118] showed that the distribution of color-coded cells appears to be highly selective. In a prestriate region of visual association cortex, which he designates V_4, Zeki found cell columns that consistently responded to certain bands of the color spectrum. Other investigators have reached similar results,[7,109] while others have questioned the idea that the color-coded columns have a restricted distribution. The most recent development in the field is the discovery, by Hubel and Livingstone, of sets of cells in areas 17 and 18 of the monkey, which are unequivocally associated with color processing.[70,82] These cells are distinguished by their staining with cytochrome oxidase, in itself an indicator of high metabolic rates. In tangential cuts of area 17, they appear to form an intricate polka-dot pattern, which Hubel and Livingstone call "the blobs." The blobs alternate regularly with groups of cells that do not stain with cytochrome oxidase and that seem to encode orientation but not color. Unlike the cells around them, the cells in the blobs are located in layer II of the cortex. But just as in the case of cells interested in orientation, these cells receive direct projections from the lateral geniculate body and in turn project to groups of cells in area 18, which also stain with cytochrome oxidase and appear as thin stripes in tangential cuts. Between these thin stripes lie thick stripes, which receive connections from the nonblob regions of area 17. In short, there is an intricate anatomic arrangement at the cellular level that may lead to an anatomic separation of color from form processing, at least in part. It is possible that the visual association areas linked to achromatopsia in humans are, at least in part, homologous to Zeki's V_4, and that the histologic and functional organization that Hubel and Livingstone have begun to uncover will reveal a highly specific region dedicated to color processing.

OTHER DISORDERS RELATED TO COLOR PROCESSING

Disorders of Color Naming

A situation entirely different from achromatopsia occurs when otherwise nonaphasic patients can experience the sensation of color and can match colors according to hue but are unable to name those colors, which apparently they perceive without difficulty. An operational definition should read as follows: *failure to name color* or to *point to colors* given their names, in the *absence of demonstrable defect of color perception or of aphasia.*

In our experience all patients with a color-naming defect have a right-homonymous hemianopia and intact color perception in the left field. The other common correlate of color-naming defect is pure alexia.[40] Only in one case to our knowledge has a color-naming defect been described without alexia, although right-homonymous hemianopia was present.[87]

Analysis of an extensive series of our own patients as well as review of the literature has led us to the conclusion that the lesion necessary for color-naming defects is located in the left hemisphere, mesially, in the transition between the occipital and temporal lobes, in a subsplenial position. The right-homonymous hemianopia, which is also present, is caused by an additional lesion either in the geniculate body, visual cortex, or optic radiations.

Regardless of the lesion that causes it, the net effect of the right field cut is to circumscribe visual information to the right visual cortex. The effect of the occipitotemporal lesion is to interfere with the ability of language areas in the left hemisphere to receive visual information related to color. It is not known whether this lesion simply disconnects color information conveyed by the corpus callosum or whether some crucial

DISORDERS OF COMPLEX VISUAL PROCESSING: AGNOSIAS, ACHROMATOPSIA, BALINT'S SYNDROME, AND RELATED DIFFICULTIES OF ORIENTATION AND CONSTRUCTION

273

step in the processing of that information takes place in the occipitotemporal transition cortex. It should be clear, however, that other types of visual information still reach the remainder of the left visual association cortices, which explains the preservation of other aspects of visual naming.

Some authors prefer to call this disorder "color agnosia." They argue that the terms "color-naming defect" and "color anomia" fail to suggest the two-way impairment that renders patients unable to name a color and also unable to point to a color when given its name. Furthermore, they indicate that the perception of color devoid of its verbal tag corresponds to a percept without meaning—that is, an agnosia. We have serious reservations about the designation "color agnosia" for this difficulty. At any rate, the reader should know that a color-naming impairment, as described here, generally implies a two-way defect, and that the method to distinguish a color-naming defect from achromatopsia includes: (1) verbal inquiry concerning the patient's experience of color, (2) testing of the ability to perform color matching, and (3) the Ishihara and Farnsworth-Munsell tests. We should note that in some patients with color anomia, especially as they recover, the magnitude of the impairment of naming colors on confrontation is greater than that of pointing to colors when given the names. This has been termed "one-way" defect. The explanation for this dissociation is not evident, but the two tasks are cognitively different. In the condition of pointing to colors when given their names, it is possible that presentation of the color name activates bilateral hemispheric structures (directly or through auditory callosal transfer), and thus permits a more prompt auditory-visual matching in the right hemisphere. In the opposite condition, given the patient's lesions, the activation caused by color stimuli is confined to the right hemisphere alone and may be less likely to arouse pertinent verbal information.

Disorders of Color Association

An entirely different process is the ability to indicate which color is associated with a specific object, using an exclusively verbal question and a verbal mode of response (e.g., requesting the patient to complete sentences such as "the color of blood is _____," or "the color of a banana is _____"). Performance of this task does not require the processing of visual information. It is reasonable to expect that patients with disturbances of the visual system will not show an impairment in such a task, whereas patients with an auditory or auditory-verbal defect may perform defectively. That is indeed the case. Although patients with aphasia do poorly in this task, patients with isolated color-naming defects perform such tasks flawlessly.

Another type of color association task consists of asking the patient to color black-and-white line drawings of common objects. The patient must select the correct pencil out of an array of colored pencils and proceed with coloring. Patients with a color-naming defect may or may not fail such a task. A study we performed some years ago persuaded us that when they do fail, the failure often results from an intriguing mechanism that consists of: (1) making the correct *verbal* choice of which color pencil to use (i.e., "this is a banana; a banana is yellow; I'll color it yellow"), but then (2) choosing the wrong pencil because the correct choice depends on a verbal-visual match (yellow name to yellow pencil). After applying the wrong color, these patients are often puzzled by the result and may be uncertain that there is indeed an error. Analysis of this kind of performance has led us to believe that the patient switches strategies and solves part of the problem at a verbal level and another at a visual level but has no means to combine the results of both approaches.[35] Such a failure does raise the question of whether the patients still have a working concept of color and hence whether they should be considered agnosic. It has been our impression that the cognitive basis of the concept of color is fragile, being principally dependent on (1) the association of a verbal tag with a given color, and (2) the association of a given color with the objects that commonly carry it (an association that must be, at least in part, verbally mediated). Thus, a functional dissociation between visual and verbal processes is likely to prevent the concomitant arousal of verbal and visual memory traces on which the attribution of meaning to a normal color perception must depend. Some support for this hypothesis comes from studies that indicate that patients with aphasia not only fail tasks involving the linguistic aspects of color but also frequently fail color-matching tasks of a nonverbal variety.[44]

DISORDERS OF SPATIAL ANALYSIS

Balint's Syndrome and Its Components

Balint's syndrome (named after the Hungarian neurologist who called attention to this symptom complex in 1909)[8] is one of the most striking disorders of spatial analysis. The eponym should be reserved for those cases that present all three major components: *visual disorientation* (also known as simultanagnosia*), *optic ataxia* (deficit of visual reaching), and *ocular apraxia* (deficit of visual scanning). Both visual disorientation and optic ataxia can occur by themselves. To the best of our knowledge ocular apraxia is always accompanied by either optic ataxia or visual disorientation. The patients may have visual field defects (when they do, they consist of inferior quadrant cuts), but these are not necessary for the appearance of Balint's syndrome. We have seen several cases of patients in whom the visual fields were intact. An operational definition of Balint's syndrome would be as follows: an *acquired* disturbance of the *ability to perceive the visual field as a whole*, resulting in the *unpredictable perception and recognition of only parts of it (simultanagnosia)*; which is accompanied by an *impairment of target pointing under visual guidance (optic ataxia)* and an *inability to shift gaze at will toward new visual stimuli (ocular apraxia)*.

The essence of the syndrome, visual disorientation (simultanagnosia), is subjective. The patient becomes unable to grasp the whole field

---

*Neither term is ideal; *simultanagnosia* is briefer but has the disadvantage of connoting a disorder of recognition.

of vision in its entirety and appears to see clearly in only a small fraction of the panorama, outside of which vision is described as hazy. But to complicate matters this fragment of useful field (usually a part of the macular representation) is not stable and moves erratically from quadrant to quadrant. This often happens in the absence of a true visual field cut, although some patients may also have inferior visual field defects. As a result of this disturbance, patients commonly fail to detect and orient to new stimuli that may appear in the periphery of the visual field, except when, by chance, the constantly changing center of their vision actually falls into the area of the field where a new stimulus appeared. Also as a result of this defect, an object that is clearly seen at a given moment may suddenly disappear from view, as the center of vision shifts. Patients often complain about objects vanishing from the scene they are inspecting. This is also why the patients are not able to report more than one or two components of the visual field at any one time. The term "simultanagosia" refers to this feature. The defect can be easily tested by asking for the description of a large array of objects or people in front of the patient, or by requesting the description of a complex scene. The Cookie Theft picture from the Boston Diagnostic Aphasia Examination[55] is an excellent means of deciding whether or not the patient can cope with the rapid analysis of a visual scene.

Some patients with visual disorientation have complained that moving objects are especially difficult to perceive, and we have noted that these patients fail to acknowledge the presence of an object moving about in the visual field. Such a defect, the converse of "visual static agnosia" mentioned in the section on pattern recognition, indicates the likelihood that the perception of movement is handled separately from that of static stimuli. But we have never seen a patient with an impairment of movement perception in isolation, nor do we know of any published instance in which the defect was not part of visual disorientation (the patient described by Zihl and associates[119] as having a "selective" disturbance of movement vision also had visual disorientation, optic ataxia, and astereopsis).

The second component of the syndrome is optic ataxia. The defect consists of an inability to point accurately to a target, under visual guidance. Patients with optic ataxia have no difficulty pointing with precision to targets in their own body or garments, using somatosensory information. Nor do they have difficulty pointing to the source of sounds. But under visual guidance there is marked difficulty, especially in reaching toward small targets and in finger pointing. The error may be as small as half an inch, and the mispointing is generally not accompanied by tremor. Optic ataxia, also known as visuomotor ataxia, can be found in isolation, in both upper limbs, and in patients who are recovering from Balint's syndrome and no longer have visual disorientation or significant ocular apraxia.[24,36] Optic ataxia can also be seen in the absence of visual disorientation or ocular apraxia, unilaterally, in the field opposite the affected hand.[100]

Ocular apraxia consists of the inability to direct gaze voluntarily toward a new stimulus that has appeared in the periphery of the visual field. In other words, whereas a normal individual will easily produce a quick saccade toward a new interesting stimulus that has appeared in the periphery of the visual field, a patient with ocular apraxia will not be able to produce this saccade even when alerted verbally that a stimulus has indeed entered his visual field. Or, if a saccade does take place, it may be directed inaccurately and not bring the new stimulus into the fovea. Patients probably fail to see anything that is not brought or maintained in optimal central vision. In his original description, Balint referred to ocular apraxia as "psychic gaze paralysis." Gordon Holmes called it "spasm of fixation."

The appearance of the full Balint's syndrome is, in our experience, strongly related to bilateral damage of the occipitoparietal region. More often than not, the lesions are the result of infarctions in the border-zone (watershed) between the anterior and posterior cerebral artery territories. Sudden and severe hypotension is perhaps the main cause of this type of cerebrovascular accident. Most of the patients who come to medical attention with Balint's syndrome are elderly individuals who are especially vulnerable to systemic blood pressure drop. Many cases appear after open heart surgery. In the vascular cases, the lesions are often quite symmetric, but they may be of different sizes. Thus, the presence of Balint's syndrome in a patient with cerebrovascular disease usually indicates a type of stroke different from the one that causes the agnosias or pure alexia. Whereas the lesions in the agnosias are ventral (occipitotemporal), those in Balint's syndrome tend to be more dorsal (occipitoparietal). Balint's syndrome can also be caused by bilateral metastases in the occipitoparietal region, even when asymmetrically placed (Fig. 4).

When the entire syndrome is present some involvement (either medially or dorsally) of area 7 is quite common, although more often than not the adjoining area 19 and sometimes part of area 39 is also involved.[36] In some patients, the lesions are far more extensive and may even encompass white matter or cortex in the lower occipital or occipitotemporal region. In such patients severe visual field defects or an additional agnosia may complicate the presentation. In general, the lesions

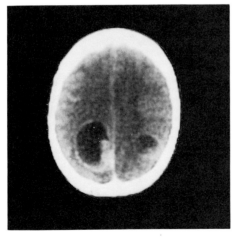

FIGURE 4. CT scan of a patient with Balint's syndrome caused by bilateral metastases from carcinoma. Note that the metastases involve functionally comparable areas in the occipitoparietal region.

DISORDERS OF COMPLEX VISUAL PROCESSING: AGNOSIAS, ACHROMATOPSIA, BALINT'S SYNDROME, AND RELATED DIFFICULTIES OF ORIENTATION AND CONSTRUCTION

responsible for Balint's syndrome are located in the occipitoparietal sector of the visual cortex. Ratcliff and co-workers[98,99] obtained results compatible with these notions in an experimental task requiring patients to touch stimuli presented in a perimeter while maintaining fixation of gaze. The authors found that patients with posterior lesions of either hemisphere performed defectively in this task and that a visual field defect was not a significant correlate.

Cursory observation of patients with Balint's syndrome may often give the examiner the false impression of dealing with a case of visual agnosia (some patients with the so-called apperceptive agnosia may have had Balint's syndrome). This may occur when the patients report that they *see* but do not *know* what the stimulus is. This may happen with any stimulus, including human faces, and often occurs with reading material. The consequence is that these patients are often mislabeled as agnosic and alexic. It is necessary to take pains to introduce the stimuli in the effective portion of the visual field and to show that the patient's real recognition performance is, in fact, preserved under certain conditions of stimulus presentation.

We have now seen instances of patients with visual disorientation (simultanagnosia) in isolation—that is, without optic ataxia. In these intriguing cases of defective spatial analysis, the patients claim not to see objects that they can point to accurately. It is possible, in those instances, that there is enough visual information to permit appropriate guidance of the hand and fingers but that the patient is unable to synthesize a visual field into a cogent whole. Patients with visual disorientation alone

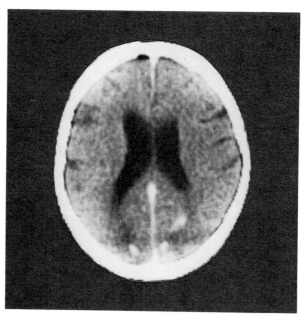

FIGURE 5. CT scan of a patient with visual disorientation but *without* optic ataxia. The high-density images are located in the vascular border zone above the calcarine fissure and correspond to gray matter enhancement (scan obtained 1 week post-onset of stroke). Autopsy confirmed the position of the lesions and the intactness of all other central visual structures (occipital, parietal, and temporal cortices, geniculate bodies, and superior colliculi).

have bilateral lesions confined to the supracalcarine cortex, thus sparing the adjoining parietal cortex (Fig. 5).[41]

The contrast between the lesions that cause the full Balint's syndrome (occipitoparietal) and those that cause visual disorientation alone (superior-occipital) suggests that these two regions may have different roles in spatial analysis and visuomotor performance. Both human data[36] and animal experiments[54,57,88−90,116] suggest that the parietal lobe, especially area 7, contains a visuomotor center capable of directing gaze toward interesting new stimuli that appear in panoramic vision and of guiding hand movements toward visual targets. On the other hand, the superior-occipital cortex (striate and peristriate) appears to contain structures and contribute to the structuring of the visual field as a whole.

Disturbance of Stereopsis

Stereopsis is the *ability to discriminate depth on the basis of binocular visual information.* Because of the physical separation of the eyes, the two-dimensional projections of three-dimensional objects occupy slightly different positions on the right and left retinas. This horizontal disparity is the source of stereopsis. Stereopsis is different from depth perception (depth perception depends on *monocular* information derived from image contours, texture, apparent size of familiar objects, interposition of near and far, linear perspectives, and monocular parallax).

The first step in the testing of stereopsis consists of ensuring that the patient has adequate visual acuity and that the ocular alignment is correct (the latter may be assessed with orthoptic techniques such as the four-diopter baseout prism and the Wirth four-dot tests). Full-field stereopsis can be tested with standard handheld tests of stereovision, available commercially. Such tests include the TNO anaglyphs (Lameris Instrumenten), which require red-green glasses, and the Titmus (Titmus Optical Company) and Randot (Stereo Optical Company) stereotests, which use polarized glasses and printed materials. Testing of stereopsis within quadrants is a difficult task that should only be undertaken in a laboratory equipped with a tachistoscope.

Carmon and Bechtoldt[31] and Benton and Hecaen[17] postulated that the right hemisphere was dominant for stereopsis. Brain-damaged patients were asked to localize random-letter stereograms to one quadrant of a background field. Performance appeared especially impaired and response time prolonged in patients with damage to the right hemisphere. But the exact cerebral localization of the lesions was not known and the age, type, and size of lesions were not considered. Several patients had malignant tumors and intracerebral hematomas, which could have increased intracranial pressure and affected brain function in a nonfocal manner. Support for the hypothesis that stereopsis could be preferentially related to right-hemisphere processing also came from a study in normal individuals, in which random-dot stereograms were presented tachistoscopically, and performance in the left hemifield proved superior. On the other hand, Julesz and associates found markedly different results in normal subjects, in whom the left and right visual fields showed equivalent stereoptic ability.[74] These results indicate that both hemispheres are capable of processing stereopsis. However, Julesz did find a significant difference between upper and lower visual fields.[74] Interestingly, Gazzaniga, Bogen, and Sperry[49] found that stereopsis was preserved after complete callosal lesions, except when the chiasm was also split. This finding indicates the existence of stereoptic ability in each hemisphere.

The visual cortex has both monocular and binocular cells.[21,22] Direct electrophysiologic evidence in the cat and monkey shows that many cells in striate and prestriate cortex respond to binocular stimuli. Animals deprived of binocular stimuli at a critical age develop predominantly monocular cortex.[65,67,71] Neuropsychologic and electrophysiologic studies show that humans also have binocular and monocular visual cortices.[77] The use of random-dot stereograms has enabled researchers to study stereopsis and confirmed the need to postulate a central processing mechanism for this function.[71] Random-dot stereograms consist of two nearly identical computer-generated stereopairs composed of black dots. They differ in that a group of these random elements, in the shape of a geometric form, have been shifted in precise fashion with respect to the identical random sequence in the other member of the stereopair. When viewed stereoscopically (one array presented to each eye), a form such as a square is seen above or below the random background. Levels of brightness, spatial and temporal disparity, order of presentation, object size, stimulus familiarity, and area of retinal stimulation influence performance.[73,101,114]

Information from animal studies is limited but does confirm that stereopsis depends on central processing and that neurons related to stereopsis are localized at a high level in the hierarchy of the visual system. Along central visual pathways, there is no known binocular interaction prior to area 17.[96] Hubel and Wiesel[66] found that binocular disparity units are found in area 18 and beyond in the macaque, but not in area 17. Four kinds of neurons involved in stereopsis were isolated in the macaque.[97]

Current research in this area is aimed at discovering the neuroanatomic sites of damage associated with astereopsis and deciding whether or not unilateral lesions can impair stereopsis. The elucidation of the phenomenon could lead to a better understanding of amblyopia in children with strabismus, who are deprived of stereoscopic stimuli during critical phases of brain development. Syndromes of monofixation and stereoblindness in individuals with normal peripheral visual apparatus could also be better understood. Testing stereopsis is crucial in individuals with occupations dependent on binocular depth discrimination, such as airline piloting, and can be used with advantage in the evaluation of patients with visual disorders caused by focal lesions of the central visual system.

Other Disorders of Spatial Analysis

The ability to judge the directional orientation of lines has been the object of several experimental studies and is a useful test of visual abilities in neurologic patients. In the test of judgment of Line Orientation,[20] the patient is presented with pairs of lines placed in a given angle and requested to point to a similarly oriented line in a different array. Defective performance in this test is strongly correlated with right posterior lesions.[16] We believe that right occipitoparietal lesions are probably the best correlate of such defects.

The ability to discriminate unfamiliar faces is another process that is dependent on intact visual perception. The widespread use of Benton and Van Allen's Facial Recognition test has shown that impaired performance is strongly associated with right hemisphere damage, especially postrolandic, but also with left-hemisphere damage in patients with fluent, postrolandic aphasias.[59] Incidentally, the well-entrenched title of the test may be misleading because the task calls for a matching of unfamiliar faces on the basis of a visual discrimination performance, and not on the basis of recognition. As noted before, prosopagnosic patients can pass this test.

DISTURBANCES OF CONSTRUCTIONAL ABILITY

Under this heading we will discuss impairments in the ability to construct the copy of a visually presented model by means of assembling blocks or by drawing. This disorder is generally known as "constructional apraxia," a term introduced by Kleist.[75] The term is commonly used but the fact that the disorder bears no relation to Liepmann's definition of apraxias has given rise to some confusion. We prefer to define these impairments of disturbances of "visuoconstructive ability" or "constructional ability" rather than as apraxias.

The performance of construction tasks presupposes normal visual acuity, the ability to perceive the several elements of the model as well as their spatial relationship, and finally adequate motor ability. Failure in any of these elementary prerequisites can lead to impairment in these tasks, so that the constructional abilities of such patients could not be tested adequately.

There are numerous ways of testing for visuoconstructive ability. Using paper and pencil, the examiner can test two-dimensional constructions by requiring the copy of line drawings. In a variation on these tasks, the examiner may request the drawing of a simple object from memory (e.g., a house, a clock, or a face) or may ask for the copy of a line model using small sticks or matches. The testing of three-dimensional construction requires an appropriate set of models and of separate building components. Benton's Three Dimensional Praxis Test is a well-standardized example.[16] The block designs subtest of the Wechsler Adult Intelligence Test is commonly used to assess constructional abilities.

Although Kleist was convinced that a disturbance of constructional ability was caused by left parietal damage, numerous studies have now confirmed that both left and right parietal lesions can cause a defect. In our experience, lesions of the left parietal region are less likely to lead to constructional defects than are lesions of the right. Furthermore, defects caused by lesions of the right parietal region tend to be different in degree and in nature. The disturbance is more profound and quite often there is a characteristic neglect of the left side of the elements being copied, assuming that the patient can copy anything at all.[11,15,43]

Most patients with constructional defect caused by a left-hemisphere lesion tend to be aphasic as well. The aphasia is of the fluent type,

DISORDERS OF COMPLEX VISUAL PROCESSING: AGNOSIAS, ACHROMATOPSIA, BALINT'S SYNDROME, AND RELATED DIFFICULTIES OF ORIENTATION AND CONSTRUCTION

281

generally associated with postrolandic lesions.[12] It is possible that the presence of a linguistic defect can contribute to the disturbance, although it is equally probable that the lesions more frequently associated with constructional praxis are located closer to the language cortices of the left hemisphere and thus determine an association of symptoms on the basis of anatomic contiguity.

The types of tasks and the degree of difficulty of the tasks that have been used in studies of visuoconstructive ability are extremely varied. Obviously, use of two-dimensional and of three-dimensional models pose different degrees of difficulty, and as do copying from actual models and copying from graphic representations of those models. These differences underscore the difficulty in comparing the results of various studies and the potential for discrepancy and controversy.

From the practical perspective of diagnosis, a deficit of constructional ability is a useful indicator of parietal dysfunction.

DISTURBANCE OF THE ABILITY TO DRESS

Numerous neurologic patients are unable to don a shirt or a coat properly, even when the specific garment is handed to them. The symptom, generally known as dressing apraxia,* is perhaps most frequent in patients with dementia or confusion, although patients with clear sensorium and preserved intellect can show the symptom as well. In those with focal lesions, the impairment is generally seen in the setting of left-sided neglect or Balint's syndrome; that is, in patients with right occipitoparietal or bilateral occipitoparietal lesions. But we must emphasize that we have never seen dressing apraxia in isolation.

A variety of mechanisms account for the impairment. For instance, patients with left-sided motor neglect will fail to swing a jacket around their backs, in a leftward direction, and will not orient the left arm toward the opening of the sleeve. Sensory neglect of the left hemispace, even in the absence of motor neglect or hemiparesis, may lead to a similar result. On the other hand, patients with simultanagnosia and optic ataxia, maintain all the automatic motions related to dressing and yet are unable to coordinate the orientation of the garment in relation to the body. When patients attempt a corrective action under visual guidance, their visuomotor impairment often does not permit a successful movement.

In theory, it seems possible that a somatosensory feedback defect, in the absence of paralysis, could lead to an impairment in dressing. But we have never seen a patient who fit that specification.

CONCLUDING REMARKS

In the past two decades there has been remarkable progress in the understanding of visual function. The neurophysiology of the central visual system, especially of the primary visual cortex, has been explored by

---

*Much the same reservations that apply to the term "constructional apraxia" apply to "dressing apraxia."

Hubel and Wiesel in a series of pathbreaking studies.[65-70] The functional exploration of the visual related parietal lobe by Mountcastle and co-workers[90] and the physiologic unraveling of the visual association cortices[117,118] constitute other benchmarks.

The split-brain experiments in animals, and the study of humans with split-brain operations[49] added to our understanding of the function and relations of the visual cortices, which had been initiated by Dejerine[46,47] and Verrey.[111] In a parallel development, the work of those and other turn-of-the-century neurologists concerned with the visual disturbances was interpreted in modern neurologic terms by Geschwind.[51,52] The revival and expansion of the work of those pioneers of the study in visual function has continued in the past decade.[2,37,39,40,84,85]

Current experience in the areas of behavioral neurology and neuropsychology suggests some principles of organization that deserve further investigation. It appears that the superior and inferior visual association cortices are associated with different functions. Thus, damage to the inferior visual cortices and to the adjoining temporal cortices has been consistently associated with disturbances of pattern recognition and of color perception, whereas damage to the superior visual cortices and the adjoining parietal cortices has been consistently related to disorders of visuospatial analysis. This implies that, at least in part, the separation of pattern recognition from spatial analytic processes occurs at a cortical level, within the central visual system (Fig. 6).

The differences in the specialization of the inferior and superior visual *association* cortices does not have any necessary relation to differences in the type of information processing in the upper and lower visual fields. In all likelihood, although information from both upper and lower fields arrives separately in the inferior and superior primary visual cortices, it then becomes available, in the form of a single map, to each of the specialized visual association areas. This means that the color information obtained from the upper *and* lower visual fields probably mixes in a ventrally located area of the inferior occipital and occipitotemporal cortex (the equivalent of Zeki's area V_4). Thus, a single, inferior occipitotemporal lesion, can cause achromatopsia in *all* of the contralateral hemifield. Ultimately, it is possible that the specialized areas are placed superiorly or inferiorly, according to the association cortex with which they interact more crucially, parietal in the instance of visuospatial processing, and temporal, in the instance of object and color vision.

Some of these formulations had been hinted at in neuropsychologic investigations[76,91] and are supported by current experimental work in animals. For example, Ungerleider and Mishkin have proposed that the mechanisms of object identification depend on a ventral path that includes visual association cortex in the temporal lobe, whereas the spatial aspects of vision may depend on a dorsal path that proceeds from dorsal visual association regions to the adjacent parietal cortex.[108] Ongoing refinements of neuroimaging technique—CT, positron emission tomography (PET), single photon emission tomography (SPET), nuclear magnetic resonance imaging (MRI)—will undoubtedly permit more elaborate clinical and experimental work related to the processing of complex visual information in the human brain (see also Chapter 1).

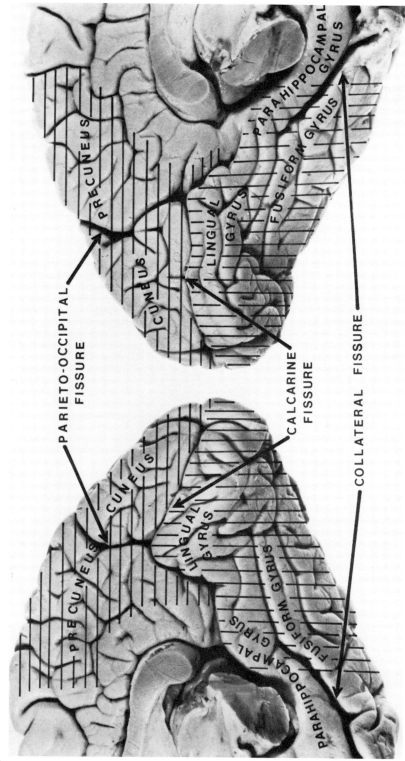

FIGURE 6. Mesial view of a human brain, showing major sulci and gyri. Patients with visual agnosia (including prosopagnosia) have lesions in bilateral structures of the occipitotemporal region (vertical hatching) comprising the lingual, fusiform, and parahippocampal gyri. Patients with Balint's syndrome have bilateral lesions in the occipitoparietal region (horizontal hatching), which is composed of the cuneus and precuneus. Unilateral lesions of either occipitotemporal region can produce visual field defects (contralaterally) and hemiachromatopsia (also contralaterally). Unilateral lesions of the left occipitotemporal region often produce the syndrome of pure alexia (in patients with left cerebral dominance for language, provided that connections of the corpus callosum are also damaged).

REFERENCES

1. AGNETTI, V, CARRERAS, M, PINNA, L, AND ROSATI, G: *Ictal prosopagnosia and epileptogenic damage of the dominant hemisphere: A case history.* Cortex 14:50–57, 1978.

2. ALBERT, ML, RECHES, A, AND SILVERBERG, R: *Hemianopic colour blindness.* J Neurol Neurosurg Psychiatry 38:546–549, 1975.

3. ANTON, F: Ueber die Selbstwahrnehmungen der Herderkrankungen des Gehirns durch den Kranken bei Rindenblindheit und Rindentaubheit. Arch Psychiatry 32:86–127, 1889.

4. ASSAL, G, EISERT, HG, AND HECAEN, H: *Analyse des resultats du Farnsworth D15 chez 155 malades atteints de lesions hemispheriques droites ou gauches.* Acta Neurologica et Psychiatrica Belgica 69:705–717, 1969.

5. ARRIGONI, G, AND DeRENZI, E: *Constructional apraxia and hemispheric locus of lesion.* Cortex 1:180–197, 1964.

6. ARSENI, C, AND BOTEZ, M: *Consideraciones sobre un caso de agnosia de las fisonomias.* Rev Neuropsiquiatr 3:157–160, 1965.

7. BAIZER, JS, ROBINSON, DL, AND DOW, BM: *Visual responses of area 18 neurons in awake, behaving monkey.* J Neurophysiol 40:1024–1037, 1977.

8. BALINT, R: *Seelenlahmung des "Schauens", optische Ataxie, raumliche Storung der Aufmerksamkeit.* Monatsschr Psychiat Neurol 25:51–81, 1909.

9. BAY, E: *Disturbances of visual perception and their examination.* Brain 76:515–550, 1953.

10. BENSON, DF, SEGARRA, J, AND ALBERT, ML: *Visual agnosia-prosopagnosia.* Arch Neurol 30:307–310, 1974.

11. BENTON, AL: *Constructional apraxia and the minor hemisphere.* Confin Neurol 29:1–16, 1967.

12. BENTON, AL: *Visuoconstructive disability in patients with cerebral disease: Its relationship to side of lesion and aphasic disorder.* Doc Ophthalmol 34:67–76, 1973.

13. BENTON, AL AND HECAEN, H: *Stereoscopic vision in patients with unilateral cerebral disease.* Neurology 20:1084–1088, 1970.

14. BENTON, AL, HAMSHER, K, AND STONE, FB: *Visual Retention Test: Multiple Choice Form.* Iowa City, Department of Neurology, University of Iowa Hospitals, 1977.

15. BENTON, AL, AND FOGEL, ML: *Three-dimensional constructional praxis.* Arch Neurol 7:347–354, 1962.

16. BENTON, AL, HAMSHER, K, VARNEY, NR, AND SPREEN, O: *Contributions to Neuropsychological Assessment.* Oxford University Press, New York, 1983.

17. BENTON, AL, AND HECAEN, H: *Stereoscopic vision in patients with unilateral cerebral disease.* Neurology 20:1084–1088, 1970.

18. BENTON, AL, LEVIN, HS, AND VAN ALLEN, MW: *Geographic orientation in patients with unilateral cerebral disease.* Neuropsychologia 12:183–191, 1974.

19. BENTON, AL, AND VAN ALLEN, MW: *Impairment in facial recognition in patients with cerebral disease.* Cortex 4:344–358, 1968.

20. BENTON, AL, VARNEY, NR, AND HAMSHER, K: *Visuospatial judgment: A clinical test.* Arch Neurol 35:364–367, 1978.

21. BLAKE, R, AND LEVINSON, E: *Spatial properties of binocular neurons in the human visual system.* Exp Brain Res 27:221–232, 1977.

22. BLAKE, R, AND CORMACK, R: *Psychophysical evidence for a monocular visual cortex in stereoblind humans.* Science 203:274–275, 1979.

23. BODAMER, J: *Die Prosop-Agnosie.* Arch Psychiatr Nervenkr 179:6–54, 1947.

24. BOLLER, F, COLE, M, KIM, Y, ET AL: *Optic ataxia: Clinical radiological correlations with the EMI scan.* J Neurol Neurosurg Psychiatry 38:954–958, 1975.

25. BORNSTEIN, B: *Prosopagnosia.* In HALPERN L (ED): *Problems of Dynamic Neurology.* Hadassah Medical Organization, Jerusalem, 1963, pp 283–318.

26. BORNSTEIN, B, STROKA, H, AND MUNITZ, H: *Prosopagnosia with animal face agnosia.* Cortex 5:164–169, 1969.

27. BORNSTEIN, B: *Prosopagnosia.* 8th International Congress of Neurology Proceedings 3:157–160, 1965.

28. BOTEZ, MI: *Two visual systems in clinical neurology: Readaptive role of the primitive system in visual agnosic patients.* Eur Neurol 13:101–122, 1975.

29. BOWEN, FP, HOEHN, MM, AND YAHR, MD: *Parkinsonism: Alterations in spatial orientation as determined by a route-walking test.* Neuropsychologia 10:355, 1972.

DISORDERS OF COMPLEX VISUAL PROCESSING: AGNOSIAS, ACHROMATOPSIA, BALINT'S SYNDROME, AND RELATED DIFFICULTIES OF ORIENTATION AND CONSTRUCTION

285

30. BRAZIS, PW, BILLER, J, AND FINE, M: *Central achromatopsia.* Neurology 31:920, 1981.

31. CARMON A, AND BECHTOLDT, HP: *Dominance of the right cerebral hemisphere for stereopsis.* Neuropsychologia 7:29–39, 1969.

32. COHN, R, NEUMANN, MS, AND WOOD, DH: *Prosopagnosia: A clinicopathological study.* Ann Neurol 1:177–182, 1977.

33. CRITCHLEY, M: *Acquired anomalies of colour perception of central origin.* Brain 88:711–724, 1965.

34. DAMASIO, AR, LIMA, PA, AND DAMASIO, H: *Nervous function after right hemispherectomy.* Neurology (Minneap) 25:89–93, 1975.

35. DAMASIO, AR, MCKEE, J, AND DAMASIO, H: *Determinants of performance in color anomia.* Brain Lang 7:74–85, 1979.

36. DAMASIO, AR, AND BENTON, AL: *Impairment of hand movements under visual guidance.* Neurology 29:170–178, 1979.

37. DAMASIO, AR, YAMADA, T, DAMASIO, H, CORBET, J, AND MCKEE, J: *Central achromatopsia: Behavioral, anatomic and physiologic aspects.* Neurology 30:1064–1071, 1980.

38. DAMASIO, AR: *Central achromatopsia.* Neurology 31:920–921, 1981.

39. DAMASIO, AR, DAMASIO, H, AND VAN HOESEN, GW: *Prosopagnosia: Anatomic basis and behavioral mechanisms.* Neurology 32:331–341, 1982.

40. DAMASIO, AR, AND DAMASIO H: *Anatomical basis of pure alexia.* Neurology 33:1573–1583, 1983.

41. DAMASIO, AR: *Visual disorientation without optical ataxia.* 1985 (in preparation).

42. DE MONASTERIO, FM, AND SCHEIN, SJ: *Spectral bandwidths of color-opponent cells of geniculocortical pathway of macaque monkeys.* J Neurophysiol 47:214–224, 1982.

43. DE RENZI, E, AND FAGLIONI, P: *The relationship between visuospatial impairment and constructional apraxia.* Cortex 3:327–342, 1967.

44. DE RENZI, E, AND SPINNLER, H: *Impaired performance on color tasks in patients with hemispheric damage.* Cortex 3:194–216, 1967.

45. DE RENZI, E, FAGLIONI, P, SCOTTI, G, AND SPINNLER, H: *Impairment of color sorting behavior after hemispheric damage: An experimental study with the Holmgren skein test.* Cortex 8:147–163, 1972.

46. DEJERINE, J: *Sur un cas de cecité verbale avec agraphie, suivi d'autopsie.* Memoires Societé Biologique 3:197–201, 1891.

47. DEJERINE, J: *Contribution a l'étude anatomo-pathologique et clinique des differentes varietés de cecité verbale.* Memoires Societé Biologique 4:61–90, 1892.

48. FREUND, S: *Zur ueber optische aphasie and seelenblindheit.* Arch Psychiatr Nervenkr 20:276–297, 371–416, 1889.

49. GAZZANIGA, MS, BOGEN, JE, AND SPERRY, RW: *Observations on visual perception after disconnexion of the cerebral hemispheres in man.* Brain 88:221–236, 1965.

50. GAZZANIGA, MS, AND SMYLIE, CS: *Facial recognition and brain asymmetries: Clues to underlying mechanisms.* Ann Neurol 13:537–540, 1984.

51. GESCHWIND, N: *Disconnexion syndromes in animals and man.* Brain 88:237–294, 1965.

52. GESCHWIND, N, AND FUSILLO, M: *Color-naming defects in association with alexia.* Arch Neurol 15:137–146, 1966.

53. GLONING, I, GLONING, K, JELLINGER, K, AND QUATEMBER, R: *A case of "prosopagnosia" with necropsy findings.* Neuropsychologia 8:199–204, 1970.

54. GOLDBERG, ME, AND ROBINSON DL: *Visual mechanisms underlying gaze: Function of the cerebral cortex.* In BAKER AND BERTHOZ (EDS): *Control of Gaze by Brain Stem Neurons, Developments in Neuroscience,* Vol 1. Elsevier North-Holland Biomedical Press, 1977, pp 469–476.

55. GOODGLASS, H, AND KAPLAN, E: *The Assessment of Aphasia and Related Disorders.* Lea & Febiger, Philadelphia, 1972.

56. GREEN, GJ, AND LESSELL, S: *Acquired cerebral dyschromatopsia.* Arch Ophthalmol 95:121–128, 1977.

57. HAAXMA, R, AND KUYPERS, HGJM: *Intrahemispheric cortical connexions and visual guidance of hand and finger movements in the rhesus monkey.* Brain 98:239–260, 1975.

58. HAMSHER, K, DES: *Stereopsis and unilateral brain disease.* Invest Ophthalmol Vis Sci 4:336–343, 1978.

59. HAMSHER, K, LEVIN, HS, AND BENTON, AL: *Facial recognition in patients with focal brain lesions.* Arch Neurol 36:837–839, 1979.

60. HECAEN, H, AND ANGELERGUES, R: *Agnosia for faces (prosopagnosia).* Arch Neurol 7:92–100, 1962.

61. HECAEN, H, AND DE AJURIAGUERRA, J: *Balint's syndrome (psychic paralysis of visual fixation) and its minor forms.* Brain 77:373–400, 1954.

62. HEIDENHAIN, A: *Beitrag zur Kenntnis der Seelenblindheit.* Monattschr Psychiatr Neurol 66:61–116, 1927.

63. HOLMES, G: *Disturbances of visual orientation.* Br J Ophthalmol 2:449–486, 506–516, 1918.

64. HOLMES, G: *Disturbances of spatial orientation and visual attention with loss of stereoscopic vision.* Arch Neurol Psychiatry 1:385, 1919.

65. HUBEL, DH, AND WIESEL, TN: *Receptive fields and functional architecture in two non-striate visual areas (18 and 19) of the cat.* J Neurophysiol 28:229–289, 1965.

66. HUBEL, DH, AND WIESEL, TN: *Cells sensitive to binocular depth in area 18 of the Macaque monkey cortex.* Nature 225:41–42, 1970.

67. HUBEL, DH, AND WIESEL, TN: *The period of susceptibility to the physiological effects of unilateral eye closure in kittens.* J Physiol (Lond) 206:419–436, 1970.

68. HUBEL, DH, AND WIESEL, TN: *Sequence regularity and geometry of orientation columns in the money striate cortex.* J Comp Neurol 158:267–294, 1974.

69. HUBEL, DH, AND WIESEL, TN: *Laminar and columnar distribution of geniculocortical fibers in the Macaque monkey.* J Comp Neurol 146:421–450, 1982.

70. HUBEL, D: *Exploration of the primary visual cortex, 1955–78.* Nature 299:515–524, 1982.

71. JULESZ, B: *Binocular depth perception without familiarity cues.* Science 145:356–362, 1964.

72. JULESZ, B: *Foundations of Cyclopean Perception.* University of Chicago Press, Chicago, 1971.

73. JULESZ, B, AND SPIVACK, GH: *Stereopsis based on vernier acuity clues alone.* Science 157:563–565, 1967.

74. JULESZ, B, BREITMEYER, B, AND KROPFL, W: *Binocular disparity dependent upper lower hemifield anisotropy and left right hemifield isotropy as revealed by dynamic randon dot stereograms.* Perception 5:129–141, 1976.

75. KLEIST, K: *Kriegsverletzungen des Gehirns in ihrer Bedutung Fur die Hirnlokalisation und Hirnpathologie.* In VON SCHJERNING, O (ED): *Handbuch der Arztlichen Erfahrung im Weltkriege,* Vol 4. Barth, Leipzig, 1923.

76. LANSDELL, HC: *Effect of extent of temporal lobe ablations on two lateralized deficits.* Physiol Behav 3:271–273, 1968.

77. LENNERSTRAND, G: *Binocular interaction studied with visual evoked responses (VER) in humans with normal or impaired binocular vision.* Acta Ophthalmol 56:628–637, 1978.

78. LEVY, J, TREVARTHEN, C, AND SPERRY, RW: *Perception of bilateral chimeric figures following hemispheric disconnection.* Brain 95:61–78, 1972.

79. LHERMITTE, F, CHAIN, F, ARON, D, LEBLANC, M, AND JOUTY, O: *Les troubles de la vision des couleurs dans les lesions posterieures du cerveau.* Rev Neurol 121:5–29, 1969.

80. LHERMITTE, J, CHAIN, F, ESCOUROLLE, R, DUCARNE, B, AND PILLON, B: *Etude anatomo-clinique d'un cas de prosopagnosie.* Rev Neurol 126:329–346, 1972.

81. LISSAUER, H: *Ein fall von Seelenblindheit nebst einem Beitrag zur Theorie derselben.* Arch Psychiatr Nervenkr 21:22–70, 1890.

82. LIVINGSTONE, MS, AND HUBEL, DH: *Anatomy and physiology of a color system in the primate visual cortex.* J Neurosci 4:309–356, 1984.

83. MACKAY, G, AND DUNLOP, JC: *The cerebral lesions in a case of complete acquired colourblindness.* Scott Med Surg J 5:503–512, 1899.

84. MEADOWS, JC: *The anatomical basis of prosopagnosia.* J Neurol Neurosurg Psychiatry 37:489–501, 1974a.

85. MEADOWS, JC: *Disturbed perception of colors associated with localized cerebral lesions.* Brain 97:615–632, 1974b.

86. MEIER, MJ, AND FRENCH, LA: *Lateralized deficits in complex visual discrimination and bilateral transfer of reminiscence following unilateral temporal lobectomy.* Neuropsychologia 3:261–272, 1965.

87. MOHR, JP, LEICESTER, J, STODDARD, LT, AND SIDMAN, M: *Right hemianopia with memory and color deficits in circumscribed left posterior cerebral artery territory infarction.* Neurology 21:1104, 1971.

88. MOLL, L, AND KUYPERS, HGJM: *Premotor cortical ablations in monkeys: Contralateral changes in visually guided reaching behavior.* Science 198:317–319, 1977.

89. MOLL, L, AND KUYPERS, H: *Role of premotor cortical areas and VL nucleus in visual guidance of relatively independent hand and finger movements in monkeys.* Exp Brain Res (Suppl) 23:142, 1975.

DISORDERS OF
COMPLEX VISUAL
PROCESSING:
AGNOSIAS,
ACHROMATOPSIA,
BALINT'S SYNDROME,
AND RELATED
DIFFICULTIES OF
ORIENTATION AND
CONSTRUCTION

287

90. MOUNTCASTLE, VB, LYNCH, JC, AND GEORGOPOULOS, A: *Posterior parietal association cortex of the monkey: Command functions for operations within extrapersonal space.* J Neurophysiol 38:871–908, 1975.

91. NEWCOMBE, F, AND RUSSELL, WR: *Dissociated visual perceptual and spatial deficits in focal lesions of the right hemisphere.* J Neurol Neurosurg Psychiatry 32: 73–81, 1969.

92. ORGASS, B, POECK, K, KERCHENSTEINER, M, AND HARTJE, W: *Visuocognitive performances in patients with unilateral hemispheric lesions.* Z Neurol 202:177–195, 1972.

93. PARKS, M: *The monofixation syndrome.* Trans Am Ophthalmol Soc 67:609–657, 1969.

94. PEARLMAN, AL, BIRCH, J, AND MEADOWS, JC: *Cerebral color blindness: An acquired defect in hue discrimination.* Ann Neurol 5:253–261, 1979.

95. PEVZNER S, BORNSTEIN, B, AND LOEWENTHAL, M: *Prosopagnosia.* J Neurol Neurosurg Psychiatry 25:336–338, 1962.

96. POGGIO, GR: *Central neural mechanisms in vision.* In MOUNTCASTLE, VD (ED): *Medical Physiology,* Vol 1. CV Mosby, St. Louis, 1980, pp 573–579.

97. POGGIO, GF, AND FISCHER, B: *Binocular interaction and depth sensitivity in striate and prestriate cortex of behaving rhesus monkey.* Neurophysiol 40:1392–1405, 1977.

98. RATCLIFF, G, AND DAVIES-JONES, GAB: *Defective visual localization in focal brain wounds.* Brain 95:49–60, 1972.

99. RATCLIFF, G, AND NEWCOMBE, F: *Spatial orientation in man: Effects of left, right and bilateral posterior lesions.* J Neurol Neurosurg Psychiatry 36:448–454, 1973.

100. RONDOT, P, AND DE RECONDO J: *Ataxie optique: Trouble de la coordination visuo-motrice.* Brain Res 71:367–375, 1974.

101. ROSS, J: *Stereopsis by binocular delay.* Nature 248:363–364, 1973.

102. ROSS, ED: *Sensory-specific and fractional disorders of recent memory in man: 1. Isolated loss of visual recent memory.* Arch Neurol 37:192–200, 1980.

103. SCOTTI, G, AND SPINNLER, H: *Colour imperception in unilateral hemisphere-damaged patients.* J Neurol Neurosurg Psychiatry 33:22–28, 1970.

104. SCHEIN, SJ, MARROCCO, RT, AND DE MONASTERIO, FM: *Is there a high concentration of color-selective cells in area V4 of monkey cortex?* J Neurophysiol 47(2):193–213, 1982.

105. SEMMES, J, WEINSTEIN, S, GHENT, L, AND TEUBER, HL: *Spatial orientation in man: 1. Analysis by locus of lesion.* J Psychol 39:227–244, 1955.

106. SERGENT, J, AND BINDRA, D: *Differential hemispheric processing of faces: Methodological considerations and reinterpretation.* Psychol Bull 89:541–554, 1981.

107. TEUBER, HL: *Alteration of perception and memory in man.* In WEISKRANTZ, L (ED): *Analysis of Behavioral Change.* Harper and Row, New York, 1968.

108. UNGERLEIDER, LG, AND MISHKIN, M: *Two cortical visual systems.* In INGLE, DJ, MANSFIELD, RJW, AND GOODALE, MD (EDS): *The Analysis of Visual Behavior.* MIT Press, Cambridge, 1982.

109. VAN ESSEN, DC, MAUNSELL, JHR, AND BIXBY, JL: *The middle temporal area in the macaque: myeloarchitecture, connections, functional properties and topographic organization.* J Comp Neurol 199:293–326, 1981.

110. VAN ESSEN, D, AND MAUNSELL, J: *Hierarchical organization and functional streams in the visual cortex.* Trends Neurosci 63:370–377, 1983.

111. VERREY, D: *Hemiachromatopsie droite absolue.* Arch Ophthalmol (Paris) 8:289–300, 1888.

112. VERREY, D (CITED BY LANDOLT, E): *De la cecité verbale.* Utrecht, 1888; AND DEJERINE, J: *Differentes varietés de cecité verbale.* Societé de Biologie (Paris) 44:64, 1892.

113. VIALET, N: *Les Centres Cerebraux de La Vision et L' Appareil Nerveux Visuel Intra-Cerebral.* Faculté de Medecine de Paris, Paris, 1893.

114. WESTHEIMER, G: *Cooperative neural processes involved in stereoscopic acuity.* Exp Brain Res 36:585–597, 1979.

115. WILBRAND, H: *Ein Fall von Seelenblindheit und Hemianopsie mit Sectionsbefund.* Deutsche Z Nervenheik 2:361–387, 1892.

116. YIN, TCT, AND MOUNTCASTLE, VB: *Visual input to the visuomotor mechanisms of the monkey's parietal lobe.* Science 197:1381–1383, 1977.

117. ZEKI, SM: *Colour coding in rhesus monkey prestriate cortex.* Brain Res 53:422–427, 1973.

118. ZEKI, SM: *Colour coding in the superior temporal sulcus of rhesus monkey visual cortex.* Pro R Soc Lond [Biol]197:195–223, 1977.

119. ZIHL, J, VON CRAMON, D, AND MAI, N: *Selective disturbance of movement vision after bilateral brain damage.* Brain 106:313–340, 1983.

Chapter 8

TEMPOROLIMBIC EPILEPSY AND BEHAVIOR*

Paul A. Spiers, M.A., Donald L. Schomer, M.D., Howard W. Blume, M.D., and M-Marsel Mesulam, M.D.

Temporal lobe epilepsy is a common condition in which complex neurologic and psychiatric phenomena are frequently seen in the same patient. Understanding the interactions between these two types of symptoms is essential for the effective management of such patients. Furthermore, the investigation of these interactions provides considerable insight into the behavioral specializations of the temporal lobe and its limbic components.

TERMINOLOGY

Classification of Epilepsies

The nomenclature for epilepsy is based on the origin of seizure activity and its associated clinical manifestation.[18] The current system has two major categories. The first of these is "primary generalized" epilepsy. In this category, the exact mechanism underlying the generation of epileptic activity is not known, but the seizures affect both hemispheres of the brain simultaneously and in a widespread manner. The actual clinical manifestations of the seizure may vary considerably and include sudden rigidity (tonic seizure), rhythmic movement (clonic seizure), a combination of both (tonic-clonic seizure), sudden loss of motor tone (atonic event or drop attack), and alterations in consciousness (e.g., absence).

The second major category in this system is "partial" epilepsy, which can be further divided into "simple partial" and "complex partial"

*The preparation of portions of this manuscript was supported by an NINCDS Teacher-Investigator Award (Dr. Blume), by NINCDS grants NS-09211 and NS-20285, and by the Javits Neuroscience Investigator Award (Dr. Mesulam). The authors also wish to express their gratitude to Gina Termine, Leah Christie, and Della Grigsby for their invaluable assistance.

subtypes. Temporal lobe epilepsy, also known as psychomotor, psycho-paretic, and limbic epilepsy, belongs to this group. In this category, all seizures are presumed to have a focal origin in some neocortical region. It is the subsequent progression of the seizure that determines its sub-classification. If it remains confined to the region of origin without pro-ducing an alteration in consciousness, it is called a simple partial seizure. "Simple," therefore, denotes that no significant involvement of con-sciousness has occurred during the event and does not refer to the quality or complexity of the experience or phenomenon that may have been evoked by the seizure. On the other hand, if consciousness is impaired during a partial seizure, it is referred to as a complex partial seizure. If the seizure progresses to involve major motor systems in a tonic, clonic, or tonic-clonic fashion, it is referred to as a partial seizure that has sec-ondarily generalized. This type of partial seizure may be very difficult to distinguish from primary generalized epilepsy.

Partial seizures on the motor sensory strip, in the prefrontal cortex region, or over occipitoparietal surfaces commonly have an associated EEG abnormality, since recording electrodes can be placed directly over these areas. Seizures confined to the medial and inferior temporal sur-face, cingulate area, orbitofrontal region, or insular cortex, on the other hand, may escape detection by standard scalp electrodes. This second group of partial seizures, which are among those of greatest interest to behavioral neurology, are therefore often difficult to detect by routine EEG.

Temporolimbic: A Descriptive Term

The term "temporal lobe epilepsy," which has gained widespread usage, has potential difficulties. This is due primarily to the heterogeneous anat-omy of the region. The temporal lobe includes not only neocortical areas where complex sensory associations are formed but also components of the limbic system such as the temporal pole, the parahippocampal region, the amygdala, and the hippocampus. The limbic components of the temporal lobe also have widespread reciprocal connections with par-alimbic regions in the insula, the orbitofrontal cortex, and the cingulate gyrus. These temporal and extratemporal constituents of the limbic sys-tem have many common anatomic, neurochemical, and behavioral fea-tures (see Chapter 1). One outcome of the intimate interconnectivity among these areas is the rapid spread of seizure activity from one com-ponent of this limbic network to the other. Therefore, it may be difficult, if not impossible, to differentiate the clinical manifestations of a seizure that originates in cingulate cortex and subsequently incorporates the amygdala from those of a seizure that begins in the hippocampus or amygdala and then spreads either to overlying temporal cortex or some other connected region. Since these relevant parts of the brain are diffi-cult to monitor by the standard placement of EEG electrodes, it may be impossible to distinguish between these two alternatives even by electro-physiologic criteria. Within this group of partial seizures, those that orig-inate in the temporal lobe, especially in its limbic parts, are the most com-mon.[122] For these reasons, we adopt the term "temporolimbic epilepsy" as a generic term for seizures that begin not only in the temporal lobe but also in other cortical components of the limbic system (e.g., posterior orbitofrontal, cingulate, and insular regions). Despite differences in the exact location of the focus these seizures share many physiologic and clinical features.

Prevalence

The overall prevalence of epilepsy in North America is approximately 0.7 percent, affecting 1.75 million people. It is estimated that another 0.3 percent of the total population have active clinical epilepsy that is undetected and untreated, reaching a total of 2.5 million affected individuals.[56] Of the total population of patients with epilepsy, between one third and one half have partial seizure disorders. For patients with partial epilepsy, more than one half of these disorders are of temporolimbic origin. Consequently, we would currently estimate that approximately 600,000 to 1 million people are affected with temporolimbic epilepsy. Even this may be an underestimation, since one epidemiologic study in Europe has suggested that as much as two thirds of the total population of patients with epilepsy may not be under treatment at any given time, one third because they have discontinued treatment and one third because they have never been treated. Given that primary generalized seizures are by nature more likely to be identified and treated, the majority of cases that escape reporting are likely to be those of partial epilepsy. It is also reasonable to assume that cases of temporolimbic epilepsy that have predominantly behavioral manifestations are particularly prone to be misdiagnosed as primary psychiatric conditions. Consequently, one could speculate that the prevalence rate for temporolimbic epilepsy in the general population may be as high as 1 percent.[122,164,173-175]

ICTAL ALTERATIONS

Some of the behavioral changes seen in conjunction with temporolimbic epilepsy are abrupt and transient and are generally considered to represent the ictal behavioral concomitants of paroxysmal seizure acitivity. The common description of a temporolimbic seizure usually emphasizes some sudden event during which there is an alteration in consciousness accompanied by automatic behaviors, such as closing a window, running in circles, taking off a shirt, or picking at clothing. Staring, deviation of the head and eyes to one side, lip smacking, some rhythmic twitching of an extremity, aimless wandering, or some nonsensical verbalization are also reported. The patient may answer questions in a manner reflecting confusion and may even become combative if restrained during such episodes.

Although this description may be appropriate for many of the seizures encountered in patients with temporolimbic epilepsy, there are many other manifestations of ictal events. In fact, certain types of seizures in this region may exclusively elicit alterations in mood and perception. Some of these ictal symptoms may even mimic the manifestations of idiopathic psychiatric disorders.[2,30,71] Even a brief catalogue of reliably identified temporolimbic seizure phenomena would include every major category of cerebral functioning.[46,98] These can be broadly categorized as sensory, motor, autonomic, hallucinatory, experiential, and emotional manifestations. Whether the patient experiences only selected symptoms or a combination of these will vary depending on the origin and spread of the seizure within the temporolimbic network.[47,169] These symptoms may be experienced in what appears to be full consciousness and may be as brief as a few seconds, persist for hours, or recur with increasing frequency over the course of several days. In some cases, the epileptic basis for the symptoms can be inferred by observing the regularity and relatively stereotyped nature of the phenomena. In others, the ictal nature of the symptoms can be corroborated on EEG, but this often requires the use of prolonged telemetric recording with special electrodes.[2,47,169]

Behavioral Aspects of Ictal Phenomena

Sensory alterations accompanying a temporolimbic seizure may include headache, focal pain, discomfort, tingling, or numbness. The feeling of something crawling on or under the skin has also been reported. These sensory alterations may be unilateral or bilateral and may follow almost any distribution, even those that may seem to defy conventional innervation patterns. It is important to keep in mind that it is the patient's felt experience at the level of the central nervous system that constitutes the sensory seizure. Dizziness or vertiginous sensations are also common. When the focus is on the left, aphasic comprehension deficits may also emerge, often accompanied by speech arrest. The neurologic examination does not necessarily reveal any objective sensory findings even during the actual seizure event.[46,70,98,172]

   Motor symptoms include complex automatisms, abnormal eye movements or nystagmus, staring, twitching, or jerking of upper or lower extremities. Head turning or visual checking, furtive scanning of the environment, or looking over the shoulder as if searching for something have all been reported and may represent, in some patients, the motor response to hallucinatory stimuli.[125] Patients sometimes spontaneously assume unusual facial expressions or body postures and may demonstrate "waxy flexibility," a condition in which a passively manipulated limb remains in the configuration to which it was last positioned.[39,149] Stuttering, slurred speech, and even complete speech arrest have also been described.[7,44,55,97,117] Transient weakness or motor paresis, sometimes resulting in suddenly falling to one side, may be an initial symptom. The distribution of the weakness may be atypical and the patient's muscle tone or reflexes may remain intact. A helpful sign is sometimes a unilateral sluggish pupillary reflex or spontaneous unilateral pupillary dilation during a temporolimbic seizure. These changes are frequently ipsilateral to the side of the seizure focus.

   Temporolimbic seizures may also have *autonomic* manifestations.[31,158,159] Patients may experience flushing or a "hot sensation" that may be bilateral or unilateral, sometimes restricted only to the head or upper torso. They may complain of shortness of breath,[70] "weight" on their chest, or of being unable to take a full breath, which, in our experience, has not been reflected by any pulmonary function abnormalities. On the other hand, patients may have frank apneic attacks with clear evidence of cyanosis. Cardiac symptoms reported by the patient may include chest pain or a feeling that the heart has stopped, is racing, or is pounding hard.[27] For example, sinus tachycardia has been documented in conjunction with seizure discharges using simultaneous EEG and ECG monitoring in selected patients.[81] Perhaps the most frequently reported symptom in this category, however, is nausea or epigastric distress.[88,99,118,172] This is usually experienced as a rising or sinking feeling, "like an elevator ride," but has also been reported to feel like a sudden urge to throw up, as if the intestines were tied in knots. Others have described the sensation of being punched in the stomach. We have even seen patients with no other clinical manifestation of seizures who came to medical attention because of cardiac arrhythmias or abdominal pain of unknown etiology and who were later found to have temporolimbic epilepsy as the cause.

Hallucinations or *illusions* as symptoms of a temporolimbic seizure may occur in virtually any sensory modality.[18,46,47,98,169] Visual phenomena such as patterns, geometric shapes, and flashing white or colored lights are common and in some cases may be restricted to one visual field. Fully formed, complex images such as of someone standing in the doorway, of a frightening personage or a demon, or of a godlike, fatherly figure have been reported. The following three phenomena should be included here: metamorphosia, the sudden distortion of a common object or person; micropsia, the illusion that objects have become smaller, further away, or suddenly out of reach; and macropsia, the illusion that objects have become larger and are closer to or "towering over" the patient. For example, a doorway may seem too small to use or bathroom fixtures may seem 10 feet tall. These may be some of the reasons patients have compared themselves to Alice in Wonderland.

Auditory hallucinations may consist of a ringing or buzzing sound or of a voice calling the patient's name or repeating several stereotyped phrases. Gustatory hallucinations are usually of a metallic or foul taste. Uncinate or olfactory phenomena are similarly unpleasant and may include odors such as ammonia, burning rubber or garbage, a decaying or fecal odor, "as if a skunk were under my nose," or a suffocating smell of garlic. Even perfumed odors are usually described as overly pungent or sickeningly sweet and are rarely experienced by the patient as positive.

Experiential manifestations of a temporolimbic seizure may include memory flashbacks ("I feel like I did as a child in the summer, falling into the lake") and illusions of familiarity or unfamiliarity.[18,46,47,98] The former is called déja vu or déjà vécu (already seen or already experienced) and consists of a sense that the events, conversation, or location at the moment have all been experienced or witnessed previously by the patient. This often leads patients to believe they are clairvoyant or prescient ("I must have known it was going to happen, because it was all familiar to me"). In the converse situation, jamais vu or jamais vécu, patients may report that they intellectually know the place or people to be familiar but that they are unable to shake the feeling that they have never seen them before. In one extreme case, the patient felt that she herself had no past and had never existed before the moment of the seizure. Though she was perfectly lucid and communicative, everything that occurred during the event was felt to be a novel experience and the patient was even unsure of her own identity. This could last for over an hour without relief and was accompanied by fear and bilateral numbness of the hands.

The sense that someone is nearby, watching over a shoulder, or outside the house (e.g., the feeling of a presence), without any accompanying visual hallucination, has also been reported.[46] Less frequent but equally well documented are symptoms such as the illusion of being possessed by the devil or by multiple personalities. The illusion that others are possessed or suddenly appear malevolent may occur. Mind-body dissociation is perhaps even more common. Patients may be reluctant to volunteer such information but when questioned may readily admit that they sometimes step outside of their own body or hover above themselves and find they are watching their own actions from the perspective of a detached observer.[47,84,98]

The *emotional* manifestations of a temporolimbic seizure disorder are most often negative in content and can usually be characterized as "forced."[18,47,169] This means that the emotion occurs suddenly and unpredictably without apparent precipitant, is often out of context or inappropriate to the patient's activities at the time, and generally ceases as abruptly as it began. Emotional manifestations reported during temporolimbic seizures have included embarrassment,[26] sadness, depression or sudden crying,[47,124] explosive laughter (gelastic epilepsy)[169] usually without the feeling of happiness, peacefulness with a sense of serenity or of "oneness with the universe," and, perhaps most commonly, fear.[47,148] The latter manifestation is typically fear of personal injury or harm from some unknown source and has been described as fear of impending doom or of death (timor mortis). There are frequently autonomic accompaniments, and the patient may feel compelled to escape, even from familiar surroundings. This may be the phenomenon that forms the basis for "running seizures" in which patients have been described to suddenly appear confused and then run away, at times covering great distances before regaining their composure at which point they are most often lost and in some unfamiliar place.

Sudden anger or increased irritability, although less frequent, can be seen either as part of the prodrome leading up to a temporolimbic seizure or may occur during the actual event.[89,102,169] In a few patients, this may take the form of explosive but undirected aggressive behavior, such as throwing objects or clawing at the wall, and the patient may become more violent if restrained. Spontaneous orgasm, pleasurable genital sensations, and sexual behaviors, including exhibitionism, have been reported in several patients.[40,119,120,144] Finally, forced thoughts, particularly those of an obsessive or compulsive nature, may occasionally be seizure related.[47,98,169] One patient with left parietotemporolimbic seizures felt compelled to engage in various motor rituals, such as touching each side of his face twice and then three times, during electrographic seizure activity. Several patients with temporolimbic seizure foci have engaged in self-mutilation. One female patient had a paroxysmal illusion of being possessed by a demon and during these episodes felt compelled to escape from her home or the hospital in order to have sexual relations with some unclean and unattractive stranger. She had acted on this impulse several times and admitted having had intercourse on at least one occasion with some unknown derelict. Prolonged confusional and fugue states, in some cases accompanied by extremely disorganized behaviors, have also been associated with ongoing seizure activity.[2,6,30,34,93,134]

It should be apparent from this review of some of the more salient ictal phenomena that the protean manifestations of temporolimbic epilepsy may be varied, complex, and difficult to differentiate from spontaneously occurring behavior. In essence, it would appear that an electrical discharge in the temporolimbic network may reproduce or evoke virtually any stimulus, memory, thought, or feeling state that the brain is capable of experiencing.

Neural Substrate of Ictal Phenomena

The diversity of alterations reported during temporolimbic seizures undoubtedly reflects the functional and anatomic complexity of the temporal lobe (see Chapter 1). Specific functional areas such as the visual cortex have direct connections with the middle and inferior temporal gyri. The primary auditory cortex in the supratemporal plane receives input from the medial geniculate nucleus and, in turn, is closely integrated with adjacent auditory association areas. Stimulation and ablation studies in both animals and humans have demonstrated the role of these more lateral regions of the temporal lobe in the perception and coding of auditory and visual information in a manner consistent with these anatomical affiliations. More medial temporal structures, meanwhile, belong to the limbic system and are involved in the control of learning, memory, affect, visceral tone, and hormonal balance (see Chapter 1). Other cortical components of the limbic system (e.g., caudal orbitofrontal cortex, cingulate cortex, and insula) that are closely interconnected with these medial temporal regions also share many of these same behavioral affiliations (see Chapter 1). One temporal lobe (usually the left) appears to be more concerned with verbal material, whereas the other is more specialized for processing nonverbal material.[85,86,127,131] Stimulation of the temporal lobe, and especially of its limbic components, has resulted in autonomic responses manifested by pupillary changes, lacrimation, salivation, pulse rate, and blood pressure changes, peristalsis, micturition, uterine contractions, and penile erections. Experimental work with bilateral symmetric lesions of medial temporal structures has resulted in amnestic syndromes, altered rage responses, hypersexuality, and changes in feeding (see Chapters 1 and 4). Consequently, it is not surprising to discover that temporolimbic epileptics manifest changes in those same behaviors that depend for their mediation and control on the integrity of temporal and limbic cortical areas.[9,46,85,86,131]

For all of the episodic phenomena described in the preceding section, patients have either shown simultaneous electrographic abnormalities or have had EEGs that were at some time characteristic of a temporolimbic seizure disorder. Electrographic observations indicate that simple visual stimuli or vertiginous experiences are likely to be associated with a discharge in the posterior temporal or parietal-occipital regions. Complex auditory hallucinations, including those of music, have been shown in certain patients to be time-locked to ictal discharges in or near Heschl's gyrus.[169] Motor and sensory phenomena in temporolimbic epilepsy generally reflect spread to the primary motor or sensory cortices, although in certain motor manifestations the supplementary motor area or frontal eye field is probably implicated. Certain investigators have also suggested that experiential, emotional, and autonomic manifestations are more likely to be associated with lateralization of the seizure foci to the right hemisphere.[54]

Important pathophysiologic correlations in temporolimbic epilepsy have come from recent studies using sterotaxically implanted intracerebral depth electrodes.[47,169] This research has shown that the experiential and emotional manifestations of temporolimbic epilepsy are most likely to be associated with seizure activity in the *limbic* regions of the temporal lobe. These ictal phenomena may include such transient events as momentary feelings of loneliness, depression, fear, the recollection of complicated scenes, or experiences that contain forced memories, perceptions, and emotions. Furthermore, temporal neocortex is apparently not a crucial neural substrate for these epileptic events. In the one study that examined this issue systematically, 3 percent of experiential, hallucinatory or emotional manifestations were associated with seizures restricted to the neocortical part of the temporal lobe.[47] In contrast, almost 30 percent were produced by discharges restricted to the amygdala and hippocampus, either alone or in combination. If the adjacent parahippocampal gyrus was also included, this accounted for 42 percent of the total number of such seizures. Therefore, almost half of all experiential events will not be associated with a seizure discharge in lateral temporal neocortex. Consequently, the electrical correlates of experiential phenomena are likely to escape detection by standard scalp EEG electrodes, even at times when the patient is actively symptomatic. Virtually identical results regarding the pivotal role of these mesial limbic structures in the production of experiential, hallucinatory, or emotional ictal manifestations were obtained by examining the effects of stimulation with intraoperative depth electrodes in an independent sample of patients.[47]

It is noteworthy that many symptoms that are known to have an epileptic origin can also lead to the diagnosis of an idiopathic psychiatric disorder. In this context, it is of interest to consider briefly those studies that have attempted to examine electrographic abnormalities in certain groups of psychiatric patients.[1] First, it has been demonstrated that as high as 20 to 30 percent of patients with schizophrenia have EEG abnormalities localized over one or both temporal lobes. In one study, greater than half of the patients with the abnormal EEGs had frankly paroxysmal activity in their records.[53,151] Patients with schizophrenia have also been reported to have reduced waveform stability in visually evoked potentials, more so over the left hemisphere, and greater power in the 20- to 30-Hz band in the left temporal derivatives. Although this might suggest preferential left-hemispheric dysfunction in this population, more wide-spread changes in multiple frequency bands and without significant hemispheric asymmetry have also been described.[1,42,72,126]

Of even greater interest are those studies in which a correlation has been shown to exist between displays of abnormal behavior and specific EEG changes. For example, high-voltage theta activity has been reported in patients during catatonic states. Recently, a carefully defined sample of schizophrenic patients and control subjects was examined to determine the presence of a specific power spectrum pattern that is thought to reflect subcortical spike activity. This pattern was never observed in the records of control subjects but occurred with considerable frequency in the schizophrenic patients. This pattern was also twice as common during epochs when the patients displayed abnormal behaviors such as catatonia, auditory hallucinations, and visual checking.[147]

There are also some related observations derived from a small number of psychotic patients who had implanted depth electrodes.[89] For example, abnormal spike and slow-wave activity from the septal region has been reported during psychotic behaviors, and limbic dysrhythmias have been correlated with episodic confusion and aggressive episodes. One patient who had been diagnosed as having chronic paranoid schizophrenia was recorded during episodes when she was delusional, confused, disoriented, experiencing auditory and visual hallucinations, and, occasionally, while showing intense rage with aggressive and destructive acts. Depth electrodes showed repeated paroxysmal bursts of high-amplitude multiphasic spiking activity in hippocampal and septal regions, with occasional temporal cortex involvement. These abnormalities were not detected by conventional scalp leads.

Although negative results have also been reported in attempting to correlate psychotic behaviors with EEG changes, in some cases this may have been due to the focal nature of these alterations and to their localization in mesial temporolimbic structures that are not readily accessible to conventional scalp EEG recording.[23,112] In addition, patients who experience the most severe behavioral problems are frequently excluded from study in facilities where sophisticated EEG technology is available, since such facilities are rarely equipped to manage these patients. Even when studies are designed to assess "psychotic" temporolimbic epileptics, these individuals are usually studied during a period when they are not acutely symptomatic.[122] Consequently, there may be a subgroup of individuals with a diagnosis of idiopathic schizophrenia who also have temporolimbic EEG abnormalities, and some of these abnormalities may be epileptic in nature.

Similar considerations exist with regard to the relationship between temporolimbic epilepsy and aggressive behavior. For example, even the widely accepted opinion that premeditated, directed, or purposeful aggressive behavior is only rarely an ictal manifestation may need to be re-evaluated.[23] One reason is that recordings capable of detecting changes in mesial temporolimbic structures which may be associated with such episodes[169] are rarely obtained during violent outbursts. Although the relationships between temporolimbic epilepsy and psychotic or aggressive behavior are extremely difficult to investigate, the theoretical, clinical, and therapeutic relevance of this area is self-evident.[79,80,145,146,161]

INTERICTAL (LONG-TERM) ALTERATIONS

Interictal Psychosis

In the preceding section, some investigations were reviewed that have attempted to discover EEG abnormalities in psychotic patients with no other clinical indication of a convulsive disorder. There is also a much more extensive literature showing that the incidence of psychosis (especially with features of paranoid schizophrenia) is unusually high in individuals who are *known* to have temporolimbic epilepsy.[142] In fact, while the incidence of schizophrenia in the general population is 1 percent, as much as 10 to 15 percent of individuals with poorly controlled temporolimbic epilepsy receive the diagnosis of schizophrenia or psychosis.[137,168] Because many of the changes observed in these patients are chronic, it was maintained that the abnormal behavior was *interictal* in nature.[43] A brief review of the salient findings in this literature over the past 15 years yields a surprisingly coherent group of factors that appear to contribute to the likelihood of developing this behavioral pattern in conjunction with temporolimbic epilepsy.

Lateralization of the epileptic focus to the left hemisphere is the most consistently reported feature in patients who have temporolimbic epilepsy associated with psychosis. This relationship is particularly strong in women who have "alien tissue" lesions (i.e., hamartomas, focal dysplasia, or small focal tumors) in relation to their seizure foci. An elevated incidence of sinistrality or ambidexterity, potentially implicating early left-hemisphere injury, has also been common to most of these reports. There is also an increased incidence of birth trauma, head injury, encephalitis, and meningitis, as well as positive findings on the neurologic examination and neuroradiologic studies. Less frequently reported factors have been a positive family history of epilepsy, suggesting some genetic predisposition. Factors such as the age at onset and the duration of the temporolimbic seizure disorder have not yielded consistent effects.[36,37,63,67,136,137,153,155]

Although temporolimbic epileptics with schizophreniform disorders have been proclaimed to be virtually indistinguishable from patients with idiopathic schizophrenia, many studies in this area have also commented on the atypical phenomenologic aspects of the "interictal" psychosis.[100,136,137,142,152,156] The psychosis is prominently characterized by hallucinatory phenomena (both auditory and visual), persecutory delusions, and ideas of reference. However, in contrast to idiopathic schizophrenia, family histories and characteristic premorbid personality disturbances are generally absent. The affect is usually preserved, the psychotic episodes are more periodic and less socially disruptive, and long-term hospitalization and treatment with major tranquilizers is not typically required. In comparison to typical schizophrenic patients, temporolimbic epileptics have also been described as more capable of establishing interpersonal rapport and better able to encapsulate their psychotic symptoms. In order to differentiate this condition from the more typical cases of schizophrenia, descriptive terms such as "schizophreniform psychosis of epilepsy" can be used. It is also necessary to realize, however, that some psychotic patients with temporolimbic epilepsy may have clinical pictures that are quite indistinguishable from idiopathic schizophrenia.[66,68]

In addition to psychotic conditions, we have observed patients who had manic-depressive illness, anorexia nervosa, and multiple personality in conjunction with their temporolimbic epilepsy. It is not clear if the incidence of these disorders will also be higher in the population of temporolimbic epileptics. However, it is conceivable that there is a subgroup of patients in each of these diagnostic categories in whom the epilepsy influences the clinical course and perhaps the genesis of the psychiatric disorder. From a practical point of view, the presence of atypical features in psychosis, affective illness, multiple personality, and even anorexia nervosa should raise the possibility of an underlying epileptic condition.

Interictal Changes in Behavior Traits

In addition to the psychiatric syndromes described above, papers published since the 1950s have also emphasized the association of specific behavioral traits with temporolimbic epilepsy.[35,62,103,152] These traits include deepened emotionality, humorlessness, hyposexualism and unusual sexual practices, anger and hostility, hyperreligiosity, enhanced philosophical preoccupation and an augmented sense of personal destiny, passivity, paranoia; moralism, guilt, obsessionalism, circumstantiality, viscosity (emotional clinging), and hypergraphia. In our own experience, one or more of these traits are frequently seen in patients with temporolimbic epilepsy, sometimes to a dramatic extent. However, we have also seen many patients with temporolimbic epilepsy who have lacked these traits altogether. There are currently two considerations associated with these traits: (1) Are these traits more common in patients who have temporolimbic epilepsy when compared with patients who have other types of epilepsy, other neuropsychiatric conditions or with normal individuals? and (2) Can these traits be used to assist in the diagnosis of temporolimbic epilepsy? Several issues relevant to these two questions are reviewed in this section.

The traits most consistently associated in the literature with temporolimbic epilepsy were recently developed into a questionnaire called the Behavior Inventory (Table 1).[10] This questionnaire was administered to patients with "left and right temporal foci," to patients with neuromuscular disorders, and to a group of normal control subjects. The results showed that the inventory significantly differentiated the patients with epilepsy from those with neuromuscular disorders and normal subjects on all 18 traits. Few of the traits, however, significantly differentiated right from left temporal lobe epileptics. A more detailed factor analysis yielded two principal components. The first of these was labeled ideative and consisted of traits such as increased rumination over religion, morality, or personal destiny, paranoid concerns, and hypergraphia. The other

TABLE 1. Characteristics Historically Attributed to Temporal Lobe Epilepsy*

Inventory Trait	Reported Clinical Observations
Emotionality	Deepening of all emotions, sustained intense affect.
Elation, euphoria	Grandiosity, exhilarated mood; diagnosis of manic-depressive disease.
Sadness	Discouragement, tearfulness, self-deprecation; diagnosis of depression, suicide attempts.
Anger	Increased temper, irritability.
Agression	Overt hostility, rage attacks, violent crimes, murder.
Altered sexual interest	Loss of libido, hyposexualism; fetishism, transvestism, exhibitionism, hypersexual episodes.
Guilt	Tendency to self-scrutiny and self-recrimination.
Hypermoralism	Attention to rules with inability to distinguish significant from minor infractions; desire to punish offenders.
Obsessionalism	Ritualism; orderliness; compulsive attention to detail.
Circumstantiality	Loquacious, pedantic; overly detailed, peripheral.
Viscosity	Stickiness; tendency to repetition.
Sense of personal destiny	Events given highly charged, personalized significance; divine guidance ascribed to many features of patient's life.
Hypergraphia	Keeping extensive diaries, detailed notes; writing autobiography or novel.
Religiosity	Holding deep religious beliefs, often idiosyncratic; multiple conversions, mystical states.
Philosophical interest	Nascent metaphysical or moral speculations, cosmological theories.
Dependence	Cosmic helplessness, "at hands of fate"; protestations of helplessness.
Humorlessness	Overgeneralized ponderous concern; humor lacking or idiosyncratic.
Paranoia	Suspicious, overinterpretative of motives and events; diagnosis of paranoid schizophrenia.

*Adapted from Bear and Fedio,[10] with permission.

was considered to be an emotive component and included such traits as elation, sadness, and emotionality. Further analysis of the data showed that patients with left-hemisphere foci tended to cluster along the ideative axis, while those with right-hemisphere foci were centered along the emotive axis.[10]

Attempts to replicate the original findings using the same Behavior Inventory have yielded inconsistent results. When psychiatric patients were included as controls and different methods of statistical analysis were utilized, epileptic patients who had clear histories and behavior abnormalities were virtually indistinguishable from the nonepileptic comparison group on the Behavior Inventory.[92] In another study, however, when a structured interview based on the inventory was used, consistent differences emerged between psychiatrically hospitalized patients with temporolimbic epilepsy and selected groups of nonepileptic psychiatric patients. However, the significant traits differed somewhat in each comparison, and only a few of the original 18 appeared to be specific to epilepsy.[11] For example, the trait of hypergraphia was useful to distinguish patients with temporolimbic epilepsy from those with schizophrenia but not from patients with affective or character disorders. In yet another study, patients with temporolimbic epilepsy differed from patients with primary generalized seizure disorders with regard to their concerns over personal destiny, dependency, paranoia, and philosophic interests.[57]

Additional studies have focused on hypergraphia, in an attempt to determine its prevalence and specificity in temporolimbic epilepsy. When sent a health-related questionnaire that required written answers, more than 50 percent of patients with focal temporal lobe epilepsy responded, whereas less than 25 percent of patients with focal nontemporal or generalized epilepsy did so. Furthermore, the mean number of words for the temporal lobe epileptics was 12 times that of the other epileptics.[130] An additional study using patients with focal temporal lobe seizures, with focal nontemporal or generalized seizures, and with mixed temporal and generalized seizure disorders failed to replicate this observation, even though some of the longest replies received did come from patients in the focal temporal lobe group.[58] In this context, there are also some very preliminary data suggesting that hypergraphia may be more frequently associated with temporolimbic foci in the right hemisphere.[123]

The tendency or even compulsion to write excessively has been apparent in many temporolimbic epilepsy patients we follow, including some of the most well-adapted and highly functioning. These patients may take creative writing courses, belong to poetry groups, or write short

TABLE 2. Poem Written by a Patient with Temporolimbic Epilepsy

Seizure

Put a gun to your head
Slowly, pull the trigger
Click
What's that mommy?
Oh, another seizure

Suppose you never knew
if the gun is loaded,
whether every time
that little flick,
that tiny discharge
in your brain
goes click
you'll come out
alive

Talk to a group
Speak to anyone
Just try to work
never knowing
if there will be
30 clicks,
or none

Life's a struggle
Everyone knows that
Click
Try it
with this extra

trick

Phyllis Tourse
April 16, 1982

stories. Some of the work they produce is obsessive, ruminative, and moralistic; some of it, however, is poignant, insightful, and well written (Table 2). It could certainly be argued that these patients do not differ from thousands of nonepileptic individuals who enjoy writing. However, it is the driven nature of the writing, its excessive quantity, and the repetitive moralistic and philosophic content that we tend to associate with temporolimbic epilepsy.

In certain individuals, the onset of hypergraphia may be quite abrupt and out of character. For example, a case has been reported in which an athletic instructor who had never had intellectual interests became consumed with the thought that he had the mission to write something important after he developed temporolimbic epilepsy secondary to temporal lobe injury during clipping of an aneurysm.[39] We ourselves have seen a middle-aged semiliterate laborer who started to keep a minute-by-minute log of his feelings and their relationship to God immediately following the development of a right temporolimbic epileptic focus subsequent to the drainage of an abscess.[84]

The relationship of these interictal behavioral traits to temporolimbic epilepsy is likely to attract a great deal of additional research. At present, the literature seems to indicate that some of these traits may have a relatively specific association with temporolimbic epilepsy.[95,157] It is even possible that a certain cluster of such traits may constitute an interictal behavior syndrome of temporolimbic epilepsy.[167] The clustering of these traits in certain patients should increase the clinician's index of suspicion for the presence of underlying temporolimbic epilepsy.[40] However, it is also abundantly clear that the absence of these traits cannot rule out temporolimbic epilepsy and that there are many individuals with one or more of these traits who do not have any epilepsy at all. Perhaps future research will uncover a relationship between specific subtypes of temporolimbic seizures and these behavioral traits.

Potential Mechanisms for Long-Term Behavioral Changes

Since they can be shown to have a temporal relationship to seizure discharges, the mechanisms for the experiential phenomena, forced emotions, autonomic discharges, and even some dissociative fugue states are relatively easy to conceptualize. However, it is much more difficult to understand how episodic seizure discharges can give rise to long-term phenomena such as psychosis, manic-depressive illness, multiple personality, and alterations of behavioral traits.

The nature of the patient's psychologic response to an unpredictable illness has been cited as one possible causative factor in the development of these long-term phenomena. Although there can be no doubt that such factors play a role, the adoption of this hypothesis would result in the prediction, not substantiated by facts, that patients with other unpredictable diseases (such as multiple sclerosis or migraines) should also manifest similar behaviors. Anticonvulsant medications have also been implicated as being responsible for these behavior alterations. This is unlikely, since we have seen many patients in whom the characteristic behavioral changes clearly preceded the diagnosis and therefore the anticonvulsant therapy.

Another hypothesis—one of special interest to behavioral neurology—is that the intermittent temporolimbic discharges directly cause the long-term behavioral changes. One possibility is that the seizures and electrical instability of temporolimbic areas could lead to an enhancement of affect-laden sensory-limbic associations.[8] This sensory-limbic *hyperconnectivity*[8] could increase the emotional significance of objects, actions, and events. "As a result of the increasing investment of the environment with limbic significance, external stimuli begin to take on great importance; this leads in turn to increased concern with philosophical, religious and cosmic matters. Since all events become charged with importance, the patients frequently resort to recording them in written form at great length and in highly charged language."[39]

This is an attractive hypothesis, consistent with the role of the limbic system in channeling drive and affect toward external stimuli (see Chapter 1). Since the evidence in the next section shows that seizure activity is usually associated with the *malfunction* or *inhibition* of the relevant regions, perhaps an even more general phenomenon is one of sensory-limbic *misconnections*. In fact, a variety of mechanisms may be operating to produce a situation in which patients with temporolimbic epilepsy are subject to unpredictable perturbations in mapping the appropriate affective tone onto experience and thought.[84] In certain patients or under certain circumstances, this disturbance may lead to an enhancement of affective associations, perhaps through a malfunctioning of inhibitory pathways. In other instances, the mapping of affect onto experience may be insufficient, inappropriate, or distorted. Depending on individual differences, this could prepare the background for disturbances that range in severity from psychosis to the accentuation of certain behavioral traits. This mechanism is also compatible with the alternative possibility that the epilepsy and the behavioral changes are *not* causally related. Instead, each could represent an independent manifestation of a common underlying factor such as temporolimbic tissue damage or neurochemical alterations. Such changes could lead to *epilepsy* by interfering with local neuronal physiology and to the *behavioral changes* by disrupting limbic-sensory interactions.

PHYSIOLOGIC CONSIDERATIONS

Membrane/Cellular Phenomena

A paroxysmal depolarization shift (PDS) of the neuronal cell membrane with an accompanying brief high-frequency train of action potentials has been considered the hallmark of focal epilepsy in animal models.[28,106] When large numbers of neurons synchronously behave in this fashion, electrical fields generated by extracellular currents can be recorded in the form of EEG spikes by surface electrodes.[5,165] These cellular membrane events correlate temporally with the spike seen on surface EEG.[5]

The neurons involved in the generation of a PDS transmit high-frequency action potentials for the duration of the PDS.[163] These action potentials trigger recurrent intrinsic inhibitory systems.[107,132] One such system that has been studied extensively is the recurrent inhibitory circuit of the hippocampus involving basket cells as the inhibitory interneuron.[13,132] The inhibitory interneuron inhibits not only the cells that give rise to the seizure but also those in the surrounding regions.[108] This inhibition is felt to be gamma aminobutyric acid (GABA) mediated and leads to a transient cell membrane hyperpolarization lasting from 200 to 250 milliseconds.[12,69,108,132] This state of hyperpolarization temporally corresponds to the slow wave potential that is seen following an epileptic spike on the surface EEG. This inhibition encompasses an area larger than the area responsible for generating the PDS itself. The EEG slow-wave potential following an epileptic spike thus suggests a state of probable hyperpolarization and relatively decreased neuronal excitability in the area of generation of the PDS as well as in the immediately surrounding regions.

The neurons involved in the generation of a PDS also transmit their action potentials through normal neuronal connections to distal sites where they can exert a net excitatory or inhibitory influence. The most common net effect recorded at distant sites so far appears to be inhibitory.[20] Thus, the secondarily involved sites are more often than not left in a state of relative hypoexcitability. Cortical areas involved in epileptic activity have also been found to induce antidromically directed discharges along thalamocortical axons.[133] Thus, a focal epileptic phenomenon may induce neural dysfunction in many different parts of the brain, especially those that are interconnected with the focus. This may explain why the behavioral manifestations associated with seizures emanating from different but interconnected components of the limbic system can have so many similarities.

Kindling

The kindling phenomenon deserves separate consideration because of its possible relevance to the behavioral changes seen with temporolimbic epilepsy. The basic study design for demonstrating kindling involves the placement of a stimulating electrode into the brain, following which brief and usually infrequent bursts of electrical stimulation are given at an intensity that is initially too low to elicit electrical or clinical seizures. Over time, however, the animal will develop first electrical phenomena that outlast the stimulation itself and then full-blown, sustained convulsions.[49,50] Eventually, spontaneous seizures will also begin to occur. The limbic parts of the brain are most susceptible to this kind of kindling.[49,51] Animal experiments are beginning to provide insight into the behavioral implications of this phenomenon. For example, in a recent report of hippocampal kindling in rats, stimulation was given until the animals developed overt convulsive events, and then it was discontinued. In association with the onset of convulsions, these animals developed alterations in their response to handling, and to the social hierarchy within their groups. Although overt convulsions ceased completely 2 weeks after stimulation was terminated, the behavioral alterations did not change or disappear. Depth electrode recording from the hippocampus, however, revealed persistent paroxysmal discharges restricted to this portion of the temporolimbic network.[83] Thus, relatively few overt convulsive events can trigger long-term physiologic changes that can chronically alter behavior. Because of its high susceptibility to kindling, this sequence of events is likely to be encountered most frequently in temporolimbic areas of the brain.

It is also possible to achieve chemical kindling through the administration of several agents that have local anesthetic properties. For example, the systemic use of such drugs as lidocaine and cocaine in subconvulsive doses and on an intermittent basis eventually leads to limbic seizures.[105,166] This may provide at least one possible explanation for the long-term behavioral changes that can be seen after abuse of cocaine or similar drugs.

It is quite likely that the phenomenon of kindling provides an important mechanism for the establishment of mirror[90] and daughter foci.[78,91] The particular susceptibility of the limbic system to kindling suggests that this spread in the form of mirror and daughter foci is likely to occur preferentially in the direction of limbic areas. The gradual accentuation of behavioral difficulties in patients who may have had life-long temporolimbic epilepsy may reflect this type of spread. The preferential spread in the direction of limbic structures may also explain why some individuals with longstanding parietal or frontal lobe foci may eventually begin to develop behavioral manifestations of temporolimbic epilepsy.

Alteration of the Endocrine Milieu

As described in Chapter 1, the limbic region and, in particular, the amygdala and hippocampus have direct and reciprocal connections with hypothalamic nuclei. Stimulation and ablation of specific amygdaloid nuclei reproducibly alter luteinizing hormone (LH), follicle stimulating hormone (FSH), thyroid stimulating hormone (TSH), adrenocorticotropic hormone (ACTH), growth hormone (GH), and prolactin levels. It is also known that there are relatively high numbers of gonadal hormone receptors on limbic neurons so that changes in hormone levels could directly influence their activity and convulsive thresholds.[101,110,116,121,150] One well-known example of this is catamenial epilepsy, the premenstrual increase of clinical events in certain women who have temporolimbic seizures.[74]

Polycystic ovarian disease, recurrent endometriosis, menstrual irregularity, infertility, and impotence have been described in patients who have temporolimbic epileptiform abnormalities on the EEG.[59-61] It is conceivable that in some of these patients the epileptiform abnormality could have preceded and perhaps caused the onset of the associated endocrinologic disturbance. Another possibility is that the altered endocrinologic state could have activated epileptiform disturbances in vulnerable individuals. There are several reports of patients with abnormal EEGs and polycystic ovarian disease, amenorrhea, and hyposexuality who have remained refractory to classic endocrinologic therapy but have responded to anticonvulsants.[135] There are also some patients who experience better seizure control in response to hormonal therapy.[16,73] The interactions between temporolimbic seizures and the endocrine system are likely to attract a great deal of attention. Advances in this area have immediate therapeutic implications. At the present, atypical endocrine abnormalities that remain refractory to treatment should raise the possibility of an underlying temporolimbic seizure disorder. Furthermore, in patients with temporolimbic epilepsy who also have hormonal abnormalities, it may be useful to consider the potential role of hormonally active agents for seizure control.

Perhaps the most impressive evidence favoring the intimate relationship between temporolimbic epilepsy and hormonal regulation is the increasing number of reports demonstrating subtle but very frequent abnormalities of neuroendocrine control in this population of patients. In a recent study and subsequent followup of untreated patients with temporolimbic epilepsy, the vast majority of female patients had an abnormal hormonal response to challenges with luteinizing hormone releasing hormone (LHRH) and chlorpromazine.[59,61] Also, frequent sampling of the various trophic hormones has suggested changes in prolactin, FSH, and LH following partial seizures.[21] Chronic hormonal changes may provide one possible mechanism for the occurrence of anorexia in some patients with epilepsy.

In addition to their well-known endocrine effects on peripheral target organs, many hypothalamic and pituitary hormones also have powerful effects on mood and behavior. The chronic influence of temporolimbic epilepsy on the endocrine milieu may therefore provide an alternative mechanism for some of the long-term behavioral alterations in these patients.[138]

DIAGNOSIS

The evaluation for a potential temporolimbic seizure disorder can occur in one of several settings. In certain cases, the history and clinical manifestations are entirely typical. For example, the patient may have had motor automatisms, olfactory hallucinations, déjà vu, jamais vu, mind-body dissociations, forced thinking or a sense of impending doom—all of which are widely accepted to be convulsive in origin.[18] There are patients, however, who will present predominantly with behavioral alterations or symptoms that may be more difficult to attribute to the seizure disorder. The eventual correlation of these symptoms with abnormal temporolimbic discharges may have to rely more heavily on specialized EEG procedures.

History and Clinical Examination

Historic information that is consistent with prior brain damage is of particular relevance to the diagnosis. Such information includes gestational distress, birth complications, febrile convulsions, intracranial infections, head trauma, delayed milestones, hyperactivity, learning disabilities, and left-handedness.

A personal history of endocrine disturbances such as those described above should increase the index of suspicion. Our initial impression is that a family history of left-handedness, autoimmune disease, and psychiatric disorders also seems to cluster in this population. This may reflect a genetic loading that increases the possibility of developing a temporolimbic seizure disorder.

A detailed account of the patient's actual experiences is helpful. This requires a careful and open-minded inquiry by a clinician who is aware of the vast spectrum of ictal and interictal behavioral alterations accompanying a temporolimbic seizure disorder. Events that occur out of context to environmental situations are more likely to have an epileptic origin, especially if they cluster in relationship to the menstrual phases or to the sleep-wake cycle. Seizures may also be specific to certain activities, such as watching television or movies where subtle photosensitive effects may be operating.[3,14,45] Specific patterns or colors may act as triggers for certain patients and seizures have also been reported that were induced by reading or listening to certain musical passages.[38,41,94,111] It is, of course, important to realize that many "normal" people may experience episodes of déjà vu, jamais vu, or even depersonalization in the course of their lives. It is the abnormal frequency of such events that leads to the suspicion of epilepsy.

On clinical examination, it is not unusual to find skeletal growth anomalies and pigmentation disturbances such as cutaneous nevi. The finding of clear reflex abnormalities, focal neurologic deficits, or an asymmetry in the growth of one side of the body would all suggest the possibility of a structural lesion in the central nervous system.

EEG Studies

The EEG is the only definitive diagnostic criterion for temporolimbic epilepsy. Routine scalp EEG must often be extended to include recordings from the inferior temporal or anterior mesial temporal regions, in order to reveal seizure activity in patients with temporolimbic epilepsy. To that end, a variety of placements have been used, including sphenoidal,[128,129] zygomatic,[141] modified anterior temporal,[140] and nasopharyngeal[22] electrodes. In our experience, the most useful of these are the modified anterior temporal and sphenoidal electrodes. In addition, the state of sleep—either spontaneous or following a night of deprivation—seems to enhance the probability of detecting abnormalities on the EEG.[82] There are, of course, many areas of the brain besides the anterior-medial temporal regions that are not well seen with standard scalp EEG electrode placements. These areas include the orbitofrontal regions, the interhemispheric (cingulate) regions, the posterior inferotemporal and parahippocampal regions, and the cortices lying within the sylvian fissures, especially the insula. Seizure activity in these "hidden" areas could become the origin of discharges in temporolimbic epilepsy. Unfortunately, abnormalities in these regions are likely to be missed not only by the standard scalp electrode placements but even when the more specialized electrodes are used. Furthermore, since epileptic discharges are transient, the period of recording may fail to overlap with periods of this abnormal activity. Consequently, a negative tracing does not rule out a temporolimbic seizure disorder. Repeated EEG studies, monitoring for prolonged periods, and recording while the patient is behaviorally symptomatic will increase the yield of detection (Fig. 1).

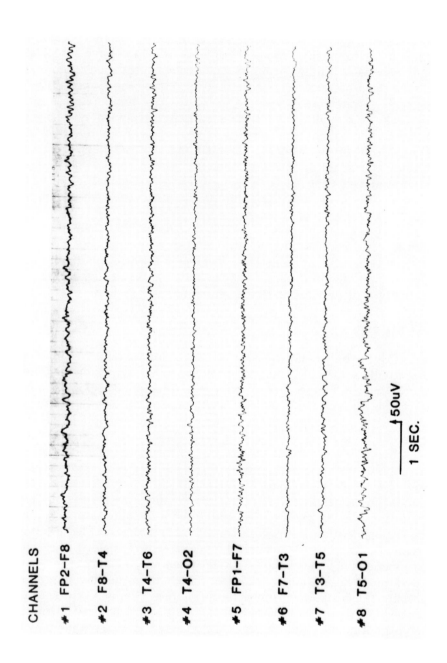

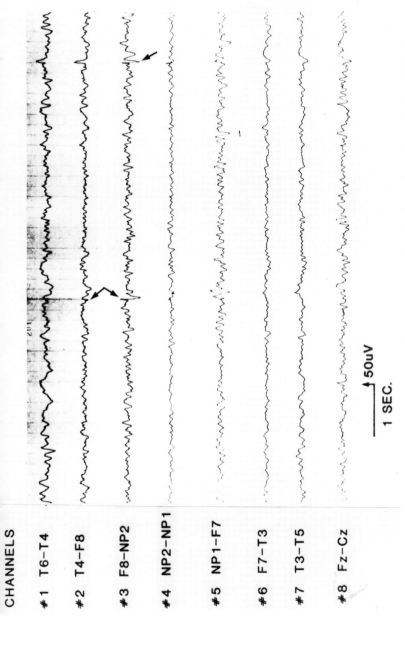

CHANNELS

#1 T6–T4

#2 T4–F8

#3 F8–NP2

#4 NP2–NP1

#5 NP1–F7

#6 F7–T3

#7 T3–T5

#8 Fz–Cz

50uV

1 SEC.

FIGURE 1. This 29-year-old female had a long history of episodic abdominal pain associated with a feeling of fear. Multiple EEGs done over many years, including some with nasopharyngeal electrodes, were reported to be normal. This tracing was the first one done while she was symptomatic. The top tracing shows the surface scalp recording while the patient was drowsy, and is normal. The bottom tracing shows recordings from nasopharyngeal electrodes (NP1 and NP2) as well as from the scalp. Localized spike and spike–slow-wave complexes from the right temporal region, maximum at the right nasopharyngeal electrode, are seen (*arrows*). Thus, obtaining a tracing during the symptomatic phase and the use of specialized electrodes both increase the yield of detection.

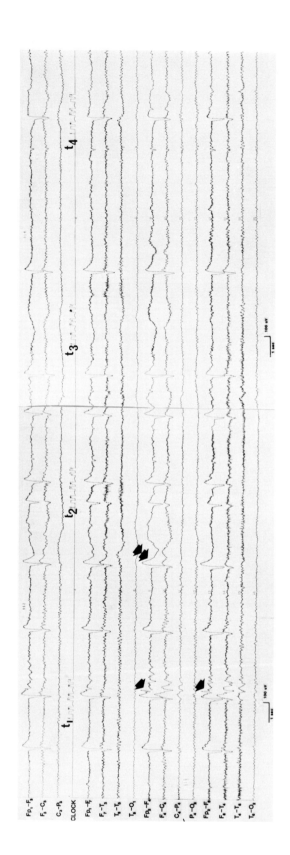

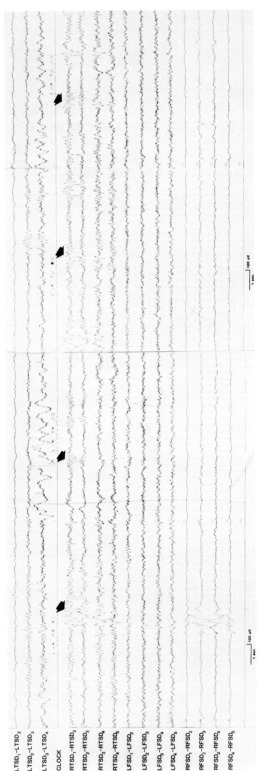

FIGURE 2. This EEG is from a 20-year-old man with intractable right frontotemporal seizures. This is a simultaneous record from standard 10 to 20 placement of scalp electrodes (*top*) and subdural (SD) strip electrodes (*bottom*). The strip electrodes are on the surface of the brain in the subdural space. The LTSD and RTSD electrodes are over the lateral-inferior portions of the left and right temporal lobes, respectively. Of these, electrode numbers 1 and 2 are most inferior, whereas 3 to 5 are more lateral. Electrodes LFSD and RFSD are over the left and right superior frontal lobes. Frontal electrode numbers 1 and 2 are most anterior, whereas 4 and 5 are more posterior, near the motor-sensory cortex. At the t_1 epoch there is an epileptic discharge seen from RFSD and RTSD strips that is also seen from F_{p2}, F_8, T_4, and F_4 on the scalp (*single arrowheads*). However, the patient was symptomatic with abdominal pain and a sense of depression on a more continuous basis. It is therefore interesting to note that there are frequent bursts of epileptic activity at $RTSD_1$, $RTSD_2$, and $RTSD_3$ at t_1, t_2, t_3, and t_4, even though no surface-scalp equivalent is noted. Thus, a behavior pattern that is "interictal" with reference to scalp electrodes may, in fact, be correlated with ongoing seizure activity. The large, bilaterally synchronous waveforms (*double arrowheads*) are eye-movement artifacts.

TEMPOROLIMBIC EPILEPSY AND BEHAVIOR

311

Even in patients who are not surgical candidates, subdural strip electrodes may be considered in order to facilitate the detection of seizures that originate in areas that are hard to sample by extracranial electrodes. It is also important to keep in mind that epileptiform activity can be present in approximately 0.5 to 2 percent of "normal" persons studied with routine or sleep-deprived EEG.[33] Therefore, the discovery of an abnormal paroxysmal EEG after repeated recordings in patients with certain behavioral disturbances need not indicate that there is an epileptic basis to this aspect of their clinical presentation. In contrast, the finding of a clear temporal relationship between a specific behavioral alteration and an epileptiform EEG abnormality is a powerful indication that the two are causally related. The demonstration of such a relationship, however, usually necessitates prolonged telemetric recordings during which the patient is asked to identify target events for subsequent correlation with the EEG tracing.

In addition to focal spike activity, there are a number of electroencephalographic findings that carry considerable significance for the diagnosis of a temporolimbic seizure disorder. The presence of a focal slow-wave abnormality, either continuous or paroxysmal, suggests the possibility of a focal structural lesion or a convulsive equivalent in which the spike discharge itself is not being recorded.[52] Studies using surgically implanted electrodes have shown that often there may be isolated or even sustained epileptic activity occurring in deep regions either without any surface reflection or with slow-wave activity as the only surface correlate (Fig. 2).

Radiologic Investigation

Computerized tomographic x-ray scans (or CT scans) have generally replaced pneumoencephalograms and angiograms for the routine screening of lesions associated with epilepsy. Tumors, hamartomas, cysts, and even arteriovenous malformations can often be detected with CT scans, particularly when intravenous contrast is given. Any patient with seizure onset at age 20 to 40 years should be suspected of having a tumor as a cause and deserves a CT scan with intravenous contrast. No more than 20 percent of such patients will eventually be found to have a tumor. Sometimes slow-growing tumors have been found in patients who have had a seizure disorder for up to 20 years.

Recent developments in imaging techniques may be particularly useful in the diagnosis of temporolimbic epilepsy. These techniques include positron emission computed tomography (PECT), single photon emission computerized tomography (SPECT), and nuclear magnetic resonance imaging (MRI) (see Chapter 10 for a description of these methods).

Generally, areas involved in epileptic discharges tend to show decreased metabolic activity or, less often, normal metabolic activity in the interictal and postictal states. Repeat PECT scans during periods of epileptic activity may show a metabolic activity higher than the adjacent "normal" brain tissue or an increase relative to the interictal state (see Fig. 3, Chapter 10). The area of abnormal metabolic activity on the PECT scan tends to be larger than the areas of epileptic discharge mapped out by EEG techniques and may represent the spreading effects of abnormal epileptic activity. It should be emphasized that the PECT scan reflects the averaged activity over many minutes and that this period may contain interictal, ictal, and postical states, making it difficult to determine the spatial and metabolic correlates of different phases in the epileptic process.[32,48,154]

SPECT scanning with I-123 iodoamphetamine has also provided useful information. The tracer binds to blood vessels in the first 2 to 3 minutes after intravenous injection and allows a three-dimensional measurements of regional cerebral physiologic activity. The short time necessary for reaching steady state permits the relatively specific investigation of the pre-ictal, ictal, and postical states. The actual scanning can be carried out as late as 4 to 6 hours after injection of the tracer, when the patient may be more tractable to the procedure. Furthermore, the equipment is presently much less elaborate and the isotopes have a relatively long half-life so that this method does not necessitate an on-site dedicated cyclotron.

SPECT scans obtained during seizure discharges revealed increased activity in areas that correspond to the electrographic focus. The epileptic focus shows relatively decreased activity on SPECT if obtained while the EEG is quiescent, especially in the immediately postical period. In some cases, the areas demonstrating alterations in metabolic activity are larger than those outlined by the EEG abnormality, probably representing the spread of electrical inhibition.[76,77]

Recent observations with the MRI scanner indicate that lesions undetectable by x-ray CT may be visualized by this technique. For example, one patient with unilateral medial temporal epileptic discharges on EEG had increased tissue density in the uncal hippocampal region of the same side, even though no abnormality was detected by conventional x-ray CT.

MANAGEMENT

Pharmacologic Approach

Once a diagnosis of temporolimbic epilepsy has been established, the primary approach to treatment is medical. The most widely accepted drug for the treatment of temporolimbic epilepsy is carbamazepine; the next is phenytoin, and the next is one of the barbiturates, which include phenobarbital, primidone, and mephobarbital. We also find that valproate sodium, the benzodiazepines, and carbonic anhydrase inhibitors can be useful in managing temporolimbic epilepsy. In addition, most epileptologists prefer single- to combination-drug therapy.[17] These basic rules in the management of focal partial epilepsy should not be taken to reflect the relative effectiveness of these anticonvulsants, because they appear to be relatively equal in this regard, depending on the individual patient's response.[17] Rather, the order of preference listed here has evolved in response to the issue of compliance and the side effects of these different medications.[24,25] A very poorly tolerated side effect is sedation. Since this correlates directly with the likelihood of patient compliance, carbamazepine and phenytoin, which are least sedative, are the preferred drugs. Carbamazepine can, however, lead to a lowering of the white blood count, and phenytoin may cause unacceptable gingival hyperplasia, especially in children. These side effects are most commonly reversible upon withdrawal of the drug.

Since most of these drugs are metabolized by the liver, the more drugs used, the greater the likelihood for drug interactions, particularly at the level of hepatic microsomal enzyme induction. Unfortunately, while the sedative properties of the various anticonvulsants are often cumulative, their combined effectiveness may prove less than additive. Consequently, when combination-drug therapies are used, sedation and other side effects such as memory or concentration deficits may increase disproportionately to the improvement in overall control. Although most reviews suggest that the majority of patients with temporolimbic epilepsy can be successfully managed with medications, the definition of "control" is not uniform. In most clinical trials, "control" is equated with a statistically significant reduction in overt seizure events.[17,104,113] "Control" as viewed by the patient, however, usually means the prevention of all clinical events, including those that are covert and perhaps predominantly experiential. This disparity in the perception of the term "control" may explain why some physicians have estimated that up to 80 percent of patients with partial seizures can be "controlled," whereas patient surveys would show that significantly fewer feel that they are "controlled." The actual statistics on drug efficacy and "control" may therefore be much less encouraging than is commonly realized.

If kindling does occur in humans, then suppression of the major manifestations of a convulsive disorder without suppression of the isolated focal epileptic electrical events might not be sufficient. Several studies have suggested that unless the EEG disturbances also resolve with treatment, patients are more likely to be refractory at a later date.[75] Consequently, the issue of human kindling needs to be addressed systematically. If its occurrence and the parameters controlling its evolution are identified, then drugs will need to be studied that can specifically prevent or delay the development of this process.

There are many gaps in our understanding of anticonvulsant drug actions. The barbiturates probably act by enhancing GABA-mediated inhibition.[109] Therefore, barbiturates are most likely to prevent the progression and spread of isolated, spontaneous epileptic discharges but will not suppress their continued occurrence at the focus. Phenytoin, on the other hand, is probably effective at the calcium ion channel level of the neuronal membrane and therefore probably inhibits the generation of activity at the focus.[171] The mechanism of action of carbamazepine, is even less well defined.[65] Second-line anticonvulsants for temporolimbic epilepsy, such as valproic acid[64] and the benzodiazepines,[29] probably achieve their effectiveness in a manner similar to the barbiturates by enhancing GABA-mediated inhibition. Although this information is not yet sufficient to dictate a more rational pharmacologic approach to individual patients, it may explain some of the possible drug interactions and side effects. For example, toxic levels of phenytoin may actually increase epileptogenic activity by increasing the calcium-dependent release of neurotransmitters.

It would appear desirable to undertake a systematic investigation of the pharmacologic approach to temporolimbic epilepsy. For example, in patients with primary generalized epilepsy with absence, long-term EEG monitoring clearly demonstrated that frequent subclinical epileptic activity altered the overall functional and intellectual capacity of affected patients. When these events were suppressed pharmacologically, the patients improved functionally.[15a,87a] Similar studies on temporolimbic epilepsy are sorely needed.

Surgical Treatment

When an adequate trial of medical treatment has failed to control a patient's epilepsy, surgical treatment should be considered. Although no firm time schedule can be imposed on an individual patient, drug trials for most major anticonvulsants with careful monitoring of serum levels can usually be accomplished over a period of 2 years. Medical control is considered inadequate when the clinical condition, or the drug therapy required for its control, interferes with the normal lifestyle of the patient. This determination must be based on careful consideration of a variety of medical, social, and psychologic factors. For example, a patient with daily brief staring episodes may consider the condition acceptable. Another patient with automatisms occurring only once every 2 weeks, on the other hand, may consider the condition unacceptable if these automatisms dramatically interfere with the patient's career. The handicap imposed by epilepsy is especially severe during the adolescent years, which are so important to social development and learning. Inadequately controlled epilepsy in this period may greatly impede a return to normal lifestyle in the future. This may favor a consideration for early surgical intervention.

The objective of surgical treatment is to remove the seizure focus. Therefore, only patients with a clearly focal onset of seizures are surgical candidates. Prior to any surgical treatment of epilepsy, a thorough evaluation must be carried out by a trained team, including a neuroradiologist, neuropsychologist, electroencephalographer, neurologist, and neurosurgeon. Although the basic format of this evaluation is similar to the investigation of any patient with temporolimbic epilepsy, a special emphasis is placed on the localization of the epileptic focus. Centers with programs in epilepsy surgery vary in their approach to the localization of epileptic areas, but most of these centers rely on scalp, sphenoidal, and other superficially placed EEG electrodes for preoperative assessment. These EEG recordings are often carried out with telemetry and video techniques in order to increase the probability of capturing epileptic events, which are correlated with the patient's behavioral alterations. If the origin of focal epileptic discharges cannot be specifically determined using these techniques, some form of implanted electrode is usually considered.

Subdural strip electrodes with multiple contact points can be placed through burr holes. They are particularly useful in recording over the convexities as well as in subfrontal and subtemporal areas. Their broad contact points give them a wide electrical field of sampling; this feature is often helpful in obtaining artifact-free recordings of relatively large areas, such as the frontal lobes, where discharges may spread so rapidly to the contralateral side that the laterality of the onset is difficult to determine from scalp recordings.

Intracerebral depth electrodes are placed with stereotaxic techniques through small holes in the skull. There is considerable variation in the methods of placement, types of electrodes, and ways in which recordings are utilized. Most centers now use flexible wire electrodes that are inserted laterally or through the brain convexities with five to nine small contact points. Intracerebral depth electrodes are particularly useful in epilepsy originating in the amygdala and hippocampus. Precision low-artifact records of these structures can often determine the laterality of onset in a unilateral medial temporolimbic focus that has a rapid contralateral spread of discharges. Depth electrodes may also be used for stimulation of contact points with small currents in the milliampere range. This may reproduce clinical epileptic events or cause electrical discharges that further assist in localizing the epileptogenic brain areas.

In order to guard against signficiant deficits of speech and memory function, most centers carry out preoperative intracarotid injections of amobarbital to each hemisphere (Wada test). During a period of a few minutes following the amobarbital injection, the ipsilateral hemisphere's function is markedly depressed. Testing of speech and memory during this interval reveals the hemispheric dominance pattern for these functions.[15,162] Since details of intracranial vasculature vary, the anatomic distribution of physiologic suppression by the amobarbital can be monitored using sequential angiography and simultaneous EEG recordings. There are two major uses of the Wada test. First, if the side to be operated on is dominant for language, then special intraoperative precautions are necessary in order to avoid a postsurgical aphasia. Second, if barbiturization of one side results in a profound amnesia (as could happen in patients who also have pathology in the contralateral medial temporal areas), then the traditional temporal lobectomy on that side is contraindicated.

During the actual surgery for cortical excision, most centers expose a wide area of cortex under local anesthesia with intravenous neuroleptic medication. This keeps the patient sedated but conscious in a way that allows intraoperative behavioral testing. Direct EEG recording of cortical activity from lateral brain surfaces is recorded with a special electrocorticography frame that attaches to the skull. EEG activity from the undersurfaces of the frontal and temporal lobes can be recorded with flexible wire electrodes, and the activity of the medial temporal structures including the amygdala and hippocampus can be recorded with depth electrodes. Surface and depth stimulation can be carried out at this time to map out the primary sensory and motor cortex and speech areas. Stimulation may also reproduce behavioral alterations associated with the patient's seizures, and these can then be correlated directly with EEG abnormalities.

Although a preliminary decision regarding the area of cortical removal is made in advance of the surgery, the final determination is dependent on pre- and post-excision electrocorticography. An anterior temporal removal in the dominant hemisphere can usually extend back to a point slightly behind the central sulcus without interfering with speech, and a dominant frontal removal must spare several centimeters around Broca's area. Some centers carry out more generous temporal removals in the dominant hemisphere when the cortical stimulation indicates a more posterior and/or superior location for speech function. The supplementary motor area in the dominant hemisphere can be removed, when necessary, with a fairly high probability of a transient speech deficit lasting from several weeks up to several months. Nondominant temporal excisions that extend back more than 7 to 8 cm result in a contralateral upper quadrantanopsia. This visual field deficit is usually not functionally significant and is often an acceptable loss if necessary for successful removal of an active epileptogenic area.

Residual areas of epileptic discharge, if small and relatively inactive, often disappear progressively during the early postoperative months or years if a major portion of the epileptogenic brain tissue is removed. This is frequently seen with suprasylvian discharges following removal of an epileptogenic temporal lobe.[115] Early experience also showed that insular cortex discharges do not continue to cause epilepsy if epileptogenic tissue below the sylvian fissure is removed.[139] There is general agreement that the amygdala and anterior hippocampus are often involved in the generation of epileptic discharges and should be removed; however, there is controversy over the involvement and removal of the more posterior hippocampal areas. It had been thought that the posterior hippocampus could be preserved, even when demonstrating active discharges, so long as the lateral temporal areas with epileptic discharges have been removed. However, some centers now believe that the amygdala and entire hippocampus back to the level of the trigone of the lateral ventricle should always be removed in patients with clinical features of "temporal lobe epilepsy," regardless of the EEG activity.[143,170] These centers believe that the epileptic discharges regularly involve these posterior hippocampal areas and that their removal with the preservation of more lateral cortical areas (even those that have active discharges) will result in a surgical reduction of seizures.

The decision to carry out a cortical excision of epileptogenic brain tissue is made only when the area involved can be removed without major functional deficit. Temporal and frontal removals can be performed in people with superior intelligence or professional employment without producing major deficits that interfere with their normal functions.[4,87] This obviously does not mean that these regions of the brain normally have no useful function. In epileptics, the areas of excised brain often show signs of chronic structural damage. These patients usually have some preoperative evidence of cognitive compromise that corresponds to this area of damage. In some cases, the patient may have already compensated for this impairment by selecting an occupation that capitalizes on their intact abilities. In others, the function of the damaged epileptic area may have shifted to adjacent or contralateral brain regions at an early age. In either instance, the ongoing epileptic activity further compromises brain function, not only at the focus but also in other distant areas. Therefore, excision of the epileptogenic focus in these patients frequently improves cognitive performance.[4] Although this has sometimes been attributed to nonspecific increases of attention and concentration, recent studies have suggested that the improvement may, in fact, be specific to those cognitive functions mediated by the homologous region contralateral to the seizure focus.[96] The mechanism responsible for this improvement may be that these regions are released from the distal inhibitory effects of the primary seizure focus.

From the perspective of seizure control, the results of cortical excision are fairly uniform for different centers. By selecting patients with epileptic areas confined to the anterior and/or medial temporal lobe, many patients will become seizure free. More widespread areas of epileptogenic tissue and suprasylvian areas of epilepsy are more difficult to treat surgically because of the limited areas that can be removed without functional deficit. Fortunately, many of the patients with focal epilepsy involving partial complex seizures have mainly anterior temporal foci. Overall, about two thirds of patients with cortical excisions show improvement in seizure frequency, with about one third becoming seizure free.[19,114,160] Some of the latter are slowly withdrawn from all medication after several years. One third have only slight or no reduction of seizure frequency. Even some of these patients, however, have an alteration in seizure pattern so that their seizures are less severe, less disruptive of their lifestyle, and more amenable to pharmacologic management.

Only a few hundred operations for temporolimbic epilepsy are available annually in North America. However, the number of potential candidates for this type of surgery vastly exceed this number. The factors contributing to the small numbers of surgical procedures for temporolimbic epilepsy have been broadly grouped into three categories.[164] The first is related to misconceptions regarding the degree of control that can be achieved with anticonvulsants. The second is related to the expense, time, and expertise required to evaluate and treat patients in this manner. Finally, the public at large, medical professionals, and even neurologists appear to have major misgivings, many of which are difficult to justify when the epilepsy is medically uncontrollable, when its focus can be identified, and when its impact on the patient's life is considerable.

Psychiatric Aspects of Management

As described earlier in this chapter, patients with temporolimbic epilepsy commonly report a number of marked behavioral and emotional symptoms. These range in severity from relatively mild characterologic problems to disabling psychotic conditions, manic-depressive illness, and dissociative states. Some of the psychiatric manifestations, especially those that can be considered ictal in nature (e.g., panic attacks, complex hallucinations, transient depersonalization), may respond to anticonvulsant agents without any need for further treatment. There is less information on the response of the long-term (interictal) behavioral traits to treatment. We have made some preliminary observations on patients who have shown distinct improvements in obsessionalism, circumstantiality, and hyposexuality in response to anticonvulsant agents. Sometimes even chronic and disabling manic depressive symptomatology has responded to treatment with phenytoin. Since the additional manifestations of epilepsy in these patients can be quite minor, it is unlikely that the alleviation of their depression is merely reactive to the improved seizure control. When such patients are treated with carbamazepine, the improvement of the depression may be more difficult to interpret, since carbamazepine has a direct antidepressant effect itself, probably because of its structural similarity to tricyclic agents.

In other patients, however, disabling characterologic, affective, and psychotic disturbances remain quite refractory to antiepileptic treatment. The customary modalities of psychiatric treatment including individual psychotherapy, group therapy, and pharmacotherapy are indicated for these patients. In fact, the vast majority of patients with temporolimbic epilepsy would greatly benefit from coordinated and simultaneous neurologic and psychiatric management. Since many neuroleptic and antidepressant drugs, as well as lithium salts, can depress seizure thresholds, these agents need to be used judiciously.

Clinicians who manage patients with temporolimbic epilepsy may find themselves unwitting participants in the mind-body controversy. Patients may be quite incredulous when told that their profound anxiety or pervasive sense of premonition is triggered by an abnormal electrical discharge from the temporal lobe. It may be necessary to explain this to the patient, since such symptomatology is usually refractory to psychologic insight and since much self-blame and pointless search for underlying antecedents can be avoided by explaining that the individual has little control over the occurrence of these discharges. "Is this the seizure or is it me?" is a most challenging question that often faces the clinician managing patients with temporolimbic epilepsy. A group therapy setting in which common experiences are shared is sometimes quite effective for dealing with these issues, especially since it dispels the sense of uniqueness associated with some of the experiences.

Psychiatric patients with an atypical psychosis, affective illness, or dissociative condition are sometimes referred to the neurologist with the suspicion of an underlying seizure disorder, even though there are no other clinical manifestations of epilepsy. In many of these patients, no evidence for seizures is uncovered despite repeated EEGs. However, there is an increasing number of such patients in whom one of the EEGs, especially when obtained with the nasopharyngeal or sphenoidal leads, may show temporolimbic paroxysmal discharges in the form of spikes or slow waves. Some of these patients improve when antiepileptics are added to the regimen; others show no such response. In patients who do not respond to the addition of one or more antiepileptics, it is not clear if these drugs should be withdrawn entirely or if they should be maintained in order to prevent the theoretical possibility of kindling. Since these patients do not have overt clinical manifestations of seizures and since their EEGs are usually minimally abnormal, a great deal of caution should be exercised before labeling them as epileptic. Perhaps the descriptive term of "temporolimbic dysrhythmia" may be more appropriate until more is understood about the natural course of these conditions.

REFERENCES

1. ABRAMS, R, AND TAYLOR, MA: *Differential EEG patterns in affective disorder and schizophrenia.* Arch Gen Psychiatry 36:1355, 1979.

2. ADEBIMPE, VR: *Complex partial seizures simulating schizophrenia.* J Am Med Assoc 237:1339, 1977.

3. ALLAN, TM, AND STEWART, RS: *Photosensitive epilepsy and driving.* Lancet 1:1125, 1971.

4. AUGUSTINE, AA, NOVELLY, R, MATTSON, R, GLASER, GH, WILLIAMSON, PD, SPENSER, D, AND SPENCER, SS: *Occupational adjustment following neurosurgical treatment of epilepsy.* Ann Neurol 15:68, 1984.

5. AYALA, GF, DICHTER, M, GUMNIT, RJ, MATUMOTO, H, AND SPENSER, WA: *Genesis of epileptic interictal spikes: New knowledge of cortical feedback systems suggests a neurophysiologic explanation of brief paroxysms.* Brain Res 52:1, 1973.

6. BALLENGER, CS, KING, DW, AND GALLAGHER, B: *Partial complex status epilepticus.* Neurology 33:1545, 1983.

7. BARATZ, R, AND MESULAM, M-M: *Adult onset stuttering treated with anticonvulsants.* Arch Neurol 38:132, 1981.

8. BEAR, DM: *Temporal lobe epilepsy: A syndrome of sensory-limbic hyperconnection.* Cortex 15:357, 1979.

9. BEAR, D: *Hemispheric specialization and the neurology of emotion.* Arch Neurol 40:195, 1983.

10. BEAR, DM, AND FEDIO, P: *Quantitative analysis of interictal behavior in temporal lobe epilepsy.* Arch Neurol 34:454, 1977.

11. BEAR, D, LEVIN, K, BLUMER, D, CHATMAN, D, AND REIDER, J: *Interictal behavior in hospitalized temporal lobe epileptics: Relationship to other psychiatric syndromes.* J Neurol Neurosurg Psychiatry 45:481, 1982.

12. BEN-ARI, Y, AND KRNJEVIC, K: *Actions of GABA on hippocampal neurons with special reference to the etiology of epilepsy.* In MORSELLI, P, LLOYD, K, LOSCHER, W, MELDRUM, B, AND REYNOLDS, E (EDS): *Neurotransmitters Seizures and Epilepsy.* Raven Press, New York, 1981.

13. BEN-ARI, Y, KRNJEVIC, K, AND REINHARDT, W: *Hippocampal seizures and failure of inhibition.* Can J Physiol Pharmacol 57:1462, 1979.

14. BINNIE, CD, DARBY, CE, AND HINDLEY, AT: *Electroencephalographic changes in epileptics while viewing television.* Br Med J 4:378, 1973.

15. BRANCH, C, MILNER, B, AND RASMUSSEN, T: *Intracarotid sodium amytal for the lateralization of cerebral speech dominance.* J Neurosurg 21:399, 1964.

15a. BROWNE, TR, PENRY, JK, PORTER, RJ, AND DREIFUSS, FE: *A comparison of clinical estimates of absence seizure frequency with estimates based on prolonged telemetred EEGs.* Neurology 24:381, 1974.

16. CHECK, JH, LUBLIN, ED, AND MANDEL, MD: *Clomiphene as an anticonvulsant drug: A case report.* Arch Neurol 39:784, 1982.

17. COATSWORTH, JJ: *Studies on the clinical efficacy of marketed antiepileptic drugs.* NINCDS Monograph No. 12, US Government Printing Office, Washington, DC, 1971.

18. COMMISSION ON CLASSIFICATION AND TERMINOLOGY OF THE INTERNATIONAL LEAGUE AGAINST EPILEPSY: *Proposal for revised clinical and electroencephalographic classification of epileptic seizures.* Epilepsia 22:489, 1981.

19. CRANDALL, P: *Postoperative management and criteria for evaluation.* In PURPURA, D, PENRY, L, AND WALTER, R (EDS): *Neurosurgical Management of the Epilepsies.* Adv Neurol 8:265, 1975.

20. CROWELL, RM: *Distant effects of a focal epileptogenic process.* Brain Res 18:137, 1970.

21. DANA-HAERI, J, TRIMBLE, MR, AND OXLEY, J: *Prolactin and gonadotrophin changes following generalized and partial seizures.* J Neurol Neurosurg Psychiatry 46:331, 1983.

22. DEJESUS, PC, AND MASLAND, WS: *The role of nasopharyngeal electrodes in clinical electroencephalography.* Neurology 20:869, 1970.

23. DELGADO-ESCUETA, AV, MATTSON, RH, KING, L, GOLDENSOHN, ES, SPIEGEL, H, MADSEN, J, CRANDALL, P, DREIFUSS, F, AND PORTER, RJ: *The nature of aggression during epileptic seizures.* N Engl J Med 305:711, 1981.

24. DELGADO-ESCUETA, AV, TREIMAN, DM, AND WALSH, GO: *The treatable epilepsies.* N Engl J Med 308: 1508, 1983.

25. DELGADO-ESCUETA, AV, TRIEMAN, DM, AND WALSH, GO: *The treatable epilepsies.* N Engl J Med 308:1576, 1983.

26. DEVINSKY, O, HAFLER, DA, AND VICTOR, J: *Embarrassment as the aura of a complex partial seizure.* Neurology 32:1284, 1982.

27. DEVINSKY, O, PRICE, BH, AND COHEN, S: *Cardiac manifestations of complex partial seizures.* Am J Med (in press).

28. DICHTER, M, AND SPENSER, WA: *Penicillin-induced interictal discharges from the cat hypocampus. I. Characteristics and topographical features. II. Mechanisms underlying origin and instruction.* J Neurophysiol 32:649, 1969.

29. DREYFUS, FE, AND SATO, S: *Clonazepam.* 2 ed. In WOODBURY, DM, PENRY, JK, AND PEPPENGER, CE (EDS): *Antiepileptic Drugs.* Raven Press, New York, 1982.

30. DRAKE, ME, AND COFFEY, E: *Complex partial status epilepticus simulating psychogenic unresponsiveness.* Am J Psychiatry 140:800, 1983.

31. DRAKE, ME: *Isolated ictal autonomic symptoms in complex partial seizures.* (Letters.) Ann Neurol 14:100, 1983.

32. ENGEL, J, KUHL, D, PHELPS, M, AND CRANDALL, P: *Comparative localization of epileptic foci in partial epilepsy by PET and EEG.* Ann Neurol 12:529, 1982.

33. ENGEL, M, LEUDERS, N, AND CHUTORIAN, A: *The significance of focal spikes in epileptic and nonepileptic children who are otherwise normal.* Ann Neurol 2:257, 1977.

34. ESCUETA, AV, BUXLEY, J, AND STUBBS, N: *Prolonged twilight state and automatisms: A case report.* Neurology 24:331, 1974.

35. FALCONER, MA: *Reversibility by temporal-lobe resection of the behavioral abnormalities of temporal-lobe epilepsy.* N Engl J Med 289:451, 1973.

36. FLOR-HENRY, P: *Schizophrenic-like reactions and affective psychoses associated with temporal lobe epilepsy. Etiological factors.* Am J Psychiatry 126:400, 1969.

37. FLOR-HENRY, P: *Lateralized temporal-limbic dysfunction and psychopathology.* Ann NY Acad Sci 280:777, 1976.

38. FORSTER, FM: *Reflex Epilepsy. Behavioral Therapy and Conditioned Reflexes.* Charles C Thomas, Springfield, IL, 1977.

39. GESCHWIND, N: *Pathogenesis of behavior change in temporal lobe epilepsy.* In WARD, AA, PENRY, JR, AND PURPURA, D (EDS): *Epilepsy.* Raven Press, New York, 1983.

40. GESCHWIND, N, SHADER, RI, BEAR, DM, NORTH, B, LEVIN, K, AND CHETHAM, D: *Behavioral changes with temporal lobe epilepsy: Assessment and treatment.* J Clin Psychiatry 41:89, 1980.

41. GESCHWIND, N, AND SHERWIN, I: *Language-induced epilepsy.* Arch Neurol 16:25, 1967.

42. GIANNITRAPANI, D, AND KAYTON, L: *Schizophrenia and EEG spectral analysis.* Electroencephalogr Clin Neurophysiol 36:377, 1974.

43. GIBBS, FA: *Ictal and non-ictal psychiatric disorders in temporal lobe epilepsy.* J Nerv Ment Dis 113:522, 1951.

44. GILMORE, RL, AND HEILMAN, R: *Speech arrest in partial seizures: Evidence of an associated language disorder.* Neurology 31:1016, 1981.

45. GLISTA, GG, FRANK, HG, AND TRACY, FW: *Video games and seizures.* Arch Neurol 40:588, 1983.

46. GLOOR, P, AND FEINDEL, W: *The temporal lobe and affective behavior.* In MONNIER, M (ED): *Physiologie des Vegetativen Nerven Systems,* Vol 2, Hippokrates Verlag, Stuttgart, 1963.

47. GLOOR, P, OLIVIER, A, QUESNEY, LF, ANDERMANN, F, AND HOROWITZ, S: *The role of the limbic system in experiential phenomena of temporal lobe epilepsy.* Ann Neurol 12:129, 1982.

48. GLOOR, P, YAMAMOTO, L, OCHS, R, FEINDEL, W, GOTMAN, J, QUESNEY, F, MEYER, E, AND THOMPSON, C: *Regional blood flow: Oxygen utilization and glucose metabolism measured by postneuron emission tomography in patients with partial epilepsy.* Abstract, 15th International Epilepsy Symposium, p 238, 1983.

49. GODDARD, GV: *Development of epileptic seizures through brain stimulation at low intensity.* Nature 214:1020, 1967.

50. GODDARD, GV: *The kindling model of limbic epilepsy.* In GERGIS, M, AND KILOH, LG (EDS): *Limbic Epilepsy and the Dyscontrol Syndrome.* Elsevier/North Holland, 1980.

51. GODDARD, GV: *The kindling model of epilepsy.* Trends in Neurosciences, 1983.

52. GOLDENSOHN, ES: *Use of the EEG for evaluation of focal intracranial lesions.* In KLASS, DW, AND HOLZ, DD (EDS): *Current Practice of Clinical Electroencephalography.* Raven Press, New York, p 307, 1979.

53. GOON, Y, ROBINSON, S, AND LEVY, S: *Electroencephalographic changes in schizophrenic patients.* Isr Ann Psychiatr Relat Discip 11:99, 1973.

54. GUPTA, AR, JEAVONS, PM, HUGHES, RC, AND COVANIS, A: *Aura in temporal lobe epilepsy: Clinical and electroencephalographic correlation.* J Neurol Neurosurg Psychiatry 46:1079, 1983.

55. HAMILTON, NG, AND MATTHEWS, T: *Aphasia: The sole manifestation of focal status epilepticus.* Neurology 29:745, 1979.

56. HAUSER, WA, ANNEGERS, JF, AND ANDERSON, VE: *Epidemiology and the genetics of epilepsy.* In WARD, AA, PENRY, JK, AND PURPURA, D (EDS): *Epilepsy.* Raven Press, New York, p 267, 1983.

57. HERMANN, BP, AND RIEL, P: *Interictal personality and behavioral traits in temporal lobe and generalized epilepsy.* Cortex 17:125, 1981.

58. HERMANN, BP, WHITMAN, S, AND ARNTSON, P: *Hypergraphia in epilepsy: Is there a specificity to temporal lobe epilepsy?* J Neurol Neurosurg Psychiatry 46:848, 1983.

59. HERZOG, AG, RUSSELL, V, VAITUKATIS, JL, AND GESCHWIND, N: *Neuroendocrine dysfunction in temporal lobe epilepsy.* Arch Neurol 39:133, 1982.

60. HERZOG, AG, SEIBEL, M, SCHOMER, D, VAITUKATIS, JL, AND GESCHWIND, N: *Temporal lobe epilepsy: An extrahypothalamic pathogenesis for polycystic ovarian syndrome.* Neurology 34:1389, 1984.

61. HERZOG, A, SEIBEL, M, SCHOMER, D, VAITUKATIS, JL, AND GESCHWIND, N: *Reproductive endocrine disturbances in women with partial seizures of temporal lobe origin.* Neurology 33:189, 1981.

62. HILL, JD: *Psychiatric disorders of epilepsy.* Med Press 20:473, 1953.

63. JENSEN, S, AND LARSEN, JK: *Psychoses in drug-resistant temporal lobe epilepsy.* J Neurol Neurosurg Psychiatry 42:948, 1979.

64. JOHNSTON, D: *Valproic acid: Update on its mechanism of action.* Epilepsia 25:51, 1984.

65. JULIEN, RM: *Carbamazepine.* In WOODBURY, DM, PENRY, K, AND PURPURA, D (EDS): *Mechanism of Action in Antiepileptic Drugs,* 2 ed. Raven Press, New York, p 543, 1982.

66. KOGEORGOS, J, FONAGY, P, AND SCOTT, DF: *Psychiatric symptom patterns of chronic epileptics attending a neurological clinic: A controlled investigation.* Br J Psychiatry 140:236, 1982.

67. KRISTENSEN, O, AND SINDRUP, EH: *Psychomotor epilepsy and psychoses: I. Physical aspects.* Acta Neurol Scand 57:361, 1978.

68. KRISTENSEN, O, AND SINDRUP, EH: *Psychomotor epilepsy and psychoses: III. Social and psychological correlates.* Acta Neurol Scand 59:1, 1979.

69. KRNJEVIC, K: *GABA-mediated inhibitory mechanisms in relation to epileptic discharges.* In JASPER, HH, AND VAN GELDER, N (EDS): *Basic Mechanisms of Neuronal Hyperexcitability.* Alan R Liss, New York, p 249, 1983.

70. LAPLANTE, P, SAINT-HILAIRE, JM, AND BOUVIER, G: *Headache as an epileptic manifestation.* Neurology 33:1493, 1983.

71. LESSER, RP, LUEDERS, H, CONOMY, JP, FURLAN, AJ AND DINNER, DS: *Sensory seizure mimicking a psychogenic seizure.* Neurology 33:800, 1983.

72. LIFSHITZ, R, AND GRADIJAN, J: *Spectral evaluation of the electroencephalogram: Power and variability in chronic schizophrenics and control subjects.* Psychophysiology 11:479, 1974.

73. LOGIN, IS, AND DREIFUSS, FE: *Anticonvulsant activity of clomiphene.* Arch Neurol 40: 525, 1983.

74. LOGOTHETIS, J, HARNER, R, MORRELL, F, AND TORRES, F: *The role of estrogens in catamenial exacerbation of epilepsy.* Neurology 9:352, 1958.

75. LOISEAU, P, DARTIGUES, JF, AND PESTRE, M: *Prognosis of partial epileptic seizures in the adolescent.* Epilepsia 24:472, 1983.

76. MAGISTRETTI, P, UREN, R, PARKER, T, ROYAL, H, FRONT, D, AND KOLODNY, G: *Monitoring of regional cerebral blood flow by single photon emission tomography of I-123-N-isopropyl-iodoamphetamine in epileptics.* Ann Radiol 26:68, 1983.

77. MAGISTRETTI, P, UREN, R, SCHOMER, D, AND BLUME, H: *Regional cerebral blood flow in ictal and interictal states in epilepsy.* Eur J Nucl Med 7:484, 1982.

78. MORREL, F: *Secondary epileptogenic lesions in man: Extent of the evidences.* In DELGADO-ESCUETA, AV (ED): *Basic Mechanisms of the Epilepsies.* Raven Press, New York. (In press.)

79. MARK, VH: *Epilepsy and episodic aggression.* Letters to the editor. Arch Neurol 39:384, 1982.

80. MARK, VH, AND SWELT, WH: *The role of limbic brain dysfunction in aggression.* In FRAZIER, SH (ED): *Aggression.* Williams & Wilkins, Baltimore, 1974.

81. MARSHALL, DW, WESTMORELAND, BR, AND SHARBROUGH, FW: *Ictal tachycardia during temporal lobe seizures.* Mayo Clin Proc 58:443, 1983.

82. MATTSON, RH, PRATT, KL, AND CALVERLY, JR: *Electroencephalograms in epileptics following sleep deprivation.* Arch Neurol 13:310, 1965.

83. MELLANBY, J, STRAWBRIDGE, P, COLLINGRIDGE, GI, GEORGE, G, RANDS, G, STROUD, C, AND THOMPSON, P: *Behavioral correlates of an experimental hippocampal epileptiform syndrome in rats.* J Neurol Neurosurg Psychiatry 44:1084, 1981.

84. MESULAM, M-M: *Dissociative states with abnormal temporal lobe EEG: Multiple personality and the illusion of possession.* Arch Neurol 38:176, 1981.

85. MILNER, B: *Visual recognition and recall after right temporal lobe excision in man.* Neuropsychologia 6:191, 1968.

86. MILNER, B: *Disorders of learning and memory after temporal lobe lesions in man.* Clin Neurosurg 19:421, 1972.

87. MILNER, B: *Psychological aspects of focal epilepsy and its neurosurgical management.* In PURPURA, D, PENRY, J, AND WALTER, R (EDS): *Neurosurgical Management of the Epilepsies.* Adv Neurol 8:299, 1985.

87a. MIRSKY, AF, AND VAN BUREN, JM: *On the nature of the absence in centrencephalic epilepsy. A study of some behavioral, electroencephalographic, and autonomic factors.* Electroencephalogr Clin Neurophysiol 18:334, 1965.

88. MITCHELL, WG, GREENWOOD, RS, AND MERSENHEIMER, JA: *Abdominal epilepsy: Cyclic vomiting as the major symptom of simple partial seizures.* Arch Neurol 40:251, 1983.

89. MONROE, RR: *Limbic ictus and atypical psychoses.* J Nerv Ment Dis 170:711, 1982.

90. MORRELL, F: *Physiology and histochemistry of the mirror focus.* In JASPER, HH, WARD, AA, AND POPE, A (EDS): *Basic Mechanisms of the Epilepsies.* Little, Brown & Co, Boston, p 357, 1969.

91. MORRELL, F, TSURU, N, HOEPPNER, TJ, MORGAN, D, AND HARRISON, WH: *Secondary epileptogenesis in frog forebrain: Effect of inhibition of protein synthesis.* Can J Neurol Sci 2:407, 1975.

92. MUNGAS, D: *Interictal behavior abnormality in temporal lobe epilepsy: A specific syndrome or nonspecific pathology.* Arch Gen Psychiatry 39:108, 1982.

93. NAKADA, T, LEE, H, KWEE, I, AND LERNER, A: *Epileptic Kluver-Bucy syndrome: Case report.* J Clin Psychiatry 45:87, 1984.

94. NEWMAN, PK, AND LONGLEY, BP: *Reading epilepsy.* (Letters to the editor.) Arch Neurol 41:13, 1984.

95. NIELSEN, H, AND KRISTENSEN, O: *Personality correlates of sphenoidal EEG foci in temporal lobe epilepsy.* Acta Neurol Scand 64:289, 1981.

96. NOVELLI, RA, AUGUSTINE, AA, GLASER, GH, WILLIAMSON, PD, SPENCER, DD, AND SPENCER, SS: *Selective memory improvement and improvement in temporal lobectomy for epilepsy.* Ann Neurol 15:64, 1984.

97. PELED, R, HARNES, B, BOROVICH, B, AND SHARF, B: *Speech arrest and supplementary motor area seizures.* Neurology 34:110, 1984.

98. PENFIELD, W, AND JASPER, H: *Epilepsy and the Functional Anatomy of the Human Brain.* Little, Brown and Co, Boston, 1954.

99. PEPPERCORN, MA, HERZOG, AG, AND DICHTER, MA: *Abdominal epilepsy: A cause of abdominal pain in adults.* JAMA 240:2450, 1978.

100. PEREZ, M, AND TRIMBLE, MR: *Epileptic psychosis: Diagnostic comparison with process schizophrenia.* Br J Psychiatry 137:245, 1980.

101. PFAFF, DW, GERLACH, JL, McEWEN, BS, FERIN, M, CARMEL, P, AND ZIMMERMAN, EA: *Autoradiographic localization of hormone-concentrating cells in the brain of the female rhesus monkey.* J Comp Neurol 170:279, 1976.

102. PINCUS, JH: *Violence and epilepsy.* N Engl J Med 305:696, 1981.

103. POND, DA: *Psychiatric aspects of epilepsy.* J Ind Med Prof 3:1441, 1957.

104. PORTER, RJ, AND PENRY, JK: *Efficacy and choice of antiepileptic. 1977. Psychology, pharmacotherapy and new diagnostic approaches.* Proceedings of the 13th Congress of the International League Against Epilepsy, September, 1977. Swets and Zeitlinger, RV, 1978.

105. POST, RM: *Lidocaine-kindled limbic seizures: Behavior implications.* In WADA, J (ED): *Kindling 2.* Raven Press, New York, p 149, 1981.

106. PRINCE, DA: *Electrophysiology of "epileptic" neurons: Spike generation.* Electroencephalogr Clin Neurophysiol 26:476, 1969.

107. PRINCE, DA: *Neurophysiology of epilepsy.* Ann Rev Neurosci 1:395, 1978.

108. PRINCE, DA, AND WILDER, BJ: *Control mechanisms in cortical epileptogenic foci: Surround inhibition.* Arch Neurol 16:194, 1967.

TEMPOROLIMBIC EPILEPSY AND BEHAVIOR

109. PRICHARD, JW: *Phenobarbital. Mechanisms of action.* In WOODBURY, DM, PENRY, JK, AND PEP-PENGER, CE (EDS): *Antiepileptic Drugs,* 2 ed. Raven Press, New York, p 365, 1982.

110. RAINBOW, TC, PARSON, B, MACLUSKY, HJ, AND MCEWEN, BS: *Estradiol receptor levels in rat hypothalamic and limbic nuclei.* J Neurosci 2:1439, 1982.

111. RAMANI, V: *Primary reading epilepsy.* Arch Neurol 40:39, 1983.

112. RAMANI, V, AND GUMNIT, RJ: *Intensive monitoring of interictal psychosis in epilepsy.* Ann Neurol 11:613, 1982.

113. RAMSEY, RE, WILDER, BJ, BERGER, JR, AND BRUNI, J: *A double blind study comparing carba-mazepine with phenytoin as initial seizure therapy in adults.* Neurology 33:904, 1983.

114. RASMUSSEN, T: *Cortical resection for medically refractory focal epilepsy: Results, lessons and questions.* In RASMUSSEN, T, AND MARINO, R (EDS): *Functional Neurosurgery.* Raven Press, New York, 1979.

115. RASMUSSEN, T: *Localizational aspects of epileptic seizure phenomena.* In THOMPSON, RA, AND GREEN, JR (EDS): *New Perspectives in Cerebral Localization.* Raven Press, New York, 1982.

116. REES, HD, AND MICHAEL, RP: *Brain cells of the male rhesus monkey accumulate 3H-testos-terone or its metabolites.* J Comp Neurol 206:273, 1982.

117. REILLY, TL, AND MASSEY, EW: *The syndrome of aphasia, headaches, and left temporal lobe spikes.* Headache 20:90, 1980.

118. REMILLARD, GM, ANDERMANN, F, AND GLOOR, P: *Water-drinking as ictal behavior in complex partial seizures.* Neurology 31:117, 1981.

119. REMILLARD, GM, ANDERMANN, F, TOSTA, GF, GLOOR, P, AUBE, M, MARTIN, JB, FEINDEL, W, GUBERMAN, A, AND SIMPSON, C: *Sexual ictal manifestations predominant in women with tem-poral lobe epilepsy: A finding suggesting sexual dimorphism in the human brain.* Neurology 33:323, 1983.

120. REMILLARD, GM, TOSTA, G, ANDERMANN, F, FEINDEL, W, GLOOR, P, AND MARTIN, JB: *Sexual aura in seizures with partial complex symptomatology.* In WADA, JA, AND PENRY, K (EDS): *Advances in Epileptology* (10th International Symposium). Raven Press, New York, 1980.

121. RENAUD, LP, MARTIN, JB, AND BRAZAU, P: *Depressant action of TRH, LH-RH and somatostatin on activity of central neurons.* Nature 255:233, 1975.

122. ROBB, P: *Focal epilepsy: The problem, prevalence and contributing factors.* In PURPURA, D, PENRY, JK, AND WALTER, RD (EDS): *Neurosurgical Management of the Epilepsies.* Raven Press, New York, p 11, 1975.

123. ROBERTS, JKA, ROBERTSON, MM, AND TRIMBLE, MR: *The lateralising significance of hypergra-phia in temporal lobe epilepsy.* J Neurol Neurosurg Psychiatry 45:131, 1982.

124. ROBERTSON, MM, AND TRIMBLE, MR: *Depressive illness in patients with epilepsy: A review.* Epilepsia 24(Suppl 2):S109, 1983.

125. ROBILLARD, A, SAINT-HILAIRE, JM, MERCIER, RET, AND BOUVIER, G: *The lateralizing and local-izing value of adversion in epileptic seizures.* Neurology 33:1241, 1983.

126. ROEMER, RA, SHAGASS, C, STRAUMANIS, JJ, AND AMADED, M: *Pattern evoked potential mea-surements suggesting lateralized hemispheric dysfunction in chronic schizophrenics.* Biol Psy-chiatry 13:185, 1978.

127. ROSENTHAL, L, AND FEDIO, P: *Recognition thresholds in the central and lateral visual fields following temporal lobectomy.* Cortex 11:217, 1975.

128. ROVIT, RL, AND GLOOR, P: *Temporal lobe epilepsy—A study using multiple basal electrodes. I. Description of method. II. Clinical EEG findings.* Neurochirurgia 3:6, 1960.

129. ROVIT, RL, GLOOR, P, AND RASMUSSEN, T: *Sphenoidal electrodes in the electrographic study of patients with temporal lobe epilepsy.* J Neurosurg 18:1511, 1961.

130. SACHDEV, HS, AND WAXMAN, SG: *Frequency of hypergraphia in temporal lobe epilepsy: An index of interictal behaviors syndrome.* J Neurol Neurosurg Psychiatry 44:358, 1981.

131. SAMUELS, L, BUTTERS, N, AND FEDIO, P: *Short term memory disorders following temporal lobe removals in humans.* Cortex 8:283, 1972.

132. SCHWARTZKROIN, P: *Local circuit considerations and intrinsic neuronal properties involved in hyperexcitability and cell synchronization.* In JASPER, H, AND VAN GELDER, N (EDS): *Basic Mechanisms of Neuronal Hyperexcitability.* Alan R. Liss, New York, p 75, 1983.

133. SCHWARTZKROIN, P, MUTANI, R, AND PRINCE, DA: *Orthodromic and antidromic effects of a cortical epileptiform focus on ventrolateral nucleus of the cat.* J Neurophysiol 38:795, 1975.

134. SHALEV, RS, AND AMIR, N: *Complex partial status epilepticus.* Arch Neurol 40:90, 1983.

135. SHARF, M, SHARF, B, BENTAL, E, AND KRYMINSKY, T: *The electroencephalogram in the inves-tigation of anovulation and its treatment with clomiphene.* Lancet 1:750, 1976.

136. SHERWIN, I: *Psychosis associated with epilepsy: Significance of the laterality of the epilepto-genic lesion.* J Neurol Neurosurg Psychiatry 44:83, 1981.

137. SHERWIN, S, PERON-MAGNAN, P, BANCAUD, J, BONIS, A, AND TALAIRACH, J: *Prevalence of psychosis in epilepsy as a function of the laterality of the epileptogenic lesion.* Arch Neurol 39:621, 1982.

138. SHUKLA, GD, SRIVASTAVA, ON, AND KATIYAR, BC: *Sexual disturbances in temporal lobe epilepsy: A controlled study.* Br J Psychiatry 134:288, 1979.

139. SILFVENIUS, H, GLOOR, P, AND RASMUSSEN, T: *Evaluation of insular ablation in surgical treatment of temporal lobe epilepsy.* Epilepsia 5:307, 1964.

140. SILVERMAN, D: *The anterior temporal electrode and the ten-twenty system.* Am J EEG Technol 5:11, 1965.

141. SINDRUP, E, THYGESEN, N, KRISTENSEN, O, AND ALVINA, J: *Zygomatic electrodes: their use and value in complex partial epilepsy.* Advances in Epileptology, International-Symposium, p 313, 1981.

142. SLATER, E, AND BEARD, AW: *The schizophrenic-like psychoses of epilepsy.* Br J Psychiatry 109:95, 1963.

143. SPENCER, D, SPENCER, S, WILLIAMSON, P, AND MATTSON, R: *The posterior hippocampal focus in complex partial epilepsy.* (Abstract #35.) Am Acad Neurosci, 1981.

144. SPENCER, S, SPENCER, D, WILLIAMSON, P, AND MATTSON, R: *Sexual automatisms in complex partial seizures.* Neurology 33:527, 1983.

145. STEVENS, JR: *Psychosis and epilepsy.* (Letters.) Ann Neurol 14:347, 1983.

146. STEVENS, JR, AND HERMANN, BP: *Temporal lobe epilepsy, psychopathology, and violence: The state of the evidence.* Neurology 31:1127, 1981.

147. STEVENS, JR, AND LIVERMORE, A: *Telemetered EEG in schizophrenia: Spectral analysis during abnormal behaviour episodes.* J Neurol Neurosurg Psychiatry 45:385, 1982.

148. STRAUSS, E, RISSER, A, AND JONES, MW: *Fear responses in patients with epilepsy.* Arch Neurol 39:626, 1982.

149. STRAUSS, E, WADA, J, AND KOSAKA, B: *Spontaneous facial expressions occurring at onset of focal seizure activity.* Arch Neurol 40:545, 1983.

150. STUMPF, WE: *Anatomical distribution of steroid hormone target neurons and circuitry in the brain.* In MATTA, M (ED): *Endocrine Functions of the Brain.* Raven Press, New York, p 448, 1980.

151. TARRIER, N, COOKE, EC, AND LADER, MH: *The EEG's of chronic schizophrenic patients in hospitals and in the community.* Electroencephalogr Clin Neurophysiol 44:669, 1978.

152. TAYLOR, DC: *Mental state and temporal lobe epilepsy: A correlative account of 100 patients treated surgically.* Epilepsia 13:727, 1972.

153. TAYLOR, DC: *Factors influencing the occurrence of schizophrenia-like psychosis in patients with temporal lobe epilepsy.* Psychol Med 5:249, 1975.

154. THEODORE, W: *Position emission tomography (PET) in partial seizure disorders.* Abstracts, 15th International Epilepsy Symposium p 100, 1983.

155. TOONE, BK, DAWSON, J, AND DRIVER, MV: *Psychoses of epilepsy: A radiological evaluation.* Br J Psychiatry 140:244, 1982.

156. TOONE, BK, GARRALDA, ME, AND RON, MA: *The psychoses of epilepsy and the functional psychoses: A clinical and phenomenological comparison.* Br J Psychiatry 141:256, 1982.

157. TRIMBLE, MR: *Personality disturbances in epilepsy.* Neurology 33:1332, 1983.

158. VAN BUREN, JM: *Sensory, motor and autonomic effects of mesial temporal stimulation in man.* J Neurosurg 18:273, 1961.

159. VAN BUREN, JM, AND AJMONE-MARSAN, C: *A correlation of autonomic and EEG components in temporal lobe epilepsy.* Arch Neurol 3:683, 1960.

160. VAN BUREN, J, AJMONE-MARSAN, C, MUTSUGU, N, AND SADOVSK, D: *Surgery of temporal lobe epilepsy.* In PURPURA, D, PENRY, J AND WALTER, R (EDS): *Neurosurgical Management of the Epilepsies.* Adv Neurol 8:155, 1975.

161. VIOLENCE AND EPILEPSY, in "Correspondence," N Engl J Med 306:298, 1982.

162. WADA, J, AND RASMUSSEN, T: *Intracarotid injection of sodium amytal for the lateralization of cerebral speech dominance: Experimental and clinical observations.* J Neurosurg 17:266, 1960.

163. WARD, AA: *The epileptic neuron: Chronic foci in animals and man.* In JASPER, HH, WARD, AA, AND POPE, A (EDS): *Basic Mechanisms of the Epilepsies.* Little, Brown & Co, Boston, p 263, 1969.

164. WARD, AA: *Perspectives for surgical therapy of epilepsy.* In WARD, AA, PENRY, JK, AND PURPURA, D (EDS): *Epilepsy.* Raven Press, New York, 1983.

165. WANG, RKS, AND TRAUB, RD: *Cellular mechanisms for neuronal synchronization in epilepsy.* In DELGADO-ESCUETA, AV (ED): *Basic Mechanisms of the Epilepsies.* Raven Press, New York. (In press.)

166. WASTERLAIN, CG, MASUOKA, D, AND JANEC, V: *Chemical kindling: A study of synaptic pharmacology.* In WADA, J (ED): *Kindling 2.* Raven Press, New York, p 315, 1981.

167. WAXMAN, SG, AND GESCHWIND, N: *The interictal behavior syndrome of temporal lobe epilepsy.* Arch Gen Psychiatry 32:1580, 1975.

168. WHITMAN, S, AND HERMANN, R: *Prevalence of psychosis in temporal lobe epilepsy.* (Letters to the editor.) Arch Neurol 40:772, 1983.

169. WIESER, HG: *Depth recorded limbic seizures and psychopathology.* Neurosci Biobehav Rev 7:427, 1983.

170. WIESER, HG, AND YASARGIL, MG: *Selective amygdalohippocampectomy as a surgical treatment of mesiobasal limbic epilepsy.* Surg Neurol 17:445, 1982.

171. WOODBURY, DM: *Phenytoin. Mechanisms of action.* In WOODBURY, DM, PENRY, JK, PEPPENGER, CE (EDS): *Antiepileptic Drugs,* ed 2. Raven Press, New York, p 209, 1982.

172. YOUNG, GB, AND BLUME, WT: *Painful epileptic seizures.* Brain 106:537, 1983.

173. ZIELINSKI, J: *Epidemiology and medical-social problems of epilepsy in Warsaw.* US Dept of HEW, Social and Rehabilitation Service, 1974.

174. ZIELINSKI, J: *Epileptics not in treatment.* Epilepsia 15:203, 1974.

175. ZIELINSKY, J: *People with epilepsy who do not consult physicians.* In JANZ, D (ED): *Epileptology.* Proceedings of the 7th International Symposium, Stuttgart: Verlag, 1976.

Chapter 9

ELECTROPHYSIOLOGY IN BEHAVIORAL NEUROLOGY*

Robert T. Knight, M.D.

Cognitive processes occur in a timeframe measured in milliseconds. For instance, up to two and one half correct decisions can be made per second, and reaction times can be as fast as 150 milliseconds.[25] At present, only electrophysiologic techniques are capable of assessing this high-speed information transfer and analysis.

Although microelectrode studies of neurons in animals has advanced our understanding of the intracortical mechanisms associated with perceptual[60,61,122] and attentional[13,62,84,85] mechanisms, these invasive techniques are rarely available for study of the physiologic mechanisms involved in human cognition. EEG and evoked potential recordings can, however, be readily used to study the neurophysiologic processes underlying human cognition. The EEG tends to measure activity generated by large ensembles of neural elements firing at varying frequencies. In general EEG recording is best suited for analysis of tonic changes in neocortical activity occurring in a time domain measured in seconds. There are some technical limitations to this method, including the poor resolution of high-frequency cortical rhythms[16,95] and the relatively long time period (seconds) needed to discern behaviorally induced changes in EEG frequencies.[40,83]

Some of the deficiencies of EEG techniques are overcome by evoked potential methods, which can detect both high- and low-frequency neural activity originating from discrete regions. Furthermore, evoked potentials can measure neural activity occurring in the millisecond time range. For these reasons, evoked potentials are better suited than EEG

*This work was supported by the Medical Research Service of the Veterans Administration and by grants from the NIAA (AG 02484). The skillful manuscript preparation of Kathryn Arvin is appreciated.

for analysis of phasic brain activity associated with high-speed mental operations. The combined application of EEG and evoked potentials can, therefore, be employed to study both tonic and phasic neural activity underlying various neuropsychologic mechanisms. Before discussing the EEG and evoked-potential literature pertinent to behavioral neurology, the cellular events that underlie the generation of scalp-recorded electrical activity will be briefly reviewed.

NEURAL GENERATORS OF EEG AND EVOKED POTENTIALS

Scalp-recorded electrical activity reflects the activation of cellular or fiber pathways engaged in sensory, motor, or cognitive processing. The potential generated by cellular elements are due to summation of graded post-synaptic activity (either IPSPs or EPSPs) on dendrites and soma of neurons.[42,68,90,125] Synchronous activity in axonal pathways also generates electrical activity that can be occasionally recorded with scalp electrodes.[119]

Evoked potential methods can record activity from cortical sources directly under the electrode (near-field response) or activity from more distant subcortical generators (far-field response). In electrophysiology, "near field" and "far field" refer simply to the distance of the neural generator from the scalp electrode. An example of near-field activity is the primary somatosensory evoked potential (SEP), recorded by electrodes in a small area over pre- and postcentral gyri contralateral to the stimulated limb. The SEP is due to summation of graded postsynaptic potentials on primary sensory cortex neurons. The isopotential gradients of the equivalent dipole source for this postcentral gyrus activation undergo marked changes over millimeters of cortex. This produces rapid changes in amplitude over centimeters of the scalp (Fig. 1A). Marked changes in amplitude or morphology of evoked potentials occurring over small distances on the scalp indicate that the generators are in cortical tissue directly under the scalp electrodes. This discrete topography of the SEP has recently been employed to map out the scalp distribution of the digit representation in the human brain.[31]

The most familiar far-field–type response is the brainstem auditory evoked response (BSAEP) generated by synchronous neural activity in brainstem auditory pathways.[11] The distant electric field of this deep dipole source is equipotential over large areas of cortex (Fig. 1B). For this reason, the BSAEP has the same amplitude at widespread frontal and central scalp sites, in contrast to the focal distribution of the SEP. In general, it is more difficult to localize the generators of far-field activity. In particular, scalp electrodes are unable to differentiate widespread synchronous cortical activity from the far field of a deep neural source, since in both cases the electrical activity is synchronous over large cortical regions. Whereas the evoked potential averaging process allows the extraction of both near-field (cortical) and the much smaller far-field (subcortical) electrical activity, EEG reflects predominantly near-field (cortical) activity.[20,21,97]

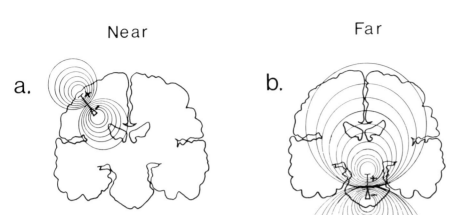

Near

Far

a.

b.

FIGURE 1. *A,* Schematic cortical dipole with isopotential contours drawn. Note rapid changes in the electric field over small regions of the cortex (near field). *B,* Identical dipole in a depth location. Note equipotentiality of the electric field at the surface (far field).

EEG

Human EEG Studies

Direct cortical recordings indicate that alpha range frequencies are synchronous over widespread cortical regions. Consequently, scalp recordings reflect alpha range frequencies fairly well. Higher-frequency cortical activity in the 20- to 50-Hz range is, however, asynchronous between adjacent cortical regions separated by as little as 1 cm. The asynchronous activity of these high-frequency dipoles tends to cancel out in a closed-field manner and is therefore attenuated at scalp electrodes.[16,95] This is unfortunate, since alpha range activity correlates with an idling state of cortical neurons, whereas high-frequency rhythms reflect more intense neural activity. In addition to this selective high-frequency attenuation, there is an overall 10- to 80-fold reduction in amplitude of all scalp-recorded activity relative to simultaneously recorded cortical activity, owing to impedance of intervening tissue.[54] This attenuation factor applies equally to EEG and evoked-potential signals.

In frequency studies, the EEG can be visually inspected or analyzed by use of the mathematical technique of fast fourier transform, in which a segment of ongoing EEG (usually from 1 to 10 seconds) is broken down into its component frequencies. This spectral analysis procedure allows one to assign a certain power or energy to different frequency bands. For example, the resting alpha range EEG may visually appear to be composed of 8- to 12-Hz activity. Mathematical frequency analysis reveals, however, that the alpha band of 8 to 12 Hz accounts for only part of the total energy content of the EEG. Beta, theta, and delta activity also account for a significant amount of the total scalp EEG energy, although these frequencies will not be readily apparent by visual inspection. Frequency analysis of a subject in deep sleep or in coma will reveal a shift in power to the lower-frequency bands; that is, the delta band will now contain most of the power. Even in these states some power will be found in higher-frequency bands. Transition to the alert, vigilant state from the resting alpha state will reveal a desynchronization of alpha and a shift in power to higher-frequency activity. This shift to higher frequencies is associated with increases in neuronal firing, metabolic activity, and regional blood flow in the underlying cortex.[64] The pioneering experiments of Moruzzi and Magoun revealed that the shift in frequency from sleep to the awake attentive state was under control of the ascending reticular activating system, and this remains as the classic example of change in tonic state of the organism.[83] The role of the reticular activating system is discussed further in Chapter 3. The use of EEG in the diagnosis of epilepsy is reviewed in Chapter 8.

Neuropsychologic Applications of EEG

Recent computer-based experiments employing fourier analysis of data from multiple scalp sites have revealed that EEG frequency changes can be focal in nature. This suggests that EEG frequency analysis may be a more valuable tool for the study of discrete neural activity than was felt possible in the past. For instance, it has been reported in humans that alpha in the left parietal lobe focally desynchronizes after right-sided stimuli only, whereas alpha in the right parietal lobe focally desynchronizes equally well after right- or left-sided stimuli. These results have been interpreted to reflect right-hemisphere dominance for attention.[52] Other studies indicate focal scalp EEG desynchronization in primary sensory regions during sensory stimulation confirming that EEG frequency changes can be quite discrete in nature.[96,108]

New methods for the display of multichannel EEG and evoked-potential data have been developed. These techniques involve converting EEG frequency or evoked-potential amplitude information into a visual code, which is then displayed on a representation of the cortical surface. In the brain electrical mapping method (BEAM) the electrophysiologic data are converted to a color code, whereas in the gray-level surface distribution method the electrical information is converted to a shade of gray scale.[10,32,88] The advantage of the BEAM technique is that it allows the topography of the EEG frequency changes to be viewed serially over time. This enhances both the amount and quality of neural information that can be examined.

The BEAM technique has been used to study both dyslexia and schizophrenia. These studies have revealed focal EEG frequency abnormalities suggesting hypofunctioning in the left temporal-parietal region of the dyslexia population with additional unexpected dysfunction in the frontal regions of the brains of these children.[33] The studies on patients with schizophrenia indicate hypofunctioning in the frontal regions in accord with cerebral metabolic studies of this population.[63,82] BEAM has also been used to map age-related changes in brain-wave activity and also to differentiate healthy individuals from those with dementia.[31a,31b]

Animal EEG Studies

Animal EEG experiments provide another perspective on cognitive processes. Focal frontal-parietal rhythms in the 35- to 45-Hz range during focused attention in the cat have recently been described with direct cortical surface recordings.[7] Similarly, a fast rhythm over the postcentral gyrus and posterior parietal cortex is seen during focused attention in the baboon and squirrel monkey.[104] Based on recordings from multiple electrodes in the olfactory bulb of the rabbit, it has been suggested that the processes of expectancy and attention are dependent on a background carrier frequency of 40- to 80-Hz activity.[36] Unfortunately, the attenuation and canceling of fast frequencies in the human scalp EEG prevents their correlation to the attention-related fast frequencies recorded in animal species. The development of the human magnetoencephalogram with its higher spatial and frequency resolution may circumvent some of these difficulties.[22]

Animal studies have also shown that whereas the surface EEG shows desynchronization to a low-voltage fast pattern during attentive behavior, simultaneous depth recordings in the hippocampus reveal synchronized 4- to 12-Hz patterns. The hippocampal activity can be further broken down into an 8- to 12-Hz amine-modulated motor component and a 4- to 7-Hz acetylcholine-modulated cognitive component.[123] These advances in the chemical and physiologic understanding of animal attention are difficult to confirm in the human, since they require recordings obtained from depth electrodes.

EVOKED POTENTIALS

Human Evoked-Potential Studies

In the 1940s, computer advances led to the development of techniques wherein small (0.1- to 10-μv) electrical signals could be reliably extracted from the background scalp EEG (10 to 100 μv). This is accomplished through the process of signal averaging, in which repetitive time-locked sensory stimuli are presented to the subject, and the brain electrical activity following each stimuli is stored in computer memory. As the stimuli are added together, the random background EEG averages to zero and the stimulus-locked activity emerges from the background as the evoked response. This methodology was immediately employed by neurophysiologists to map out the topography of the primary sensory cortices in many animal species.[17,135]

These stimulus evoked potentials have subsequently developed into powerful clinical tools for the evaluation of neural sensory pathways from the peripheral receptor to the primary cortex.[66,117] Recording of auditory, somatosensory, and visual evoked potentials has become a routine part of most clinical neurophysiology departments. These sensory evoked potentials are referred to as exogenous, since the stimulus that is delivered to the subject determines the amplitude, latency, and morphology of the evoked response that will be recorded. These sensory evoked potentials have not convincingly been shown to be altered by any mental activity, be it attention, stimulus recognition, or memory.[27,99] In fact, normal sensory evoked potentials are accurately recorded in comatose patients lacking structural damage to brainstem auditory or primary somatosensory pathways,[43] thus further dissociating them from active mental activity.

A series of major breakthroughs occurred in the 1960s and early 1970s, when researchers discovered several new types of human brain potentials that were generated during the cognitive operations associated with attention, orientation, and stimulus detection.[57,120,128] These potentials have subsequently been referred to as endogenous potentials in that they are stimulus independent and are generated solely as a function of the mental processing allocated to the stimulus. For example, endogenous potentials are increased in size when near threshold sensory stimuli are detected. These observations dissociate endogenous potentials from the exogenous "hard-wired" sensory evoked potentials which decrease in size with decreasing stimulus strength.[56] In fact, large endogenous potentials are generated even when a subject detects that a stimulus has not occurred.[110]

These endogenous brain potentials occur in a timeframe of 50 to 500 msec poststimulus and are direct physiologic markers of the neuropsychologic events underlying human cognitive processing. Unlike microelectrode studies, they allow for a noninvasive electrophysiologic analysis of cognitive operations in humans. Two of the most widely studied and robust endogenous brain potentials are the selective attention negativity and the P300.

It has been shown that when a subject attends to a stimulus—whether auditory, tactile, or visual—a negative electrical shift is generated in response to the attended stimulus.[27,50,57] The negative shift begins at about 60 msec poststimulus and lasts for up to 500 msec. This endogenous brain potential greatly augments the "hard-wired" sensory evoked potential generated by the attended stimulus. The integrated area under this shift is referred to as the processing negativity or the selective attention effect (Nd) in the evoked-potential literature.[49,86]

The P300 endogenous brain potential (P for positive, 300 because of its typical 300-msec poststimulus latency) is generated whenever infrequently occurring sensory stimuli are correctly detected or when unexpected and highly deviant stimuli are delivered to a subject.[14,30,59] The P300 reflects both the stimulus evaluation and orientation capability of the subject and does not require an overt motor response for its generation. For these reasons it is a valuable tool in the analysis of behavioral syndromes wherein verbal or motor output might be impaired.[100]

Other endogenous brain potentials have been recorded in humans. Since they have not proved as reliable as the Nd and the P300, they will be briefly described. The contingent negative variation (CNV) is generated during the period between a warning stimulus and a stimulus to which the person must respond. This potential may reflect the motivational state of the subject.[8] Recent studies suggest, however, that the CNV may in fact be the summation of two other responses: the post-P300 negativity generated by detection of the warning stimulus and the readiness potential (Bereitschaftspotential), which precedes the motor response.[103] The readiness potential may be generated by supplementary motor regions.[26]

Linguistic studies employing evoked-potential methods have not revealed consistent findings, owing to a number of technical and methodological difficulties.[38] Recently, however, a negativity has been described that peaks at 400 msec poststimulus and is generated by semantically incongruous words.[73] This phenomenon may prove useful in the evaluation of aphasic symptoms.

Selective Attention Evoked Potential

Attention involves the selection of a particular stimulus or channel of stimuli for further processing. Since attention to a task is at the entry zone to many cognitive operations, it is a critical brain mechanism for the understanding of ongoing mental activity. Many major neurologic syndromes including dementias, confusional states, frontal and nondominant parietal syndromes, childhood hyperkinesis, autism, and schizophrenia have prominent attentional disturbances (see Chapter 3). Understanding of the physiologic substrates of attention promises major advance in the analysis of these disorders.

One key question is the level of the central nervous system at which attended stimuli are selected.[55,111,136] Two divergent theories have been proposed. In one schema attended stimuli are selected at an early stage in processing—for instance, at the cochlear level in the auditory system. The opposite view holds that external events are fully analyzed before one stimulus is selected for further processing. In the first view selective attention might begin within the first few milliseconds, whereas in the latter case selective processing might not begin for several hundred milliseconds. At present, there is no human evidence to support this proposal of early filtering of sensory inputs. In the human studies of auditory and somatosensory selective attention to be reviewed, the first recordable difference between attended and nonattended stimuli occurs at 60 to 80 msec poststimulus. This is long after the peripheral sensory volley has reached primary cortical receiving areas.[27,99]

The first reports of electrophysiologic evidence for selective processing of attended stimuli in humans were provided in the early 1970s.[57,58] The experiments involved the presentation of two channels of brief randomized tones delivered to the left and right ears at high rates. The subjects were instructed to attend to one ear and count the occurrence of rare, slightly longer-duration tones in that ear. The behavioral strategy of high stimulus rate and difficult detection of the slightly longer tone forced attention preferentially to one channel of input. Typical results from this type of experiment are seen in Figure 2.

As can be seen in this figure, when subjects attend to the left-ear stimuli (solid line) there is a negative electrical shift whose onset is at 60 to 80 msec and whose duration is several hundred milliseconds. When the subject shifts attention to right-ear tones and we average the response to the now unattended left-ear tones still being presented, this negativity disappears (dotted lines). The shaded area in Figure 2 represents the endogenous brain potential generated by the subjects' attention

ATTENTION EFFECT

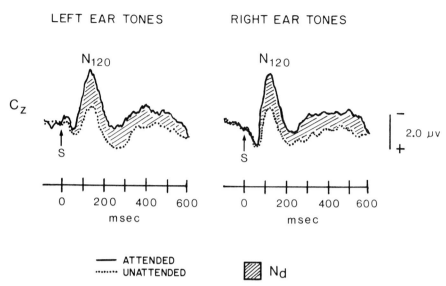

FIGURE 2. Evoked responses to left and right ear tones when they are attended *(solid line)* and unattended *(dotted line)*. Note the negative electrical enhancement of the attended stimulus *(cross-hatched area)*. Each tracing is the average waveform of 13 subjects. All recordings from C_z.

to the stimulus; this is called the Nd, as defined previously. The auditory selective attention effect has been replicated in many laboratories and investigators have extended this phenomenon to the somatosensory and visual modalities in which similar negative shifts to stimuli in attended channels of input have been found.[27,50,124]

A key feature of these experiments is that the tonic arousal level of the subject stays constant throughout the experiment. This eliminates nonspecific effects of arousal level, which are known to alter evoked response amplitude and morphology.[39]

Of particular importance is the finding that the Nd is modality specific in scalp topography. For instance, when the patient attends to the left hand, the Nd is maximal over right parietal cortex, whereas this shifts to the parietal-occipital cortex when the patient attends to visual stimuli. These findings of attention-related negative shifts in the parietal association region of humans is paralleled by observations of attention specific unit activity in the parietal association cortex in monkeys performing directed attention tasks.[84]

The amplitude of the vertex potential is doubled during the early phase of the Nd (60 to 200 msec). The vertex potentials are the prominent negative evoked responses peaking at 100 to 150 msec poststimulus. They are generated by auditory, somatosensory, or visual stimuli that require no cognitive processing.[15,24,42] Of interest are observations that the vertex potentials are markedly attentuated by lesions involving temporoparietal association cortex.[41,69]

Neuropsychologic Applications of the Selective Attention Effect (Nd)

As also discussed in Chapter 3, studies on animals and humans have demonstrated an important role for the prefrontal cortex in regulating attentional process.[6,9,23,35,51,65,105–107,118] Projections from frontal cortex to the thalamus, limbic cortex, reticular formation, posterior association cortex, and limbic forebrain[1,44,78,94,129] have been implicated in sensory gating and alerting functions in animals and humans.[37,76,81,89,111] In general, these effects are mediated by a net inhibitory prefrontal output to cortical and subcortical regions[2,4,9,34,77,109]

The selective attention effect (Nd) has been studied in humans with focal left or right prefrontal lesions.[70] Normal subjects generate the expected electrical sign of enhanced negativity to attended stimuli when focusing attention to either left- or right-ear stimuli. Subjects with left prefrontal lesions generate moderately diminished attention-related negativity to stimulation on either side. Subjects with right prefrontal lesions generate a normal attention shift to right-ear stimuli; however, there is a complete inability to generate attention-specific processing negativity when attending to the left-ear tones (Fig. 3).

As described in Chapter 3, these findings converge with other observations indicating a critical role for prefrontal cortex in the control

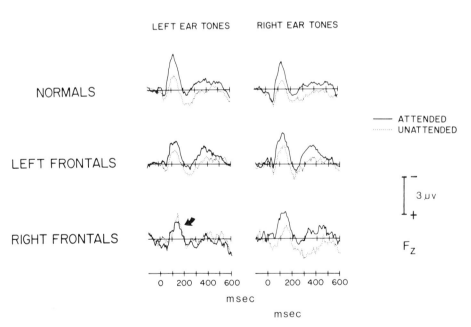

FIGURE 3. Attention effect for control subjects and patients with unilateral prefrontal lesions. Normal subjects generate a processing negativity to the selectively attended ear. Left frontals generate a diminished but equal attention effect for each ear. Right frontals reveal a complete hemi-inattention to left ear stimuli (*large arrow*). All recordings from C_z.

ELECTROPHYSIOLOGY
IN BEHAVIORAL
NEUROLOGY

of attentional systems and further indicate a hemisphere asymmetry in this attentional capacity.[53,80,107,127]

Whether the prefrontal modulation of the processing negativity is due to control of subcortical or posterior association cortex generators remains to be determined. The auditory findings also need to be extended to the somatosensory and visual modality, thus testing whether the hemispheric attention asymmetry found in the auditory system is a general property for all sensory modalities.

The attention effect paradigm has been applied to schizophrenic populations and to hyperkinetic children, two conditions with known clinical deficits of attention. Both these syndromes show pronounced defects in the ability to generate processing negativity to attended stimuli.[3,74]

Evoked-potential studies have provided other interesting data on patients with hemispatial neglect. For example, patients with evoked-potential evidence of damage to primary auditory or somatosensory regions do not necessarily show neglect.[41,47,72,131] Hemispatial neglect is apparent only when there is evoked-potential evidence of parietal or prefrontal association cortex damage.[41,70,79] These findings confirm the clinical observations that the neglect syndrome is not simply due to sensory loss.

P300 Evoked Potential

When an infrequent sensory event (referred to as the target) is correctly detected by a subject, a large, P300 waveform (maximal over parietal regions) is generated. This is usually preceded by a modality-specific N200 wave.[87,102] An undetected target does not generate an N200-P300 complex. The P300 wave has the same distribution for auditory, somatosensory, or visual stimuli, suggesting a common neural generator.[113] This potential was initially felt to reflect widespread activity in parietal association cortex, possibly related to a transient decrease in mesencephalic reticular formation-induced negative electrical tone in the neocortex.[28,29] However, depth studies in humans undergoing surgical procedure for intractable complex partial seizures have not shown polarity inversion of the P300 when passing through parietal cortex.[132] This is strong evidence against a parietal cortex source for P300, since a cortical dipole source should reveal polarity inversion when a electrode passes through it. Subsequent studies, again in temporal lobectomy patients, have revealed endogenous potentials in limbic regions with the P300 latency, which show polarity inversion in this area.[48,115] Neuromagnetic studies in humans have also indicated a limbic source for the P300 wave.[92] The widespread scalp distribution of the P300 may therefore be due to a subcortical source generating a far-field response at the scalp.

The N200 component is usually elicited as a complex with the P300 wave, but, in contrast, this potential has a modality-specific scalp distribution. For instance, it is largest over parieto-occipital regions for visual stimuli, central regions for auditory stimuli, and parietal regions for somatosensory stimuli, indicating that it is generated in modality-specific association cortex. Recent work indicates that the negativity in the N200 latency range may be further separable into two negativities, one associated with passive and the other with active signal detection.[87]

The parameters that control the latency, amplitude, and morphology of the N200-P300 complex have been extensively studied. In essence, more difficult tasks involving detection of weak, randomly occurring, and infrequent stimuli elicit the largest N200-P300 complexes. In contrast, the amplitude of "hard-wired" sensory evoked potentials is directly proportional to stimulus magnitude. Furthermore, exogenous potentials habituate at short interstimulus intervals, whereas the amplitude of the N200-P300 complex remains unaffected in signal detection experiments with stimuli presented as fast as three per second.[133,134] This provides additional evidence in support of a direct relationship to high-speed cognitive activity.

The N200-P300 complex is also generated by presentation of highly novel stimuli to a subject who has not been forewarned of their occurrence. This effect has been shown for both visual and auditory stimuli.[18,19,71] This *novelty* N200-P300 complex is felt to reflect a CNS component of the orienting reflex.[75,101,112,114,116] The N200-P300 waveform to novel, unexpected, orienting-type stimuli (1) occurs earlier, (2) is more phasic in character, and (3) has a more frontocentral scalp distribution in contrast to the *target detection* N200-P300, which tends to be predominantly parietal in distribution.

An example of these two varieties of N200-P300 complexes is shown in Figure 4. Repetitive stimuli (referred to as standards) elicit the classic vertex N100 response. Interspersed among the standard stimuli are tones of slightly different pitch (designated as targets). When subjects are asked to detect the frequency shifts, they generate a large P300 with maximal amplitude over the parietal lobe to the correctly detected targets. When rare, highly deviant unexpected complex tones (designated as novels) are delivered to the subject, a large P300 with maximal amplitude over the frontal lobe is generated. This novelty P300 occurs approximately 50 msec earlier and is more phasic in morphology than is the target detection N200-P300.

Both of these types of N200-P300 responses are sufficiently large (5 to 20 μv) that the response to a single target or novel stimulus can occasionally be seen in the ongoing EEG, particularly when the EEG is in a low-voltage desynchronized pattern. Figure 5 shows the occurrence of N200-P300 complexes to individual stimuli in four different subjects.

Whether the two varieties of N200-P300 reflect unitary or different brain mechanisms remains to be determined. The most parsimonious explanation is that of a single generator for both complexes. A generator in limbic or diencephalic structures related to the organism's orienting capability would seem most likely. Both the target and novelty N200-P300 complexes may reflect activity of the cerebral detector proposed by Sokolov.[114] The target P300 may reflect active and the novelty P300 passive engagement of the orienting system. The target P300 would occur slightly later, owing to the intervening decision processes required to identify the stimulus as a target; whereas highly deviant unexpected stimuli would engage the novelty P300 generator without the need of intervening decision processes.

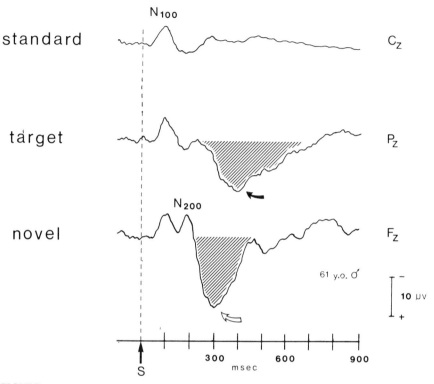

FIGURE 4. Repetitive 400 HZ background stimuli elicit the classic vertex N100 response (*top*, C$_z$). Correctly detected infrequently occurring 450 Hz tones elicit the classic parietal maximal target P300 response (*solid arrow, middle*, P$_z$). Rare, unexpected, highly deviant complex tones elicit the frontocentral maximal novelty P300 response (*open arrow, bottom*, F$_z$).

Animal microelectrode studies indicate that the cerebral mismatch detector may be found in limbic regions, which have also been suggested to generate P300 in humans.[93,126] Studies employing patient groups with neocortical or limbic lesions should help answer many of the questions concerning the P300 generator.

Neuropsychologic Applications of P300

The P300 wave has been shown reliably to increase in latency and diminish in amplitude with aging.[45] Interestingly, it shows the opposite effect from infancy to adulthood, when a shortening of latency and an increase in amplitude have been reported.[19] The P300 effect has recently been applied to the analysis of neuropsychologic syndromes. The target P300 was initially shown to increase in latency and decrease in amplitude with dementing syndromes of diverse etiology.[46,121] It has been proposed that these P300 changes may be used to differentiate true dementia from depression-related dementia (in which the P300 would be expected to be normal). The P300 is also diminished in chronic alcoholic patients with evidence or cortical atrophy.[5]

SINGLE TRIAL P$_{300}$

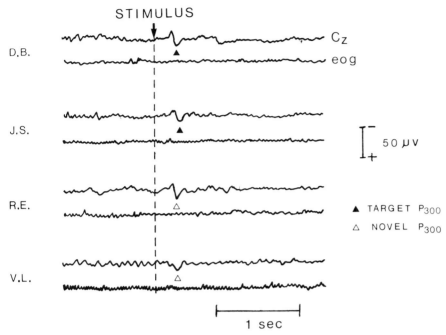

FIGURE 5. P300 responses in the ongoing EEG to target or novel stimuli in four subjects. Note that the electro-oculogram shows no artifact generating eye movement. All responses recorded from C$_z$.

Since electrophysiologic studies reveal a net inhibitory prefrontal output to cortical, subcortical, and limbic structures,[2,69,70,111] prefrontal damage might produce a chronic disinhibition of neural inputs to both sensory and limbic circuits.[4] This would fatigue the orienting system and diminish the neural reaction to any truly unexpected novel event, perhaps accounting for the observations of decreased orienting response activity (decreased galvanic skin response, decreased EEG desynchronization) in patients with prefrontal lesions.[67,76]

P300 experiments provide evidence of orientation abnormalities in patients with prefrontal lesions. Both simple target stimuli (450-Hz tones) and highly deviant, unexpected novel auditory stimuli were presented to control subjects and to patients with unilateral prefrontal damage (Fig. 6). Controls generated the expected P300 to correctly detected targets and an earlier, more phasic, frontocentrally distributed P300 to the unexpected novel stimuli. Prefrontal patients generate normal P300 responses to the detected targets. However, they have markedly decreased P300 responses to novel stimuli. In fact, the novel P300 responses in the patients are similar to their target P300 in amplitude and scalp topography. This provides electrophysiologic evidence of the decreased spontaneous orientation capabilities with respect to novelty in these patients.[71]

It is of interest that the abnormal attention effects and P300 orientation defects can be detected in patients with unilateral prefrontal

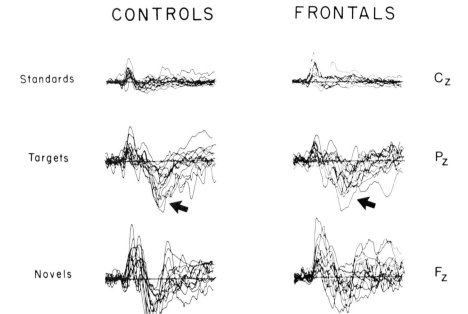

CONTROLS   FRONTALS

Standards   Cz

Targets   Pz

Novels   Fz

10μv   10 μv
300 600 900   300 600 900
msec   msec
S   S

FIGURE 6. Vertex responses (*top*, C$_z$), target P300 responses (*middle*, P$_z$) and novelty P300 responses (*bottom*, F$_z$) in control subjects and patients with unilateral prefrontal lesions. There is no difference in the vertex or target P300 responses between the two groups (*solid arrowheads*). The prefrontal patients have a markedly decreased P300 to the novel stimulus when compared with the control subjects (*open arrowheads*). Each tracing is the evoked response of an individual subject.

pathology who show minimal additional neuropsychologic deficits. Since patients with prefrontal damage are known to have cognitive deficits that can be elusive (see Chapter 1), these electrophysiologic tools offer an alternative means for the diagnosis and clinical investigation in such patients.

Animal Models of P300

In contrast to the area of exogenous potentials, remarkably little animal information is available on the neural sources of the cognition-related processing negativity or the N200-P300 complex. Recently, attempts have been made to develop P300 models in both the cat[12,130] and the rat.[91] The use of restraints and anesthetics in these animal experiments, however, necessitates caution in comparing the results with those of human P300 experiments. The use of highly deviant stimuli in attempts to elicit novelty N200-P300 complexes might allow free field animal experiments, thus circumventing the need for restraints and anesthetics.

CONCLUSIONS

Developments in computer techniques have increased the capability of handling large amounts of neurophysiologic data. These methods have been directly applied to the study of human EEG and evoked-potential data. Fourier analysis of EEG data from multiple scalp electrodes is increasingly being employed to study regional changes in cortical activity during various cognitive tasks. Recent discoveries of evoked potentials generated during specific cognitive operation are also being extensively applied to the study of the neural events underlying mental activity.

The majority of these human electrophysiologic studies have centered on the study of normal populations. Recently, however, these methods have been profitably employed in the study of patient groups with various neuropsychologic syndromes. These approaches have produced data in accord with behavioral observation and in addition have generated findings, which complement traditional neuropsychologic analysis.

REFERENCES

1. ADEY, WR, AND MEYER, M: *An experimental study of hippocampal afferent pathways from prefrontal and cingulate area in the monkey.* J Anat 86:58–74, 1952.
2. ALEXANDER, GE, NEWMAN, JD, AND SYMMES, D: *Convergence of prefrontal and acoustic inputs upon neurons in the superior temporal gyrus of the awake squirrel monkey.* Brain Res 116:334–338, 1976.
3. BARIBEAU-BRAUN, J, PINCTON, TW, AND GOSSELIN, J: *Schizophrenia: A neurophysiological evaluation of abnormal information processing.* Science 219:874–876, 1983.
4. BARTUS, RT, AND LEVERE, TE: *Frontal decortication in rhesus monkeys: A test of the interference hypothesis.* Brain Res 119:233–248, 1977.
5. BEGLEITER, H, PORJESZ, B, AND TENNER, M: *Neuroradiological and neurophysiological evidence of brain deficits in chronic alcoholics.* Acta Psychiatr Scand (Suppl 286)62:3–14, 1980.
6. BIANCHI, L: *The functions of the frontal lobes.* Brain 18:497–522, 1895.
7. BOUYER, JJ, MONTARON, MF, AND ROUGEUL, A: *Fast fronto-parietal rhythms during combined focused attentive behavior and immobility in cat: Cortical and thalamic localizations.* Electroencephalogr Clin Neurophysiol 51:244–252, 1981.
8. BOYD, EH, BOYD, ES, AND BROWN, LE: *Precentral cortex unit activity during the M-wave and contingent negative variation in behaving squirrel monkeys.* Exp Neurol 75:535–554, 1982.
9. BRUTKOWSKI, S: *Functions of prefrontal cortex in animals.* Physiol Rev 45:721–746, 1965.
10. BUCHSBAUM, MS, RIGAL, F, COPPOLA, R, CAPELLETTI, J, KING, C, AND JOHNSON, J: *A new system for gray-level surface distribution maps of electrical activity.* Electroencephalogr Clin Neurophysiol 53:237–242, 1982.
11. BUCHWALD, JS: *Generators.* In MOORE, EJ (ED): *Basis of Auditory Brain-Stem Evoked Responses.* Grune & Stratton, 1983, pp 157–196.
12. BUCHWALD, JS, AND SQUIRES, NS: *Endogenous auditory potentials in the cat: A P300 model.* In WOODY, CD (ED): *Conditioning.* Plenum Publishing, pp 503–515.
13. BUSHNELL, MC, GOLDBERG, ME, AND ROBINSON, DL: *Behavioral enhancement of visual responses in monkey cerebral cortex. I. Modulation in posterior parietal cortex related to selective visual attention.* J Neurophysiol 46:755–787, 1981.
14. CALLAWAY, E, TUETING, P, AND KOSLOW, S: *Event-Related Brain Potentials in Man.* Academic Press, New York, 1978.
15. CELESIA, GG: *Organization of auditory cortical areas in man.* Brain 99:403–414, 1976.
16. COOPER, R, WINTER, AL, CROW, HJ, AND WALTER, WG: *Comparison of subcortical, cortical and scalp activity using chronically indwelling electrodes in man.* Electroencephalogr Clin Neurophysiol 18:217–228, 1965.
17. CORWIN, JT, BULLOCK, TH, AND SCHWEITZER, J: *The auditory brain stem response in five vertebrate classes.* Electroencephalogr Clin Neurophysiol 54:629–641, 1982.

18. COURCHESNE, E, HILLYARD, SA, AND GALAMBOS, R: *Stimulus novelty, task relevance and the visual evoked potential in man.* Electroencephalogr Clin Neurophysiol 39:131–143, 1975.

19. COURCHESNE, E: *The maturation of cognitive components of the event-related brain potential.* In GAILLARD, AWK, AND RITTER, W (EDS): *Tutorials in ERP Research: Endogenous Components.* North-Holland Publishing, Amsterdam, 1983, pp 329–344.

20. CREUTZFELD, OD, WATANABE, S, AND LUX, HD: *Relations between EEG phenomena and potentials of single cortical cells. I. Evoked responses after thalamic and epicortical stimulation.* Electroencephalogr Clin Neurophysiol 20:1–8, 1966.

21. CREUTZFELD, OD, WATANABE, S, AND LUX, HD: *Relations between EEG phenomena and potentials of single cortical cells. II. Spontaneous and convulsoid activity.* Electroencephalogr Clin Neurophysiol 22:19–37, 1966.

22. CUFFIN, BN, AND COHEN, D: *Comparison of the magnetoencephalogram and electroencephalogram.* Electroencephalogr Clin Neurophysiol 47:132–146, 1979.

23. DAMASIO, A, DAMASIO, H, AND CHUI, HC: *Neglect following damage to frontal lobe or basal ganglia.* Neuropsychologia 18:123–132, 1980.

24. DAVIS, H, OSTERHAMMEL, RA, AND WEIR, CC: *Slow vertex potentials: Interactions among auditory, tactile, electric, and visual stimuli.* Electroencephalogr Clin Neurophysiol 33:537–545, 1972.

25. DEBECKER, J, AND DESMEDT, JE: *Maximum capacity for sequential one-bit auditory decisions.* J Exp Psychol 83:366–372, 1970.

26. DEECKE, J, AND KORNHUBER, HH: *An electrical sign of participation of the mesial supplementary motor cortex in human voluntary finger movement.* Brain Res 159:473–476, 1978.

27. DESMEDT, JE, AND ROBERTSON, D: *Differential enhancement of early and late components of the cerebral somatosensory evoked potentials during forced-paced cognitive tasks in man.* J Physiol (Lond) 271:761–782, 1977.

28. DESMEDT, JE, AND DEBECKER, S: *Wave form and neural mechanism of the decision P350 elicited without pre-stimulus CNV or readiness potential in random sequences of near threshold auditory clicks and finger stimuli.* Electroencephalogr Clin Neurophysiol 47:648–670, 1979a.

29. DESMEDT, JD, AND DEBECKER, J: *Slow potential shifts and decision P350 interactions in tasks with random sequences for near-threshold clicks and finger stimuli delivered at regular intervals.* Electroencephalogr Clin Neurophysiol 47:671–679, 1979b.

30. DESMEDT, JE, *Scalp-recorded cerebral event-related potentials in man as point of entry into the analysis of cognitive processing.* In SCHMITT FO, WORDEN, FG, ADELMAN, G, AND DENNIS, SD (EDS): *The Organization of the Cerebral Cortex.* MIT Press, Cambridge, MA, 1981, pp 441–473.

31. DUFF, T: *Topography of scalp recorded potentials evoked by stimulation of the digits.* Electroencephalogr Clin Neurophysiol 49:452–460, 1980.

31a. DUFFY, FH, ALBERT, MS, MCANULTY, G, AND GARVEY, AJ: *Age-related differences in brain electrical activity of healthy subjects.* Ann Neurol 16:430–438, 1984.

31b. DUFFY, FH, ALBERT, MS, AND MCANULTY, G: *Brain electrical activity in patients with presenile dementia of the Alzheimer type.* Ann Neurol 16:439–448, 1984.

32. DUFFY, FH, BURCHFIEL, JL, AND LOMBROSO, CT: *Brain electrical activity mapping (BEAM): A method for extending the clinical utility of EEG and evoked potential data.* Ann Neurol 5:309–321, 1979.

33. DUFFY, FH, DENCKLA, MB, BARTELS, PH, AND SANDINI, G: *Dyslexia: Regional differences in brain electrical activity by topographic mapping.* Ann Neurol 1:412–420, 1979.

34. EDINGER, HM, SIEGEL, A, AND, TROIANO, R: *Effect of stimulation of prefrontal cortex and amygdala on diencephalic neurons.* Brain Res 97:17–31, 1975.

35. FERRIER, D: *Functions of the Brain.* Smith & Elder, London, 1886.

36. FREEMAN, WJ: *Models of dynamics of neural populations.* Electroencephalogr Clin Neurophysiol 34:9–18, 1978.

37. FUSTER, JM: *The Prefrontal Cortex.* Raven Press, New York, 1980.

38. GALAMBOS, R, BENSON, P, SMITH, TS, SCHULMAN-GALAMBOS, C, AND OSIER, H: *On hemispheric differences in evoked potentials to speech stimuli.* Electroencephalogr Clin Neurophysiol 39:279–283, 1975.

39. GALAMBOS, R, AND HILLYARD, SA: *Electrophysiological approaches to human cognitive processing.* NRP Bull 20:145–265, 1981.

40. GEVINS, AS, ZEITLIN, GM, DOYLE, JC, YINGLING, CD, SCHAFFER, RE, CALLAWAY, E, AND YEAGER, CL: *Electroencephalogram correlates of higher cortical functions.* Science 203:665–668, 1979.

41. GIBLIN, DR: *Somatosensory evoked potentials in healthy subjects and in patients with lesions of the nervous system.* Ann NY Acad Sci 112:93–142, 1964.

42. GOFF, WR, ALLISON, T, AND VAUGHAN, HG JR: *The functional neuroanatomy of event-related potentials.* In CALLAWAY, E, TUETING, P, AND KOSLOW, SH (EDS): *Event-Related Brain Potentials in Man.* Academic Press, New York, 1978, pp 1–79.

43. GOLDIE, WD, CHIAPPA, KH, YOUNG, RR, AND BROOK, EB: *Brainstem auditory and short latency somatosensory evoked response in brain death.* Neurology 31:248–256, 1981.

44. GOLDMAN, PS, AND NAUTA, WJH: *Columnar distribution of corticocortical fibers in the frontal association, limbic, and motor cortex of the developing rhesus monkey.* Brain Res 122:393–413, 1977.

45. GOODIN, DS, SQUIRES, KC, HENDERSON, BH, AND STARR, A: *Age-related variations in evoked potentials to auditory stimuli in normal human subjects.* Electroencephalogr Clin Neurophysiol 44:447–458, 1978.

46. GOODIN, DS, SQUIRES, KC, AND STARR, A: *Long latency event-related components of the auditory evoked potential in dementia.* Brain 101:635–648, 1978.

47. GREEN, JB, AND HAMILTON, WJ: *Anosognosia for hemiplegia: Somatosensory evoked potential studies.* Neurology 26:1141–1144, 1976.

48. HALGREN, E, SQUIRE, NK, WILSON, CL, ROHRBAUGH, JW, BABB, TL, AND CRANDALL, PH: *Endogenous potentials generated in the human hippocampal formation and amygdala by infrequent events.* Science 210:803–805, 1980.

49. HANSEN, JC, AND HILLYARD, SA: *Endogenous brain potentials associated with selective auditory attention.* Electroencephalogr Clin Neurophysiol 49:277–290, 1980.

50. HARTER, MR, AND GUIDO, W: *Attention to pattern orientation: Negative cortical potentials, reaction time and the selection process.* Electroencephalogr Clin Neurophysiol 49:461–475, 1980.

51. HEILMAN, KM, AND VALENSTEIN, E. *Frontal lobe neglect in man.* Neurology 22:660–664, 1972.

52. HEILMAN, KA, AND VAN DEN ABELL, T: *Right hemisphere dominance for attention: The mechanism underlying hemisphere asymmetries of inattention (neglect).* Neurology 30:327–330, 1980.

53. HEILMAN, KM, AND WATSON, RT: *Mechanisms underlying the unilateral neglect syndrome.* In WEINSTEIN, EA, AND FRIEDLAND, RP (EDS): *Advances in Neurology,* Vol 18. Raven Press, New York, 1977, pp 93–106.

54. HENDERSON, CJ, BUTLER, SR, AND GLASS, A: *The localization of equivalent dipoles of EEG sources by the application of electric field theory.* Electroencephalogr Clin Neurophysiol 39:117–130, 1975.

55. HERNANDEZ-PEON, R, SCHERRER, H, AND JOUVET, M: *Modification of electric activity in cochlear nucleus during attention in unanesthesized cats.* Science 123:331–332, 1956.

56. HILLYARD, SA, SQUIRES, KC, BAUER, JW, AND LINDSEY, PN: *Evoked potential correlates of auditory signal detection.* Science 172:1357–1360, 1971.

57. HILLYARD, SA, HINK, RF, SCHWENT, VL, AND PICTON, TW: *Electrical signs of selective attention in the human brain.* Science 182:177–180, 1973.

58. HILLYARD, SA, AND PICTON, TW: *Event-related brain potentials and selective information processing in man.* In DESMEDT, JE (ED): *Progress in Clinical Neurophysiology.* (Vol 6: Cognitive Components in Cerebral Event-Related Potentials and Selective Attention.) Karger-Basel, 1979, pp 1–52.

59. HILLYARD, SA, AND KUTAS, M: *Electrophysiology of cognitive processing.* Ann Rev Psychol 34:33–61, 1983.

60. HUBEL, DH: *Exploration of the primary visual cortex.* Nature 299:515 524, 1982.

61. HUBEL, DH, AND WIESEL, TN: *Receptive fields, binocular interaction and functional architecture in the cat's visual cortex.* J Physiol (Lond) 160:106–154, 1962.

62. HYVARINEN, J, PORANEN, A, AND JOKINEN, Y: *Influence of attentive behavior on neuronal response to vibration in primary somatosensory cortex of the monkey.* J Neurophysiol 43:870–882, 1980.

63. INGVAR, DH, AND FRANZEN, G: *Distribution of cerebral activity in chronic schizophrenia.* Lancet 2:1484–1486, 1974.

64. INGVAR, DH, SJOLUND, B, AND ARDO, A: *Correlation between dominant EEG frequency, cerebral oxygen uptake and blood flow.* Electroencephalogr Clin Neurophysiol 41:268–276, 1976.

65. JACOBSEN, CF: *Functions of frontal association area in primates.* Arch Neurol Psychiat 33:558–569, 1935.

66. JEWETT, DL, ROMANO, MN, AND WILLISTON, JS: *Human auditory evoked potentials: Possible brainstem components detected in the scalp.* Science 167:1517–1518, 1970.

67. KIMBLE, DP, BAGSHAW, MH, AND PRIBRAM, KH: *The GSR of monkeys during orienting and habituation after selective partial ablations of the cingulate and frontal cortex.* Neuropsychologia 3:121–128, 1965.

68. KLEE, M, AND RALL, W: *Computed potentials of cortically arranged populations of neurons.* J Neurophysiol 40:647–666, 1977.

69. KNIGHT, RT, HILLYARD, SA, WOODS, DL, AND NEVILLE, HJ: *The effects of frontal and temporal-parietal lesions on the auditory evoked potential in man.* Electroencephalogr Clin Neurophysiol 50:112–124, 1980.

70. KNIGHT, RT, HILLYARD, SA, WOODS, DL, AND, NEVILLE, HJ: *The effects of frontal cortex lesions on event-related potentials during auditory selective attention.* Electroencephalogr Clin Neurophysiol 52:571–582, 1981.

71. KNIGHT, RT: *Decreased response to novel stimuli after prefrontal lesions in man.* Electroencephalogr Clin Neurophysiol 54:9–20, 1984.

72. KRAUS, N, OZDAMAR, O, HIER, D, AND STEIN, L: *Auditory middle latency responses (MRLs) in patients with cortical lesions.* Electroencephalogr Clin Neurophysiol 54:275–287, 1982.

73. KUTAS, M, AND HILLYARD, SA: *Reading senseless sentences: Brain potentials reflect semantic incongruity.* Science 207:203–205, 1980.

74. LOISELLE, DL, STAMN, JS, MAITINSKY, S, AND WHITE, SC: *Evoked potential and behavioral signs of attentive dysfunctions in hyperactive boys.* Psychophysiology 17:193–201, 1980.

75. LYNN, R: *Attention, Arousal and the Orientation Reaction.* Pergamon Press, 1966.

76. LURIA, AR, AND HOMSKAYA, ED: *Frontal lobes and the regulation of arousal process.* In MOSTOFKSY, DI (ED): *Attention: Contemporary Theory and Analysis.* Appleton-Century-Crofts, East Norwalk, CT, 1970.

77. MALMO, RB: *Interference factors in delayed response in monkeys after removal of frontal lobes.* J Neurophysiol 5:295–308, 1942.

78. MARKOWITSCH, HJ, AND PRITZEL, M: *Comparative analysis of prefrontal learning functions in rats, cats, and monkeys.* Psychol Bull 84:817–837, 1977.

79. MAUGUIERE, F, BRECHARD, S, PERNIER, J, COURJON, J, AND SCHOTT, B: *Anosognosia with hemiplegia: Auditory evoked potential studies.* In COURJON, J, MAUGUIERE, F, AND REVOL, M (EDS): *Clinical Applications of Evoked Potentials in Neurology.* Raven Press, New York, 1982, pp 271–278.

80. MESULAM, M-M: *A cortical network for directed attention and unilateral neglect.* Ann Neurol 10:309–325, 1981.

81. MILNER, B: *Effects of different brain lesions on card sorting.* Arch Neurol 9:100–110, 1963.

82. MORSTYN, R, DUFFY, FH, AND McCARLEY, RW: *Altered topography of EEG spectral content in schizophrenia.* Electroencephalogr Clin Neurophysiol 56:263–271, 1983.

83. MORUZZI, G, AND MAGOUN, HW: *Brain stem reticular formation and activation of the EEG.* Electroencephalogr Clin Neurophysiol 1:455–473, 1949.

84. MOUNTCASTLE, VB: *The world around us: Neural command functions for selective attention.* Neurosci Res Prog Bull (Suppl) 14:1–47, 1976.

85. MOUNTCASTLE, VB, ANDERSON, RA, AND MOTTER, BC: *The influence of attentive fixation upon the excitability of the light-sensitive neurons of the posterior parietal cortex.* J Neurosci 1:1218–1235, 1981.

86. NAATANEN, R: *Processing negativity: An evoked-potential reflection of selective attention.* Psychol Bull 92:605–640, 1982.

87. NAATANEN, R, AND GAILLARD, AWK: *The orienting reflex and the N2 deflection of the event-related potential (ERP).* In GAILLARD, AWK, AND RITTER, W (EDS): *Tutorials in Event-Related Potential Research: Endogenous Components.* North-Holland Publishing, 1983, pp 119–141.

88. NAGATA, K, MIZUKAMI, M, ARAKI, G, KAVASE, T, AND HIRANO, M: *Topographic electroencephalographic study of cerebral infarction using computed mapping of the EEG.* Journal of Cerebral Blood Flow and Metabolism 2:79–88, 1982.

89. NAUTA, WJH: *The problem of the frontal lobe: A reinterpretation.* J Psychiatr Res 8:167–187, 1971.

90. NUNEZ, PL: *Electrical Fields of the Brain.* Oxford University Press, 1981.

91. O'BRIEN, JH: *P300 wave elicited by a stimulus-change paradigm in acutely prepared rats.* Physiol Behav 28:711–713, 1982.

92. OKADA, YC, KAUFMAN, L, AND WILLIAMSON, SJ: *The hippocampal formation as a source of the slow endogenous potentials.* Electroencephalogr Clin Neurophysiol 55:417–426, 1983.

93. O'KEEFE, J, AND NADEL, L: *The Hippocampus as a Cognitive Map.* Clarendon Press, Oxford, 1978.

94. PANDYA, DN, AND SELTZER, B: *Association area of the cerebral cortex.* Trends in Neuroscience 5:386–390, 1982.

95. PFURTSCHELLER, G, AND COOPER, R: *Frequency dependence of the transmission of the EEG from cortex to scalp.* Electroencephalogr Clin Neurophysiol 38:93–96, 1975.

96. PFURTSCHELLER, G, AND ARANIBAR, A: *Event-related desynchronization detected by power measurements of scalp EEG.* Electroencephalogr Clin Neurophysiol 42:817–826, 1977.

97. PFURTSCHELLER, G, BUSER, P, LOPES DA SILVA, FH, AND PETSCHE, H: *Rhythmic EEG Activities and Cortical Functioning.* Elsevier/North-Holland, 1980.

98. PICTON, TW, HILLYARD, SA, KRAUSE, HI, AND GALAMBOS, RG: *Human auditory potentials. I. Evaluation of components.* Electroencephalogr Clin Neurophysiol 36:179–190, 1974.

99. PICTON, TW, AND HILLYARD, SA: *Human auditory evoked potentials. II. Effects of attention.* Electroencephalogr Clin Neurophysiol 36:191–200, 1974.

100. RAGOT, R, DEROUESNE, C, RENAULT, B, AND LESEVRE, N: *Ideomotor apraxia and P300: A preliminary study.* In COURJON, J, MAUGUIERE, F, AND REVOL, M: *Clinical Application of Evoked Potentials in Neurology.* Raven Press, New York, 1982, pp 263–269.

101. RITTER, W, VAUGHAN, HG, JR, AND COSTA, LD: *Orienting and habituation to auditory stimuli: A study of short term changes in averaged evoked responses.* Electroencephalogr Clin Neurophysiol 25:550–556, 1968.

102. RITTER, W, SIMSON, R, VAUGHAN, HG, JR, AND MACHT, M: *Manipulation of event-related potential manifestations of information processing stages.* Science 218:909–911, 1982.

103. ROHRBAUGH, JW, AND GAILLARD, AWK: *Sensory and motor aspects of the contingent negative variation.* In GAILLARD, AWK, AND RITTER W: *Tutorials in Event Related Potential Research: Endogenous Components.* North-Holland Publishing, Amsterdam, 1983, pp 269–310.

104. ROUGEL, A, BOUYEN, JJ, DEDET, L, AND DEBRAY, O: *Fast somatoparietal rhythms during combined focal attention and immobility in baboon and squirrel monkey.* Electroencephalogr Clin Neurophysiol 46:310–319, 1979.

105. ROLAND, PE, SKINHOJ, E, AND LASSEN, NA: *Focal activities of human cerebral cortex during auditory discrimination.* J Neurophysiol 45:1139–1151, 1981a.

106. ROLAND, PE: *Somatotopical tuning of postcentral gyrus during focal attention in man: A regional cerebral blood flow study.* J Neurophysiol 46:744–754, 1981.

107. ROLAND, PE: *Cortical regulation of selective attention in man. A regional cerebral blood flow study.* J Neurophysiol 48:1059–1078, 1982.

108. SHARPLESS, S, AND JASPER, H: *Habituation of the arousal reaction.* Brain 79:655–680, 1956.

109. SIEGEL, A, EDINGER, H, AND LOWENTHAL, H: *Effects of electrical stimulation of the medical aspect of the prefrontal cortex upon attack behavior in rats.* Brain Res 66:467–479, 1974.

110. SIMSON, R, VAUGHAN, HG, JR, AND RITTER, W: *The scalp topography of potentials associated with missing visual or auditory stimuli.* Electroencephalogr Clin Neurophysiol 40:33–42, 1976.

111. SKINNER, JE, AND YINGLING, CD: *Central gating mechanisms that regulate event-related potentials and behavior.* In DESMEDT, JE (ED): *Attention, Voluntary Contraction and Event-Related Cerebral Potentials.* Prog Clin Neurophysiol Vol 1. Karger, Basel, 1977, pp 30–69.

112. SNYDER, E, AND HILLYARD, SA: *Long-latency evoked potentials to irrelevant, deviant stimuli.* Behav Biol 16:319–331, 1976.

113. SNYDER, E, HILLYARD, SA, AND GALAMBOS, R: *Similarities and differences among the P3 waves to detected signals in three modalities.* Psychophysiology 17:112–122, 1980.

114. SOKOLOV, EN: *Higher nervous functions: The orienting reflex.* Ann Rev Physiol 25:545–580, 1963.

115. SQUIRES, NK, HALGREN, E, WILSON, C, AND CRANDALL, P: *Human endogenous limbic potentials: Cross modality and depth/surface comparisons in epileptic subjects.* In GAILLARD, AWK, AND RITTER, W (EDS): *Tutorials in Event-Related Potential Research: Endogenous Components.* 1983, pp 217–232.

116. SQUIRES, NK, SQUIRES, KC, AND HILLYARD, SA: *Two varieties of long-latency positive waves evoked by unpredictable auditory stimuli in man.* Electroencephalogr Clin Neurophysiol 38:387–401, 1975.

117. STARR, A: *Sensory evoked potentials in clinical disorders of the nervous system.* Ann Rev Neurosci 1:103–127, 1978.

118. STEIN, S, AND VOLPE, BT: *Classic "parietal" neglect syndrome after subcortical right frontal lobe infarction.* Neurology 33:797–799, 1983.

119. STEINSCHNEIDER, M, AREZZO, J, AND VAUGHAN, HG, JR: *Speech evoked activity in the auditory radiations and cortex of the awake monkey.* Brain Res 252:353–365, 1982.

120. SUTTON, S, BRAREN, M, ZUBIN, J, AND JOHN, ER: *Evoked-potential correlates of stimulus uncertainty.* Science 150:1187–1188, 1965.

121. SYNDULKO, K, HANSCH, EC, COHEN, SN, PEARCE, JW, GOLDBERG, Z, MONTAN, B, TOURTELLOLTE, WW, AND POTVUM, AR: *Long-latency event-related potentials in normal aging and dementia.* In COURJON, J, MAUGUIERE, F, AND REVOL, M: *Clinical Applications of Evoked Potentials in Neurology.* Raven Press, New York, 1982, pp 279–285.

122. SZENTAGOTHAI, J: *The module concept in cerebral cortex architecture.* Brain Res 95:475–496, 1975.

123. VANDERWOLF, CH, AND ROBINSON, TE: *Reticulo-cortical activity and behavior: A critique of the arousal theory and a new synthesis.* The Behavioral and Brain Sciences 4:459–514, 1981.

124. VAN VOORHIS, S, AND HILLYARD, SA: *Visual evoked potentials and selective attention to points in space.* Perception and Psychophysics 22:54–62, 1977.

125. VAUGHAN, HG, JR, RITTER, W, AND SIMPSON, R: *Neurophysiological consideration in event-related potential research.* In GAILLARD, WK, AND RITTER, W (EDS): *Tutorials in Event-Related Potential Research: Endogenous Components.* North-Holland Publishing, Amsterdam, 1983, pp 1–7.

126. VINOGRADOVA, OS: *Functional organization of the limbic system in the process of registration of information: Facts and hypothesis.* In ISAACSON, RL, AND PRIBRAM, KH: *The Hippocampus,* Vol 2: *Neurophysiology and Behavior.* Plenum Press, New York, 1975, pp 3–67.

127. WADA, JA, CLARKE, R, AND HAMM, A: *Cerebral hemispheric asymmetry in humans.* Arch Neurol 32:239–246, 1975.

128. WALTER, WGR, COOPER, R, ALDRIDGE, VJ, McCALLUM, WC, AND WINTER, AL: *Contingent negative variation: An electric sign of sensorimotor association and expectancy in the human brain.* Nature 203:380–384, 1964.

129. WELCH, K, AND STUTEVILLE, P: *Experimental production of unilateral neglect in monkeys.* Brain 81:341–347, 1958.

130. WILDER, MB, FARLEY, GR, AND STARR, A: *Endogenous late positive componenet of the evoked potential in cats corresponding to P300 in humans.* Science 211:605–607, 1981.

131. WILLIAMSON, PD, GOFF, WR, AND ALLISON, T: *Somatosensory evoked responses in patients with unilateral cerebral lesions.* Electroencephalogr Clin Neurophysiol 28:566–575, 1970.

132. WOOD, CC, ALLISON, T, GOFF, WR, WILLIAMSON, PD, AND SPENCER, DD: *On the neural origin of P300 in man.* In KORNHUBER, HH, AND DEECKE, L: *Progress in Brain Research Vol 54. Motivation, Motor and Sensory Processes of the Brain: Electric Potentials, Behavior and Clinical Use.* Elsevier/North Holland, 1980, pp 51–56.

133. WOODS, DL, COURCHESNE, E, HILLYARD, SA, AND GALAMBOS, R: *Recovery cycle of event-related potentials in multiple decision tasks.* Electroencephalogr Clin Neurophysiol 50:335–347, 1980.

134. WOODS, DL, HILLYARD, SA, COURCHESNE, E, AND GALAMBOS, R: *Electrophysiological signs of split-second decision making.* Science 207:655–657, 1980.

135. WOOLSEY, CN, MARSHALL, WH, AND BARD, P: *Representation of cutaneous tactile sensibility in the cerebral cortex of the monkey as indicated by evoked potentials.* Bull Johns Hopkins Hosp 70:399–440, 1942.

136. WORDEN, FG, AND MARSH, JT: *Amplitude changes of auditory potentials evoked at cochlear nucleus during acoustic habituation.* Electroencephalogr Clin Neurophysiol 15:866–881, 1963.

Chapter 10

IMAGING REGIONAL BRAIN PHYSIOLOGY IN BEHAVIORAL NEUROLOGY*

Ruben C. Gur, Ph.D.

THE TECHNIQUES

In the 1970s, a number of techniques have been developed for the imaging of metabolic activity in the living human brain. Only initial steps have been taken to integrate these new methods into the mainstream of behavioral neurosciences. This chapter describes some of these preliminary applications and illustrates how the developing techniques may be used to open new horizons for research in behavioral neurology.

Many of the techniques for the in vivo measurement of human brain physiology are an outgrowth of the Kety-Schmidt technique for measuring whole-brain flow and metabolism.[24] Using the Fick principle, cerebral blood flow (CBF), as well as cerebral glucose metabolism (CMRgl) and oxygen metabolism (CMRO2), can be calculated with considerable precision. The two major limitations of this technique are that it is invasive (requiring carotid and jugular catheterization) and that it is unable to provide information on the regional distribution of blood flow and metabolic activity. However, no other technique yields more accurate quantitation of whole-brain physiologic parameters, and values obtained with this technique still serve as a standard against which values obtained with other techniques are matched.

The Xenon Clearance Techniques

Lassen, Ingvar, and colleagues[28] applied the 133-xenon tracer methodology for measuring regional cerebral blood flow (rCBF). This method

*Supported by grants NS 14867, MN 36891, and MH 30456 and by the Spencer Foundation. I thank Raquel E. Gur, Marek-Marsel Mesulam, Walter D. Obrist, Martin Reivich, and Sushma S. Trivedi for their comments.

requires delivery of the radioisotope to the brain, and its clearance is measured with scintillation detectors. Initially, the 133-Xe was administered by intracarotid injection, enabling measurement of clearance in one hemisphere at a time. These limitations have been overcome by Obrist and co-workers'[34] 133-xenon inhalation technique, which is noninvasive and provides simultaneous rCBF measures in both hemispheres. The patient inhales radioactive gas mixed with air for about a minute. Xenon diffuses rapidly in blood and enters brain tissue. Algorithms for calculating a number of physiologic indices have been developed.[35] These include (1) the flow of the fast clearing compartment (f1), which presumably reflects gray matter flow (also referred to as fg); (2) the initial slope (IS) of the clearance curve, which corresponds to the slope at time zero of an equivalent bolus injection and represents a noncompartmental index of gray matter flow; and (3) the average flow (gray and white matter), equivalent to the height-over-area measure when the integration is carried out to 15 minutes of clearance (CBF-15). In addition, the percentage of counts emanating from the fast-clearing compartment can be calculated (w1 or wg). This is presumed to reflect, in normal brains, the percentage of gray matter. Thus, the method provides physiologic as well as limited anatomic information. Instruments for measuring rCBF with this technique are now commercially available, and their application in research and clinical practice is increasing. However, the imaging is two dimensional and does not adequately sample the medial and ventral parts of the brain. Furthermore, up to 15 minutes of clearance are needed for measuring rCBF, during which time the subject needs to be engaged in the relevant task.

Positron Emission Computerized Tomography (PECT)

PECT combines principles of the Kety-Schmidt technique with those of computerized tomography. Metabolically significant molecules are tagged with positron emitting radioisotopes and injected intravenously. The distribution of the tagged molecule in the brain can be determined because the emitted positrons annihilate when they collide with circulating electrons, and the two gamma photons, emitted in each annihilation, travel in opposite directions. The coincidental emissions are recorded by scintillation detectors, and a three-dimensional representation of the distribution of activity in the brain is reconstructed (Fig. 1). Thus far, 18-F-fluoro-d-2-deoxyglucose (FDG) has been used most extensively as a tracer for cerebral glucose metabolism.[40] Oxygen metabolism can be measured using inhalation of oxygen-15, and blood flow can be measured with oxygen-15–labeled water infused intravenously. Labeling of brain amines also is in progress, and images of the distribution of dopamine have already been generated.[8]

The widespread use of PECT is presently limited by the cost of the instrumentation and by the fact that a dedicated cyclotron is necessary to produce the short-lived isotopes. Furthermore, the FDG method requires 30 to 40 minutes to reach steady state. This means that the final distribution that has been imaged is a composite reflection of this entire

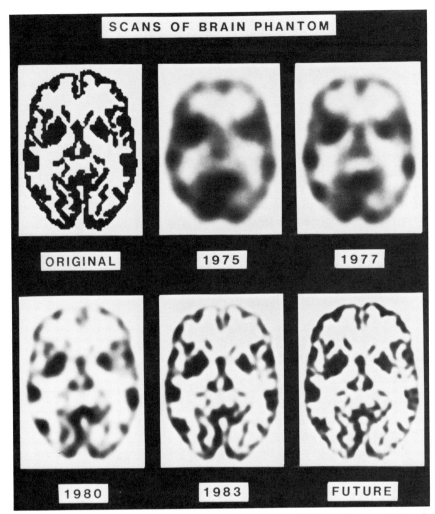

FIGURE 1. A phantom scan showing the effects of improved resolution on the expected quality of the PET images. (From Michael Phelps, UCLA, with permission.)

period. This creates a potential problem, since it is difficult to engage a subject in a uniform behavioral state for such an extended period.

Single-Photon Emission Computed Tomography (SPECT)

Tomographic techniques have been developed for imaging the three-dimensional distribution of cerebral blood flow (CBF) in conjunction with administration of single-photon (gamma ray) emitting isotopes. These techniques may provide more affordable substitutes for PECT in the measurement of CBF, because the relevant isotopes have much longer half-lives and do not necessitate an on-site cyclotron. However, at present the resolution of SPECT used with 133-xenon is relatively low, and the iodoamphetamine tracer (I-123) produces considerable radiation and has a long washout period. This poses difficulties for repeated studies or studies with normal individuals. However, SPECT scanning with I-123 iodoamphetamine offers a unique potential advantage: the tracer reaches its final distribution in the brain essentially after 2 to 3 minutes, and this distribution remains stable for several hours. Thus, it is possible to inject the tracer while the patient is engaged in a certain behavior in the neuropsychology laboratory and then to take the patient to the scanner for imaging.

Stable Xenon Enhancement for X-ray CT

Inhaled stable (nonradioactive) xenon rapidly diffuses in brain tissue. The gas has attenuation characteristics that permit measurement by sequential x-ray CT scans. Such sequential scanning may be used for quantitation and high-resolution imaging of regional CBF.[10] The cost of applying this technique is low, because it is used in conjunction with a conventional CT scanner. However, only one plane can be obtained per study, and the radiation exposure to the head from repeated scanning limits the number of studies that can be performed on one individual. A specific disadvantage of this technique for behavioral research is that, in the concentrations needed, xenon produces analgesic and mood-elevating effects.

Magnetic Resonance Imaging (MRI)

The use of powerful magnets for producing resonance of nuclei with changes in the polarity of the magnetic field has been a recent development that enables brain imaging without the use of radioactivity. Proton MRI devices are already in use in a number of clinical laboratories and are generating images of brain anatomy.[5] The resolution of these images varies with the device, but the overall quality of images is impressive. In particular, the separation of gray and white matter is superior to x-ray CT, and MRI images are not subject to beam-hardening effects that compromise the quality of conventional CT images near bone. Phosphorus nuclear magnetic resonance (31-P NMR) has been developed for measurement of pH and of relative concentrations of mobile phosphorus compounds involved in energy metabolism. This technique is still in its early development, but it has the potential of providing data that will be of physiologic significance.

Which Technique Will Prevail?

Within the span of a decade the technology of in vivo measurement and imaging of brain anatomy and physiology has expanded from one technique that enabled determination of whole-brain metabolism and blood-flow (the Kety-Schmidt technique) to a range of methods for two-dimensional and three-dimensional determinations of these parameters. It is unlikely that one of these techniques will replace the others. Each of the techniques can be improved, but each also has its own inherent limitations and is distinguished by the kind of information it can optimally provide. Currently MRI yields strictly anatomic information even though it has the potential for mapping regional metabolic activity. Of the techniques for measuring brain physiology, PECT seems most versatile, because it enables measurement of metabolic rates for glucose and oxygen, local cerebral blood flow, as well as the distribution of brain amines. Furthermore, it provides three-dimensional information, with reasonable resolution. Its major disadvantage is the high cost entailed by complex and cumbersome instrumentation and the need for a dedicated cyclotron. In addition, the procedures for quantitation of the physiologic parameters are still controversial and are in need of standardization. The 133-xenon clearance techniques and SPECT offer much more affordable alternatives, since they involve simpler instrumentation, do not require a dedicated cyclotron, and use commercially available isotopes. Other advantages of the 133-xenon clearance techniques are much lower radiation exposure; rapid clearance of isotope, which permits repeated measurements within a single session; and the capability of bedside measurement. They also have well-developed algorithms for data quantitation with good reproducibility. The disadvantage is that only two-dimensional information can be obtained and only on cerebral blood flow. SPECT provides three-dimensional information, but its resolution with currently available isotopes is not as good as that of PECT. The stable xenon x-ray CT yields three-dimensional information with good spatial resolution, but the radiation exposure from repeated x-ray CT scans prohibits the measurement of flow in more than one or two planes, and the required concentration of xenon produces changes in mental state.

APPLICATION TO NEUROPSYCHIATRIC CONDITIONS

The Normal Resting Pattern

An assessment of the effects of brain pathology requires comparison of normal patterns with specific clinical disorders. The normal pattern of regional brain metabolism and blood flow has been studied with the 133-xenon technique and the PECT. Although initial studies with the 133-xenon technique reported greater overall flows in the right hemisphere,[3] subsequent work both with this technique[12,34,41] and with PECT,[9,37] indicated that the normal brain is characterized by overall symmetry in blood flow and metabolism. The 133-xenon studies also show a "hyperfrontal pattern" of blood flow,[22] meaning that anterior flows are consistently higher than posterior flows. PECT scans likewise show increased resting metabolism and blood flow in frontal lobe regions, especially under conditions of sensory deprivation, but they also show peaks in flow for occipital lobe regions. The latter regions are not adequately sampled by the two-dimensional 133-xenon technique.

Cerebrovascular Disease

The 133-xenon technique has revealed bilaterally reduced rCBF in stroke patients, even when the disease is unilateral.[41,45] This "diaschisis" phenomenon suggests remote effects on brain regions in the contralateral hemisphere. It is as yet unclear how well focal reduction in rCBF can be demonstrated with the xenon clearance techniques. Since the rCBF indices are calculated on the basis of clearance rates, reduced rCBF in the area of an infarct may be obscured by normal flows in surrounding tissue. Thus, a detector would "see through" an area where there is no flow.[20] Furthermore, collateral circulation may normalize flow rate in the area of infarct. Resolution of these issues must await further study.

The PECT method has introduced marked advances in imaging the physiologic consequences of stroke. For example, PECT studies suggest that the region of reduced metabolism and flow after an infarct is considerably larger than the volume of anatomic destruction depicted by x-ray CT.[27] This can be seen clearly in Figure 2. This raises the possibility that the lesion seen on CT scan is not necessarily the only structural correlate for the observed behavioral deficit (see Chapter 5, section on subcortical aphasia). Furthermore, some brain regions that have been subjected to ischemia may suffer a sufficient depression of metabolism to result in behavioral changes, even though the tissue changes may not be of a magnitude that can be detected by CT scans. The value of PECT scanning in such situations is obvious.

Epilepsy

The physiologic measures may have their greatest potential for characterizing regional brain abnormalities in diseases such as epilepsy that do not usually manifest anatomic destruction detectable by x-ray CT.

Consistent with Penfield and co-workers'[36] observation of increased CBF during induced epileptic seizures in humans and animals, increased rCBF was noted in epileptics studied with the 133-xenon clearance technique.[43] Furthermore, it has been noted that in most patients there is a decrease of blood flow in the region of the focus during the interictal period, while there is an increase in rCBF in patients with extensive interictal spiking.[21]

PECT studies have been performed both in focal and nonfocal epilepsy. Kuhl and associates[26] studied patients with partial epilepsy who were candidates for focal surgery. Local cerebral glucose metabolism was measured by FDG PECT scans, and local cerebral blood flow was imaged with 13N-ammonia labeling. Interictal glucose metabolism and blood-flow were reduced in regions corresponding to the epileptic foci, whereas ictal scans showed increased metabolism and blood flow in the regions of EEG abnormalities. Recently, increased labeling over the seizure focus has also been shown with SPECT scans obtained during a seizure (see Chapter 8). As discussed in Chapter 8, these studies raise the exciting possibility that PECT and SPECT scans can provide useful information for localizing the seizure focus, especially in surgical candidates, in a way that could obviate the need for invasive depth electrodes (Fig. 3).[6]

Interictal hypometabolism in the area of the focus may explain some of the chronic psychologic deficits characteristic of epilepsy (see Chapter 8). Observations with the PECT scans also show that the area of hypometabolism can exceed the electrical focus. In fact, the exact distribution of interictal hypometabolism may provide considerable insight into the pattern of interictal behavioral disturbances.[15]

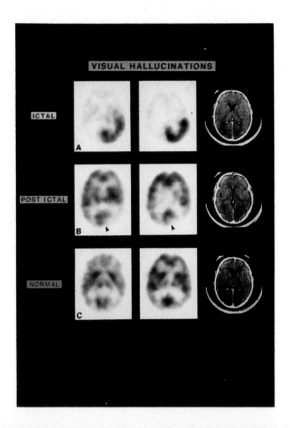

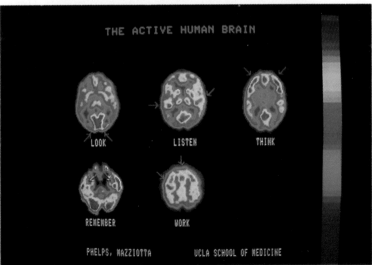

FIGURE 3. *Top*, PECT scan of an epileptic patient with ictal visual hallucinations. Note that the same region that is hypermetabolic ictally becomes hypometabolic in the postictal state.[6] (Courtesy of Michael Phelps, UCLA, with permission.) *Bottom*, PECT scans of normal subjects during different types of cognitive stimulation. (From Michael Phelps, UCLA, with permission.)

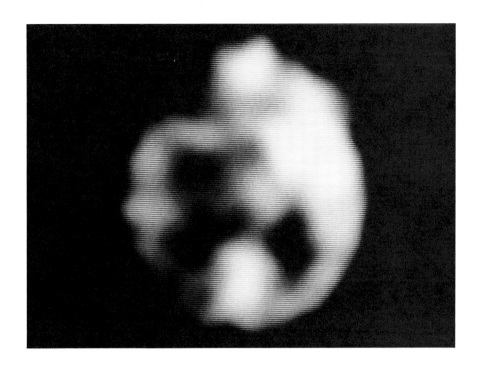

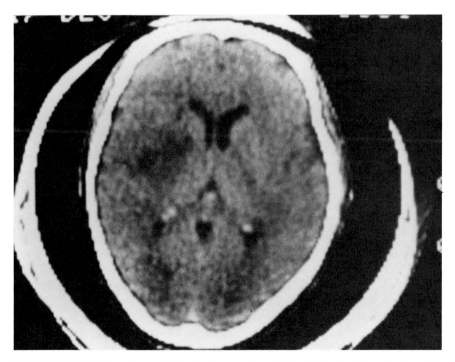

FIGURE 2. PECT *(top)* and x-ray CT *(bottom)* scans of a patient with stroke. Note that the wide-spread metabolic depression shown by PECT is more extensive than the size of the lucency shown on x-ray CT. (From Kushner, M, Alavi, A, Reivich, M, Dann, R, Burke, A, and Robinson, G: *Contralateral cerebellar hypometabolism following cerebral insult: A positron emission tomographic study.* Ann Neurol 15:425–434, 1984.

Dementia

Alavi[1] and Farkas[7] and their associates have reported FDG PECT data suggesting globally reduced glucose metabolism in demented patients, and rCBF measurements with the 133-xenon technique[33] have also suggested a reduced blood flow in this condition. However, these studies did not examine regional physiologic abnormalities in relation to the behavioral deficits, and the control samples were inadequately matched to the patients.

A recent report of patients with the clinical features of Alzheimer's disease suggested that there are areas that exhibit hypometabolism when compared with the same areas in normal individuals, especially in the superior parietal and adjacent temporal cortices. The sites of maximum hypometabolism agreed with the major behavioral deficits in individual patients. However, only five control subjects were studied.[4] The findings are certainly tantalizing, but their interpretation must await further study with adequate samples of controls matched for age, education, and overall premorbid intellectual functioning. However, such initial results do suggest that the PECT scanner may be very useful in identifying tissue hypofunction in dementia even when other neurodiagnostic investigations remain normal. Similar developments with the x-ray CT scanner are discussed in Chapter 11. Furthermore, these initial results also raise the possibility that the specific pattern of PECT hypometabolism in an individual patient may reflect that patient's unique pattern of behavioral deficits and the corresponding distribution of pathology. If confirmed, such correlations would be extremely useful in differentiating subtypes of dementia and for individually tailoring different therapeutic procedures.

Most dementia occurs in old age. Therefore, in order to determine if demented individuals vary from normal control subjects, it is necessary to acquire normative data on aging. In general, such studies show that aging is associated with decreased rCBF and glucose metabolism.[19,27] However, it is not yet known if this simply reflects a diminution of cerebral mass that occurs with age or if it indicates that remaining neurons show a decline in metabolic activity.

Psychiatric Disorders

The study of psychiatric disorders may benefit particularly from the application of the physiologic techniques. Even in major psychoses, there exist neither pronounced x-ray CT abnormalities nor consistent EEG patterns. However, behavioral alterations are severe enough to merit the hypothesis that regional brain function is disturbed in these conditions.

Studies with the Kety-Schmidt technique have indicated no differences in blood flow between schizophrenics and matched control subjects.[25] Initial studies with the 133-xenon technique showed reduced overall flows in schizophrenics,[30] as well as in depressives.[31] However, these differences may have been the result of variability in age, sex, education, or pCO_2 between patients and normal subjects. Gur and co-workers used the 133-xenon technique in three samples of psychiatric patients: medicated schizophrenics,[16] medicated depressives,[17] and unmedicated schizophrenics.[18] In each study patients were compared with matched control subjects and tested under standardized resting procedures and during cognitive activity. Consistent with the results of the Kety-Schmidt technique, no statistically significant differences were found in overall resting flows between patient and control groups.

Kety and associates[25] pointed out that the lack of differences in overall metabolism and blood flow between psychiatric patients and control subjects does not rule out regional differences. Indeed, initial studies with the 133-xenon technique[22] and with PECT[2,7] reported an absence of the normal "hyperfrontal pattern" in schizophrenics. Gur and co-workers[18] found a significant difference between unmedicated schizophrenics and matched control subjects in the hemispheric asymmetry of resting flows. Unlike control subjects who had symmetric flows, unmedicated schizophrenics had higher left-hemisphere flows. Furthermore, this abnormality was more pronounced in the severely disturbed patients. This finding was interpreted as supporting a hypothesis, derived from behavioral observations, that schizophrenia is characterized by left-hemispheric overactivation.[13] Recently, Sheppard and associates[44] reported a well-controlled PECT study of CBF in schizophrenics using the ^{15}O-labeled water infusion technique. The only significant abnormality in schizophrenics was increased left-hemispheric flows, a finding consistent with Gur and associates' results using the 133-xenon technique.

THE ACTIVATED BRAIN VERSUS THE IDLING BRAIN

Almost all of the metabolic mapping studies in neuropsychiatric conditions have compared the resting brain state of patients with that of matched control samples. It is conceivable, however, that even greater and more specific effects can be obtained if the comparisons are based on activation by specific tasks. For example, if the resting brain state of a patient with early dementia is compared with that of normal subjects, one might obtain only minor differences, if any at all. However, consistent physiologic differences could become detectable if the comparisons are done during activation by tasks that are most impaired in the neuropsychologic assessment of that person. This possibility is demonstrated in Figure 4, which shows the topographic display of rCBF measures using the 133-xenon technique in patients with left- and right-hemispheric stroke, during the performance of a verbal and a spatial task.

Activation strategies have also helped in highlighting abnormalities of brain activity in psychiatric conditions such as schizophrenia.[16,18]

INVESTIGATING BRAIN-BEHAVIOR RELATIONS IN NEUROLOGICALLY INTACT VOLUNTEERS

Since neural activity produces increased energy metabolism and since metabolic activity is, in turn, highly coupled with CBF,[38] the blood flow measures could be used in experimental research on brain-behavior relationships in which human subjects are engaged in specific tasks. Supporting the feasibility of such investigations, increased rCBF was demonstrated with the 133-xenon technique for simple sensory stimulation,

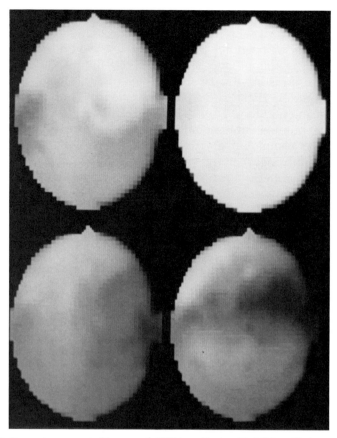

FIGURE 4. Computer-generated images of rCBF in patients with left (n = 11, *top row*) and right hemispheric (n = 9, *bottom row*) strokes. The images were obtained during the performance of a verbal *(left)* and a spatial *(right)* task. Note the focal relative hypoperfusion in the left hemisphere for the verbal task in patient with left-hemispheric stroke, and the focal hypoperfusion in the right for the spatial task in patients with right-hemispheric stroke. (Left is on left, right on right.)

as well as during the execution of motor movements.[29] For example, visual stimulation produced increased rCBF in the area of primary and association visual cortex, and auditory stimulation increased rCBF in the auditory cortex. Rhythmic movement of the eyes produced bilateral increase in rCBF in areas corresponding to the "frontal eye fields." Rhythmic movement of the hand, fingers, lips, and toes activated the respective contralateral motor areas.[29] Similar findings were demonstrated with PECT. Greenberg and co-workers[9] studied the effects of unilateral sensory stimulation on local glucose metabolism using FDG. Contralateral increase in metabolic rates was found in primary and secondary visual cortices during visual stimulation and in regions corresponding to auditory cortex during auditory stimulation. Phelps and associates[37] reported an effect of stimulus complexity: metabolism in response to complex scenes was higher than that in response to an alternating checkerboard pattern. Furthermore, white light activated only the primary visual cortex, whereas a complex scene also activated visual association cortex. These findings were replicated and extended in a systematic series of studies by the UCLA group (reviewed by Mazziotta[32]).

The sensitivity of the physiologic techniques to metabolic changes produced by sensory stimulation suggests that they can also be used to map more complex cognitive functions. This was initially examined by Ingvar and Risberg,[23] using the intracarotid 133-xenon technique. Increased rCBF was found during cognitive activity, compared with the resting baseline of rCBF. Risberg and Ingvar[42] measured rCBF in the dominant hemisphere in a group of chronic alcoholics. A visually presented reasoning task and an auditory digit-span-backward task were administered. Compared with resting rCBF, the rCBF was observed to be increased in the occipital, temporo-occipital, parietal, and frontal regions for the former task, and in the anterior frontal, prerolandic, and posterior temporal regions for the latter task. With the same technique, Carmon and co-workers[3] found greater increase in the left hemispheres that were studied in patients performing a verbal memory task, compared with patients who were listening to music. By contrast, greater increase occurred for right hemispheres in patients who were listening to music, compared with patients performing the same verbal task.

With the 133-xenon inhalation technique, which permits bilateral determination of rCBF, Risberg and associates[41] compared hemispheric rCBF during rest with laterality of flow during performance of a verbal analogies task and a spatial task (Street's incomplete figures). The subjects were right-handed volunteers who were rewarded for good performance. The rewarded subjects showed greater left-hemispheric increase for the verbal task and greater right-hemispheric increase for the spatial task. A group of subjects who were not rewarded for performance showed much weaker effects. Gur and Reivich[12] administered the same tasks to a group of 36 right-handed males. The main finding was that whereas the verbal task produced highly significant lateralized increases in rCBF in the expected left hemisphere, the spatial task produced greater right-hemispheric increase in 17 of the 36 subjects and reverse asymmetry in 17 subjects (the remaining two subjects showed the same laterality of rCBF as during rest). This suggests that the figure completion task could be solved either by right-hemispheric or by left-hemispheric strategies, whereas the verbal task was more "hardwired" to the left hemisphere. However, the evidence that right-hemispheric lesions produce greater impairment on the completion test suggests that right-hemispheric strategies are more essential and perhaps more effective. In support of this possibility, the subjects who showed greater right-hemispheric increase for the spatial task performed better than subjects who showed greater left-hemispheric increase. In a subsequent study with 62 young subjects,[14] the spatial completion test was substituted by a spatial line-orientation task developed by Benton (see Chapter 2). The task is highly sensitive to right-hemispheric lesions and shows a left visual field superiority in tachistoscopic studies. This time, both the verbal and the spatial task produced significant lateralized increase in rCBF. Furthermore, the effects were influenced by handedness and sex, factors reputed to affect the direction and degree of hemispheric specialization (Fig. 5).

The findings with rCBF were replicated in a PECT study.[11] The verbal and spatial tasks produced the expected increased metabolism in the

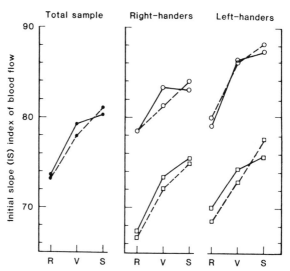

FIGURE 5. Hemispheric rCBF in 62 young normals during rest (R) and during activation with a verbal (V) and a spatial (S) task. Left *(solid lines)* and right *(dashed lines)* hemispheric values are shown for the total sample, and for right- and left-handed males (□) and females (○). (From Gur et al,[14] with permission.)

left and right hemispheres, respectively. This effect was especially significant in the superior temporal and inferior parietal regions (including Wernicke's area).

Thus far, studies of psychologic factors influencing regional brain physiology have focused on the effects of cognitive tasks. However, these techniques may also be suitable for the study of conation and affect. Attention is a central conative function whose anatomic substrates have recently been studied in a systematic fashion. Single-cell recordings demonstrated the existence in the inferior parietal region of neurons that are selectively activated by visual fixation and pursuit of motivationally relevant objects. The connectivity of area PG in the monkey brain, which is in the caudal part of the inferior parietal lobe, has been studied extensively by Mesulam and colleagues using the horseradish peroxidase technique. Based on this connectivity, Mesulam has suggested that attention to sensory targets may depend on the coordinated activity of a neural network that includes the inferior parietal lobule, cingulate cortex, and frontal eye fields (see Chapter 3).

In humans, there is a distinct hemispheric asymmetry in that unilateral neglect occurs much more frequently following right-hemispheric lesions. To test the hypothesized asymmetry in the attentional network, we have analyzed the PECT scans of subjects who were given sensory stimuli in the left and right visual fields in order to examine metabolic asymmetries in primary sensory cortices. Greater metabolic activity in the contralateral hemisphere was demonstrated in these subjects for the primary sensory cortices.[9] However, the experimental design has also permitted an examination of the effects of an attentional task. To ensure central fixation, the subjects were given an attentional task in which a light-emitting diode was placed in central fixation. The light source dimmed occasionally, and the subjects were told to pay attention to this light source and indicate when dimmings occurred. The PECT scans of these stimulated subjects were compared with scans obtained from control subjects who were studied while their eyes were blindfolded and their ears plugged. These control subjects were not given an attentional task. Three regions were selected for determination of local glucose metabolism: inferior parietal (IP), superior temporal (ST) lobe, and cerebellum (CR). The IP was the target region of interest, the ST was selected as a control region proximal to the IP, and the CR was selected as a control region that has not been implicated in attentional functions. The results showed greater right-hemispheric metabolism in stimulated subjects specifically observed for IP, with no significant lateralization effects in ST and CR. The attentional network was examined in further detail in a subsequent study, using a scanner with better resolution. In addition to the three regions evaluated in the first study, metabolic rates were calculated for the cingulate gyrus, the frontal eye fields, and the prefrontal cortex, as well as the calcarine cortex and several control regions. The results were replicated for the IP, with subjects attending to stimuli showing significantly higher right-hemispheric laterality of metabolism than was shown in control subjects. In addition, the same effects were obtained for the frontal eye fields, the prefrontal cortex, and the calcarine cortex, but not in the cingulate gyrus or any of the control regions.[39]

Studies that aim to investigate brain-behavior relationships tacitly assume that the organization of all brains is approximately equivalent. However, this may be an unwarranted assumption. For example, there are consistent differences in the resting pattern of blood flow in relationship to sex and handedness.[14] Furthermore, it is possible that individuals with special talents in diverse areas (e.g., mathematics, music, athletics) have systematic differences in brain physiology that underlie these special skills. The discovery of such differences promises to be an exciting area of research.

LIMITATIONS AND PITFALLS

The emphasis in this chapter is on the potentials of the new techniques. This should not be taken to imply that the techniques are free from artifacts and pitfalls or that the neuroscientist interested in pursuing their potential need only acquire the instruments and launch the research. Like any new methodology, the isotopic techniques demand attention to many technical problems that may render the results meaningless and even misleading when overlooked. It would be impossible to enumerate here the artifacts that are already recognized, and it is quite possible that additional artifacts are yet to be discovered. Anyone contemplating the use of these techniques in research or clinical practice is advised not only to become acquainted with the technical literature but also to consult with centers that are currently using the methods and have experience in their application. One example may help illustrate the issue.

There are commercially available machines for measuring rCBF with the 133-xenon technique, and it appears that the operator need only push a few buttons, place the subject or patient in the scanner, and receive a printed output of rCBF values. However, in all likelihood a number of the probes will be aimed at the sinuses, and clearance curves from these most likely contain artifacts reflecting the xenon concentrations in the air passages rather than in the brain. The algorithms for calculating rCBF may fail to detect these artifacts, and only visual inspection of the clearance curve by someone experienced in the detection of such artifacts can determine whether the curve can be corrected or whether the particular probe should be deleted. Failure to correct or delete such probes could invalidate the study.

Another major pitfall, due in part to the high cost of using the physiologic techniques, is reliance on small samples and failure to include control subjects and procedures. Few of the studies reported herein have included more subjects than variables, and statistical analysis in this situation is tenuous. This practice is particularly risky for behavioral research, which uses "soft" and multivariate data. Many of the questions that are of utmost interest will not be answered until the methodology of research in this field is improved.

Aside from the technical and methodologic pitfalls, there are theoretical issues that are yet to be resolved. A glaring example is the issue of what effects inhibitory activity may have on metabolism and bloodflow. In all the studies discussed in the preceding sections, the assumption has been that increased metabolism signifies activation of brain regions and, conversely, that increased activity of a particular brain region will be reflected in increased neural activity and hence metabolism and blood flow.[38] The possible role of inhibitory activity and the manner in which it could be studied with this technology are yet to be examined.

Finally, physiologic data need to be integrated with anatomic information. Thus far, this has been done with rather imprecise procedures, and localization of anatomic regions on the PECT images, as well as for the other physiologic techniques, is only approximate. Computer algorithms are currently being developed to permit better precision in localizing anatomic regions. This will be invaluable for future behavioral research.

REFERENCES

1. ALAVI, A, REIVICH, M, FERRIS, S, ET AL: Regional cerebral glucose metabolism in aging and senile dementia as determined by 18-F-deoxyglucose and positron emission tomography. Exp Brain Res 5(Suppl):187–195, 1982.

2. BUCHSBAUM, MS, INGVAR, DH, KESSLER, R, ET AL: Cerebral glucography with positron tomography. Arch Gen Psychiatry 39:251–259, 1982.

3. CARMON, A, LAVY, S, GORDON, H, AND PORTNOY, Z: Hemispheric differences in rCBF during verbal and nonverbal tasks. In INGVAR, DH, AND LASSEN, NA (EDS): Brain Work. Munksgaard, Copenhagen, 1975, pp 414–423.

4. CHASE, TN, FOSTER, NL, FEDIO, P, ET AL: Regional cortical dysfunction in Alzheimer's disease as determined by positron emission tomography. Ann Neurol 15(Suppl):170–175, 1984.

5. DOYLE, FH, GORE, JC, PENNOCK, JM, ET AL: Imaging of the brain by nuclear magnetic resonance. Lancet 2:53–57, 1981.

6. ENGEL, J, KUHL, DE, PHELPS, ME, AND CRANDALL, PH: Comparative localization of epileptic foci in partial epilepsy by PCT and EEG. Ann Neurol 12:529–537, 1982.

7. FARKAS, T, FERRIS, SH, WOLF, AP, ET AL: 18-F-2-deoxy-2-fluoro-D-glucose as a tracer in the positron emission tomographic study of senile dementia. Am J Psychiatry 139:352–353, 1982.

8. GARNETT, ES, NAHMIAS, C, AND FIRNAU, G: *Central dopaminergic pathways in hemiparkinsonism examined by positron emission tomography.* Can J Neurol Sci 11(Suppl):174–179, 1984.

9. GREENBERG, JH, REIVICH, M, ALAVI, A, ET AL: *Metabolic mapping of functional activity in man with 18F-fluorodeoxyglucose technique.* Science 212:678–680, 1981.

10. GUR, D, YONAS, H, HERBERT, D, ET AL: *Xenon enhanced dynamic computed tomography: Multilevel cerebral blood flow studies.* J Comp Assist Tomogr 5:334–340, 1981.

11. GUR, RC, GUR, RE, ROSEN, AD, ET AL: *A cognitive-motor network demonstrated by positron emission tomography.* Neuropsychologia 21:601–606, 1983.

12. GUR, RC, AND REIVICH, M: *Cognitive task effects on hemispheric blood flow in humans.* Brain Lang 9:78–93, 1980.

13. GUR, RE: *Left hemisphere dysfunction and left hemisphere overactivation in schizophrenia.* J Abnorm Psychol 87:226–238, 1978.

14. GUR, RC, GUR, RE, OBRIST, WD, ET AL: *Sex and handedness differences in cerebral blood flow during rest and cognitive activity.* Science 217:256–261, 1982.

15. GUR, RC, SUSSMAN, NM, ALAVI, A, ET AL: *Positron emission tomography in two cases of childhood epileptic encephalopathy (Lennox-Gastaut syndrome).* Neurology 32:1191–1195, 1982.

16. GUR, RE, SKOLNICK, BS, GUR, RC, ET AL: *Brain function in psychiatric disorders: I. Regional cerebral blood flow in medicated schizophrenics.* Arch Gen Psychiatry 40:1250–1254, 1983.

17. GUR, RE, SKOLNICK, BS, GUR, RC, ET AL: *Brain function in psychiatric disorders: II. Regional cerebral blood flow in medicated depressives.* Arch Gen Psychiatry 41:695–699, 1984.

18. GUR, RE, GUR, RC, SKOLNICK, BE, ET AL: *Brain function in psychiatric disorders: III. Regional cerebral blood flow in unmedicated schizophrenics.* Arch Gen Psychiatry (in press).

19. HAGSTADIUS, S, AND RISBERG, J: *The effects of normal aging on rCBF during resting and functional activation.* rCBF Bulletin 6:116–120, 1983.

20. HALSEY, JH: *Is there clinical value in the measurement of rCBF?* rCBF Bulletin 1:5–6, 1981.

21. INGVAR, DH: *Regional cerebral blood flow in focal cortical epilepsy.* Stroke 4:359–360, 1973.

22. INGVAR, DH, AND FRANZEN, G: *Distribution of cerebral activity in chronic schizophrenia.* Lancet 2:1484–1486, 1974.

23. INGVAR, DH, AND RISBERG, J: *Increase of regional cerebral blood flow during mental effort in normals and in patients with focal brain disorders.* Exp Brain Res 3:195–211, 1967.

24. KETY, SS, AND SCHMIDT, CF: *The nitrous oxide method for the quantitative determination of cerebral blood flow in man: Theory, procedure and normal values.* J Clin Invest 27:476–483, 1948.

25. KETY, SD, WOODFORD, RB, HARMEL, MH, ET AL: *Cerebral blood flow and metabolism in schizophrenia.* Am J Psychiatry 104:765–770, 1948.

26. KUHL, DE, ENGEL, J, PHELPS, ME, AND SELIN, C: *Epileptic pattern of local cerebral metabolism and perfusion in man determined by emission computed tomography of 18 FDG and 13 NH3.* Ann Neurol 8:348–360, 1980.

27. KUHL, DE, PHELPS, ME, KOWELL, AP, ET AL: *Effects of stroke on local cerebral metabolism and perfusion: Mapping by emission computed tomography of 18-FDG and 13-NH3.* Ann Neurol 8:47–60, 1980.

28. LASSEN, NA, AND INGVAR, GH: *Radioisotopic assessment of rCBF.* Prog Nucl Med 1:376–409, 1972.

29. LASSEN, NA, ROLAND, PE, LARSEN, B, ET AL: *Mapping of human cerebral functions: A study of the regional cerebral blood flow pattern during rest, its reproducibility and the activations seen during basic sensory and motor functions.* Acta Neurol Scand (Suppl)56:262–263, 1977.

30. MATHEW, RJ, DUNCAN, GC, WEINMAN, ML, ET AL: *Regional cerebral blood flow in schizophrenia.* Arch Gen Psychiatry 39:1121–1124, 1982.

31. MATHEW, R, MEYER, JS, FRANCIS, DJ, ET AL: *Cerebral blood flow in depression.* Am J Psychiatry 137:1449–1450, 1980.

32. MAZZIOTTA, JC, AND PHELPS, ME: *Human sensory stimulation and deprivation: Positron emission tomographic results and strategies.* Ann Neurol 15(Suppl): 48–49, 1984.

33. OBRIST, WD, CHIVIAN, E, CRONQVIST, S, AND INGVAR, D: *Regional cerebral blood flow in senile and presenile dementia.* Neurology 20:315–322, 1970.

34. OBRIST, WD, THOMPSON, HK, WANG, HS, ET AL: *Regional cerebral blood flow estimated by 133-Xe inhalation.* Stroke 6:245–256, 1975.

35. OBRIST, WD, AND WILKINSON, WE: *The noninvasive Xe-133 method: Evaluation of CBF indices.* In BES, A, AND GERAUD, G (EDS): *Cerebral Circulation.* Elsevier, Amsterdam, 1979.

36. PENFIELD, W, VON SANTHA, K, AND CIPRIANI, A: *Cerebral blood flow during induced epileptiform seizures in animals and man.* J Neurophysiol 2:257–267, 1939.

37. PHELPS, ME, KUHL, DE, AND MAZZIOTTA, JC: *Metabolic mapping of the brain's response to visual stimulation: Studies in humans.* Science 211:1445–1448, 1981.

38. REIVICH, M: *Blood flow metabolism couple in brain.* In PLUM, F (ED): *Brain Dysfunction in Metabolic Disorders.* Raven Press, New York, 1974.

39. REIVICH, M, GUR, RC, AND ALAVI, A: *Positron emission tomographic studies of sensory stimuli, cognitive processes and anxiety.* Human Neurobiology 2:25–33, 1983.

40. REIVICH, M, KUHL, DE, WOLF, A, ET AL: *The 18-F-fluorodeoxyglucose method for the measurement of local cerebral glucose utilization in man.* Circ Res 44:127–137, 1979.

41. RISBERG, J, HALSEY, JH, BLAUENSTEIN, VW, ET AL: Bilateral measurements of the rCBF during mental activation in normals and in dysphasic patients. In HARPER, AM, ET AL (EDS): *Blood flow and metabolism in the brain.* Churchill Livingstone, London, 1975.

42. RISBERG, J, AND INGVAR, DH: *Patterns of activation in grey matter of the dominant hemisphere during memorizing and reasoning.* Brain 96:737–756, 1973.

43. SAKAI, F, MEYER, JS, NARITOMI, H, ET AL: *Regional cerebral blood flow and EEG in patients with epilepsy.* Arch Neurol 35:648–657, 1978.

44. SHEPPARD, G, GRUZELIER, J, MANCHANDA, R, ET AL: *15-O positron emission tomographic scanning in predominantly never-treated acute schizophrenic patients.* Lancet 2:1448–1452, 1983.

45. SLATER, R, REIVICH, M, GOLDBERG, H, ET AL: *Diaschisis with cerebral infarction.* Stroke 8:684–690, 1977.

Chapter 11

QUANTITATIVE APPROACHES TO COMPUTERIZED TOMOGRAPHY IN BEHAVIORAL NEUROLOGY*

Margaret A. Naeser, Ph.D.

This chapter discusses the application of quantitative computerized tomography (CT) scan analysis techniques to three areas of behavioral neurology: (1) aphasia, (2) normal aging, and (3) dementia (presenile and senile dementia of the Alzheimer type). Emphasis is directed toward understanding how quantitative CT scan information may be useful in the diagnostic process within these areas of behavioral neurology.

ORIENTATION TO THE SEMIAUTOMATED COMPUTER PROGRAM, ASI-I, USED IN QUANTIFYING LESION SIZE

The semiautomated computer program called Automated Slice Information I (ASI-I) yields quantitative information regarding the relative percentage of hemispheric tissue damage (infarct size).[33,47] The picture elements (pixels) for any given CT slice are represented on a 256 × 256 X-Y coordinate matrix, which can be displayed on a cathode ray tube (CRT) viewer screen. The information for relevant pixels can be accessed with the help of a floating cursor, which can be used to delineate areas of interest on the CRT screen. (For more information regarding basic CT scan terminology, see work by McCullough.[39]) The software used to run the ASI-I program is programmed into the computer that is part of the CT scanner system. The ASI-I program is run sometime after the scan is

*This research was supported in part by the Medical Research Service of the Veterans Administration, in part by USPHS Grants NS06209 and NS07615, and in part by the National Institute on Aging Grant PO1-AG-0226.

The author would like to thank Dr. Marilyn Albert, Janis Fang, Hope Heller, Susan Lo Castro, Carole Palumbo, and Denise Stiassny for assistance in preparation of the manuscript, Charles Foltz for artwork, and Suzanne Ruscitti and Roger Ray for typing the manuscript.

QUANTITATIVE
APPROACHES TO
COMPUTERIZED
TOMOGRAPHY IN
BEHAVIORAL
NEUROLOGY

363

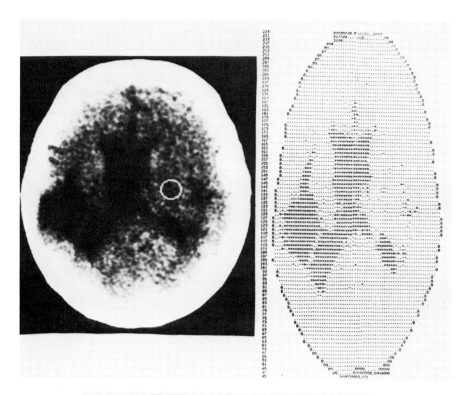

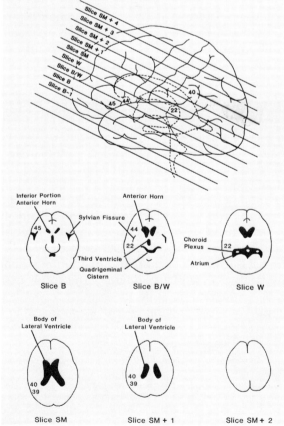

completed and takes approximately 1 hour of technician time per scan to complete. The steps involved are as follows:

1. When comparing one hemisphere to the other, the ASI-I program requires the user to indicate, with the cursor on the CRT screen, the X-Y coordinates of the midline structures of the brain (interhemispheric fissure, septum pellucidum, and so on), so that the right hemisphere (RH) can be distinguished from the left hemisphere (LH).
2. A 169-pixel healthy tissue sample (HTS) is taken in the hemisphere contralateral to the lesion, just lateral to the ventricles (usually mixed white and gray matter) on each slice as visualized at the CRT (Fig. 1, *top left*). The X-Y coordinates for the "center" of the HTS are entered into the computer.
3. The ASI-I program is executed, and the mean CT density number of the HTS at a given slice is then compared with the mean CT density number of four-pixel samples in the entire LH, then RH, of that slice with a *t* test. Each four-pixel sample that is significantly lower in mean CT density number (closer to cerebrospinal fluid [CSF] or water level) is represented by a probability value marker ($-$, $=$, or *) referring to the p <0.05, <0.01, or <0.001 levels, respectively, and printed out (Fig. 1, *top right*). The remaining tissue four-pixel samples that are not significantly lower than the HTS are printed out with periods. On the Ohio Nuclear scanners, CSF is near zero, and normal white or gray matter is 20 to 50 Hounsfield Units (HU) (-1000, air; $+1000$, bone). The significantly low CT number areas represent ventricle, fissures, and also infarcted tissue. *LoPix* is the abbreviation for significantly low four-pixel samples. In patients who have infarctions in only one hemisphere, the difference in %LoPix between the two hemispheres provides a measure of infarct size. ASI-I analysis of CT scans of normal subjects has shown that there is approximately only 1 percent difference between the left and right hemisphere %LoPix.[47]

Our programs have shown that hemisphere size can range from 6000 to 11,000 pixels, depending on the patient's head size and the slice level observed.[47] Therefore, the amount of tissue damage in *percentage* is more meaningful than the raw number of lesion pixels or their total area.[47,48]

FIGURE 1. *Top,* Sample ASI-I printout *(right)* of CT scan at the roof of the third ventricle slice level with left temporal lobe infarct in Wernicke's area *(left)*. The percentage of significantly low CT number pixels in the right hemisphere was 13 percent; left hemisphere, 29 percent; difference, 16 percent (left-hemisphere infarct size). The white circle represents the approximate location of the right hemisphere, healthy tissue sample (HTS). From Naeser et al,[47] with permission.
Bottom, Lateral and cross-sectional views of the brain that show relationship of cortical language areas to shape of ventricles on CT scan slices. Broca's area is 44 and 45; Wernicke's area, 22; supramarginal gyrus, 40; and angular gyrus, 39.

APHASIAS

For aphasia research purposes, each CT slice is labeled with respect to the hypothetical cortical language areas that are present on that slice (Fig. 1, *bottom*). Slice B is marked by frontal horns that are oval-shaped: part of Broca's area (pars triangularis, area 45) is immediately lateral to the left frontal horn. Slice B/W is marked by frontal horns that are butterfly shaped and by the presence of the third ventricle: part of Broca's area (pars opercularis, area 44) is immediately lateral to the left frontal horn and part of Wernicke's area (midportion of superior temporal gyrus, area 22) is lateral to the third ventricle. Slice W is marked by the characteristic shape of the occipital horns: part of Wernicke's area (posterior superior temporal gyrus, area 22) is anterolateral to the junction of the atrium with the left occipital horn. Slices SM and SM + 1 are marked by the presence of the bow-shaped bodies of the lateral ventricles: the supramarginal gyrus (area 40) is lateral to the posterior half of the body of the lateral ventricle. The angular gyrus (area 39) is posterior to the supramarginal gyrus. Several other reports contain further information regarding vascular supply to CT scan areas and the CT correlates of aphasia.[3−6,9−12,14,20,22,27,29,30,35,37,38,40,41,44,46,49,50]

The ability to quantify infarct size and its careful anatomic delineation lead to a better understanding of the relationship between the structural characteristics of the lesion and the clinical detail of the resultant aphasia in stroke cases.[41,47,48] The most reliable lesion size information is always obtained from CT scans performed at least 2 months or more after the stroke. Scans obtained earlier than this often do not show the entire extent of the lesion. The ASI-I program was run on each of the five language CT slices (B, B/W, W, SM, SM + 1) for each of the aphasia cases presented here. All patients were tested and diagnosed for the aphasia type with the Boston Diagnostic Aphasia Exam (BDAE).[26]

Transcortical Motor Aphasia versus Broca's Aphasia

Figure 2 shows the CT scans of a patient with mild transcortical motor aphasia *(top)* and one with moderately severe Broca's aphasia *(bottom)*. The patient with the transcortical motor aphasia (TCM) had a mild right hemiparesis, which cleared within the first week. The patient with Broca's aphasia had a dense right hemiplegia, affecting the arm more than the leg, which did not clear and persisted for at least 7 years. The CT scan for the TCM case was done 6 months after the stroke (Fig. 2, *top*) and shows no lesion in Broca's area at slice B. The CT scan for the patient

---

FIGURE 2. CT scans of patient with mild transcortical motor aphasia *(top)* versus patient with moderate Broca's aphasia *(bottom)*. This TCM case has small lesion in Broca's area (slice B/W only) and larger lesion in the frontal lobe, superior to Broca's area at slices W, SM, SM + 1, and SM + 2. The Broca's aphasia case has large cortical lesion in Broca's area (slices B and B/W), which extends deep across the frontal lobe to the lateral border of the left frontal horn *(arrows)*. This Broca's aphasia lesion also includes lower motor and sensory cortex for face and deep pyramidal tract in periventricular white matter on slice SM *(arrow)*. Percentage lesion size on each slice is noted in lower left corner.

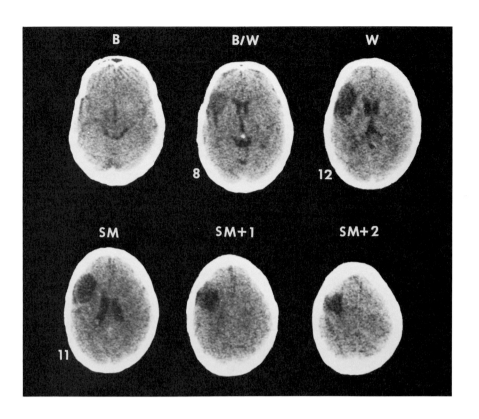

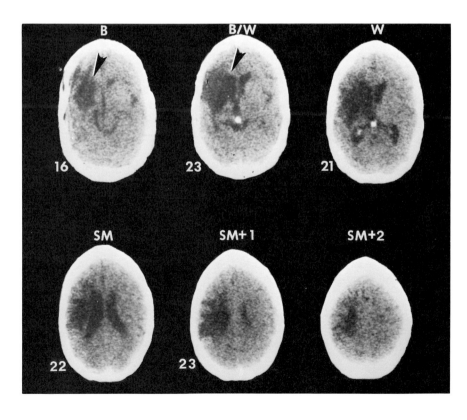

QUANTITATIVE
APPROACHES TO
COMPUTERIZED
TOMOGRAPHY IN
BEHAVIORAL
NEUROLOGY

367

with Broca's aphasia, however, was done 7 years after the stroke (Fig. 2, *bottom*) and shows a large lesion (LH %LoPix − RH %LoPix = 16) on slice B, which included Broca's area. The lesion extended deep across the frontal lobe to the lateral border of the left frontal horn *(arrow)*. On slice B/W, the CT scan of the TCM case showed a small lesion (8 percent) in Broca's area that did not extend across the frontal lobe to the lateral border of the left frontal horn. There was additional lesion, extension into the basal ganglia and internal capsule on slices B and B/W in the CT scan of the Broca case, but not in that of the TCM case. The largest lesion for the TCM case was present on slice W (12 percent) in the frontal lobe, superior to Broca's area. The remainder of the lesion for the TCM case extended into the more dorsal left frontal lobe at slices SM (11 percent), SM+1, and SM+2. The lesion in the patient with Broca's aphasia continued to enlarge dorsally on each succeeding slice: slice W, 21 percent; slice SM, 22 percent; slice SM+1, 23 percent. This more superior part of the lesion (slices SM and SM+1) extended into the motor and sensory cortex areas for the face, as well as into the anterior parietal lobe.

The motor cortex for the face is well represented on the surface at slice SM. Also, many deep pyramidal tract fibers for the arm and leg are represented in periventricular white matter adjacent to the body of the lateral ventricle on slice SM.[52] Hence, it is not surprising that the patient with Broca's aphasia, with a large lesion in these areas on slice SM, had a permanent hemiplegia and slow, poorly articulated speech.

Conduction Aphasia versus Wernicke's Aphasia

Figure 3 shows the CT scans of a patient with a mild conduction aphasia *(top)* and of a patient with Wernicke's aphasia *(bottom)*. The patient with the conduction aphasia initially had a mild right hemiparesis, which rapidly cleared. The one with Wernicke's aphasia had a very mild right hemiparesis (affecting the arm more than the leg) and right visual field deficit, both of which persisted. Hemiparesis is rare in Wernicke's aphasia even though it was present in a mild form in this case. It was probably due to deep lesion extension into the posterior limb of the internal capsule at slice B/W.

The CT scan for the patient with the conduction aphasia was obtained 4 months after the stroke (Fig. 3, *top*), and shows no lesion at slice B/W or W. Small lesions were present in the supramarginal/angular gyrus areas of the parietal lobe on slices SM (5 percent) and SM+1 (5 percent), with further extension into the parietal lobe on slice SM+2. The white matter bundle, which extends from the inferior parietal lobule and Wernicke's area toward Broca's area, is designated as the arcuate fasciculus (see Chapter 5). The extension of the lesion deep to the supramarginal gyrus on slices SM and SM+1 suggests that the arcuate fasci-

FIGURE 3. CT scans of patient with mild conduction aphasia *(top)* versus patient with moderate-to-severe Wernicke's aphasia *(bottom)*. This conduction aphasia case has small supramarginal/angular gyrus lesion at slices SM, SM+1, and SM+2. The Wernicke's aphasia case has larger Wernicke's area lesion at slices B/W and W and supramarginal gyrus lesion at slices SM and SM+1.

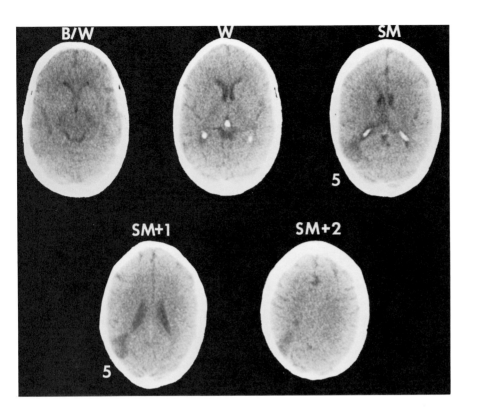

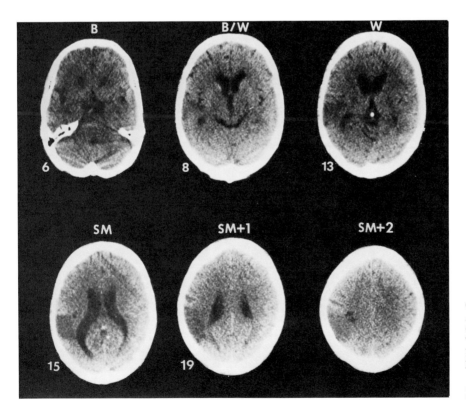

QUANTITATIVE
APPROACHES TO
COMPUTERIZED
TOMOGRAPHY IN
BEHAVIORAL
NEUROLOGY

culus could have been lesioned in this patient with conduction aphasia. The CT scan for the patient with Wernicke's aphasia, however, obtained a similar length of time after the stroke (Fig. 3, *bottom*), shows lesions in Wernicke's area and in the temporal isthmus on slices B/W (8 percent) and W (13 percent). For the patient with Wernicke's aphasia, the larger temporal lobe lesion on slices B/W and W continued into the supramarginal gyrus region of the parietal lobe on slice SM (15 percent) and into the supramarginal/angular gyrus region on slice SM+1 (19 percent). Thus, the CT scan of the patient with Wernicke's aphasia, who had a comprehension deficit, also revealed a large lesion in Wernicke's area on slices B/W (8 percent) and W (13 percent), whereas that of the patient with conduction aphasia, who had good comprehension, revealed no lesion there.

Future aphasia studies with lesion size and lesion site information will focus on the yet "unclassified" aphasia cases, including those with unusual cortical and/or subcortical lesion patterns, as well as those with two or three lesions or bilateral lesions. Long-term information on treatment programs and recovery patterns for the various lesion size and lesion site patterns is another area for future aphasia research. The relationship of right-hemisphere lesions to other aspects of language behavior, such as prosody, is also an area that will attract research in the future (see Chapter 6).

ORIENTATION TO THE SEMIAUTOMATED COMPUTER PROGRAM, ASI-II, USED IN QUANTIFYING VENTRICLE AND SULCAL SIZE IN NORMAL AGING AND DEMENTIA

The semiautomated computer program called Automated Slice Information II (ASI-II) is similar to the ASI-I program in that it computes several quantitative measures for a given CT slice. However, it does not use a *t* test to define low CT density number areas.[33] Whereas the ASI-I program considers each four-pixel sample as either "normal tissue" (not significantly low) or "CSF" (significantly low), based on the *t* test results, the ASI-II program takes into account the fact that some pixels may not be homogeneous and that they may contain part brain and part CSF. This is especially true in sulcal areas and around the borders of the ventricles. A partially volumed pixel is more accurately referred to as a partially volumed *voxel* (a voxel is a three-dimensional pixel). The ASI-II CSF *volume* analysis for a given slice always reveals less low-density (CSF-like) *volume* (e.g., 11 percent) than the ASI-I CSF *area* analysis (e.g., 21 percent) for the same slice.[42]

The ASI-II program uses a previously determined ideal CT density number difference between normal brain tissue and CSF for the given CT scanner in use. (For Ohio Nuclear Delta 2010 at the Boston Veterans Administration Medical Center, the ideal CT density number difference was 23.8 HU.) This ideal difference is then used to determine which four-pixel samples are totally composed of brain, which are totally composed of CSF, and which are combinations of partially volumed brain and CSF. The ratio of actual-to-ideal difference constitutes the CSF fraction for a four-pixel sample (Fig. 4, *left*).

Control - Age 60

Semi-automated Computer Analysis

Bodies of Lateral Ventricles Slice (#5)

Percent CSF Volume = 6.317

L = 6.646   R = 5.983

ASI-II Program   Sample Partially-Volumed Pixel :

10 mm

BRAIN   6 mm

VENTRICLE   4 mm

"Known average CT number difference" between tissue and CSF was 13.

CSF fraction of pixel was computed as follows:

$$\frac{\text{Mean CT number of 169-pixel HTS (30) minus mean CT number of partially-volumed 4-pixel sample (24.8)} = 5.2}{13}$$

$\frac{5.2}{13} = .4$ i.e. CSF = .4 of pixel and tissue = .6 of pixel

FIGURE 4. *Left*, Explanation of ASI-II program which computes percentage CSF and percentage brain in partially volumed pixels (voxels). *Right*, Sample ASI-II program printout for bodies of lateral ventricles slice (number 5) for normal control subject, age 60 (see Fig. 5, *top*).

QUANTITATIVE
APPROACHES TO
COMPUTERIZED
TOMOGRAPHY IN
BEHAVIORAL
NEUROLOGY

371

Control – Age 60

CT Scan Slices

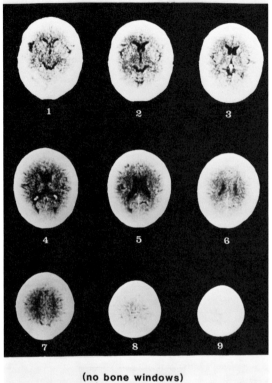

(no bone windows)

MEDIAL SEGMENT—SLICE AT BODIES OF LATERAL
VENTRICLES

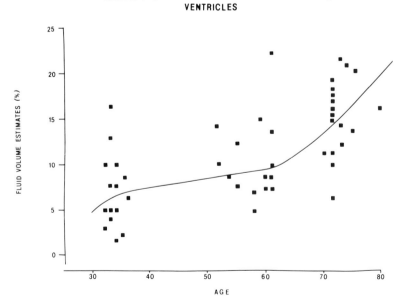

As in ASI-I, the mean CT density number of a 169-pixel healthy tissue sample (HTS) (Fig. 1) is first taken for each subject and is used as a baseline to determine which four-pixel samples throughout that slice are composed of brain, CSF, or partially volumed voxels. The normal aging control cases and Alzheimer's dementia cases presented in this chapter were scanned on the Ohio Nuclear Delta 2010 with slices 7 mm in thickness. Hence, the voxel size was 1 mm × 1 mm × 7 mm.

An ASI-II printout is shown in Figure 4, *right.* This is the printout for slice number 5 (slice SM) for the same 60-year-old normal control subject shown at the top of Figure 5. Figure 4, *right,* shows that the percent CSF volume on slice number 5 was 6.3 percent. On the ASI-II printout, the completely volumed voxels with only brain tissue are shown as periods. The partially volumed voxels that contain 0.9 CSF are shown as number signs (#); 0.7 to 0.9 CSF as asterisks (*); 0.5 to 0.7 CSF as plus signs (+); 0.3 to 0.5 CSF as minus signs (−); and 0.1 to 0.3 CSF as colon signs (:). More information on the ASI-II program is available elsewhere.[1,2,42,58,59]

Material presented in the following sections on quantitative CT scan data in cases of normal aging and of Alzheimer's dementia is largely restricted to that obtained with the ASI-I and ASI-II programs. Following the advent of CT scanning in 1973, numerous papers were published about linear measurements of CT scans. Discussion of linear measurements is kept to a minimum in this chapter, however, because in studies for which *both* quantitative and linear CT data were available, more significant differences as a result of age, or between patients and normals, were consistently observed when the quantitative CT data were used.[1,23,42] More recently, a few papers about dementia research have been published in which ventricle size was computed directly at the CRT with the use of a light pen to trace the ventricle contours.[13,23] This quantitative method may also become more widely used in the future.

CT SCAN FINDINGS IN NORMAL AGING

ASI-II data on CT scans from 123 normal subjects (male and female), aged 23 to 88 years, showed that ventricle size remained stable until about age 60, after which time it began to increase.[58] Figure 5, *bottom,* shows that a similar curve is obtained when ventricle size only at the bodies of the lateral ventricles (slice number 5) is examined. We also found that this slice was the single most meaningful one to examine with the ASI-I and ASI-II programs, in order to discriminate between patients with Alzheimer's dementia and normal controls.[1,2]

Sulcal size does not increase dramatically in normal aging until after age 60.[58] Sulcal prominence does not appear to be a useful predictor of cognitive function in normal subjects.[17] One study, however, did observe that enlargement of the left Sylvian fissure may be correlated with

←

FIGURE 5. *Top,* CT scan of normal control subject, age 60. *Bottom,* Ventricle size in normal aging from age 30 to 80 (males) from the Boston Normative Aging Study. Graph represents ASI-II ventricle size data for the bodies of the lateral ventricles slice. Note that ventricle size increases after age 60.

decreased intellectual performance.[53] Additional studies with quantitative CT scan information and neuropsychologic test information are necessary to further define the relationship between mental status changes and CT scan findings in normal aging.

An age-related decrease in the mean CT density number at the level of the centrum semiovale, in patients from 23 to 88 years of age, has been reported in a study from Palo Alto.[59] The age-related changes in the attenuation values could reflect alterations in white matter that are either primary or secondary to cortical degeneration. The Boston Normative Aging Study, however, has not yet observed a decrease in mean CT density numbers with age (n = 65, ages 30 to 80). The discrepancy in the findings of the Palo Alto and Boston studies may be due to a gender difference in the two populations. The Palo Alto group consisted of 74 women and 49 men; the Boston group consisted of 65 men. Additional large studies with separate male and female data are necessary to examine the possible relationship between aging and decreasing mean CT density number in white matter.

There have been very few CT scan studies published on aging in males versus females, and more research in this area is necessary.[28,56–58] Factors that may influence the mean CT density numbers of white and gray matter include bone thickness and calcium content of the cranial bones.[56,59] A study from Japan has recently shown that the calcium content in the cranial bone in men is fairly constant from the second to the seventh decades; whereas for women there is an increase from age 20 through 49, then a decrease from the fifth to the seventh decades.[56] Hence, it is possible that mean CT density numbers in white matter in normal females in the fifth to seventh decades could be lower in part owing to reduced bone CT numbers and hence to reduced bone artifact on the CT scan. The majority of normal cases examined in the Palo Alto study were women (especially in the older age range, for example, retired teachers). Hence, it is possible that the lower white matter CT numbers in the older ages in that study were due to the lower bone attenuation values in the older female population. These are all complex areas that deserve further research.

CT SCAN FINDINGS IN ALZHEIMER'S TYPE OF DEMENTIA

Use of Ventricle Size in the Diagnosis of Presenile Dementia of the Alzheimer Type

In previous studies that have used subjective ratings and/or linear measurements to investigate ventricle size and sulcal prominence on CT scans, a great deal of overlap was observed between the Alzheimer group and the age-matched controls, even when group differences were found to be significant.[8,15,18,31,32,34,51,53]

Eight patients with presenile dementia of the Alzheimer type (PDAT) (six females and two males) and 12 male age-matched controls (ages 53 to 64) were examined in the Boston study.[1] The diagnosis of Alzheimer's disease was reached by excluding other causes of dementia.

These patients were free of hypertension and did not have a history of head trauma, alcoholism, or severe psychiatric illness. They were all in the relatively early stages of the disease. These characteristics are common to all the patient groups that are subsequently discussed in this chapter. The patients had been manifesting changes in cognition for one to five years and were living at home. The normal aging subjects had been followed for the previous 18 years and had no history of alcoholism, psychiatric illness, learning disabilities, severe head trauma, epilepsy, hypertension, chronic lung disease, kidney disease, coronary artery disease, or cancer.

Gross visual inspection of CT scans from control cases and demented patients usually reveals no obvious differences (Fig. 6, *top*). Analysis of the ASI-II data for the CT slices for the eight dementia and 12 control cases revealed a significant difference (p <0.005) between the two groups in ventricle size at the bodies of the lateral ventricles (slice number 5) slice level (Fig. 6, *bottom*). The bottom of Figure 6 shows that only one dementia case overlapped into the control range (less than 8 percent CSF volume), and only one control case overlapped into the dementia range (greater than 8 percent CSF volume). A discriminant function analysis using only CSF volume estimate at slice number 5 correctly discriminated 88.89 percent of the subjects—all but one dementia patient and all but one control. Hence, the results from this study suggest that quantitative analysis of ventricle size at the bodies of the lateral ventricle slice may be useful in the diagnosis of PDAT.[1]

Although ventricle size was excellent in discriminating PDAT cases (below 65 years) from their controls, it was not as useful in discriminating patients with senile dementia of the Alzheimer type (greater than 65 years) from their controls.[2] This may reflect the presence of an age-related increase in ventricular size after the age of 60 (Fig. 5, *bottom*). This may make it difficult to discriminate superimposed dementia-related changes in ventricular size in the senile age group.

The relative usefulness of linear measurement of ventricle size at slice number 5 (Fig. 5, *top*) in discriminating the PDAT cases from the controls has also been investigated.[1] The linear measurements were taken as shown in Figure 7, *top*. Ventricle/brain ratios were calculated at the anterior, intermediate, and posterior portions of the bodies of the lateral ventricles and then were used to estimate total ventricular area.[1] All linear measurements were done with "bone windows," which allow better definition of the border of the inner table of the skull.[55] Analysis of the linear measurement data revealed a significant difference between the two groups (p <0.004). The linear ventricle size data for the eight presenile dementia cases (X = 0.043, SD = 0.009) and the 12 controls (X = 0.030, SD = 0.008) are plotted in Figure 7, *bottom*. Two control cases overlapped into the presenile dementia range. Hence, the linear data were not as accurate as the ASI-II data in discriminating presenile dementia cases versus control subjects: the linear data correctly classified 78.95 percent of the subjects, and two in each group were misclassified; the ASI-II data correctly classified 88.89 percent of the subjects, and only one in each group was misclassified.[1] It remains to be seen if larger studies will confirm the superiority of the ASI-II approach.

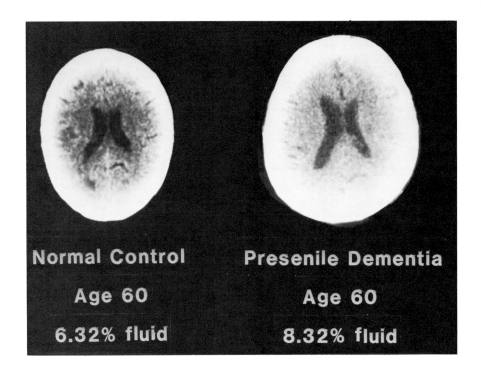

Normal Control
Age 60
6.32% fluid

Presenile Dementia
Age 60
8.32% fluid

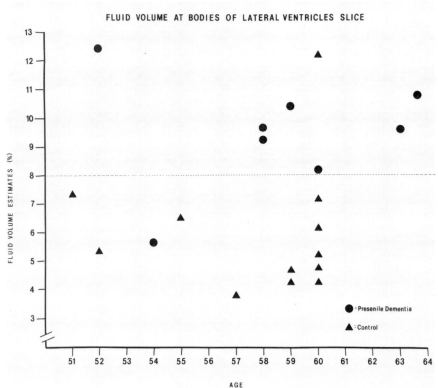

FLUID VOLUME AT BODIES OF LATERAL VENTRICLES SLICE

For the PDAT cases, there were significant correlations (p <0.01 to p <0.001) between ASI-II ventricle size data and neuropsychologic test scores (−0.59 to −0.74), as well as between linear ventricle size data and neuropsychologic test scores (−0.55 to −0.60).[1] The significant correlations were with the Wechsler Memory Scale (test of immediate memory with verbal and nonverbal stimuli), verbal fluency (ability to generate a list of words), proverb interpretation (degree of abstract thinking), and clock drawing (accuracy in drawing a clock to command). Future research may delineate on the CT scan more specific areas of brain damage in patients with Alzheimer's disease and may relate these to specific neuropsychologic deficits. Considerable progress in this direction is also being made with the use of positron emission tomography (see Chapter 10).

Use of Mean CT Density Numbers in the Diagnosis of Senile Dementia of the Alzheimer Type

The ASI-I program also provides the mean CT density number for the non-CSF tissue areas on that CT scan slice.[33,42] Figure 8, *top*, for example, shows the ASI-I printout for slice number 5 from a 76-year-old patient with SDAT. The percentage of CSF area was 28.5, and the mean CT density number of the remaining frontoparietal tissue was 35.7 HU. (On the Ohio Nuclear Scanner, water was 0, and white and gray matter was 20 to 50 HU.) Before any statistical comparisons could be made between mean CT density numbers for remaining tissue in the patients with SDAT and those of the control subjects, however, it was necessary to adjust for head size. This is done because mean CT density numbers increase near bone.[16] Therefore, head size alone will alter the mean CT density numbers.[58,59] For example, lower white and gray matter mean CT density numbers are observed on CT slices with greater total number of pixels (larger head size).[59] The ASI-I and ASI-II printouts provide data on the raw number of total pixels for each CT slice exclusive of bone. Adjustment for head size is then done in a manner previously used by Zatz, Jernigan, and Ahumada in the Palo Alto Normal Aging Study.[59]

We used mean CT density to compare normal control subjects with patients who had SDAT. The mean CT density number of the non-CSF tissue at slice number 5 for 13 SDAT cases was 34.3 HU (SD, 2.9), and for 18 normal control cases it was 39.4 HU (SD, 2.4) (p <0.00001).[2] The head size adjusted mean tissue CT density numbers at slice number 5 are shown in Figure 8 for the 13 SDAT cases and the 18 control subjects.

FIGURE 6. *Top,* Percentage CSF volume (ASI-II data, primarily ventricle size) at the bodies of the lateral ventricles slice (number 5) for the normal control subject, age 60 (see Fig. 5, *top*) and a patient with PDAT, also age 60. The 8.32 percent for the PDAT case is closer to the larger size (greater than 8 percent) observed with all but one of the other PDAT cases versus the smaller size (less than 8 percent) with all but one of the normal control cases.

Bottom, CSF volume (ASI-II data, primarily ventricle size) at the bodies of the lateral ventricles slice, for eight patients with PDAT and 12 control subjects. Note that the ventricle size is larger for the PDAT cases. Only one of eight PDAT cases falls within the normal control range (less than 8 percent), and only one of 12 normal controls falls within the PDAT range (greater than 8 percent). Ventricle size at the bodies of the lateral ventricles slice level may be useful in the diagnosis of PDAT.

QUANTITATIVE
APPROACHES TO
COMPUTERIZED
TOMOGRAPHY IN
BEHAVIORAL
NEUROLOGY

377

Linear Measurement
Multiple (x3) Bodies of
Lateral Ventricles /
Intracranial Width Ratio = 0.027

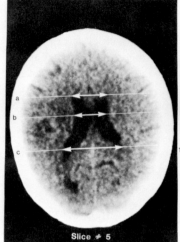

0.940/3.464 = 0.271

0.825/3.659 = 0.225

1.583/3.508 = 0.451

a x b x c = 0.027

Slice # 5

Control
Age 60
(bone window)

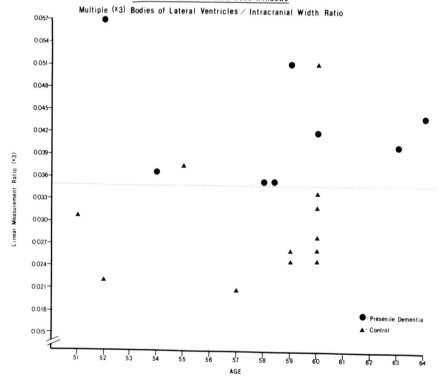

Linear Measurements with Bone Windows
Multiple (x3) Bodies of Lateral Ventricles / Intracranial Width Ratio

● : Presenile Dementia
▲ : Control

This shows that only 4 of 13 SDAT cases were in the normal range in this study.[2]

When a discriminant function analysis was performed on the head size adjusted mean CT density numbers for slice number 5, 77.42 percent of the subjects were correctly classified (three SDAT patients and four control subjects were misclassified). When additional quantitative CT scan information was added into the stepwise discriminant function analysis (e.g., fluid volume estimate at the maximum width of the third ventricle slice), 93.55 percent of the subjects were correctly classified (one SDAT patient and one control subject were misclassified). These preliminary results suggest that early in the course of the disease, CT scan information may be useful in the diagnosis of SDAT.[2] In this same group of studies,[1,2] mean CT density numbers did not differentiate normal individuals from demented individuals in the presenile age range (PDAT), although this had been observed to differentiate patients with PDAT from age-matched control subjects in a previous study.[45]

METHODOLOGIC CONSIDERATIONS

The area of research on mean CT density numbers in normal aging and dementia is very new, existing only since 1980,[2,42,45,59] and conflicting reports are present in the literature.[7,24,25,43,54] Studies that have not observed a decrease in mean CT density numbers in SDAT cases have usually not controlled for head size and have examined mean CT density numbers in cursor samples of less than 169 pixels.[24,54] A study of SDAT cases by Bondareff and co-workers,[7] however, which used cursor samples up to 388 pixels, did reveal a significant decrease in mean CT density numbers for the SDAT cases when these were compared with those of age-matched control subjects. Areas where this decrease was found included medial temporal lobe, anterior frontal lobe, and the head of the caudate.

Factors that complicate research with mean CT density numbers include the effect of bone calcium content on mean CT density numbers, photon noise,[36] and the manufacturer changes in the hardware on each new generation of scanner. For example, in our original study in 1980,[45] the mean CT density numbers differentiated the 14 PDAT and seven SDAT cases from their respective controls. Our current study (1982) did not find a significant difference in the mean CT density numbers in the PDAT cases versus those of their controls[1] but did find a significant difference in the mean CT density numbers between the SDAT cases and their controls.[2] Why the 1980 finding was not replicated in the 1982

←

FIGURE 7. *Top*, Method used for linear measurement of ventricle size at the bodies of the lateral ventricles slice level with bone windows. Note the increased clarity of the border of the inner table of the skull on this slice (number 5) with bone window versus that on slice (number 5) without bone window in Figure 5, *top* (same normal control case and same CT scan but no bone window). *Bottom*, Linear measurement data for ventricle size at the bodies of the lateral ventricles slice for the same eight patients with PDAT and 12 control subjects in Figure 6, *bottom* (ASI-II data). Two normal cases are within the PDAT range and two PDAT cases are on the borderline for the normal cases. There are fewer distinct differences between the two groups with the linear measurement data for ventricle size compared with the ASI-II data for ventricle size.

Dementia Case – Age 76,F 7yr. Hx.

Body of Lateral Ventricle Slice Level

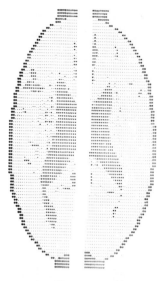

ASI-I Data

Slice Size: 18,060 Pixels

Percent CSF Area: 28.5%

Mean CT Density Number

of Remaining Tissue: 35.7HU

Semi-Automated Computer Analysis
CT Density of Remaining Tissue at Bodies of Lateral Ventricles Slice

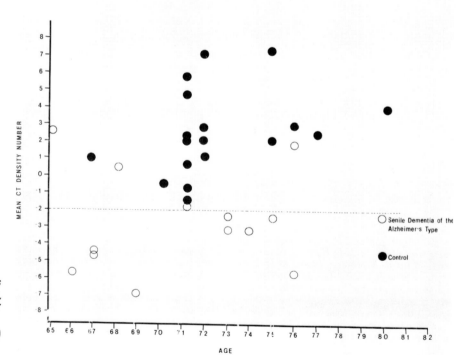

study with the PDAT cases is not known; however, the CT scanners were different in each of the two studies. In the 1980 study the Ohio Nuclear Delta 50 CT scanner with slices 13 mm thick was used, and in the 1982 study the Ohio Nuclear Delta 2010 CT scanner with slices 7 mm thick was used. The rapidly changing hardware on the current CT scanners also makes it difficult to do meaningful chronologic studies with mean CT density numbers. Every time a new scanner is installed, new baseline control data for mean CT density numbers must be obtained for that scanner.

Future CT scan research in the area of normal aging and dementia will be enhanced by more-powerful, high-resolution CT scanners, which have increasingly stable CT numbers with less machine drift and artifact. The potential for diagnostic capability with these scanners in the area of Alzheimer's disease is obvious.[1,2,7,21] At present, it appears that quantitative area and volumetric, not linear, methods of CT scan analysis will yield more meaningful information.

Future CT scan research in normal aging and dementia may center on examining mean CT density numbers in specific parts of the brain (right and left frontal, temporal or parietal lobe, thalamus, and so on) and on correlating these with specific neuropsychologic deficits, as has recently been done successfully in a few select patients with Alzheimer's disease through PET scanning.[19] CT scan research in this area could thus potentially contribute toward a better understanding of brain-behavior interrelations and assist in the diagnosis of degenerative brain diseases such as Alzheimer's dementia.

REFERENCES

1. ALBERT, MS, NAESER, MA, LEVINE, HL, AND GARVEY, J: *Ventricular size in patients with presenile dementia of the Alzheimer's type.* Arch Neurol 41:1258–1263, 1984.

2. ALBERT, MS, NAESER, MA, LEVINE, HL, AND GARVEY, J: *Mean CT density numbers in patients with senile dementia of the Alzheimer's type.* Arch Neurol 41:1264–1269, 1984.

3. ALEXANDER, MP, AND LOVERME, SR: *Aphasia following left hemispheric intracerebral hemorrhage.* Neurology 30:1193–1202, 1980.

4. ALEXANDER, MP, AND SCHMITT, MA: *The aphasia syndrome of stroke in the left anterior cerebral artery territory.* Arch Neurol 37:97–100, 1980.

5. ALTEMUS, LR, ROBERSON, GH, MILLER, FC, AND PESSIN, M: *Embolic occlusion of the superior and inferior divisions of the middle cerebral artery with angiographic-clinical correlation.* Am J Roentgenol 126:576–581, 1976.

6. BERMAN, SA, HAYMAN, LA, AND HINCK, VC: *Correlation of CT cerebral vascular territories with function: 1. Anterior cerebral artery.* Am J Roentgenol 135:253–257, 1980.

7. BONDAREFF, W, BALDY, R, AND LEVY, R: *Quantitative computed tomography in dementia.* Arch Gen Psychiatry 38:1365–1368, 1981.

FIGURE 8. *Top,* Sample ASI-I program printout for bodies of the lateral ventricles slice for a patient with SDAT, age 76. Mean CT density number for remaining tissue (35.7 HU) is for entire slice, including primarily frontal and parietal lobe tissue. *Bottom,* Head size adjusted mean CT density numbers for remaining frontal and parietal lobe tissue at the bodies of the lateral ventricles slice for 13 patients with SDAT and 18 normal control subjects. The mean CT density numbers were lower (closer to CSF levels) for the SDAT cases. Only four of 13 SDAT cases were in the normal control range (n = 18). Head size adjusted mean CT density numbers for frontoparietal areas at the bodies of the lateral ventricles slice level may be useful in the diagnosis of SDAT.

8. BRINKMAN, SD, SARWAR M, LEVIN, HS, AND MORRIS, HH: *Quantitative indexes of computed tomography in dementia and normal aging.* Neuroradiology 138:89–92, 1981.

9. DAMASIO, H: *Cerebral localization of the aphasias.* In SARNO, MT (ED): *Acquired Aphasia.* Academic Press, New York, 1981, pp. 27–50.

10. DAMASIO, H: *A computed tomographic guide to the identification of cerebral vascular territories.* Arch Neurol 40:138–142, 1983.

11. DAMASIO, H, AND DAMASIO, AR: *The anatomical basis of conduction aphasia.* Brain 103:337–350, 1980.

12. DAMASIO, AR, DAMASIO, H, RIZZO, M, VARNEY, N, AND GERSH, F: *Aphasia with nonhemorrhagic lesions in the basal ganglia and internal capsule.* Arch Neurol 39:15–20, 1982.

13. DAMASIO, H, ESLINGER, P, DAMASIO, AR, RIZZO, M, HUANG, HK, AND DEMETER, S: *Quantitative computed tomographic analysis in the diagnosis of dementia.* Arch Neurol 40:715–719, 1983.

14. DEARMOND, SJ, FUSCO, MM, AND DEWEY, MM: *Structure of the Human Brain.* Oxford University Press, New York, 1976.

15. DELEON, MJ, FERRIS, SH, GEORGE, AE, REISBERG, B, KRICHEFF, II, AND GERSHON, S: *Computed tomography evaluations of brain-behavior relationships in senile dementia of the Alzheimer's type.* Neurobiol Aging 1:69–79, 1980.

16. DICHIRO, G, BROOKS, RA, DUBAL, L, ET AL: *The apical artifact: Elevated attenuation values toward the apex of the skull.* J Comput Assist Tomog 2:65–70, 1978.

17. EARNEST, MP, HEATON, RK, WELKINSON, WE, AND MANKE, WF: *Cortical atrophy, ventricular enlargement and intellectual impairment in the aged.* Neurology 29:1138–1143, 1979.

18. FORD, CV, AND WINTER, JW: *Computerized axial tomograms and dementia in elderly patients.* J Gerontol 36:164–169, 1981.

19. FOSTER, NL, CHASE, TN, FEDIO, P, PATRONAS, NJ, BROOKS, RA, AND DICHIRO, G: *Alzheimer's disease: Focal cortical changes shown by positron emission tomography.* Neurology 33:961–965, 1983.

20. FREEDMAN, M, ALEXANDER, MP, AND NAESER, MA: *Anatomical basis of transcortical motor aphasia.* Neurology 34:409–417, 1984.

21. FREEDMAN, M, KNOEFEL, JE, NAESER, MA, AND LEVINE, HL: *CT scanning in Aging Dementia.* In ALBERT, M (ED): *Clinical Neurology of Aging.* Oxford University Press, New York.

22. GADO, M, HANAWAY, J, AND FRANK, R: *Functional anatomy of the cerebral cortex by computed tomography.* J Comput Assist Tomog 3:1–19, 1979.

23. GADO, M, HUGHES, CP, DANZIGER, W, CHI, D, JOST, G, AND BERG, L: *Volumetric measurements of the cerebrospinal fluid spaces in demented subjects and controls.* Radiology 144:535–538, 1982.

24. GADO, M, DANZIGER, W, CHI, D, HUGHES, CP, AND COBEN, LA: *Brain parenchymal density measurements by CT in demented subjects and normal controls.* Radiology 147:703–710, 1983.

25. GEORGE, A, DELEON, M, FERRIS, S, KRICHEFF, II: *Parenchymal CT correlates of senile dementia (Alzheimer's disease): Loss of grey-white matter discriminibility.* Am J Neuroradiol 2:205–213, 1981.

26. GOODGLASS, H, AND KAPLAN, E: *The Assessment of Aphasia and Related Disorders.* Lea & Febiger, Philadelphia, 1972.

27. HANAWAY, J, SCOTT, WR, AND STROTHER, CM: *Atlas of the Human Brain and the Orbit for Computed Tomography.* Warren H Green, St Louis, 1972.

28. HATAZAWA, J, ITO, M, YAMAURA, H, AND MATSUZAWA, T: *Sex difference in brain atrophy during aging: A quantitative study with computed tomography.* J Am Geriatr Soc 30:235–239, 1982.

29. HAYMAN, LA, BERMAN, SA, AND HINCK, VC: *Correlation of CT cerebral vascular territories with function: II. Posterior cerebral artery.* Am J Neuroradiol 2:219–225, 1981.

30. HIER, DB, DAVIS, KR, RICHARDSON, E, AND MOHR, JP: *Hypertensive putaminal hemorrhage.* Ann Neurol 1:152–159, 1977.

31. HUGHES, CP, AND GADO, M: *Computed tomography and aging of the brain.* Radiology 139:391–396, 1981.

32. JACOBY, RJ, LEVY, R, AND DAWSON, JM: *Computed tomography in the elderly. 1. The normal population.* Br J Psychiatry 136:249–255, 1980.

33. JERNIGAN, T, ZATZ, LM, AND NAESER, MA: *Semiautomated methods for quantitating CSF volume on cranial computed tomography.* Radiology 132:463–466, 1979.

34. KASZNIAK, AW, GARRON, DC, FOX, JH, BERGEN, D, AND HUCKMAN, M: *The relationship of cerebral atrophy, EEG slowing, age and education to cognitive functioning in older patients suspected of dementia.* Neurology 29:1273–1279, 1979.

35. KERTESZ, A, HARLOCK, W, AND COATES, R: *Computer tomographic localization, lesion size and prognosis in aphasia and nonverbal impairment*. Brain Lang 8:34–50, 1979.

36. LEVI, C, GRAY, JE, MCCULLOUGH, EC, AND HATTERY, RR: *The unreliability of CT numbers as absolute values*. Am J Neuroradiol 139:443–447, 1982.

37. MATSUI, T, AND HIRANO, A: *An Atlas of the Human Brain for Computerized Tomography*. Igaku-Shoin, Tokyo and New York, 1978.

38. MAZZOCCHI, F, AND VIGNOLO, LA: *Localization of lesions in aphasia: Clinical-CT scan correlation in stroke patients*. Cortex 15:627–654, 1979.

39. MCCULLOUGH, EC: *Factors affecting the use of quantitative information from a CT scanner*. Radiology 124:99–107, 1977.

40. MOHR, JP, PESSIN, MS, FINKELSTEIN, S, FUNKENSTEIN, HH, DUNCAN, GW, AND DAVIS, KR: *Broca's aphasia: Pathologic and clinical*. Neurology 28:311–324, 1978.

41. NAESER, MA: *CT scan lesion size and lesion locus in cortical and subcortical aphasias*. In KERTESZ, A (ED): *Localization in Neuropsychology*. Academic Press, New York, 1983, pp. 63–119.

42. NAESER, MA, ALBERT, MS, AND KLEEFIELD, J: *New methods of CT scan diagnosis of Alzheimer's disease: Examination of white and grey matter mean CT density numbers*. In CORKIN, S, DAVIS, KL, GROWDON, JH, USDIN, E, AND WURTMAN, RJ (EDS): *Alzheimer's Disease: A Review of Progress* (Vol 19, Aging). Raven Press, New York, 1982, pp 63–78.

43. NAESER, MA, ALBERT, MS, LEVINE, HL, AND GEBHARDT, C: *Published invited commentary on paper by George, DeLeon, Ferris et al*. Am J Neuroradiol 2:211–212, 1981.

44. NAESER, MA, ALEXANDER, MP, HELM-ESTABROOKS, N, LEVINE, HL, LAUGHLIN, SA, AND GESCHWIND, N: *Aphasia with predominantly subcortical lesion sites—Description of three capsular/putaminal aphasia syndromes*. Arch Neurol 39:2–14, 1982.

45. NAESER, MA, GEBHARDT, C, AND LEVINE, HL: *Decreased computerized tomography numbers in patients with presenile dementia—Detection in patients with otherwise normal scans*. Arch Neurol 37:401–409, 1980.

46. NAESER, MA, AND HAYWARD, RW: *Correlation between CT scan findings and the Boston Diagnostic Aphasia Exam*. Neurology 28:545–551, 1978.

47. NAESER, MA, HAYWARD, RW, LAUGHLIN, S, BECKER, JMT, JERNIGAN, T, AND ZATZ, LM: *Quantitative CT scan studies in aphasia, Part II: Comparison of the right and left hemispheres*. Brain Lang 12:165–189, 1981.

48. NAESER, MA, HAYWARD, RW, LAUGHLIN, S, AND ZATZ, LM: *Quantitative CT scan studies in aphasia, Part I: Infarct size and CT numbers*. Brain Lang 12:140–164, 1981.

49. NIELSEN, JM: *Agnosia, Apraxia, and Aphasia: Their Value in Cerebral Localization*. Hafner Publishing, New York, 1936, pp. 119–120.

50. PALACIOS, E, FINE, M, AND HENGHTON, V: *Multiplanar Anatomy of the Head and Neck for Computed Tomography*. John Wiley & Sons, New York, 1980.

51. ROBERTS, MA, AND CAIRD, FI: *Computerized tomography and intellectual impairment in the elderly*. J Neurol Neurosurg Psychiatry 39:986–989, 1976.

52. ROSS, ED: *Localization of the pyramidal tract in the internal capsule by whole brain dissection*. Neurology 30:59–64, 1980.

53. SOININEN, H, PURANEN, M, AND RIEKKINEN, PJ: *Computed tomography findings in senile dementia and normal aging*. J Neurol Neurosurg Psychiatry 45:50–54, 1982.

54. WILSON, RS, FOX, JH, HUCKMAN, MS, BACON, LD, AND LOBICK, JJ: *Computed tomography in dementia*. Neurology 32:1054–1057, 1982.

55. WOLPERT, SM: *The ventricular size on computed tomography*. J Comput Assist Tomogr 1:222–226, 1977.

56. YAMADA, K, ENDO, S, YOSHIOKA, S, HATAZAWA, J, YAMAURA, H, AND MATSUZAWA, T: *Age-related changes of the cranial bone mineral: A quantitative study with computed tomography*. J Am Geriatr Soc 30:756–763, 1982.

57. YAMAURA, H, ITO, M, KUBOTA, K, AND MATSUZAWA, T: *Brain atrophy during aging: A Quantitative study with computed tomography*. J Gerontol 35:492–498, 1980.

58. ZATZ, LM, JERNIGAN, TL, AND AHUMADA, AJ: *Changes on computed cranial tomography with aging: Intracranial fluid volume*. Am J Neuroradiol 6:1–11, 1982.

59. ZATZ, LM, JERNIGAN, TL, AND AHUMADA, AJ: *White matter changes in cerebral computed tomography relating to aging*. J Comput Assist Tomogr 6:19–23, 1982.

QUANTITATIVE
APPROACHES TO
COMPUTERIZED
TOMOGRAPHY IN
BEHAVIORAL
NEUROLOGY

INDEX

385

Allocortex, 2, *4*, 8, *8*, 9, 35, *46*
  cell differentiation, 3–5
Allocortical structures, 2
Alternating Sequences Test, 77t, 78t, 80, 81,
  108, 129
Alzheimer's disease, 41. *See also* Presenile
    dementia, Alzheimer type; Senile de-
    mentia, Alzheimer type
  amnesia and, 128, 174
  anomic aphasia and, 212
  apraxia and, 224
  CT scan findings in, 374–379
  dementias of, 36, 83, 113–114, *115*
  hypometabolism in, 354
  memory function, 36, 85, *88*, 89t, 184
American Indian Sign, 231–232
AMERIND. *See* American Indian Sign
Amnesia(s), 32
  alcohol and, 179
  Alzheimer's disease and, 128, 174
  animal behavior data, 181–184
  anterograde, 83, 172, 176, *177*, 180, *183*,
    184
  bilateral limbic involvement, 170–171,
    172–178
  clinical features of, 170–172
  disconnection, 178
  distraction and, 85
  drugs causing, 179
  epileptic surgery and, 316
  generalized, 19
  global, 170, 180, 267
  herpes simplex encephalitis, due to, 173,
    173
  hippocampal damage, 39, 172, 173, 174,
    183
  Korsakoff's, 83, 85, 129, 169–170, 174–
    176, *175*, 177–178, 184–187
  material-specific verbal, 73, 179
  modality-specific, 18
  psychogenic, 181
  retrograde, 83, 171, 174, 179, 180, 186,
    189
  selective, 178
  transient global, 170, *177*, 180
  unknown anatomy, 178–180
  visual, 109
Amnestic states, 39–40, 46. *See also* Amne-
    sia(s); Memory loss
  limbic system involvement, 47, 170–171,
    172–178
  progressive, *20*
Amnestic syndromes, 32, 172–181
  clinical features, 170–172
Amobarbital, 316
Amorphosynthesis, 152
Amygdala, 37–39
  affect and, 37
  atrophy, *183*
  behavioral patterns, 37–39
  connectivity, 10, 38
  damage to, 31
  stimulation of
    endocrine balance and, 39
  surgical removal, 317. *See also* Amyg-
      dalectomy

Amygdalectomy, 32, 37, 38
Amygdaloid complex, 2, *3*
Anarithmetria, 224–225. *See also* Acalculia;
    Calculation abilities, impairment of
Angular gyrus syndrome, 26–27, 225–226.
    See also Gerstmann's syndrome
Anomia, 26, 178
  aphasic, 212–213, 216
    location of pathology, 220t
    varieties of, 220t
  callosal, 222
  disconnection, 222
  semantic, 221
  transient, 46
  visual, 18, 109
  word-evocation, 220–221
  word-production, 220–221
  word-selection, 221
Anomic aphasia, 210t, 212–213
Anorexia nervosa, 299
  epilepsy and, 306
Anosognosia, 27, 150, 158, 247
Anterior alexia, 217
Anterograde amnesia, 83, 172, 176, *177*,
    180, *183*, 184
Anticholinergic drugs, 130
Anticholinesterase agents, 187
Anticonvulsant drugs, 305
  effectiveness, 313–314, 315, 318–319
  effects, 302, 313–314
Antidepressants, 130, 319
  tricyclic, 251, 252
Antiepileptic drugs, 319. *See also* Anticon-
    vulsant drugs
Antihistamines, 130
Antiparkinsonian drugs, 130
Anton's syndrome, 266
Anxiety, intractable, 45
Apathy, 33, 82, 83
  confusional states and, 129
Aphasia(s), 48, 201–203
  Alzheimer's disease and, 212
  anomic, 210t, 212–213, 216
  Broca's, 90, 91, 92, 93, 202t, 203–204,
    209, 217, 223, 366–368, *366–367*
  Broca's vs. transcortical, 366–368, *366–*
    367
  calculations and, 94, 224
  conduction, 202t, 205–206
    vs. Wernicke's, 368–370, *368–369*
  crossed, 161, 227
  CT studies, 366–370
  defined, 193, 195–196
  depression and, 228–229
  emotional accompaniments of, 228–229
  fluent, 90, 195, 196, 197, 208
  fluent-nonfluent classifications, 90, 196,
    197
  global, 202t, 208–209
  historic background, 194–195
  left-hemisphere damage and, 50
  mixed transcortical, 210t, 211–212
  naming objects and, 91, 220–222
  neurobehavioral problems of, 228, 229
  nonfluent, 90, 196, 197
  postsurgical, 316

psychosocial problems of, 228, 229
pure word deafness, 202t, 207
recovery from, 229–231
  factors affecting, 230t
relatively preserved repetition, 209–214,
  210t
striatal lesions, 41
subcortical, 213–214
syndromes, 201–214
  definition of, 195–196
  localization of, 203, 226–228, 366–
  370
testing, 77t, 90–92, 117, 118, 196–199
therapy for, 231–232
transcortical motor, 24, 209–211, 210t,
  366, 368
transcortical sensory, 91, 210t, 211, 216,
  221, 227–228
transcortical vs. Broca's, 366–368, 366–
  367
Wernicke's, 22, 26, 27, 90, 91, 129, 202t,
  204–205, 207–208, 211, 216–217,
  221, 227–228, 242, 248, 252
  vs. conduction, 368–370, 368–369
with repetition disturbances, 202t, 203–
  209
word-finding impairment and, 220–222
Aphasic acalculia, 224
Aphasic anomia, varieties, 220t, 220–222
Aphasiology, 195
Aphemia, 202t, 206–207
Apraxia, 14, 150, 199, 222–224. *See also*
  Praxis
Alzheimer's disease and, 224
buccofacial, 93
callosal, 223
constructional, 281–282
defined, 92, 222
dressing, 101, 282
facial, 204
ideational, 94, 108, 223–224
ideomotor, 92, 204, 223, 242
limb-kinetic, 223
ocular, 275, 276, 277
tests for, 92–94, 117, 199
to verbal command, 204, 206, 209
Aprosodias, 241, 246t. *See also* Prosody
affective prosody
  comprehension, 247
  repetition, 246–247
bedside evaluation, 246–247
functional-anatomic correlates, 247–250
gestural kinesics comprehension, 247
gesturing, 246
global, *245*, 248–249, 251
mixed transcortical, *245*, 249–250, 251
motor, *245*, 247–248, 250, 251
pure affective deafness and, 250
right brain infarctions and, 243, 244,
  245
sensory, *245*, 248
spontaneous affective prosody, 246
transcortical motor, 249
transcortical sensory, *245*, 249
Archicortex, 2. *See also* Hippocampal
  formation

Arousal
attention and, 126, 135–136
disorders, 129
levels of, 137
neural activity and, 48, 49
Arousal/wakefulness assessment, 76, 77t
ASI-I, 363–365, 373, 377, *380–381*
  printout, Wernicke's area infarct, *364–365*
ASI-II, 370–373, 375, *376–377*, 377, *378–379*
  program diagram, *371*
Association areas
auditory unimodal, 6, 13, 19–22, 44, 109,
  200, 228
heteromodal, 7, *8*, 9, 10, 22–23, 24–30,
  45–46, 139
  prefrontal, 27–30
  temporoparietal, 25–27
modality-specific, 6–7, 15–24, 138
motor, 23–24, 44
somatosensory unimodal, 6, 13, 22–23
unimodal, 6–7, 9, 10, 15–24, 44–45, 138
visual unimodal, 15–19
Association isocortex, 6–7
Astereopsis, 280. *See also* Stereopsis, disturbance of
Ataxia, 174
optic, 275, 276, 278, *278*, 282
visuomotor, 276
Attention. *See also* Attentional matrix
affect and, 126
arousal and, 126, 135–136
aspects of, 126
assessment, 76, 77t, 78–82, 117
defined, 125–126
directed, 24, 103, 126
  hemispheric dominance for, 159–161
  neglect and, unilateral, 127, 140–150
disorders, 140–150
  confusional states as, 127–134
  in visual space, 19
expectancy and, 331
extrapersonal space distribution deficits,
  26
neglect and, unilateral
  causes and course of, 149–150
  clinical picture, 140–142
  motivational aspects of, 149
  motor aspects of, 145–148
  perceptual aspects of, 142–145
neural activity during, 330, 331
neural network for distribution of, 156–
  158
phylogenetic corticalization of, 155–156
process of, 126–127
responses, 153
span
  amnesia and, 172, 174
  digit, 81
  nonverbal, 81
spatial distribution of, 53
  tests, 77t, 101–103, 118
tasks, trimodal, 139
Attentional disturbances, 29, 80, 81, 94, 127,
  260. *See also* Attention, disorders
hemispatial, 144, 155. *See also* Neglect,
  unilateral

Brodmann nomenclature, 22
Brown-Peterson Technique, 77t, 85
BSAEP. *See* Brainstem auditory evoked response
Buschke-Fuld selective reminding procedure, 77t, 85

CALCULATION abilities
    impairment of, 94, 224–225. *See also* Acalculia; Anarithmetria
    tests for, 77t, 94–95, 118
Callosal anomia, 222
Callosal apraxia, 223
Cancellation tests, verbal and nonverbal, 77t, *102*, 103, 142, *143*, 145, *146*, 147, *148*, 149
Carbamazepine, 313, 314, 319
Carbonic anhydrase inhibitors, 313
Card Sort Test, Wisconsin, 78t, 105–107, 108
Catatonia, 297
Categorical abstraction, nonverbal, 105
Category naming difficulties, 222
Caudate, 40, 41
CBF. *See* Cerebral blood flow
Central alexia, 27, 216–217
Cerebral blood flow, 347–348, 349–350
    alcoholism and, 357
    attentional tasks and, 358–359
    auditory stimulation and, 356
    brain-behavior relations and, 355–359
    cognitive abilities and, 357–358
    in psychiatric patients, 354–355
    measuring, 347–348
    regional (rCBF), 347–348, 351–352, 354, 355–357, *356*, *358*, 359–360
    visual stimulation and, 356
Cerebral connectivity. *See* Connectivity patterns
Cerebral cortex, 2. *See also* Cortex
    allocortical structures, 2
    corticoid areas, 2. *See also specific areas*
    idiotypic, *4*, 7, *8*
    maps of, 2
    paralimbic areas (mesocortex), 3–6, 8, *8*, 30–31
    regional variations, 2
    types, *4*
Cerebral glucose metabolism
    measuring, 347, 348, 351, 352, 354, 356, 359
Cerebral localization, 57–58. *See also* Hemispheric dominance; Hemispheric specialization
Cerebropathia toxaemia psychica, 174. *See also* Korsakoff's amnesia
Cerebrovascular disease, imaging techniques in, 351–352
Channel-mediated functions, 47–49
ChAT. *See* Enzyme choline acetyl transferase
Chlorpromazine, 306
Cholinergic agonists, 187
Cholinergic antagonists, 179
Cholinergic innervation, 35, 36, 47, 184
    impairment of, 174, 177

Cholinergic neurons, 48, 49
Cholinergic transmission
    interference with, 130–131
Cholinomimetics, 49
Chronotaraxis, 176
Cingulate complex, 5, *6*. *See also* Paralimbic areas
Cingulate lesions, 177–178
Cingulotomies, 33
Clonidine, 187
Clumsiness, right-sided, 24
CMRgl. *See* Cerebral glucose metabolism
CMRO2. *See* Oxygen metabolism
CNV. *See* Contingent negative variation
Cocaine, 305
Cognitive abilities
    amnesia and, 170, 172, 174
    cerebral blood flow and, 357–358
Cognitive functioning of elderly, 111–113
Color
    association disorders, 274–275
    naming disorders, 215, 273–274
    recognition, 215
    vision
        loss of, 269
        impairment, 17–18, 215
        perception disorders, 110, 268–273
Colored Progressive Matrices, 78t, 107
Comatose patients, evoked potentials recording of, 331–332
Commissurotomy, 51–52
    callosal, 178, 280
Communication. *See also* Comprehension, spoken and written language; Language; Speech
    emotional aspects, 56–57, 59, 240–241, 243, 251–252
    nonverbal
        aprosodias, 246–250
        gestures and, 231–232, 241–242, 244
        kinesics, 241–242
        prosody, 239–241
        right hemisphere and, 242–245
    paralinguistic aspects of, 104
        right hemisphere and, 56–57
    tests, 77t, 104
    written, 56, 198–199, 204, 218, 227
Complex perceptual tasks, tests, 77t, 95–98
Comprehension
    disturbances, 26, 211–212, 228, 242
        aphasia and, 370
        syntactic, 228
    reading, 215–218. *See also* Reading
    spoken language, 198, 204, 227–228
    testing, 77t, 91–92, 198–199, 227
    written language, 117, 198–199, 204, 218, 227. *See also* Reading
Computer programs, semiautomated
    ASI-I, 363–365. *See also* ASI-I
    ASI-II, 370–373. *See also* ASI-II
Computerized tomography, 213, 214, 283, 348
    in Alzheimer's dementia, 374–379
    in aphasias, 366–370
    in normal aging, 373–374
    methodologic considerations, 379–381

Computerized tomography—*Continued*
  positron emission (PET), 17, 160, 312, 313, *349*
  quantitative approaches to, 363–381
  single photon emission (SPECT), 312, 313
  x-ray, *20, 173, 177, 270, 277, 278,* 312, 313, *350, 351, 352, 353, 354, 364–369, 372, 373, 376, 377, 378–381. See also* X-ray imaging
Concentration, 76
  disturbances of, 29, 72
  span, 140. *See also* Attention
  tests, 76, 77t, 81
Conduction aphasia, 202t, 205–206
  vs. Wernicke's, 368–370, *368–369*
Confabulation, 171, 174, 175, 177, 186
Confrontation naming, 198, 204, 212–213, 220. *See also* Naming
Confusional state(s)
  acute, 89, 92, 94, 127–129
  agitated, 27
  agraphia and, 92, 218
  alcohol and, 129, 130
  apraxia and, 94, 224
  attention disorders and, 127–134
  barbiturates and, 129, 130
  causes and mechanisms, 130–132
  chronic, 130
  clinical picture, 127–129
  detection, 132–134
  differential diagnosis, 129–130
  epilepsy and, 131, 295, 297
  global, 27, 174. *See also* Korsakoff's amnesia
  metabolic, 89, 89t, 128, 130
  postoperative, 131
  toxic, 128, 130
  vigilance in, 78
  vulnerability, 132–134
Connectivity patterns. *See also specific anatomic areas*
  amygdala, 10, 38
  auditory, 21
  behavioral specialization and, 8–11
  frontal eye fields, 155
  horizontal, 9, 11
  human and monkey brains, 156
  limbic, 35, 36, 37, 39, 47–49
  modality-specific association nuclei, 44–45
  parietal lobe, 152, *152*
  PG area, 152, 358
  striatum, 40, 157, 158
  thalamic, 45, 157, 158
  thalamocortical, 43–44
  vertical, 9–10
Consciousness
  alterations in, 290, 291
  impairment, 176
  levels, attention and, 135–137
Constructional abilities
  disturbances, 26, 27, 129, 150, 281–282
  tests, 77t, 98–100, 118
Constructional apraxia, 281–282. *See also* Constructional abilities, disturbances
Contingent negative variation, 332

Continuous Performance Test, 76, 77t, 78
Contour discrimination, 97, 260
Convulsions, 304–305, 306. *See also* Epilepsy; Seizures
Copying tests, 77t, 99, *99,* 100, 118, *133,* 260, 281
Corsi Block Test, 77t, 81
Cortex
  allocortex, 2, *4, 5, 8*
  association, 6–7
    auditory,19–22
    motor, 23–24
    somatosensory, 22–23
    visual, 15–19
  attention and, 138–140
  behavioral patterns in. *See specific areas of cortex*
  cerebral. *See* Cerebral cortex
  heteromodal association, 24–30
  homotypical isocortex, *4,* 6–7, *8*
  idiotypic, *4, 7, 8*
  neural connectivity and behavioral specialization, 8–11
  paralimbic areas (mesocortex), 3–6, 8, *8*
  piriform, 2, 3, 36–37. *See also* Primary olfactory cortex
  prefrontal, 27–30, 335–336
  primary auditory, 12–13
  primary motor, 14
  primary olfactory, 2, 34–35
  primary somatosensory, 13
  primary visual ,11–12
  regional variations, 2
  topographic maps, 2, 15
  types, *4*
  unimodal association, 15
Cortical blindness, 266
Cortical deafness, 13, 207–208
Cortical dipole, 328, *329*
Cortical excision, 317–318
Cortical hypometabolism, 214. *See also* Hypometabolism
Cortical language areas, 199–200, *200,* 201
Cortical mantle, 2, *3*
Cortical maps, 2, 15
Cortical zones, human brain, *8*
Corticoid areas, 2. *See also* Basal forebrain structures; Cortical mantle; Cortical zones, human brain
Corticoreticular interaction, 137–138
CSF, CT analysis, 370, *371,* 373, 375, *376–377, 380–381*
CT. *See also* Computerized tomography
  density numbers, diagnosis of senile dementia, 377–379
  x-ray scans, *20, 173, 177, 270, 277, 278,* 312, 313, *350, 351, 352, 353, 354, 364–369, 372, 373, 376, 377, 378–381*
Curiosity, impaired, 134

DAILY living activities
  assessment, 78t, 115
  impairments, 108
  testing of, 75

Dyspraxia, sympathetic, 93, 223
Dysprosody, 196, 203, 241
Dysrhythmia, temporolimbic, 320

ECT. *See* Electroconvulsive therapy
EEG. *See* Electroencephalogram
Elderly. *See also* Aging
  confusional states and, 134
  dementia in, 78t, 111–115
  forgetfulness in, 180–181
  reasoning and, 105
  tests for dementia in, 78t, 113
Electroconvulsive therapy, 179, 187
Electroencephalogram(s)
  abnormalities
    epilepsy and, 296–297, 314, 352
    schizophrenia and, 296, 297
  attention and, 134–136, 160
  diagnosis of confusional states, 132
  epilepsy diagnosis, 307–312
  evoked potentials, 328
  frequency analysis, 329
  neural generators, 328
  neuropsychologic applications of, 330
  studies
    animal, 330–331
    epilepsy and, 307, *308–311*, 312
    human, 329–330
Electrophysiologic techniques, 327–328
  EEG, 327–331
  evoked potentials, 327, 328, 331–340, 341
Emotional expression, encoding and decoding, 54–55, 56
Emotional gesturing, 246, 246t, 247, 248, 249
Emotional manifestations of epilepsy, 38, 251, 294–295
Emotional reactivity, and septal lesions, 36
Emotional states, 33–34
  sensory processing and, 11
Emotions. *See also* Affect; Emotional expression; Emotional states
  amygdala and, 37–38
  aphasia and, 228–229
  communication and, 56–57, 251–252
  epilepsy and, 38, 251, 294–295
  extreme displays, 36, 251–252, 254
  feeling, 54, 55
  inappropriate, 53
  negative, 53
  paralimbic areas and, 32–33
  prosody and, 54, 240–241, 244
  right hemisphere and, 53–56, 251–252
Encephalopathies
  metabolic, 89, 89t, 114
  toxic, 128, 130
  toxic-metabolic, 128, 130, 131, 139, 149
  Wernicke's, 46, 174
Endocrine
  alterations, epilepsy and, 305–306
  control, regulation of, 35, 39
Endogenous brain potentials, 332–334
  P300, 332, 336–340
  selective attention effect, 332, 333–336
Environmental sounds, identifying, 22, 97
Enzyme choline acetyl transferase, 184

Epilepsy
  behavioral change in, 298–302, 318–319
  classification of, 289–290
  compulsive writing, 299, 300, 301, *301*, 302. *See also* Hypergraphia; Graphorrhea
  confusional states and, 130, 131
  diagnosis
    EEG studies, 307–312
    history and clinical examination, 306–307
    radiologic investigation, 312–313
  EEG abnormalities in, 296–297, 305–306
  EEG studies, 307, *308–311*, 312
  emotional manifestations, 38, 251, 294–295
  experiential manifestations, 293–294
  gelastic, 294
  graphorrhea and, 219
  hypergraphia and, 299, 300, 301–302
  ictal alterations, 291–292
    behavioral aspects of, 292–295
    neural substrate of, 295–298
  imaging techniques in, 352
  interictal alterations
    behavior traits, 299–302
    behavioral changes, potential mechanisms, 302–303
    psychosis, 298–299
  management
    pharmacologic, 313–315, 319
    psychiatric aspects of, 318–320
    surgical treatment, 315–318
  memory disorders and, 170, 179, 180, 316
  physiologic considerations
    endocrine milieu alterations, 305–306
    kindling, 304–305
    membrane/cellular phenomena, 303–304
  prevalence of, 291
  psychoses and, 291, 296, 297–298, 300, 302, 303, 318–319
  radiologic investigation, 312–313, 352
  symptoms, 292–295
  temporolimbic, 35, 38, 39, 55, 290–291
Epileptic seizures, 130, 131. *See also* Epilepsy; Seizures
Equipotentiality principle, 57
Euphoria, 53, 252
Evoked potentials
  EEG neural generators and, 328
  human studies, 331–333
  neuropsychologic applications 335–336, 338–340
  P300, 336–338
  selective attention, 333–334
Expectancy, attention and, 331
Experiential manifestations of epilepsy, 293–294
Exploration and scanning behavior, 24
Exploratory behavior, 154, 155, 157
Exploratory behavior deficits, 147
Extinction, sensory, 142–144, 150
Extrapyramidal disturbances, 42, 223
Eye movements, 154
  saccadic, programming, 42

Lesions—*Continued*
   temporo-occipital, 18
   temporoparietal, 26
   thalamic, 44, 155–156, 176
   unilateral temporal, 178
   unimodal, 15
   unimodal output channel, 15
   visual association cortex, 19, 109, 110, 263, 271
   Wernicke's area, *368–369*, 370
LGN. *See* Lateral geniculate nucleus
Lidocaine, 305
Limb movements, impairment, 14
Limb-kinetic apraxia, 223
Limbic connection in neglect, 153
Limbic connectivity, *8, 9*, 35, 36, 37, 47
Limbic cortex, *8, 9*
   behavioral roles, 9
   thalamic nuclei, 45–46
Limbic disconnections, 178
Limbic striatum, 40–41, 47
Limbic structures
   allocortical, 34. *See also* Allocortex
   amnesias, involvement in, 47, 170–171, 172–178
   amygdala, 37–39
   behavioral specializations, 35–40
   hippocampal formation, 39–40
   hypothalamic connections and, 9, 10
   memory and, *182*, 187
   piriform cortex, 36–37
   septal nuclei, 35–36
   substantia innominata, 35–36
Limbic system
   components, 46–47
   concept of, 46–47
   drive and affect and, 303
   emotional experience and, 252, *254*
   epilepsy and, 303–305,
   lesions, 178, 181
Linguistic functions, 50, 51
Lithium salts, 319
Lobectomies
   right-sided prefrontal, 29
   temporal, 316–318
     bilateral medial, 172
     left, 51
     neural activity in, 336
     right, 51, 55, 57
     unilateral, 178
Logorrhea, 219

M1 REGION, 14. *See also* Motor areas, primary; Motor cortex, primary
M2 region, 24. *See also* Sensory-motor areas, secondary; Supplementary motor area
Macropyramidal cortex, 7
Magnetic resonance imaging, *265*, 350
   nuclear, 283, 312, 313, 350
Magnetoencephalogram, 331
Manic-depressive illness, 55, 299, 302, 318–319
Manual exploration, 77t, 103, 147
Mattis Dementia Rating Scale, 78t, 113
Maze learning, visual, 178

Maze performance, 51
Medial geniculate nucleus, 44
Medial temporal lesions, 39–40, 172–174
   neurosurgical intervention, 172
Melodic Intonation Therapy, 231
Melody identification, 22, 52
Memorization. *See also* Learning
   impairment, nonlinguistic perceptual material, 51
   stages of, 83
Memorizing process
   acquiring, 185
   holding, 184
Memory. *See also* Learning
   Alzheimer's disease and, 36, 85, *88*, 89t, 184, 187
   cerebral blood flow and, 357
   complex somatosensory, 22
   contextual, 263. *See also* Memory, incidental
   deficits, 31–32
     anterograde, 83, 171, 176
     confusional states and, 129
     retrograde, 83, 171, 176
     visual recognition, 17
   disorders, 49, 169. *See also* Amnesia(s).
     amnesias, selective, 178
     amnesias of unknown anatomy, 178–180
     amnesias with bilateral limbic involvement, 172–178
     benign senescent forgetfulness, 180–181
     epilepsy and, 170, 179, 180
     psychogenic amnesias, 181
     transient global amnesias, 170, *177*, 180
     treatment, 187
   flashbacks, 33
   function, 35–36
   functions, assessment, 77t, 83–90, 116
   gap, 180
   impairment
     assessment, 77t, 83–90, 116, 171
     olfactory, 35
     verbal, 176, *177*
   incidental, 86, *86*, 87, *87*, *88*
   interference
     proactive, 84, 171
     retroactive, 84
   limbic zone and, 9
   loss, 36, 171
     material specific, 178
   neural activity and, 48–49
   nonverbal, 86, *86*, 90
     tests, 84–87, 109
     tonal, 52, 97
   olfactory/gustatory, 85
   organization
     memorizing process, 184–185
     remembering process, 186
     storing process, 185–186
   paralimbic area and, 31–32
   processes, hemispheric asymmetry, 51
   quotient, 83–84, 171
   retrieval, 88
   rote, 87

Neuropeptides, 184, 187
Neurophysiologic process, study of, 327. *See also* Electrophysiologic techniques
Neuropsychiatric conditions
  imaging technique applications, 351–355
  prominent attentional deficits and, 161–162
Neuropsychologic tests, choosing, 74, 111
Neurosurgery
  medial temporal lobe lesions, 172
  treatment of epilepsy, 315–318
Neurotransmitters, 184
  interference with action, 130
  release, 315
NMR. *See* Nuclear magnetic resonance
Nonaphasic dysarthria, 24
Nonfluent aphasia, 196, 197. *See also* Aphasia(s)
Nonlinguistic perceptual skills, 51–53
Nonverbal auditory agnosias, 22, 208. *See also* Agnosia(s), auditory, nonverbal auditory
Nonverbal communication, 104. *See also* Communication, paralinguistic aspects of
  right hemisphere and, 242–245
Nonverbal memory, 52, 86, *86*, 90, 109. *See also* Memory, nonverbal
Nonverbal modality-specific cognitive disturbances, 19
Nonverbal perceptual processing, 96–98, 107
Nonverbal reasoning, 105, 107
Nonverbal recognition defect, 262
Nonverbal tonal memory, 52, 97
Noradrenalin, 137
Noradrenergic agents/drugs, 137, 161
Noradrenergic innervation, 136
Noradrenergic neurons, 48
Noradrenergic pathways, 161
Norepinephrine
  Korsakoff's syndrome and, 176, 184, 187
Novelty P300 response, 337, *338–340*
Nuclear magnetic resonance, 283, 312, 313, 350
  imaging, *265*, 350
Nucleus(i)
  accumbens, 40, 41
  lateral geniculate, 44
  medial geniculate, 44
  thalamic, 43–46
  thalamic reticular, 46
  ventroposterior lateral, 44
  ventroposterior medial, 44
Nystagmus, 174

OBJECT agnosia, 18, 75, 109, 178
Object and form sorting tasks, 78t, 105
Object identification, 97, 107
Object identification tasks, 17
Object naming, 261
  difficulties, 222
Object recognition, 261, 266, 283
Object recognition tasks, 17

Objects
  distinguishing, 17, 37, 261, 283
  emotional significance, 303
  motivational relevance, 26, 36
Obsessive compulsive syndromes, 33
Occipital lobe. *See also* Visual association areas; Visual cortex
  lesions, 264t, 266, 267
  testing, 110
Occipitoparietal lesions, 277, 279, 281, 282
Occipitotemporal lesions, 264t, 266, 269, 271, 273, 277
Ocular apraxia, 275–277
Ocular scanning disturbances, 18
Olfaction, paralimbic areas and, 34–35
Olfactocentric paralimbic belt, 5, *6*
Olfactory cortex, primary, 2, 34–35. *See also* Piriform cortex
Olfactory input, piriform, 36–37
Olfactory tubercle, 40
Ophthalmoplegia, 174
Opiates, confusional states and, 129, 130
Optic aphasia, 266
Optic ataxia, 275, 276, 278, *278*, 282
Orbitofrontal cortex, caudal, 5, *6*. *See also* Paralimbic areas
Orbitofrontal lesions, 31, 32
Organic brain syndromes, acute, 128
Orientation
  abnormalities, 339
  deficits, 156,
  directional, judgment of, 97, 280
  motor aspects of, 154, 157
  neglect and, 156
  spatial, disorders of, 267–268
  temporal, tests of, 112, 112t
Oxygen metabolism, measuring, 347, 348, 351
Oxytocin, 184

31-P NMR. *See* Phosphorus nuclear magnetic resonance
P300 evoked potential, 332, 336–338
  animal models of, 340
  neuropsychologic applications of, 338–340
Pain perception/sensation, loss of, 33, 44
Paleocortex, 2. *See also* Primary olfactory cortex
Palilalia, 90
Pallidum, 42
Panic attacks, 38, 318
Pantomime, 231–232, 241–242
Parahippocampal gyrus, 5, *6*. *See also* Paralimbic areas
Paralexia, 217
Paralimbic areas, 3–6, 8, *8*, 30–31
  affect and, 32–33
  behavioral specializations of, 30–35
  connectivity, 10
  drive and affect and, 32–33
  functions, 9
  gustation in, 34–35
  higher autonomic control and, 33–34
  hippocampocentric, 5, *6*, 30
  lesions, 32, 33
  memory and, 31–32

Somesthetic-limbic disconnection, 33
Sound patterns, language, 200–201
Sounds
  environmental, identification of, 22, 97
  nonverbal, 208
Spastic dysarthria, 213
Spatial analysis disorders, 280–281
  Balint's syndrome, 18, 23, 27, 110, 266, 275–279, 282, *284*
  stereopsis, 279–280
Spatial disorientation, 23, 27. *See also* Topographic orientation disorders
Spatial distribution of attention, 53. *See also* Attention, neglect and, unilateral
  tests, 77t, 101–103
Spatial relationships of objects, impairment of, 26
Specialization, behavioral patterns, neural connectivity and, 8–11
SPECT. *See* Single-photon emission computerized tomography
Speech. *See also* Communication; Language, spoken; Prosody
  arrest, 292
  disordered, 193–194, 207. *See also* Aphasia
  empty, 197, 220
  evaluation 90–91
  fluent/nonfluent classification, 90, 196, 197
  linguistic aspects of, 56, 90
  prosodic features, 239–241
  prosody testing, 104, 246–247
  spontaneous, 196–197, 227
Speech therapy, 231
Speech-related motor programs, impairment, 24
SPET. *See* Single-photon emission tomography
Splenium, sex differences in, 52
Split-brain, 50. *See also* Commissurotomy
Split-brain patients, experiments, 160, 252, 265, 283
Spoken language
  comprehension of, 198, 227–228
  impairment, syndromes of, 201–214
  repetition of, 197, 201
State-mediated functions, 47–49
Stereoblindness, 280
Stereopsis, disturbance of, 279–280
Stereovision, standard tests, 279
Sternberg, Paradigm of, 77t, 89
Stimulus-response bond, breakdown of, 30–35
Striatal lesions, 41–42
Striate cortex, 4, 11–12. *See also* Visual cortex, primary
Striatum, 40–42
  dopaminergic input, 40
Stroop Interference Test, 77t, 79, 81, 119, 129, 140
Subcortical aphasia, 213–214, 213t
Substantia innominata, 2, 3, 4, 35–36
  behavioral patterns, 35–36

Supplementary motor area, 14, 24. *See also* Sensory motor areas, secondary; M2 area
Syntactic alexia, 217
Syntactic comprehension impairment, 91, 228
Syntactic language, 203, 204

TACTILE discrimination learning deficits, 31–32
TCM. *See* Transcortical motor aphasia
TCS. *See* Transcortical sensory aphasia
Temporal lobe
  lesions, 19, 109, 370
  testing, 109
Temporal pole, 5, 6. *See also* Paralimbic areas
Temporal visual association areas, 15, 17. *See also* Visual association areas
  damage, 17–19
Temporal visual neurons, 16
Temporolimbic epilepsy, 35, 38, 39, 55, 290–291. *See also* Epilepsy
  traits, 300t
Temporo-occipital area, lesions, 18, 264t
Temporoparietal heteromodal association areas, behavioral patterns, 25–27
Temporoparietal lesions, 26
Thalamic nuclei
  major groups of, 43
  modality-specific association cortex, 44–45
  non–modality-specific, 45–46
  primary, 44
  reticular, 46
Thalamic relay, 137–138
Thalamocortical connectivity patterns, 43–44
Thalamotomies, bilateral dorsomedial, 45, 176
Thalamus, 43. *See also* Thalamic nuclei; Thalamic relay
  attention and, 137–138
Three Words–Three Shapes Test, 77t, 86, *86*, 87, *87*, 88, *88*, 89, 117
Timbre discrimination, 22, 97
Token Test, 77t, 91
Topographic orientation disorders, 267–268
Toxic psychosis, 128
Trail-Making Test, 77t, 79–80, *80*, 81, 140
Transcortical motor aphasias, 24, 209–211, 210t, 366, *366*, *367*, 368
Transcortical motor aprosodia, 249
Transcortical sensory aphasias, 91, 210t, 211, 216, 221, 227–228
Transcortical sensory aprosodia, *245*, 249
Transient global amnesia, 170, *177*, 180
Tricyclic antidepressants, 251, 252

UNIMODAL association areas
  auditory, 19–22
  motor, 23–24
  sensory input and, 9
  somatosensory, 22–23

INDEX

403

Unimodal association areas—*Continued*
thalamic nuclei, 44–45
vertical connectivity, 10
visual, 15–19
Unimodal association cortex, 25. *See also* Unimodal association areas
Unimodal association isocortex, 6–7. *See also* Isocortex; Unimodal association areas

V1 AREA, 14. *See also* Striate cortex; Visual cortex, primary
V2 AREA, 14. *See also* Sensory-motor areas, secondary
Valproic acid, 313, 314
Vasopressin, 184, 187
VAT. *See* Visual Action Therapy
Ventricle and sulcal sizes, quantifying, 370–373, 374–377
Ventricle size
in normal aging, 373
linear measurement of, *378–379*
Ventroposterior lateral nucleus, 44
Ventroposterior medial nucleus, 44
Verbal auditory agnosia, 208
Vigilance, 76, 78, 160
attention and, 126, 128, 129
confusional states and, 78, 129
testing, 76, 77t, 81, 129
Visual Action Therapy, 232
Visual agnosia. *See* Agnosia(s), visual
Visual association areas, 15–19
focal damage, 269, *270,* 271, 272
language learning and, 200–201
unimodal, 6, 15–19
Visual cortex, primary, behavioral patterns, 11–12
Visual deficits, 16, 19. *See also* Blindness; Visual information processing; Visual neglect; Visual object agnosia; Visual processing disorders, complex; Visual reaching disturbances; Visual scanning defect
Visual disorientation, 275, 276, 278–279, *278*
Visual information processing, 17
interference with, 273–274
Visual maze learning, 178
Visual neglect, 154. *See also* Neglect
Visual object agnosia, 266
Visual processing disorders, complex
color perception, 268–273
color processing-related, 273–275
constructional ability, 281–282
dressing ability, 282
pattern recognition, 259–268
spatial analysis, 275–281
Visual processing impairments, 20. *See also* Visual processing disorders, complex
Visual reaching disturbances, 18, 275. *See also* Optic ataxia
Visual Retention Test, 112, 112t
Visual scanning defect, 275. *See also* Ocular apraxia

Visual unimodal association areas, 6, 15–19. *See also* Peristriate area
Visual-limbic disconnections, 18
Visual-limbic interactions, 16
Visual-somatosensory interactions, 18
Visual-spatial acalculia, 224
Visual-spatial agraphia, 219
Visual-verbal disconnection, 18, 19, 216, 267, 275
Visual-Verbal Test, 78t, 105, *106,* 119
Visuoconstructive ability, 281–282
Visuomotor ataxia, 276. *See also* Optic ataxia
Visuospatial deficits, 18, 27, 101. *See also* Visual processing disorders, complex; Visual-spatial acalculia; Visual-spatial agraphia
Visuospatial disorientation, 110
Visuospatial orientation, 22
Visuospatial processing, 16
Visuospatial reasoning, 107
Visuospatial tasks impairment, 51
Vocalizations, affective, gestures and, 37
VPL. *See* Ventroposterior lateral nucleus
VPM. *See* Ventroposterior medial nucleus

WADA test, 73, 316
Wakefulness
attention and, 126, 136
Wakefulness/arousal assessment, 76–82, 77t
Wechsler Adult Intelligence Scale, 78t, 111, 112
Arithmetic Subtest, 77t, 94
Block Design Subtest, 51, 77t, 98, *98,* 99, 118, 281
Comprehension Subtest, 77t, 105
Similarities Subtest, 77t, 105
Wechsler Memory Scale, 77t, 83–84, 111, 171, 187
Wernicke's aphasia, 22, 26, 27, 90, 91, 129, 202t, 204–205, 207–208, 211, 216–217, 221, 227–228, 242, 248, 252
conduction vs., 368–370, *368–369*
Wernicke's area, 21, 22, 26
ASI-I printout of, *364–365*
language learning and, *200,* 201
lesions, *368–369,* 370
Wernicke's encephalopathy, 46, 174
Western Aphasia Battery, 77t, 90, 111
Wide Range Achievement Test, Arithmetic Subtest, 77t, 94
Wisconsin Card Sort Test, 78t, 105–107, 108
Word blindness, 18, 48, 267. *See also* Alexia
Word deafness, 21, 26, 202t, 207–208, 227
Word-evocation anomia, 220–221
Word-finding impairment, 220–222. *See also* Anomia; Naming difficulties, disorders
Word-list generation test, 77t, 79, 81, 198
Word-production anomia, 220–221
Word-selection anomia, 221
Writing, compulsive, 301, *301,* 302. *See also* Graphorrea; Hypergraphia